Aerodynamics: Design and Applied Principles

Aerodynamics: Design and Applied Principles

Edited by
Russell Mikel

MURPHY & MOORE
www.murphy-moorepublishing.com

Published by Murphy & Moore Publishing,
1 Rockefeller Plaza,
New York City, NY 10020, USA

ISBN: 978-1-63987-025-7

Cataloging-in-Publication Data

Aerodynamics : design and applied principles / edited by Russell Mikel.
p. cm.
Includes bibliographical references and index.
ISBN 978-1-63987-025-7
1. Aerodynamics. 2. Airplanes--Design and construction. 3. Aeronautics. I. Mikel, Russell.
TL570 .A37 2022
629.132 3--dc23

For information on all Murphy & Moore Publications
visit our website at www.murphy-moorepublishing.com

MURPHY & MOORE

Contents

Preface

The main aim of this book is to educate learners and enhance their research focus by presenting diverse topics covering this vast field. This is an advanced book which compiles significant studies by distinguished experts in the area of analysis. This book addresses successive solutions to the challenges arising in the area of application, along with it; the book provides scope for future developments.

Aerodynamics is the sub-field of fluid dynamics and gas dynamics. Aerodynamics is a discipline associated with the study of motion of air and its interaction with the solid objects such as airplane. The concepts of mathematical analysis, wind tunnel experimentation, empirical approximations, and computer simulations are widely used in aerodynamics. Aerodynamics focuses on the issues like compressible flow, turbulence and boundary layers. The two branches of aerodynamics are incompressible aerodynamics and compressible aerodynamics. Incompressible aerodynamics deals with the subsonic flow, whereas compressible aerodynamics deals with transonic, supersonic and hypersonic flow. This book aims to shed light on some of the principles of aerodynamics and the recent researches in this field. It also elucidates new techniques and their applications in a multidisciplinary manner. It is appropriate for students seeking detailed information in this area as well as for experts.

It was a great honour to edit this book, though there were challenges, as it involved a lot of communication and networking between me and the editorial team. However, the end result was this all-inclusive book covering diverse themes in the field.

Finally, it is important to acknowledge the efforts of the contributors for their excellent chapters, through which a wide variety of issues have been addressed. I would also like to thank my colleagues for their valuable feedback during the making of this book.

Editor

Preface

1

Aerodynamic Performance of the Three-Dimensional Lumped Flexibility Bionic Hovering Wing

Zhu Jianyang, Jiang Lin, and Hou Yu

School of Machinery and Automation, Wuhan University of Science and Technology, Wuhan 430081, China

Correspondence should be addressed to Jiang Lin; jlxyhjl@163.com

Academic Editor: Linda L. Vahala

Understanding the effect of flexibility on the aerodynamic characteristics of the wing is one of the most important considerations for successfully designing a flapping wing micro aero vehicle (FMAV). This paper aims at providing a systematic synthesis on the flexibility effects on the hovering performance of the bionic wing based on the numerical analysis approach. We construct a novel three-dimensional bionic wing, which has lumped flexibility at the root, and develop an iterative coupling program to simulate the interaction between the flexible wing and fluid. The effects of flexibility on the hovering performance of the three-dimensional flapping wing are investigated, and the results indicate that the best performance of the wing is achieved when the wing flaps at resonance and has the density close to the natural insect wing. The feasibility of using lumped flexibility wings driven by a simple harmonic flapping for designing efficient FMAV is also concluded in this study.

1. Introduction

Flapping wing aero vehicle has become a popular research topic in recent years because it has better aerodynamic performance than the fixed-wing and rotary-wing aero vehicle under low Reynolds number flight regime [1–3]. Understanding the effect of flexibility on the aerodynamic characteristics of the wing is one of the most important considerations for successfully designing this type of vehicle. Therefore, many studies have been carried out to investigate this problem, and considerable understandings have been achieved [4–7].

The lumped flexibility wing for insect flight was first proposed by Ishihara et al. in [8]. In their study, the two-dimensional wing was modeled by a rectangular plate, and the wing's flexibility was lumped into the spring at the leading edge. It was found that the studied wing can generate sufficient lift force to support the insect's weight. Therefore, they concluded that the pitching motion can be passive in the dipteran's flapping flight. Later, they again experimentally and numerically [9, 10] investigated the performance of the three-dimensional lumped flexible wing, and it was also found that if the wing has appropriate flexibility, the lumped flexible wing can generate sufficient lift force to support the corresponding insect weight.

Michelin and Llewellyn Smith [11] studied the influence of the flexibility on the performance of a heaving wing. They found that the maximum values of the mean thrust and efficiency are obtained when the resonance occurs between the forcing frequency and a natural frequency of the system. Masoud and Alexeev [12] conducted a numerical study to focus on the aerodynamic characteristics of a flexible planar hovering wing which was flapped at resonance. They again concluded that large-amplitude resonance flapping wing can drastically enhance aerodynamic lift and efficiency of the wing under low Reynolds number. Moore [13] also found that dramatic performance improves when the wing is driven near resonance and deteriorates when it is driven at higher frequencies. Zhu et al. [14, 15] studied the aerodynamic performance of a 2D flexible wing in forward flight by solving the N-S equations coupled with structural dynamic equation and the flapping motion of the wing. They concluded that if the wing flapped at a frequency close but below the wing's natural vibration

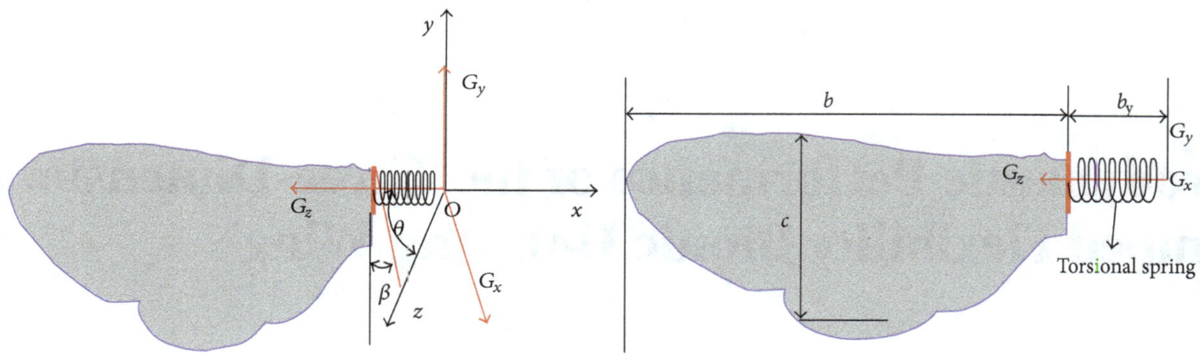

FIGURE 1: The schematic of the wing planform.

frequency, the flexibility can increase the aerodynamic force and the energy efficiency.

On the other hand, Chen et al. [16] performed a study to investigate the aerodynamic performance of a flapping and passively rotating insect wing by using experimental and computational methods. A novel wing which had a single flexible hinge allowing passive pitching under load was devised in their study. The results indicated that the wing's kinematic and flapping efficiency depends on the stiffness of a passive compliant hinge. It was not the resonance wing but the stiffness wing which had larger lift force production. Cho et al. [17] developed a computational framework to simulate a three-dimensional flapping flexible wing under simultaneous pitching and plunging motion, and the results indicated that the flexibility can be beneficial for the wing with moderate flexibility. Yeh and Alexeev [18] carried out a numerical study to investigate the performance of the plunging elastic plates with aspect ratios ranging from 0.5 to 5. They showed that the maximum velocity occurs near the first natural frequency regardless of aspect ratio, whereas the maximum economy occurs away from the first natural frequency.

In summary, the flexibility is beneficial for the flapping wing as shown in the above literatures; however, a full understanding on the mechanism of how the flexibility affects the aerodynamic characteristics of the wing is still not achieved, especially the best performance of the wing flapping at resonance or not. To fill in these gaps, a novel three-dimensional bionic wing which has lumped flexibility is constructed, and an iterative coupling program to simulate the interaction between the flexible wing and fluid is developed. Then the details of the lift force, hovering efficiency, and vortex of the flexible bionic wing are studied.

2. The Novel Three-Dimensional Bionic Wing Model

Bumblebee is one of the typical insects which have amazing flying capabilities. The shape, structure, and aerodynamic characteristics of this insect's wing have been studied extensively [19–21]. In this paper, the planform of the bionic bumblebee wing is selected similar to the experimental model studied by Nagai et al. [22] for comparison. The schematic of the wing planform is shown in Figure 1, where two coordinate systems are also defined: one is the inertial coordinate system (OXYZ) and the other is the wing-fixed coordinate system ($OG_xG_yG_z$). The symbol $b = 0.100$ m is the span, $b_r = 0.024$ m is the distance from wing base to the flapping y-axis, and c is the reference chord which is defined as a length at two-thirds of the span location ($c = 0.01575$ m). The area of the wing S is fixed at a value of 2.984×10^{-3} m^2, and the thickness of the wing is fixed at $S/(50b)$ for comparison.

In contrast to the previous study that consider the chordwise and spanwise flexibility of the wing, the flexibility of the wing in this paper is mimicked as the passive pitching motion by connecting a rigid wing to a torsional spring; therefore, the flexibility is just lumped at the root of the wing, and the simplified wing is in reasonable agreement with the natural insect wing [23].

The flapping motion θ of the wing which is defined as the angle between the axis G_x and the z-axis follows the simple sinusoidal function:

$$\theta(t) = \frac{\theta_m}{2} \sin(2\pi f t), \tag{1}$$

where θ_m is the flapping amplitude and f is the flapping frequency.

The passive pitching angle β which is defined as the angle between the axis G_y and the y-axis can be decided by the equation

$$J\ddot{\beta} + K\beta = Q, \tag{2}$$

where K is the torsion stiffness, Q the pitching torque generated by the aerodynamic force, and J is the inertia moment of the wing which can be defined as

$$J = \sum_i \Delta m_i r_i^2, \tag{3}$$

where m_i is the element mass and r_i is the distance from the element mass to the pitching axis G_x.

3. Parameters of Bionic Hovering Wing

To clearly elaborate the aerodynamic characteristics of the lumped flexibility wing, three dimensionless parameters are defined, namely, the Reynolds number Re, density ratio ρ^*, and frequency ratio f^*. They can be defined as

$$\mathrm{Re} = \frac{U_{\mathrm{ref}} c}{\nu},$$
$$\rho^* = \frac{\rho_s}{\rho}, \tag{4}$$
$$f^* = \frac{f}{f_s},$$

where $U_{\mathrm{ref}} = \pi f \Theta_m (2b/3 + b_r)$ is the maximum of flapping velocity, ν is the fluid kinematic viscosity, ρ is the fluid density, ρ_s is the wing density, and f_s is the wing vibration frequency which can be calculated as

$$f_s = \frac{\sqrt{K/J}}{2\pi}. \tag{5}$$

Note that the fluid field characteristics are determined by Re and the flexibility of the wing is characterized by f^* and ρ^*.

The coefficient of lift, drag, and energy consumption are described as follows:

$$C_L = \frac{F_L}{0.5\rho U_{\mathrm{ref}}^2 S},$$
$$C_D = \frac{F_D}{0.5\rho U_{\mathrm{ref}}^2 S}, \tag{6}$$
$$C_P = \frac{Q_Y \dot{\theta}(t)}{0.5\rho U_{\mathrm{ref}}^3 S},$$

where F_L is the lift, F_D the drag, Q_y the aerodynamic torque of wing about the y-axis, and $\theta(t)^{\bullet}$ is the derivative of $\theta(t)$. The hovering efficiency η then can be described as

$$\eta = \frac{\overline{C_L}}{\overline{C_P}}, \tag{7}$$

where $\overline{C_L}$ and $\overline{C_P}$ are the mean lift and energy coefficient, respectively.

4. The Coupling Program to Simulate the Interaction between Fluid and Flexible Wing

Considering the bumblebee's flight environment in nature, the Reynolds number of the hovering wing is at the range of 102–104; therefore, the flow field of the wing is assumed to be unsteady, laminar, and incompressible with constant thermophysical properties, and the governing equations of this flow field are described by the 3D unsteady incompressible laminar N-S equations as follows:

$$\nabla \cdot V = 0,$$
$$\frac{\partial V}{\partial t} + V \cdot \nabla V = -\nabla P + \frac{1}{\mathrm{Re}} \nabla^2 V, \tag{8}$$

where V is the nondimensional fluid velocity and P is the nondimensional fluid pressure.

To solve (8), we implement a numerical solver (Fluent 6.3) based on a finite volume method. The third-order MUSCL scheme [24] is used for spatial interpolation, and a first-order implicit algorithm is employed for time discretization. Meanwhile, the coupling between the pressure and the velocity is achieved by means of the PISO [25] algorithm. The flapping and passive pitching motion of the wing are controlled and achieved by the local remeshing dynamic mesh technique (see Fluent 6.3 Tutorial Guide).

In order to solve the governing equation of the passive pitching motion of the wing, the Newmark method is applied to discretization derivative of passive pitching angle, and then (2) can be rewritten as

$$\left(1 + \gamma \Delta t^2 \frac{K}{J}\right) \beta^{t+\Delta t} = \frac{\bar{Q}^{t+\Delta t}}{J} \Delta t^2 + \left[2 - \left(\frac{1}{2} - 2\gamma + \delta\right) \Delta t^2 \frac{K}{J}\right] \beta^t + \left[-1 - \left(\frac{1}{2} + \gamma - \delta\right) \Delta t^2 \frac{K}{J}\right] \beta^{t-\Delta t}, \tag{9}$$

where

$$\bar{Q}^{t+\Delta t} = \gamma Q^{t+\Delta t} + \left(\frac{1}{2} - 2\gamma + \delta\right) Q^t + \left(\frac{1}{2} - \gamma - \delta\right) Q^{t-\Delta t}, \tag{10}$$

where $\beta^{t+\Delta t}$, β^t, and $\beta^{t-\Delta t}$ are the pitching angle of the next step time, current moment, and last step time, respectively. $Q^{t+\Delta t}$, Q^t, and $Q^{t-\Delta t}$ are the aerodynamic torque of the next step time, current moment, and last step time, respectively. Δt is the iteration time step and γ and δ are the control parameters, which are fixed as $\gamma = 1/6$ and $\delta = 1/2$.

An iterative coupling method is developed to simulate the interaction of fluid, flapping, and passive pitching motion of wing, where the fluid field based on finite volume method and passive pitching based on Newmark method are solved separately at each time step. First, the fluid field is solved, then the aerodynamic torque on the wing is obtained, and then the pitching angle of flexibility wing can be determined under the aerodynamic torque using Newmark method, which is embedded in Fluent using the UDF. In the next time step, the fluid field is solved with the new location of the wing updated by using the dynamic mesh technique. In this way, the fluid and the flapping and passive pitching motion are coupled so that the fluid-structure interaction is taken into account properly. The details of the coupling process are illustrated in Figure 2.

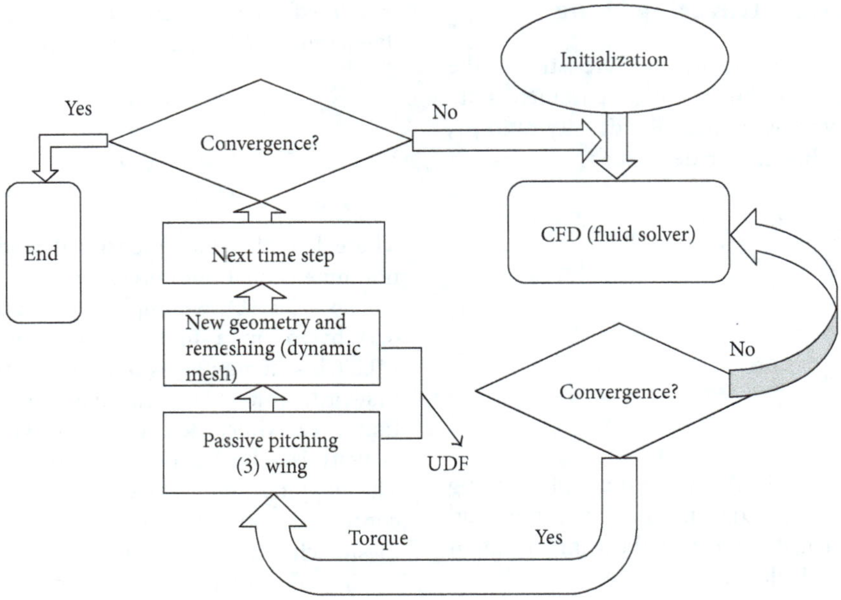

FIGURE 2: The schematic of coupling method.

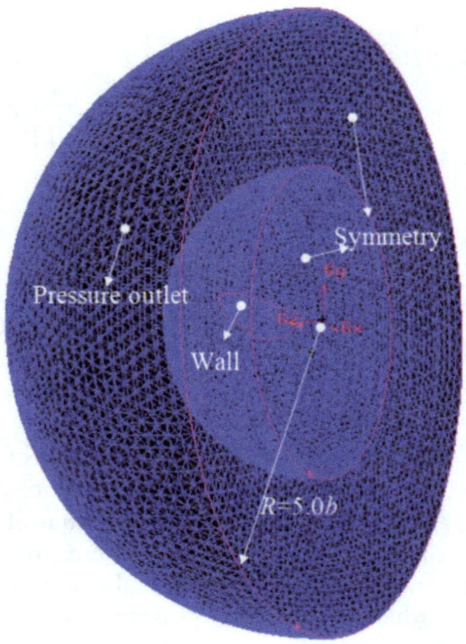

FIGURE 3: The computational grid and boundary conditions.

5. Simulation Setup and Method Validation

5.1. Simulation Setup. The computational domain has a hemispherical shape, and the radius of the hemisphere is fixed at a value of $5b$ as shown in Figure 3. A uniform unstructured mesh system is employed, where the hemispherical shape computational domain contains an inner domain and an outer domain. Tetrahedron cells are used to cluster to the wing in the inner domain, and the density of the tetrahedron cells are decreased from the inner domain to the outer domain. The no-slip wall boundary condition is applied at the bionic hovering wing surface. Meanwhile, pressure-outlet boundary condition which has zero static pressure is imposed on the outer boundary, and symmetry boundary condition is applied on the right computational boundary to save the simulation cost.

5.2. Method Validation. A grid sensitivity study is carried out first to evaluate the independence of the numerical result on the mesh size. The unsteady flow fields of a flexible wing were computed under conditions $\rho^* = 1000$, $f^* = 1.50$, and $\mathrm{Re} = 2.347 \times 10^3$ with three different grid systems. The sizes of involved grid systems were as follows: 3.5×10^5 tetrahedron cells around the wing, the first layer cell spacing above the wing $0.040b$ (system A); 2.3×10^5 tetrahedron cells around the wing, the first layer cell spacing above the wing $0.060b$ (system B); and 1.9×10^5 tetrahedron cells around the wing, the first layer cell spacing above the wing $0.070b$ (system C). Each grid was computed for 6 periods, and iteration time step was fixed at $0.001T$. Only the comparison of passive pitching angle β at different resolutions is given, because the aerodynamic performance of the developed lumped flexibility wing is mainly determined by β. The results in Figure 4 show that the difference of passive pitching angle β of the grid systems A and B are very small (no more than 1%). Therefore, grid system B is sufficiently dense and it is employed for the next simulations.

Three different iteration time steps $0.05T$, $0.001T$, and $0.0005T$ ($T = 1/f$) are employed to again simulate the unsteady flow field of a flexible wing with $\rho^* = 1000$, $f^* = 1.50$, and $\mathrm{Re} = 2.347 \times 10^3$ to validate the time-discretization independence. The results in Figure 5 show that the difference of the passive pitching angle of the three iteration time steps are very small (no more than 2%), which indicates that the iteration time step with $t = 0.001T$ is sufficiently small, and it is employed for the next simulations.

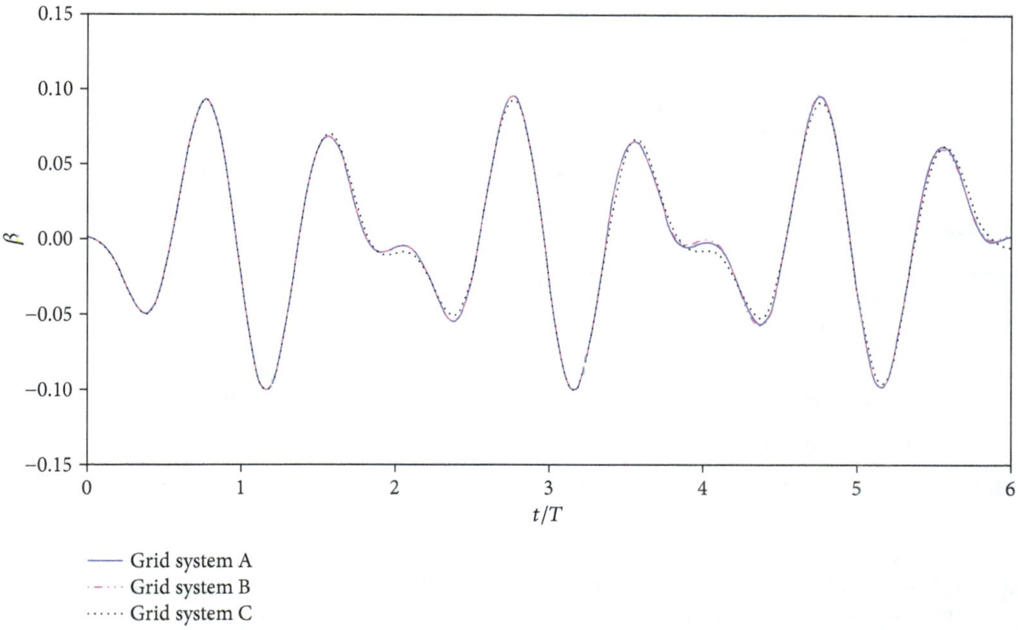

FIGURE 4: The time variation of passive pitching angle β under three different grid systems.

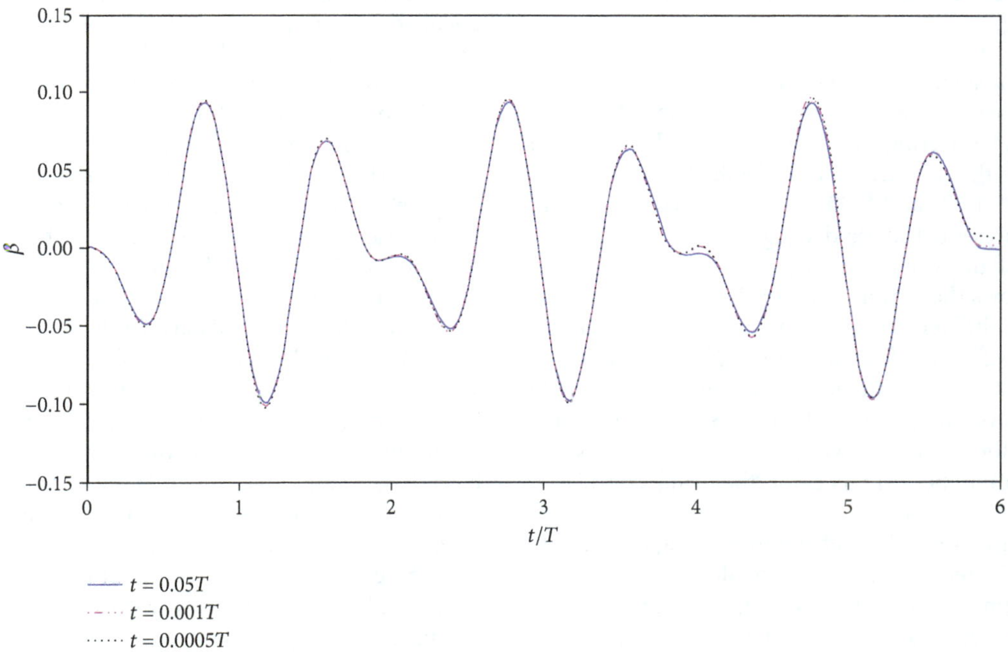

FIGURE 5: The time variation of passive pitching angle β under three different iteration time steps.

A typically bionic bumblebee hovering wing which was experimentally studied by Nagai et al. [22] is employed to validate the reliability of the present numerical method for simulating the unsteady flow around 3D hovering wing. The mean drag and lift coefficients are −0.00 and 0.46, respectively, in the present numerical computation, and in Nagai et al.'s experimental study, these two data are 0.00 and 0.49, respectively. The comparison of instantaneous drag and lift coefficient are also illustrated in Figure 6. It is again found that our computational results agree with the experimental results very well.

Validation of the reliability of the developed weak coupling method for simulating the fluid-flapping wing interaction problem had been demonstrated in our previous study. The details of the validation process can be found in [14, 26].

6. Results and Discussion

To study the aerodynamic performance of the three-dimensional (3D) lumped flexibility bionic hovering wing, the results of the 3D rigid flapping wing need to be employed

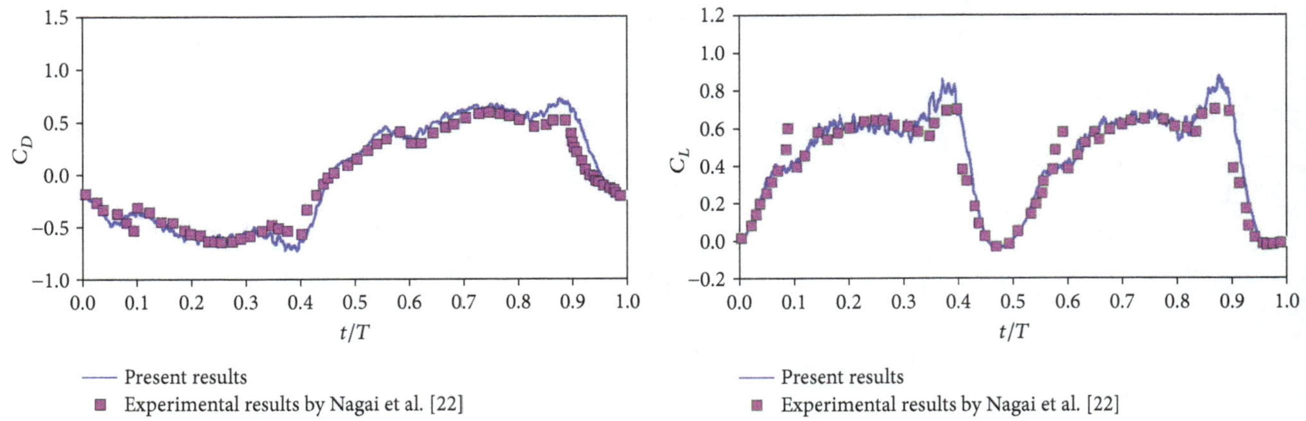

(a) Drag coefficient

(b) Lift coefficient

FIGURE 6: Comparison of the drag and lift coefficient of the wing obtained by the present numerical method and literature experimental data.

for comparison. The 3D rigid and lumped flexibility bionic hovering wing are performed with $\theta_m = \pi/3$, $f = 120$ Hz, and Re = 2347, and the parameters of the lumped flexibility wing are characterized by frequency ratio f^* and density ratio ρ^*. The details of f^* as well as ρ^* on the lift and hovering efficiency of the wing will be analyzed in the following.

6.1. Varying Frequency Ratio. To study the effect of frequency ratio f^* on the aerodynamic performance of the 3D lumped flexibility bionic hovering wing, we fix the density ratio $\rho^* = 1000$, and the frequency ratio with $f^* = 0.00$, 0.40, 0.50, 0.67, 1.00, 1.50, and 2.00 are considered. Note that $f^* = 0.00$ is represented as rigid wing, and as f^* increases, the flexibility of the wing increases.

Figure 7 shows the performance of the 3D lumped flexibility wing with different f^*. It is seen from Figure 7(a) that the mean lift coefficients first increase as f^* increases when f^* is smaller than 1.00 and reach a maximum at $f^* = 1.00$; however, they drop sharply when f^* crosses over 1.00, which indicates that the flexibility wing with $f^* = 1.00$ has the best lift generation performance for the wing considered in this section.

It is clear in Figure 7(b) that the mean drag coefficients increase as f^* increases when f^* is smaller than 0.67 and reach a maximum value 0.00 at $f^* = 0.67$; however, they drop sharply when f^* crosses over 0.67. It is also found in this figure that the mean drag coefficients have negative values expect for the wing with $f^* = 0.67$, which indicates that the thrust are generated for the other flexible wings.

It is obvious in Figure 7(c) that the minimum mean energy coefficient is obtained when the flexible wing with $f^* = 1.00$, and in comparison with the rigid wing, the mean energy coefficient of this wing is decreased by 36.27%. However, for the other flexible wing, the frequency ratio f^* has little effect on the mean energy coefficient; they have almost identical values.

It is seen from Figure 7(d) that a similar variation trend of hovering efficiencies and lift coefficients with frequency ratio f^* are observed. The energy efficiencies increase with the frequency ratio when the ratio is smaller than 1.00 and reach a maximum at $f^* = 1.00$, which indicates that the best hovering performance of the wing is achieved when the flexible wing with $f^* = 1.00$. However, they drop sharply when f^* crosses over 1.00 and even a negative value is observed, which indicates that the hovering performance is deteriorated. This conclusion is different with the results obtained in our previous 2D lumped flexible wing study [27], at which we concluded that the best performance of the wing is achieved when the wing with frequency ratio f^* is close to but less than 1.00.

To analyze the mechanism of how the flexibility enhances the hovering performance of the lumped flexibility bionic wing details, Figure 8 shows the time variation of lift coefficients, drag coefficients, and energy coefficients of the above considered wings. It is obvious in this figure that after six flapping cycle simulations, periodic lift, drag, and energy coefficients have been established. It is also observed in this figure that the flexible wing with $f^* = 1.00$ not only has smaller drag and energy coefficient amplitude than the other flexible wings but also has smoother time variation trend (with smaller curve slope), while for the lift coefficient, it has a larger amplitude, which is the reason why the flexible wing with $f^* = 1.00$ has a larger mean lift and smaller energy coefficient as shown in Figure 7.

Figure 9 plots the time variation of passive pitching angle β of the flexible wing with $f^* = 1.00$ and 1.50, and the time variation of flapping motion θ is also presented for comparison. It is obvious in this figure that f^* influences β significantly. For the flexible wing with $f^* = 1.00$, the peak negative angle takes place at a phase angle of 0.86π, while for flapping angle θ, the peak negative angle happens at a phase angle of 1.01π; moreover, the passive pitching angle β of the flexible wing with $f^* = 1.00$ also has a larger amplitude than the flapping angle θ. While for the flexible wing with $f^* = 1.50$, the passive pitching angle has visibly smaller amplitude than the flapping angle θ, and the variation cycle of β is two times that of the flapping motion θ. Those findings could be useful to explain the phenomenon why the flexible wing with $f^* = 1.00$ and 1.50 has different lift, drag, and energy coefficients as shown in Figure 8.

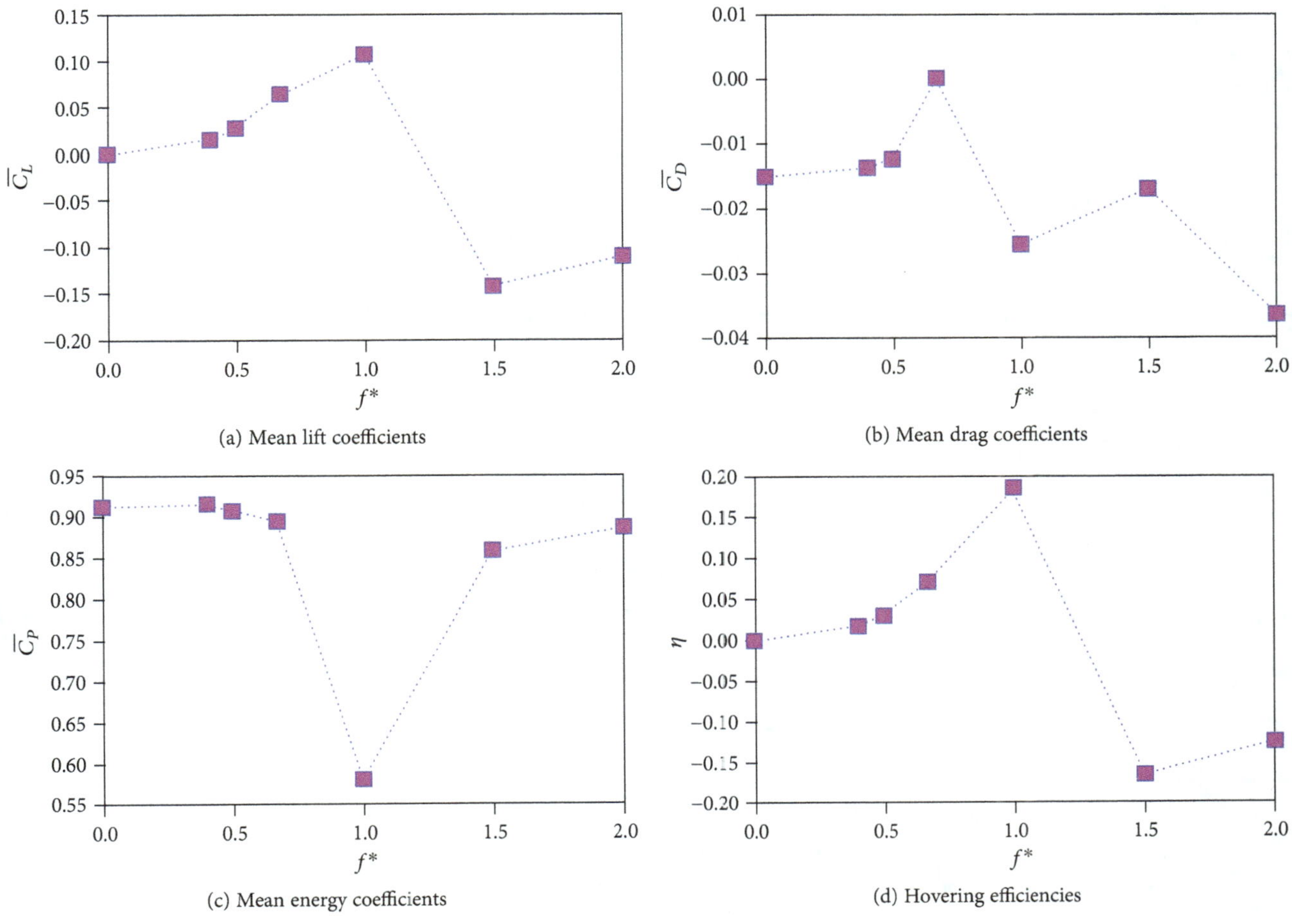

(a) Mean lift coefficients

(b) Mean drag coefficients

(c) Mean energy coefficients

(d) Hovering efficiencies

FIGURE 7: The performance of the 3D lumped flexibility wing with different f^*.

In order to investigate why the flexible wing with $f^* = 1.00$ and 1.50 possesses different performance with the rigid wing, Figures 10 and 11 show the vortex of the three considered wings at $t = 11.30T$ and $11.70T$; for each time, three different wing span positions are investigated ($z = 3.11c$, $4.70c$, and $6.29c$). Obviously, it is seen in Figure 10 that similar vortex patterns near the wing surface are observed for the rigid wing and flexible wing with $f^* = 1.00$ and 1.50; however, for the rigid wing and flexible wing with $f^* = 1.50$, a visibly separated vortex is observed. It is also found in this figure that different vortex patterns are observed at wake of the three considered wings; a jet (in the negative Y direction) is observed for the flexible wing with $f^* = 1.00$, while it is not for the rigid wing and flexible wing with $f^* = 1.50$.

In Figure 11, a comparison with the rigid wing and flexible wing with $f^* = 1.50$ is seen, in addition to the jet in the negative Y direction; the wake capture mechanism which is considered as the reason for the flapping wing having high-lift generating capacity is also observed for the flexible wing with $f^* = 1.00$, which is the reason why the lumped flexible wing with $f^* = 1.00$ has better hovering performance than the rigid wing and flexible wing with $f^* = 1.50$. It is also found in this figure that the flexible wing with $f^* = 1.50$ has the weakest vortex near the wing surface, which leads the wing to have the worst aerodynamic performance.

6.2. Varying Density Ratio. To study the effect of density ratio ρ^* on the aerodynamic performance of the 3D lumped flexibility bionic hovering wing, we fix the frequency ratio $f^* = 1.00$ which has the best hovering performance of the studied wing in the last section, and the density ratio with $\rho^* = 750$, 1000, 1250, 1500, 1750, and 2000 are considered.

Figure 12 plots the performance of the 3D lumped flexibility wing with different ρ^*. Three interesting phenomena can be concluded from this figure. Firstly, the density ratio has little effect on the mean drag coefficients, they have almost identical value at zero. Secondly, the mean energy coefficients of the wing monotonously decrease as ρ^* increases. Thirdly, the maximum mean lift coefficient as well as hovering efficiency is achieved when the wing has $\rho^* = 1750$, which indicates that the flexible wing with $\rho^* = 1750$ has better hovering performance than the other considered wing in this section. According to the study on the properties of insect wings by Combes and Daniel [28], the density ratio ρ^* of insect wing is typically of the order $O(10^3)$, which is consist with our findings.

In order to analyze the mechanism of how the density ratio influences aerodynamic performance of the 3D lumped flexibility bionic hovering wing, three typically density ratios $\rho^* = 750$, 1750, and 200 are considered.

Figure 13 shows the time variation of lift coefficients, drag coefficients, and energy coefficients of the above three

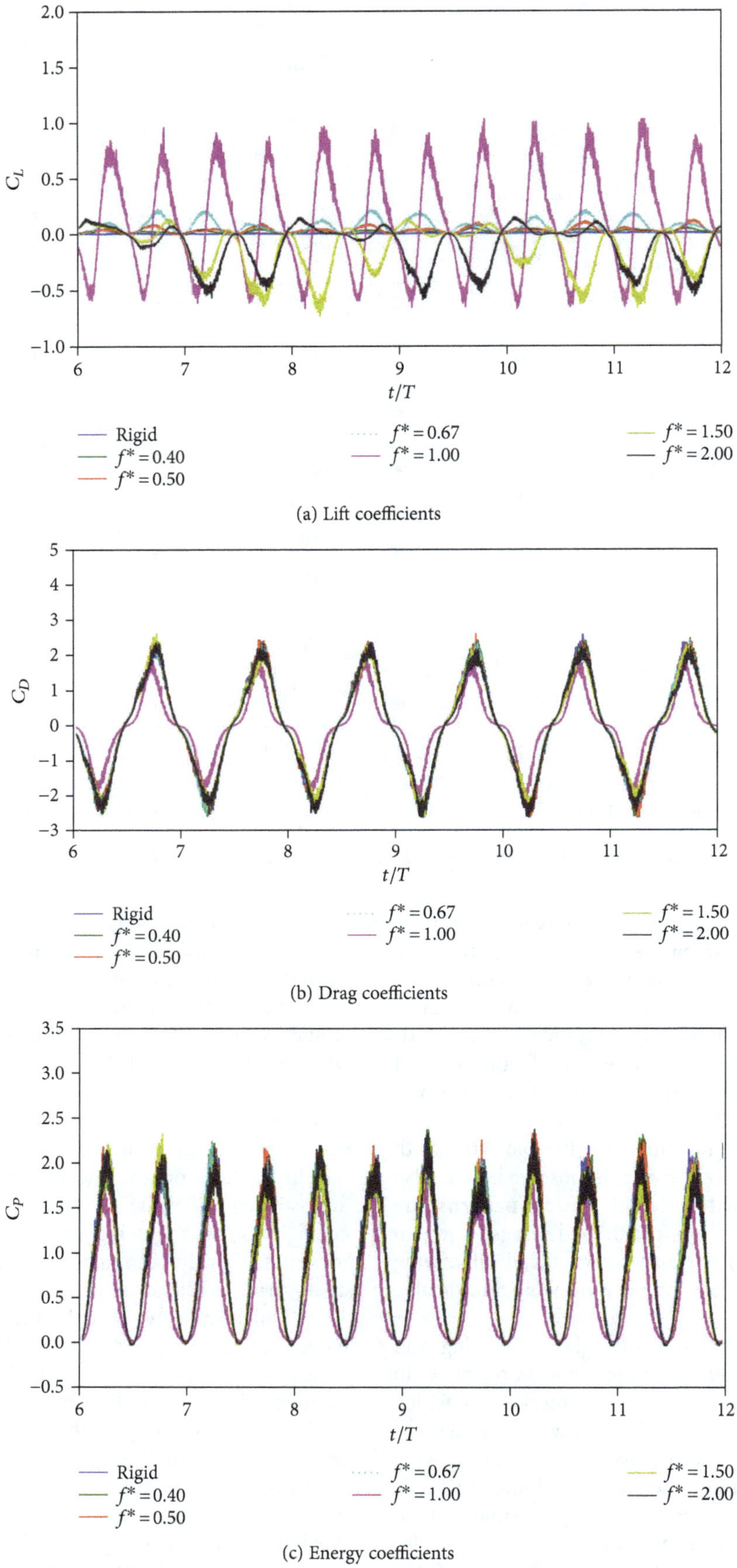

FIGURE 8: Time variation of the lift coefficients, drag coefficients, and energy coefficients of the 3D lumped flexibility wing with different f^*.

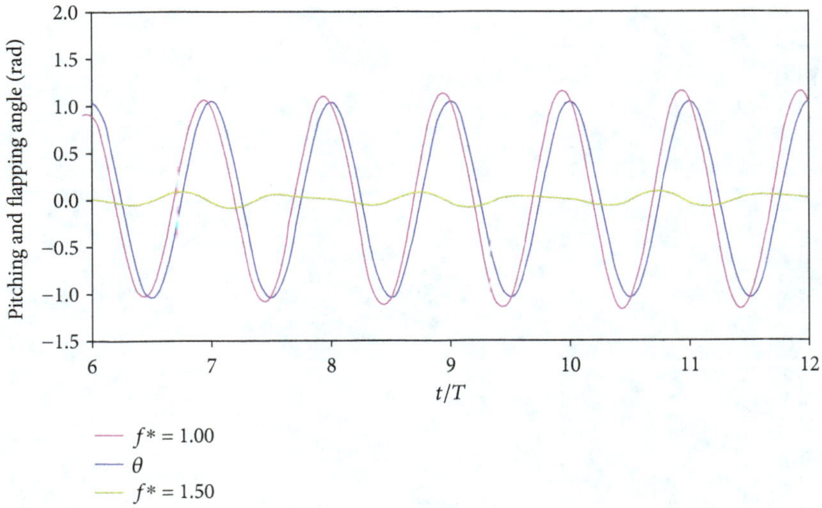

FIGURE 9: Time variation of passive pitching angle of flexible wings with f^* = 1.00 and 1.50 and flapping angle θ.

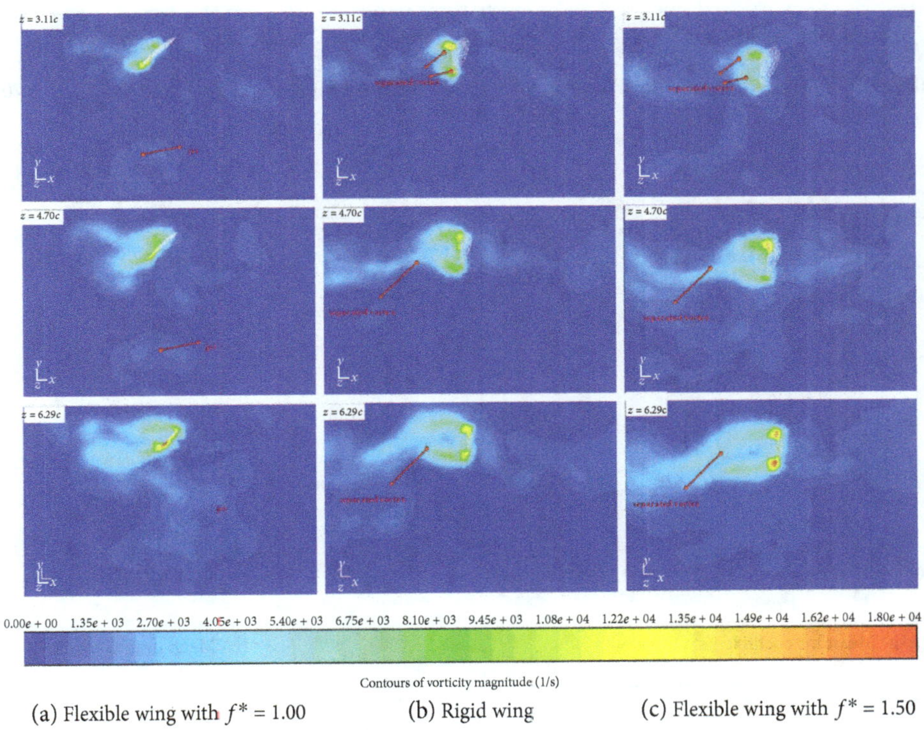

(a) Flexible wing with f^* = 1.00 (b) Rigid wing (c) Flexible wing with f^* = 1.50

FIGURE 10: The vortex of the rigid and flexible wing with f^* = 1.00 and 1.50 at $t = 11.30T$ and $z = 3.11c$, 4.70c, and 6.29c.

considered wings. Although a similar time variation trend of the lift coefficients are generated, there are phase differences between these three flexible wings, and the amplitude of the lift coefficient of the flexible wing with ρ^* = 1750 is slightly larger than that of the flexible wing with ρ^* = 750; on the contrary, the flexible wing with ρ^* = 2000 results in the flexible wing with ρ^* = 1750 to have larger mean lift coefficient and the flexible wing with ρ^* = 2000 to have smaller mean lift coefficient, as shown in Figure 12(a). Obviously, the drag coefficient of the flexible wing with ρ^* = 1750 and 2000 have almost identical values; however, the flexible wing with ρ^* = 750 has larger amplitude almost during the whole flapping cycle; moreover, it also has rougher time variation trend, which indicates that the flexible wing with ρ^* = 1750 and 2000 have better hovering stability. For the energy coefficient, the flexible wing with ρ^* = 2000 has slightly smaller value than the flexible wing with ρ^* = 1500 almost during the whole flapping cycle, while the flexible wing with ρ^* = 750 has a larger value than the flexible wing with ρ^* = 1500, which results in the mean energy coefficient of the flexible wing decreasing with ρ^* increasing, as shown in Figure 12(c).

Figure 14 plots the time variation of passive pitching angle β of the above three relation wings. It is clear in this figure that there are phase differences between these three

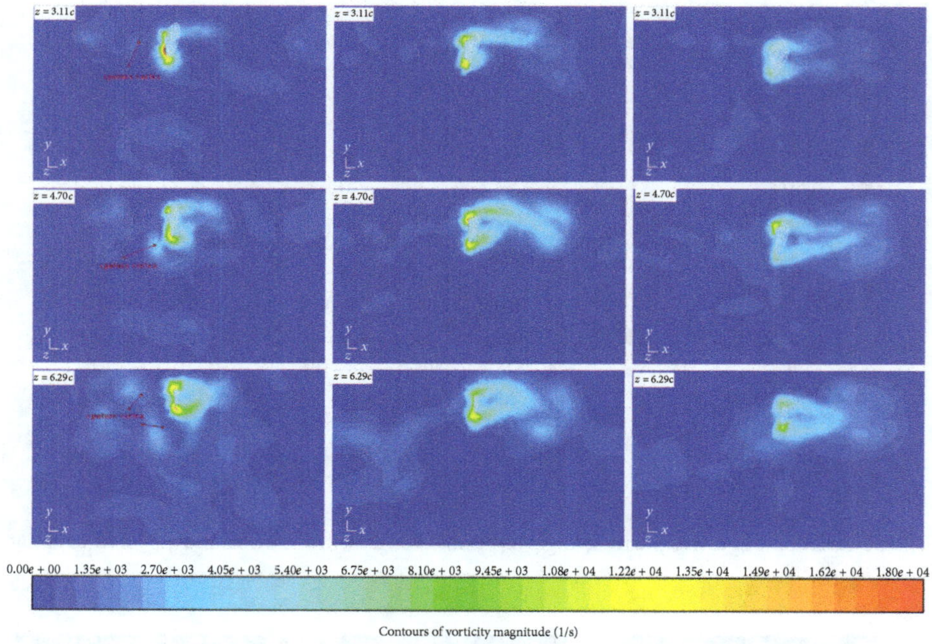

(a) Flexible wing with $f^* = 1.00$ (b) Rigid wing (c) Flexible wing with $f^* = 1.50$

FIGURE 11: The vortex of the rigid and flexible wing with $f^* = 1.00$ and 1.50 at $t = 11.70T$ and $z = 3.11c$, 4.70c, and 6.29c.

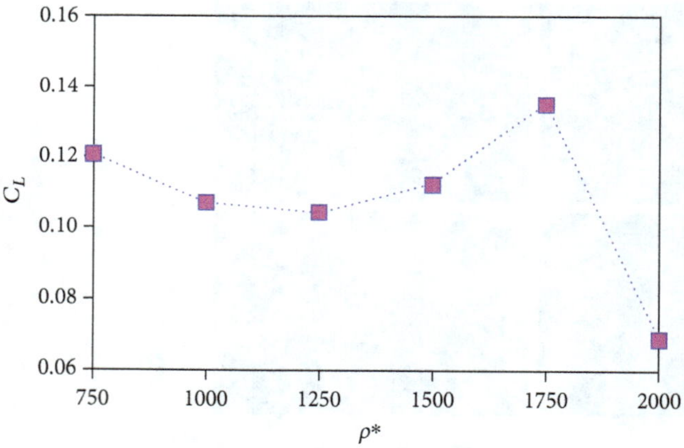

(a) Mean lift coefficients

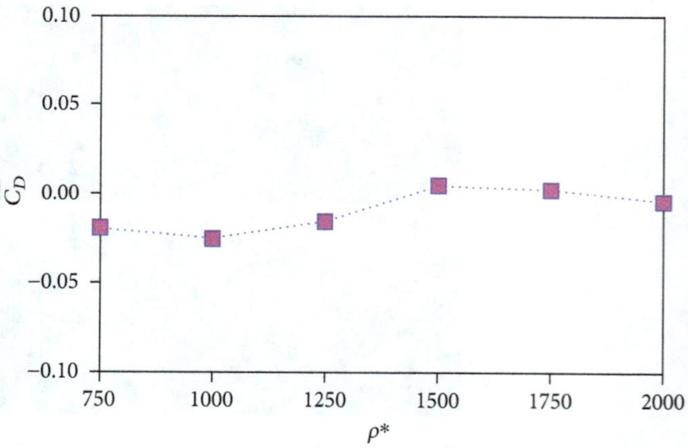

(b) Mean drag coefficients

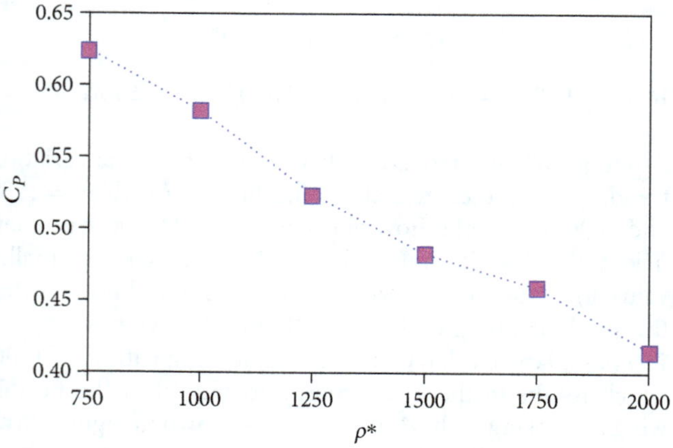

(c) Mean energy coefficients

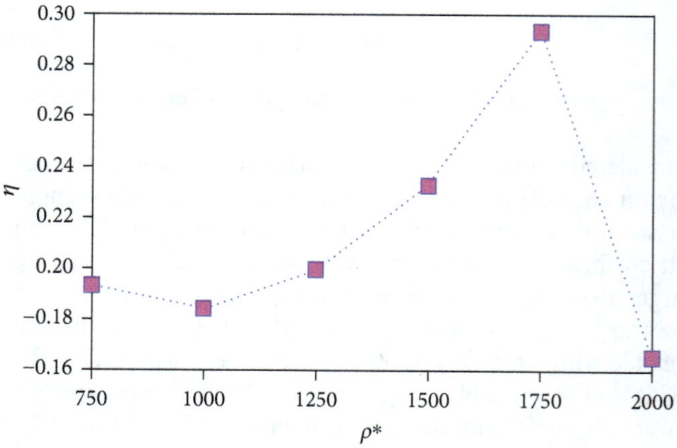

(d) Hovering efficiencies

FIGURE 12: The performance of the 3D lumped flexibility wing with different ρ^*.

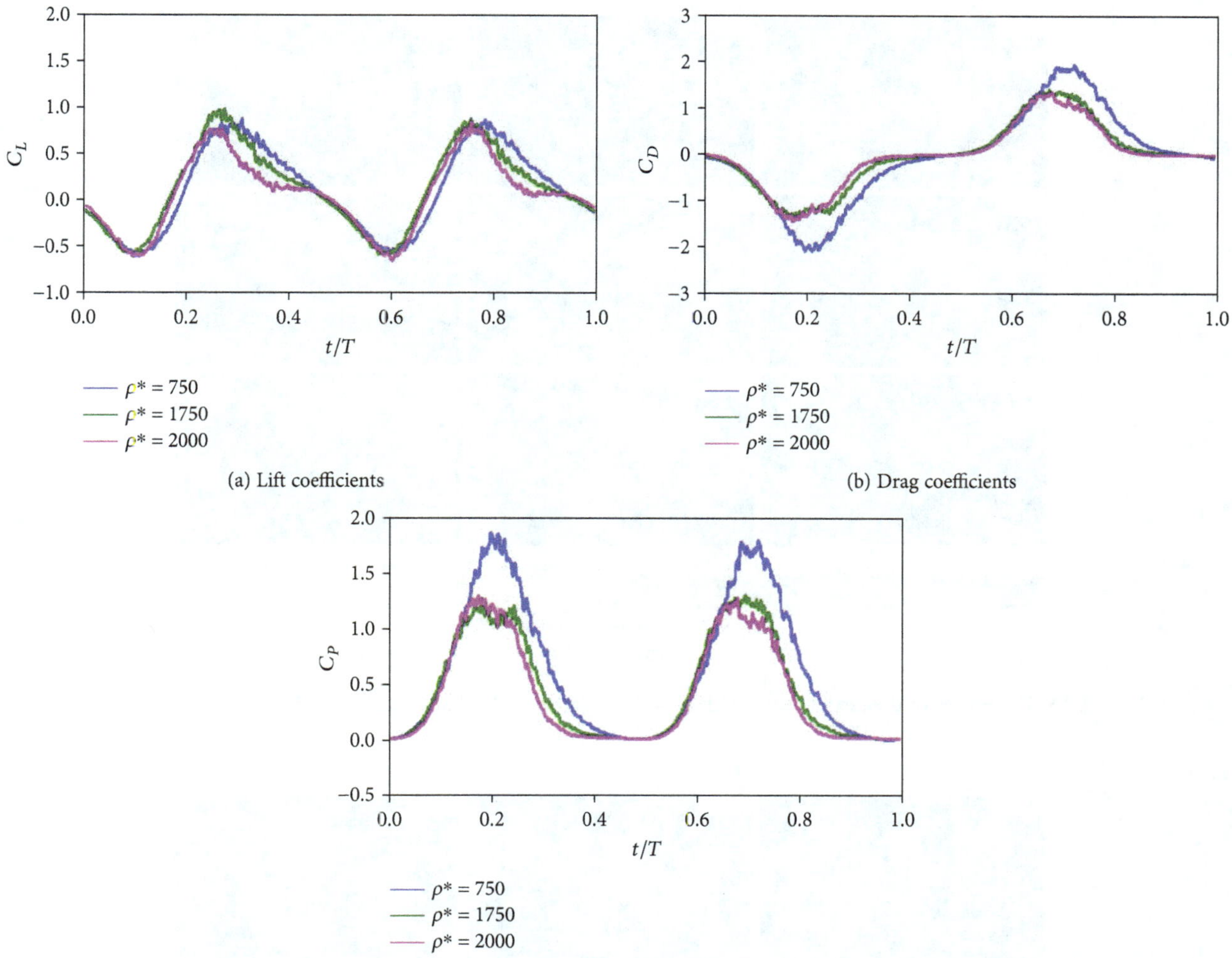

(a) Lift coefficients

(b) Drag coefficients

(c) Energy coefficients

Figure 13: Time variation of the lift coefficients, drag coefficients, and energy coefficients of the 3D lumped flexibility wing with $\rho^* = 750$, 1750, and 2000.

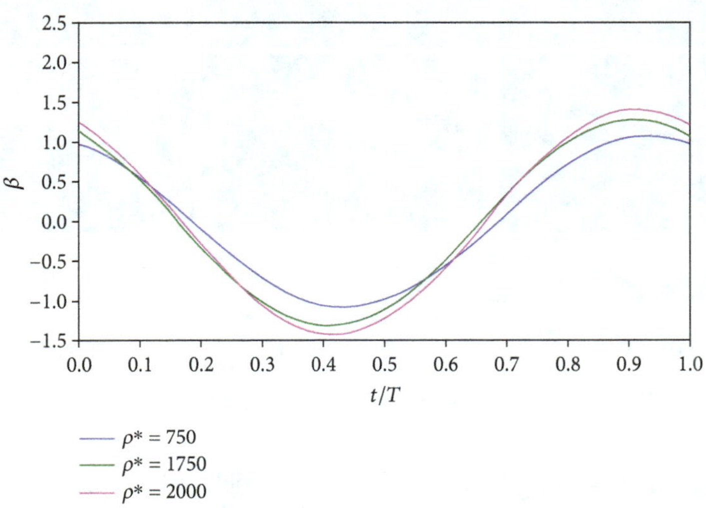

Figure 14: Time variation of passive pitching angle of flexible wings with $\rho^* = 750$, 1750, and 2000.

angles, and the amplitude of β increases with the increase in ρ^*, which may influence the vortex generated around the wing.

Figures 15 and 16 show the vortex of the flexible wing with $\rho^* = 750$, 1750, and 2000 at $t = 0.30T$ and $0.70T$, and for each of time, again three different wing

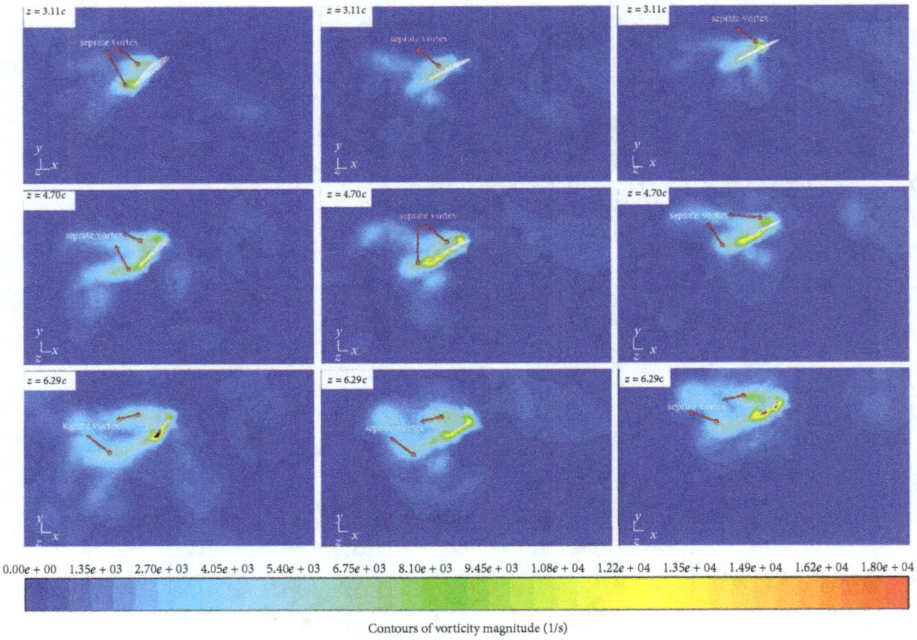

(a) $\rho^* = 750$ (b) $\rho^* = 1750$ (c) $\rho^* = 2000$

FIGURE 15: The vortex of the flexible wing with $\rho^* = 750$, 1750, and 2000 at $t = 0.30T$, $z = 3.11c$, $4.70c$, and $6.29c$.

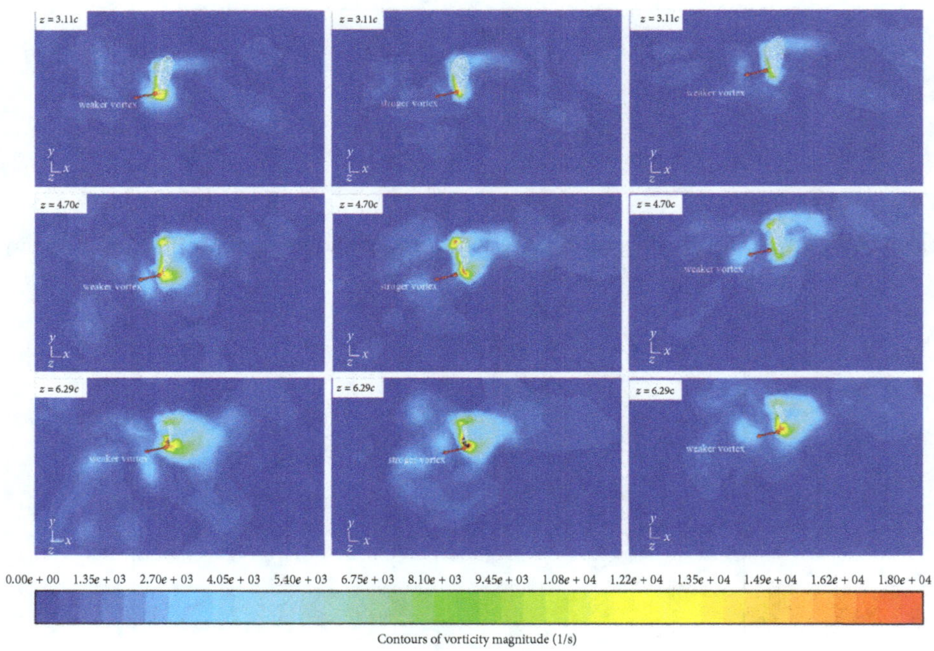

(a) $\rho^* = 750$ (b) $\rho^* = 1750$ (c) $\rho^* = 2000$

FIGURE 16: The vortex of the flexible wing with $\rho^* = 750$, 1750, and 2000 at $t = 0.70T$, $z = 3.11c$, $4.70c$, and $6.29c$.

span positions are investigated ($z = 3.11c$, $4.70c$, and $6.29c$). It is seen from Figure 15 that similar vortex patterns near the wing surface are observed for the three considered wings, from root to tip; the strength of the vortex is increased; however, the separate vortex of the flexible wing with $\rho^* = 1750$ is closer to the wing surface than the other two flexible wings, which indicates that the flexible wing with $\rho^* = 1750$ has an enhancement delay stall mechanism.

It is obvious in Figure 16 that the strongest vortex is generated around the flexible wing with $\rho^* = 1750$. Moreover, complex vortex patterns are observed under the wake of the flexible wing with $\rho^* = 750$, and the weakest vortex is generated around the flexible wing with $\rho^* = 2000$, which indicates that the flexible wing with $\rho^* = 750$ needs more energy to perform the wing hovering, and the flexible wing with $\rho^* = 2000$ generates the smallest aerodynamic force, as shown in Figure 13.

7. Conclusion and Future Work

In this paper, a novel three-dimensional bionic wing which has lumped flexibility at the root is constructed; a numerical experiment is carried out to investigate the flexibility effects on the hovering performance of this novel bionic wing, where the incompressible Navier-Stokes (N-S) equations coupled with passive pitching motion by aerodynamic torque is solved. The flow field and aerodynamic forces are analyzed for a different frequency ratio f^* and density ratio ρ^* wings, and the results show that the f^* and ρ^* influence the hovering characteristics of the wing greatly. For the wing with different frequency ratio f^*, the best performance of the wing is achieved when the wing flaps at resonance ($f^* = 1.00$). For the wing with different density ratio ρ^*, it is found that the flexible wing with $\rho^* = 1750$ has the best hovering performance, which agrees with the insect's density ratio in nature. In addition, delay stall and wake capture mechanic are observed around the wing with appropriate frequency ratio f^* and density ratio ρ^* ($f^* = 1.00$, $\rho^* = 1750$), therefore leading to better performance characteristics.

The results presented in this paper are based on lumped flexibility wing. Future work will simplify the bumblebee's wing as a 3D flexible plate.

Conflicts of Interest

The authors declare that they have no conflicts of interest.

Acknowledgments

This work was supported by the National Natural Science Foundation of China. (Project no. 51505347) and the Key Laboratory of Metallurgical Equipment and Control of Education Ministry, Wuhan University of Science and Technology Foundation (2015B07).

References

[1] F. Leys, D. Reynaerts, and D. Vandepitte, "Outperforming hummingbirds' load-lifting capability with a lightweight hummingbird-like flapping-wing mechanism," *Biology Open*, vol. 5, no. 8, pp. 1052–1060, 2016.

[2] V. M. Mwongera and Department of Aerospace Engineering, University of Bristol, UK, "A review of flapping wing MAV modelling," *International Journal of Aeronautical Science & Aerospace Research*, vol. 2, no. 2, pp. 27–36, 2015.

[3] W. Shyy, H. Aono, S. K. Chimakurthi et al., "Recent progress in flapping wing aerodynamics and aeroelasticity," *Progress in Aerospace Sciences*, vol. 46, no. 7, pp. 284–327, 2010.

[4] M. Vanella, T. Fitzgerald, S. Preidikman, E. Balaras, and B. Balachandran, "Influence of flexibility on the aerodynamic performance of a hovering wing," *Journal of Experimental Biology*, vol. 212, no. 1, pp. 95–105, 2009.

[5] S. Shkarayev and R. Kumar, "Kinematics and inertial effects in locust flapping wings," *Experimental Mechanics*, vol. 56, no. 2, pp. 245–258, 2016.

[6] C. Zhang and C. Rossi, "A review of compliant transmission mechanisms for bio-inspired flapping-wing micro air vehicles," *Bioinspiration & Biomimetics*, vol. 12, no. 2, article 025005, 2017.

[7] S. Tobing, J. Young, and J. C. S. Lai, "Effects of wing flexibility on bumblebee propulsion," *Journal of Fluids and Structures*, vol. 68, pp. 141–157, 2017.

[8] D. Ishihara, T. Horie, and M. Denda, "A two-dimensional computational study on the fluid–structure interaction cause of wing pitch changes in dipteran flapping flight," *Journal of Experimental Biology*, vol. 212, no. 1, pp. 1–10, 2009.

[9] D. Ishihara, T. Horie, and T. Niho, "An experimental and three-dimensional computational study on the aerodynamic contribution to the passive pitching motion of flapping wings in hovering flies," *Bioinspiration & Biomimetics*, vol. 9, no. 4, article 046009, 2014.

[10] D. Ishihara and T. Horie, "Passive mechanism of pitch recoil in flapping insect wings," *Bioinspiration & Biomimetics*, vol. 12, no. 1, article 016008, 2016.

[11] S. Michelin and S. G. Llewellyn Smith, "Resonance and propulsion performance of a heaving flexible wing," *Physics of Fluids*, vol. 21, no. 7, article 071902, 2009.

[12] H. Masoud and A. Alexeev, "Resonance of flexible flapping wings at low Reynolds number," *Physical Review E*, vol. 81, no. 5, article 056304, 2010.

[13] M. N. J. Moore, "Analytical results on the role of flexibility in flapping propulsion," *Journal of Fluid Mechanics*, vol. 757, pp. 599–612, 2014.

[14] J. Zhu, C. Zhou, C. Wang, and L. Jiang, "Effect of flexibility on flapping wing characteristics under forward flight," *Fluid Dynamics Research*, vol. 46, no. 5, article 055515, 2014.

[15] J. Zhu and C. Zhou, "The aerodynamic performance of flexible wing in plunge," *Journal of Mechanical Science and Technology*, vol. 28, no. 7, pp. 2687–2695, 2014.

[16] Y. Chen, N. Gravish, A. L. Desbiens, R. Malka, and R. J. Wood, "Experimental and computational studies of the aerodynamic performance of a flapping and passively rotating insect wing," *Journal of Fluid Mechanics*, vol. 791, pp. 1–33, 2016.

[17] H. Cho, N. Lee, J. Y. Kwak, S. J. Shin, and S. Lee, "Three-dimensional fluid–structure interaction analysis of a flexible flapping wing under the simultaneous pitching and plunging motion," *Nonlinear Dynamics*, vol. 86, no. 3, pp. 1951–1966, 2016.

[18] P. D. Yeh and A. Alexeev, "Effect of aspect ratio in free-swimming plunging flexible plates," *Computers & Fluids*, vol. 124, pp. 220–225, 2016.

[19] R. J. Bomphrey, G. K. Taylor, and A. L. R. Thomas, "Smoke visualization of free-flying bumblebees indicates independent leading-edge vortices on each wing pair," *Experiments in Fluids*, vol. 46, no. 5, pp. 811–821, 2009.

[20] C. P. Ellington, "The aerodynamics of hovering insect flight. III. Kinematics," *Philosophical Transactions of the Royal Society B: Biological Sciences*, vol. 305, no. 1122, pp. 41–78, 1984.

[21] C. P. Ellington, "The aerodynamics of hovering insect flight. IV. Aeorodynamic mechanisms," *Philosophical Transactions of the Royal Society B: Biological Sciences*, vol. 305, no. 1122, pp. 79–113, 1984.

[22] H. Nagai, K. Isogai, T. Fujimoto, and T. Hayase, "Experimental and numerical study of forward flight aerodynamics of insect flapping wing," *AIAA Journal*, vol. 47, no. 3, pp. 730–742, 2009.

[23] A. R. Ennos, "The importance of torsion in the design of insect wings," *Journal of Experimental Biology*, vol. 140, no. 1, pp. 137–160, 1988.

[24] S. Camarri, M. V. Salvetti, B. Koobus, and A. Dervieux, "A low-diffusion MUSCL scheme for LES on unstructured grids," *Computers & Fluids*, vol. 33, no. 9, pp. 1101–1129, 2004.

[25] T. Wang, C. G. Gu, B. Yang, and J. G. Huang, "PISO algorithm for unsteady flow field," *Journal of Hydrodynamics*, vol. 2, p. 14, 2003.

[26] J. Zhu, H. Huang, and H. Shen, "Self-starting aerodynamics analysis of vertical axis wind turbine," *Advances in Mechanical Engineering*, vol. 7, no. 12, 2015.

[27] C. Zhou and J. Zhu, "The aerodynamic performance of a 2D lumped flexible airfoil in forward flight," *Proceedings of the Institution of Mechanical Engineers Part G: Journal of Aerospace Engineering*, vol. 228, no. 8, pp. 1363–1374, 2014.

[28] S. A. Combes and T. L. Daniel, "Flexural stiffness in insect wings I. Scaling and the influence of wing venation," *Journal of Experimental Biology*, vol. 206, no. 17, pp. 2979–2987, 2003.

2

An Improved Scale-Adaptive Simulation Model for Massively Separated Flows

Yue Liu, Xiaorong Guan, and Cheng Xu

School of Mechanical Engineering, Nanjing University of Science and Technology, Nanjing 210094, China

Correspondence should be addressed to Xiaorong Guan; gxr@njust.edu.cn

Academic Editor: William W. Liou

A new hybrid modelling method termed improved scale-adaptive simulation (ISAS) is proposed by introducing the von Karman operator into the dissipation term of the turbulence scale equation, proper derivation as well as constant calibration of which is presented, and the typical circular cylinder flow at Re = 3900 is selected for validation. As expected, the proposed ISAS approach with the concept of scale-adaptive appears more efficient than the original SAS method in obtaining a convergent resolution, meanwhile, comparable with DES in visually capturing the fine-scale unsteadiness. Furthermore, the grid sensitivity issue of DES is encouragingly remedied benefiting from the local-adjusted limiter. The ISAS simulation turns out to attractively represent the development of the shear layers and the flow profiles of the recirculation region, and thus, the focused statistical quantities such as the recirculation length and drag coefficient are closer to the available measurements than DES and SAS outputs. In general, the new modelling method, combining the features of DES and SAS concepts, is capable to simulate turbulent structures down to the grid limit in a simple and effective way, which is practically valuable for engineering flows.

1. Introduction

Large-scale separations coupled with chaotic, nonlinear phenomena are often strongly pronounced in aerodynamic and industrial turbulent flows; the wide investigations for accurately predicting such flows have proven one of the most challenging CFD tasks, which is mainly related to the complex boundary layer and the troublesome broad turbulence spectra [1, 2]. Thus, at least a partial resolution of the turbulence spectrum has been the main topic of turbulence modelling for technical simulations in the last decades. In recent years, the hybrid RANS/LES methods are of great interest to both academic institutions and industry applications with the motivation that it is powerless to solve the problems at hand when using the LES or RANS method alone [3–5]. As the continuing theoretical progress and from a practical standpoint, many scholars recognize that whether a turbulence model can predict turbulent structures relies on the available eddy viscosity level when using the eddy viscous method. Following the LES method, a typical hybrid LES/RANS model termed detached eddy simulation (DES) was firstly proposed by Spalart et al. [6]. It is well known that the DES concept combines the advantages of RANS and LES methods in a simple and effective way by using the grid spacing as an adjusting scale, which has been proven promising for engineering flows, and a variety of hybrid approaches have since appeared [3, 7]. However, the DES models are not without problems [8, 9], and the main stumbling stone for DES is the ambiguous response to the grid refinement. When exceeding a critical level of grid refinement near walls, the standard DES formulation may negatively impact the RANS performance and arise some issues, such as, gray area, modeled stress depletion (MSD), and grid-induced separation (GIS) [10]. Various remedial measures such as DDES [11, 12], IDDES [13–15], and zonal DES [16] have been conducted to protect the boundary layer from the DES limiter; nevertheless, the remedies do not entirely avoid the issues appearing either conservative or inefficient [10, 17].

In contrast to DES which employs grid size as the deciding scale to produce LES content for unstable flows, the concept of scale-adaptive simulation (SAS) characterized with a second-order von Karman length scale (L_{vk}) was firstly conducted by Menter et al. [18]. The higher derivative reflects the local flow physics and gradually adjusts the resolution to the resolved scales with grid independency, which is thereby viewed the main advantage of SAS over the other hybrid methods. Again because of grid independence, the SAS is often treated as an advanced unsteady RANS (URANS) [3]. Since the first one-equation KE1E-SAS was introduced, several transformations based on different RANS background have been conducted by the Menter team [19–21]. In recent years, the SAS models have been extensively validated in numerous test cases and continuously undergone certain evolution, which dramatically shows its potential in the prediction of actual turbulent flows [22–24].

At this point, two appealing hybrid methods regarding DES and SAS have been briefly reviewed. It was expected that the LES solutions are still too expensive for industrial flows with high Reynolds number while the feasible RANS/LES approaches will dominate turbulence modelling in the next few decades [10, 17]; thus, the modification and improvement for the existing hybrid methods are very essential. For this perspective, we naturally intend to combine the formulations of DES and SAS in this paper. On one hand, the simple modelling mechanism using the ratio of modelled scale to grid scale enables DES method to be the most widely used hybrid method. But the strict grid requirement of which is difficult to cover the entire boundary layer [25]. On the other hand, the L_{vk} scale with concept of scale-adaptive enables the solution smoothly varying from RANS region to LES-like region, which allows easier grid generation especially in the complex applications. Nevertheless, the current SAS mechanism by an additional source term is troublesome and may be less efficient to obtain a convergent resolution. Furthermore, the von Karman scale and grid scale are proportional to the shear layer thickness at near-wall region, while at the separated region, both of them are proportional to the resolved scales and responsible for the generation of spectral content with LES-level viscosity [21, 26]. It thereby means the two scales are theoretically compatible with each other. In fact, there have been several reformulations by introducing L_{vk} scale into the one-equation SA-DES model [27, 28], while an even more interesting example is the "turbulence-resolving RANS" (TRANS) approach based on the two-equation k-ω model [29]. The TRANS model uses the ratio between strain and vorticity (S/Ω) as the spatial operator rather than the high derivative, which also allows the reproduction of unsteady content and makes SAS more intuitive, so the model is practical but without a clear physical mechanism between the eddy viscosity and resolved scales.

In this paper, a new hybrid model named improved scale-adaptive simulation (ISAS) is proposed by replacing the grid scale of the DES with the von Karman scale from SAS. The derivation and the constant calibration of the ISAS are described in the following sections. As for the model validation, the typical bluff body flow over a circular cylinder at Re = 3900 was selected mainly due to the rich flow physics and various references regarding both experimental reports and numerical studies. The ISAS performance was tested by numerical comparisons with available experimental data on both fine and coarse grids, and the main objective is to verify the capability of the underlying method in predicting the massively separated flows. It is worth noting that all the hybrid formulations for the following simulations are based on the two-equation shear stress transport (SST) model [30], which is demonstrated in robustly and accurately predicting the boundary layers with adverse pressure gradients.

2. Turbulence Modelling

2.1. SST Model Formulation. Benefiting from k-ω model in calculating the viscous sublayer as well as k-ε model in better dealing with the fully turbulent flows away from walls, Menter's two-equation SST model combines the best features of standard k-ω and k-ε models, which is very popular in the prediction of aeronautics flows and treated as a main platform for the hybrid formulations. In this paper, the target SST model is the version of Menter [31], and the two independent transport equations of which are written as follows:

$$\begin{aligned}\frac{\partial \rho k}{\partial t} + \nabla\cdot(\rho\mathbf{U}k) &= \tilde{P}_k - \beta^*\rho k\omega + \nabla\cdot[(\mu+\sigma_k\mu_t)\nabla k],\\ \frac{\partial \rho\omega}{\partial t} + \nabla\cdot(\rho\mathbf{U}\omega) &= \frac{\alpha}{\upsilon_t}\tilde{P}_k - \beta\rho\omega^2 + \nabla\cdot[(\mu+\sigma_\omega\mu_t)\nabla\omega]\\ &\quad + (1-F_1)\frac{2\rho\sigma_{\omega_2}}{\omega}\nabla k\nabla\omega,\end{aligned} \tag{1}$$

where k is the turbulence kinetic energy, ω is the specific dissipation rate, ρ is the density, $\mathbf{U}$ is the velocity vector, and μ is the dynamic viscosity.

All model parameters φ are obtained by an applicable compromise of k-ω and k-ε constants via $\varphi = F_1\varphi_1 + (1-F_1)\varphi_2$, in which the blending function F_1 is defined as

$$F_1 = \tanh\left\{\left\{\min\left[\max\left(\frac{\sqrt{k}}{\beta^*\omega d}, \frac{500\mu}{d^2\rho\omega}\right), \frac{4\rho\sigma_{\omega_2}k}{\mathrm{CD}_{k\omega}d^2}\right]\right\}^4\right\}, \tag{2}$$

with $\mathrm{CD}_{k\omega} = \max(2\rho\sigma_{\omega_2}(1/\omega)(\partial k/\partial x_i)(\partial\omega/\partial x_i), 10^{-10})$ and d is the distance from the nearest solid surface. The adopted model constants are $\beta^* = 0.09$, $\alpha_1 = 0.5556$, $\alpha_2 = 0.44$, $\beta_1 = 0.075$, $\beta_2 = 0.0828$, $\sigma_{k_1} = 0.85$, $\sigma_{k_2} = 1.0$, $\sigma_{\omega_1} = 0.5$, and $\sigma_{\omega_2} = 0.856$.

The turbulent eddy viscosity is expressed as

$$\mu_t = \frac{\alpha_1\rho k}{\max(\alpha_1\omega, SF_2)}, \tag{3}$$

where S is the scalar invariant of the strain rate (S_{ij}), $S = \sqrt{2S_{ij}S_{ij}}$, $\alpha_1 = 0.31$, and F_2 is a second blending function defined as

$$F_2 = \tanh\left\{\left[\max\left(\frac{2\sqrt{k}}{\beta^*\omega d}, \frac{500\mu}{d^2\rho\omega}\right)\right]^2\right\}. \tag{4}$$

It is important to note that a limit function is used in recent production terms to fairly avoid the build-up issue in stagnation regions. The production terms are expressed as follows:

$$P_k = \mu_t S^2, \quad \tilde{P}_k = \min(P_k, 10 \times \beta^* \rho k\omega). \tag{5}$$

2.2. DES Model Formulation. The transformation of the DES term to the SST model is firstly introduced by Strelets [32] and simplified by Menter et al. [17]. Here, the SST-DES model of Menter is viewed as the standard DES formulation, and the distinguishing factor over the background SST model lies in a hybrid function (F_{DES}) directly acting on the dissipation term of the k-scale equation, which is expressed as

$$\beta^* \rho k\omega \rightarrow \beta^* \rho k\omega \times F_{\mathrm{DES}}, \quad F_{\mathrm{DES}} = \max\left(\frac{l_{k\omega}}{C_{\mathrm{DES}}\Delta}, 1.0\right), \tag{6}$$

where C_{DES} is the model constant and is equal to 0.61, as the limiter should only be active in the k-ε region, $l_{k\omega}$ is the turbulent length scale defined by $l_{k\omega} = \sqrt{k}/(\beta^* \omega)$, and Δ is the local grid scale reading $\Delta = \max(\Delta x, \Delta y, \Delta z)$.

As shown above, in DES, the switch from RANS to LES-like modes depends on the ratio between the turbulent length scale (modelled) and grid scale (resolved). When near walls, $l_{k\omega} < C_{\mathrm{DES}}\Delta$ and $F_{\mathrm{DES}} = 1.0$, the model function is passive and therefore suitable for the stable flows in a RANS mode, whereas, when typically in the separating regions with refinement grids, $l_{k\omega} > C_{\mathrm{DES}}\Delta$ and $F_{\mathrm{DES}} > 1.0$, a LES-like mode is gained.

2.3. SAS Model Formulation. The formulation of the SST-SAS model differs from the pure SST model by the additional source term (Q_{SAS}) in the ω-scale equation [20]:

$$Q_{\mathrm{SAS}} = \max\left[\rho \xi \kappa S^2 \left(\frac{L_{k\omega}}{L_{\mathrm{vk}}}\right)^2 - C \cdot \frac{2\rho k}{\sigma_\varphi} \max\left(\frac{|\nabla \omega|^2}{\omega^2}, \frac{|\nabla k|^2}{k^2}\right), 0\right], \tag{7}$$

where similar to DES, $L_{k\omega}$ is the modelled turbulence length scale and reads $L_{k\omega} = \sqrt{k}/(C_\mu^{1/4}\omega)$, while the distinctive von Karman length scale (L_{vk}) is achieved by $L_{vk} = \kappa S/|\nabla^2 U|$. The model constants are $C = 2.0$, $\xi = 3.51$, $\kappa = 0.41$, $\sigma\varphi = 0.67$, and $C_\mu = 0.09$.

It is worth mentioning that the second derivative (U'') is used instead of the third derivative, appearing more sufficient in the 3-D simulation [21, 22, 33]. The von Karman length scale (L_{vk}) composed of the first and second velocity derivatives is not sensitive to grid efforts but dynamically adjusts the resolution to the spectra content. Thus, it is an appealing element that has a strong theoretical foundation and enables flow mode to smoothly vary from RANS to LES-like mode. Moreover, when in the stable region, $L_{k\omega} < L_{\mathrm{vk}}$ and $Q_{\mathrm{SAS}} = 0$, returning a pure RANS mode, while when in the unstable region, $L_{k\omega} > L_{\mathrm{vk}}$ and $Q_{\mathrm{SAS}} > 0$ and the mode will turn to LES-like mode.

2.4. ISAS Model Formulation. To combine the DES and SAS concepts based on the SST model, a simple way is to replace the grid scale in (6) with the von Karman scale.

As known, using Reynolds averaging method, the standard two-equation models such as SST always produce a turbulent length scale (L) proportional to the thickness of the boundary layer (δ), which is suitable for the stable shear flows near walls but depresses the resolved scales in the unsteady regions. When using the hybrid methods such as DES and SAS, the relation between L and δ in the RANS regions is still [20]

$$L \sim \delta, \tag{8}$$

whereas in the separated regions, the grid scale Δ and the von Karman scale L_{vk}, respectively, used in DES and SAS allow the solution to adapt to the resolved structures in a classical LES-like behavior and gradually reach to convergent solutions. The current relations between different length scales can be expressed as

$$\mathrm{DES}: (l_{k\omega} \sim C_1 \Delta) \ll \delta, \quad \mathrm{SAS}: (L_{k\omega} \sim C_2 L_{\mathrm{vk}}) \ll \delta, \tag{9}$$

where the model constants C_1 and C_2 are of the order of one. Moreover, considering the equilibrium assumptions (balance between production and destruction of the turbulence energy), the eddy viscosity of each method is of the following relation [21, 26]:

$$\nu_t^{\mathrm{DES}} = \left(\beta^{*3/4} C_{\mathrm{DES}} \cdot \Delta\right)^2 S, \tag{10}$$

$$\nu_t^{\mathrm{SAS}} = \left(\sqrt{\left(\frac{\beta}{C_\mu} - \alpha\right)/(\kappa \xi_2)} \cdot L_{\mathrm{vk}}\right)^2 S. \tag{11}$$

It is shown that both the two formulations have the similar structure as the LES viscosity:

$$\nu_t^{\mathrm{LES}} = (C_s \cdot \Delta)^2 \cdot S, \tag{12}$$

where the C_s value is also of the order of one and thus the Δ and L_{vk} scales are well compatible with each other and it is theoretically acceptable to the replacement.

Through the brief derivations above, the new proposed hybrid function employs von Karman scale instead of grid scale in (6), and the dominant factor (F_{ISAS}) of which is reformulated as

$$F_{\mathrm{ISAS}} = \max\left(\frac{l_{k\omega}}{C_{\mathrm{ISAS}} L_{\mathrm{vk}}}, 1.0\right), \tag{13}$$

where C_{ISAS} is the control parameter with the value of 1.67 calibrated optimal in the next section.

With the modification in (13), the new model can effectively adjust the turbulent fluctuations by directly operating on the k equation like DES, and more importantly, the L_{vk} scale enables the model to dynamically provide the eddy viscosity with the concept of scale-adaptive; thus, the underlying method is named improved scale-adaptive simulation (ISAS).

2.5. ISAS Model Calibration. In (10), the DES constants $\beta^* = 0.09$, $C_{\mathrm{DES}} = 0.61$. In (11), the SAS parameter

$\sqrt{((\beta/C_\mu)-\alpha)/(\kappa\xi_2)}$ appears in [21] associated with high wave number damping. We just calibrate the whole part $\sqrt{((\beta/C_\mu)-\alpha)/(\kappa\xi_2)}$ in the DHIT case like that done in [34]. Finally, we found 0.42 is feasible for the proportional relation. From the above derivation, we get the scale relation in the following:

$$\nu_t^{DES} = \left(\beta^{*3/4} C_{DES}\cdot\Delta\right)^2 S \approx (0.1\Delta)^2 S,$$
$$\nu_t^{SAS} = \left(\sqrt{\left(\frac{\beta}{C_\mu}-\alpha\right)/(\kappa\xi)}\cdot L_{vk}\right)^2 S \approx (0.42 L_{vk})^2 S. \quad (14)$$

Assuming that SAS and DES models produce the same energy dissipation, we can obtain $(0.1\Delta)^2 S = (0.42L_{vk})^2 S$ and $\Delta = 4.2L_{vk}$. If simply replace the grid scale in (6) with von Karman scale and avoid excessive dissipation, an approximate range $C_{ISAS} \le C_{DES} * 4.2 = 2.56$ will be conducted. Meanwhile, one similar relation based on (11) and (12) has been built in [21] with an optimal $C_s = 0.11$. We assume the SAS method matches the same energy spectrum of LES and the equation $\Delta = 3.82L_{vk}$ can be gained. Again, roughly use L_{vk} instead of Δ in (6) and avoid more dissipative action, and we can get $C_{ISAS} \le C_{DES} * 3.82 = 2.33$. It is very close between the two rough results.

To obtain the optimal value of C_{ISAS}, a recommended way is to calculate the simple decaying homogeneous isotropic turbulence (DHIT), for example, the classical test case performed by Comte-Bellot and Corrsin [35], just as that did for most LES and hybrid models calibrations [19, 29, 32, 36]. The experimental equipment was simplified to a cubical box with periodic boundary conditions for the ISAS simulations, and a uniform grid with $32\times32\times32$ points was used. Moreover, the numerical strategies such as the initial conditions and splitting schemes are drawn from the similar DES calibrations [29, 32]. Figure 1 shows the spectra of the resolved kinetic energy for the selected DHIT case at the normalized time $t = 2$; meanwhile, the ISAS outputs with different C_{ISAS} and the line with −5/3 slope are also given for comparison. As displayed, the energy spectra change as using different C_{ISAS} values; concretely, the lower the value, the less the dissipation. However, the curve with $C_{ISAS} = 1.67$ best matches the measurement and provides the spectral slope close to the −5/3 near the cutoff wave number, and the declared C_{ISAS} value fairly fits the rough estimate above.

Through the simple validation in the freely decaying isotropic turbulence, it turns out that, with a proper constant, the ISAS model allows for the formulation of the turbulence spectrum, which is contrary to URANS but similar to DES and SAS. As a result, the C_{ISAS} value 1.67 is proven optimal, and more comprehensive validation will be presented in the next section.

3. Generic Validation of ISAS Model

Flow over a circular cylinder is a typical case of bluff body flows, which is often involving complex vortex motion and

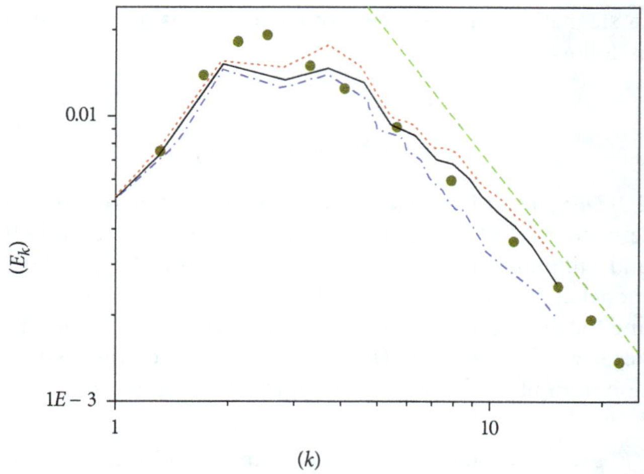

Figure 1: Comparison of energy spectra between ISAS simulations and DHIT data.

rich physics such as separation and reattachment. Because of the simple shape and representative flow phenomena, the cylinder flow is often favorable for the validation of turbulence models. As mentioned above, the flow over a circular cylinder at Re = 3900 (corresponding to cylinder diameter and freestream velocity) is selected for the thorough validation of the proposed ISAS model, mainly due to the large amount of experimental and numerical data in this particular subcritical Reynolds number.

3.1. Numerical Solution

3.1.1. Grid Generation and Boundary Conditions. All three-dimensional simulations were performed on a multiblock grid extruding from the cross-sectional plane, displayed in Figure 2(a), which is overall of hexahedral volumes, and an O-type mesh is designed near walls to keep better orthogonality. With the goal of discussing the grid-independent ability of the adopted models, it should be noted that two grids were, respectively, generated, which differ from each other in the expansion ratios for the RANS region and wake region. Specifically, as shown in Figure 2(b), a growth rate of 1.04 is used in RANS region for the relatively fine grid while a rate of 1.07 is used for the coarse one, other than that, the grid spacing of $0.15D$ and $0.2D$ is, respectively, used for the two grids in the wake region. As a result, a total of 225×158 nodes were distributed in the x-y plane of the fine grid; meanwhile, 166×126 nodes were adapted to the coarse grid. Finally, to confirm the y^+ values well below 1.0 at the wall, the minimum wall grid spacing is set to $2\times10^{-3}D$ [37], and the growth rate of both grids is below 1.1 to keep enough nodes in boundary layer.

Concerning the boundary conditions, to prevent the undesirable effort from inflow and outflow, a fixed velocity field (0.05 Ma) was used at inflow plane and a distance of $20D$ was developed from the cylinder wall, while a

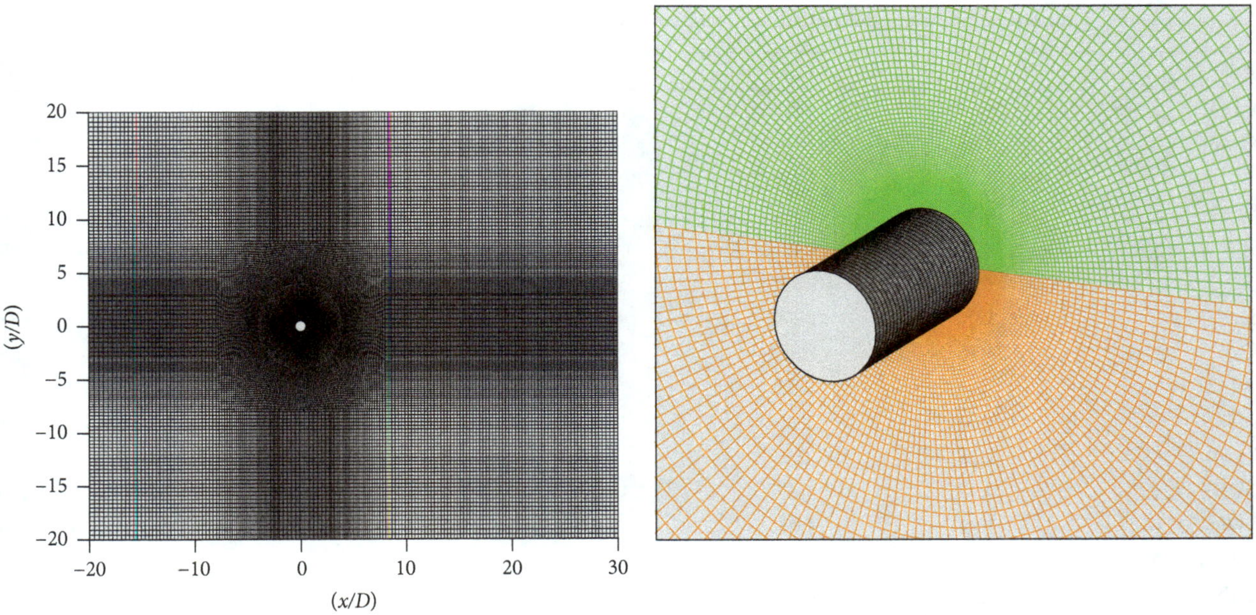

(a) x-y plane

(b) Zoom of the near-wall region

Figure 2: The design of cross-sectional grid (a) overview and (b) near-wall region (top, fine grid; bottom, coarse grid).

nonreflecting condition with the 30D distance was set for the outflow plane. The inviscid wall condition was applied to both the top and bottom sides, while the space between each side and cylinder wall is also 20D. As for the spanwise direction with a periodic assumption, the proper πD extent and normal 50 nodes were adopted for both fine and coarse grids aiming at fully reproducing the three-dimensionality [38–40].

3.1.2. Numerical Strategies. All the solutions presented in this paper are based on a 3-D incompressible solver. The space discretization of the governing equations was performed using the popular second-order finite volume method (FVM), and a second-order implicit backward method was used for time integration coupled with the dynamically adjustable stepping technique [41], while the transient PISO algorithm was employed for the pressure-velocity coupling. Concerning the convective terms which mostly affect the solving precision, both the velocity and turbulent terms were approximated using a high-resolution NVD Gamma scheme [42]. Finally, the linear systems were solved with the Gauss-Seidel method. It is worth mentioning that, to achieve stable and convergent calculations for every time step, a local Courant number close to 0.5 was adopted as well as an accuracy of 10^{-6} for pressure and 10^{-7} for other dependent variables.

It is well known that periodical vortex shedding exists in the cylinder flow at current Re, and sufficiently long physical time integration is very essential for the accurate predictions of the average surface forces. Therefore, in order to get reliable statistics, a total time of about 60 shedding cycles ($T = 300D/U_\infty$) was taken for each simulation, and the final time-spatial averaged data were recorded over approximate 50 cycles with the initial 10 cycles ignored. The focused time is more than 40 shedding cycles suggested by Franke and Frank [43].

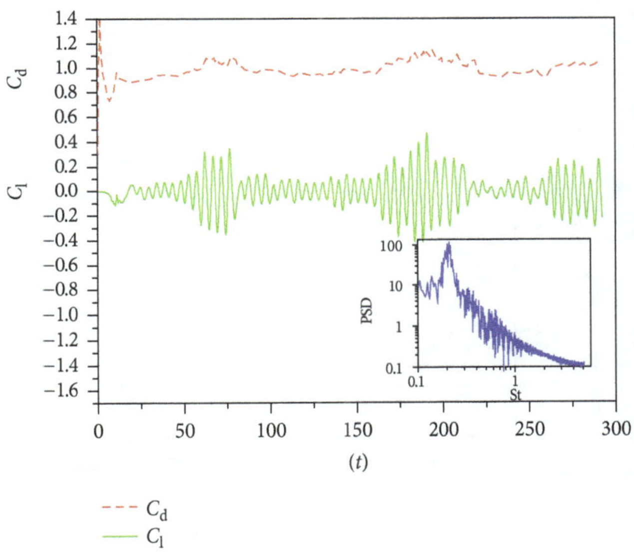

Figure 3: Time history of force coefficients using ISAS model and fine grid (lift, C_l; drag, C_d).

Figure 3 shows the time history of drag (C_d) and lift (C_l) coefficients for the ISAS simulation performed on the fine grid. As shown, the lift amplitude is unstable but appears periodical wave, while the drag magnitude appears strong fluctuation near an average value. Moreover, the Strouhal number (St) corresponding to shedding frequency is also given using a fast Fourier transform (FFT) to the lift signal with time. In addition, the mean force coefficients as well as Strouhal numbers of all simulations are listed in Table 1.

3.2. Results and Discussions. In this section, all numerical results with available experimental data are presented and detailed comparisons are conducted between ISAS,

TABLE 1: Summary of the statistical flow quantities for the flow over a cylinder flow at Re = 3900.

Case	C_d	L_r/D	$-U_{min}/U_{\infty}$	$-C_{pb}$	St	θ_{sep}
Exp. [44]	—	1.51	0.34	—	0.21	—
Exp. [46]	0.98	—	—	0.94	—	—
ISAS1	1.00	1.496	0.296	0.915	0.214	87.24
DES1	0.95	1.857	0.308	0.878	0.207	86.95
SAS1	1.01	1.485	0.285	0.964	0.212	87.90
ISAS2	1.03	1.348	0.264	1.002	0.208	88.10
DES2	1.06	1.271	0.287	1.037	0.203	88.47
SAS2	1.06	1.213	0.294	1.044	0.205	88.68
DNS [48]	—	1.41	0.291	—	0.20	—
LES [38]	1.04	1.35	0.37	0.94	0.21	88
LES-SMAG [41]	1.18	0.9	0.26	0.8	0.19	89
LES-THE [41]	0.97	1.67	0.27	0.91	0.21	88
SST-DES [37]	1.01	1.46	0.29	0.89	0.20	86.4
SST-PANS [37]	1.06	1.20	0.28	0.96	0.20	87.1
SA-DES, [40]	1.20	0.85	0.3	1.08	0.22	89.3
SA-DDES [49]	0.97	1.37	0.28	0.97	0.22	88.3
SA-IDDES [40]	1.02	1.42	0.28	0.89	0.22	87
V^2-f-DES [40]	0.99	1.68	0.4	0.83	0.21	86.4

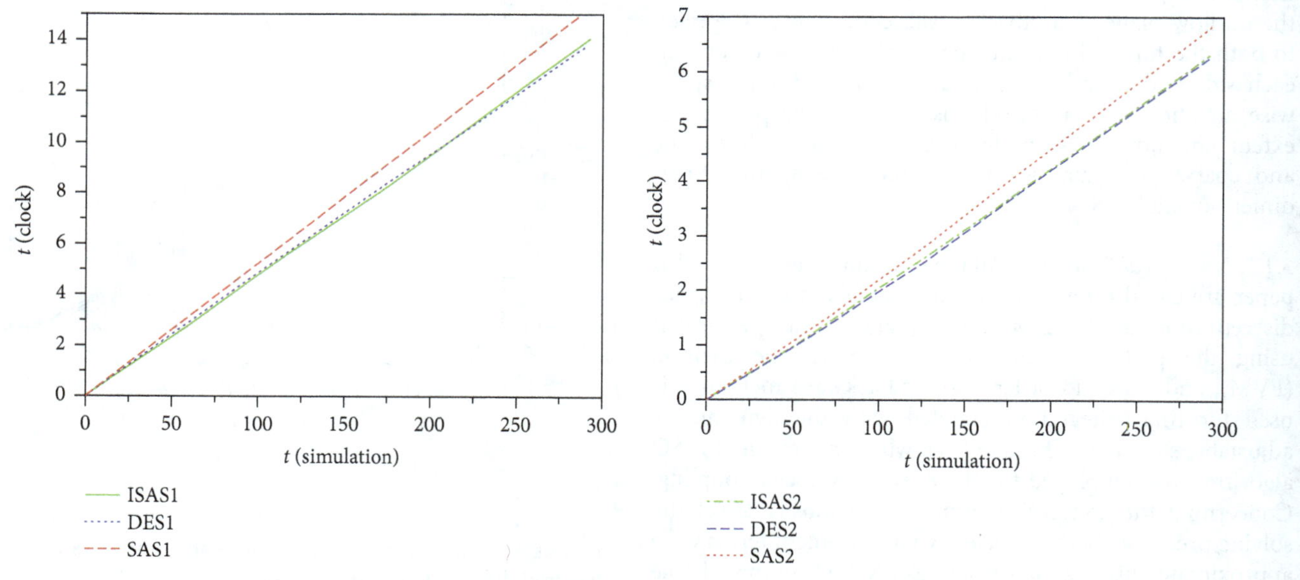

(a) Fine grid

(b) Coarse grid

FIGURE 4: Computational efficiency of ISAS, DES, and SAS.

DES, and SAS models. It is worth mentioning that the following terminologies will be used for the subsequent simulations with the goal of briefly and avoiding confusion: the results based on the fine grid for the three models are, respectively, named ISAS1, DES1, and SAS1 while those based on the coarse grid are named ISAS2, DES2, and SAS2.

3.2.1. Comparison in the Computational Efficiency. To compare the efficiency of different models, all simulations were performed on the same workstation with 40 cores. It should be emphasized that all the simulations used the same numerical setup and the iteration in every time step was ensured convergent; thus, the model efficiency just depends on the solving speed of the turbulent equations.

The computational efficiency of ISAS, DES, and SAS methods in both fine and coarse grids is discussed in Figure 4, in which the longitudinal axis is the flow time normalized with a ratio of U_{∞}/D while the transverse axis is the local clock time normalized with one hour (h). As shown, the six relations between flow time and clock time appear to be oblique lines and, respectively, with different

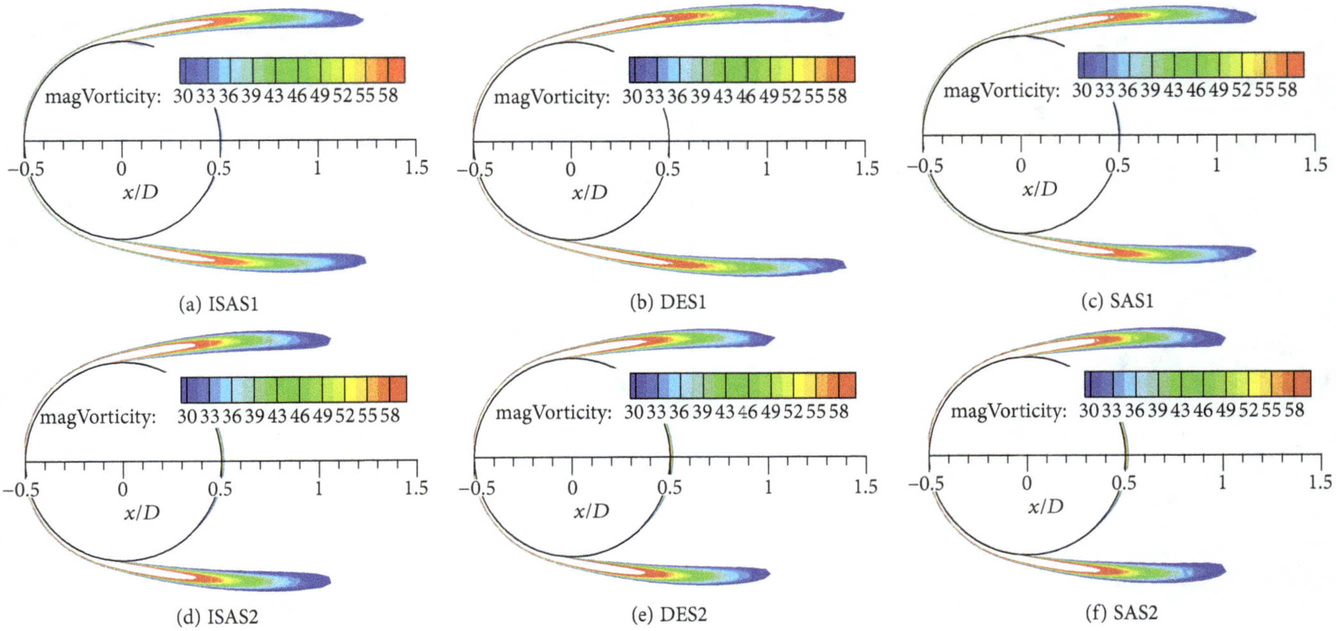

FIGURE 5: Separating shear layers displayed by the mean vorticity magnitude.

slopes. Moreover, the efficiency of ISAS appears very close to DES but drastically superior to SAS, and this trend is more obvious over flow time, which is reasonable due to different model formulations. The work mechanism using the additional source term Q_{SAS} in SAS is more complex than that only using the ratio of turbulent length scale to the von Karman length scale in ISAS, which leads the SAS equations more expensive to be solved, while with a similar formulation, there is a tiny gap between ISAS and DES efficiency though the L_{vk} scale is somewhat more complicated than grid scale.

3.2.2. Flow Statistics. It is well known that the thin shear layers with about one cylinder diameter length are the distinct feature of the cylinder flow at current Re = 3900 [38]. As observed in Figure 5, all the simulated shear layers are displayed by the mean vorticity magnitude. Both on the fine and coarse grids, all the three comparative models well represent the development of the laminar shear layers originated from the solid surface, which is especially benefited from the acceptable capacity of SST model in calculating the attached boundary layer. However, through a further inspection, the length relations among the results are DES1 > ISAS1 > SAS1, meanwhile, ISAS2 > DES2 > SAS2, mainly due to the different grid sensitivity and adjusting mechanism. The shear layers from fine grid are longer than those from coarse grid while the DES1 result from the fine grid is noticeably longest. From Figure 5, we can visibly observe the diverse model performances in simulating the shear layers but the exact length of shear layers still needs quantitative validation.

Figures 6 and 7 depict the time and space averaged velocity data extracted from different simulations; the available experimental data at same Reynolds number are listed below:

(1) PIV (particle image velocimetry) data measured by Parnaudeau et al. [44]

(2) HWA (hot-wire anemometry) data measured by Ong and Wallace [45]

Figure 6 shows the mean streamwise velocity along the wake centerline. As shown for clarity in Figure 6(a), the DES1 profiles from fine grid appear a terribly delayed development of the recirculation region by comparison with measurements, and the direct features for which are reflected on a larger recirculation length (L_r/D, the distance between the cylinder surface and the peak with 0 streamwise velocity) and a slower velocity recovery after the recirculation region. Nevertheless, the accurate recirculation size and fastest velocity recovery of ISAS1 give the best agreement with the experimental data. Finally, the SAS1 results give a similar recirculation length as ISAS1 but with larger discrepancy in the velocity peak and recovery. The similar results can be observed in Figure 7 which demonstrates the concrete velocity profiles in the near wake region. Concerning the DES1 results on the fine grid (Figures 7(a) and 7(b)), it shows a delayed development of the mean streamwise velocity from the U shape to V shape and the transverse velocity from antisymmetrical shape to declining shape. Meanwhile, the ISAS1 result gives a better recirculation size and a faster velocity recovery than the SAS1 result. When using the coarse grid, all the simulated accuracy is reasonably reduced as shown in Figures 6(b), 7(c) and 7(d). The accuracy of ISAS, DES, and SAS models is nearly at the same level compared to the measurements, but the ISAS2 results are concretely more accurate than those of the other two methods, though underestimate the velocity peak in Figure 6(b).

Combining the analysis of Figures 5–7, the fact is revealed that the reproduction of the recirculation region is specifically associated with the development of the shear layers. Thus, the recirculation data such as region size and velocity profiles are applicable to validate the accuracy of the shear layers. More concretely, with extremely longer

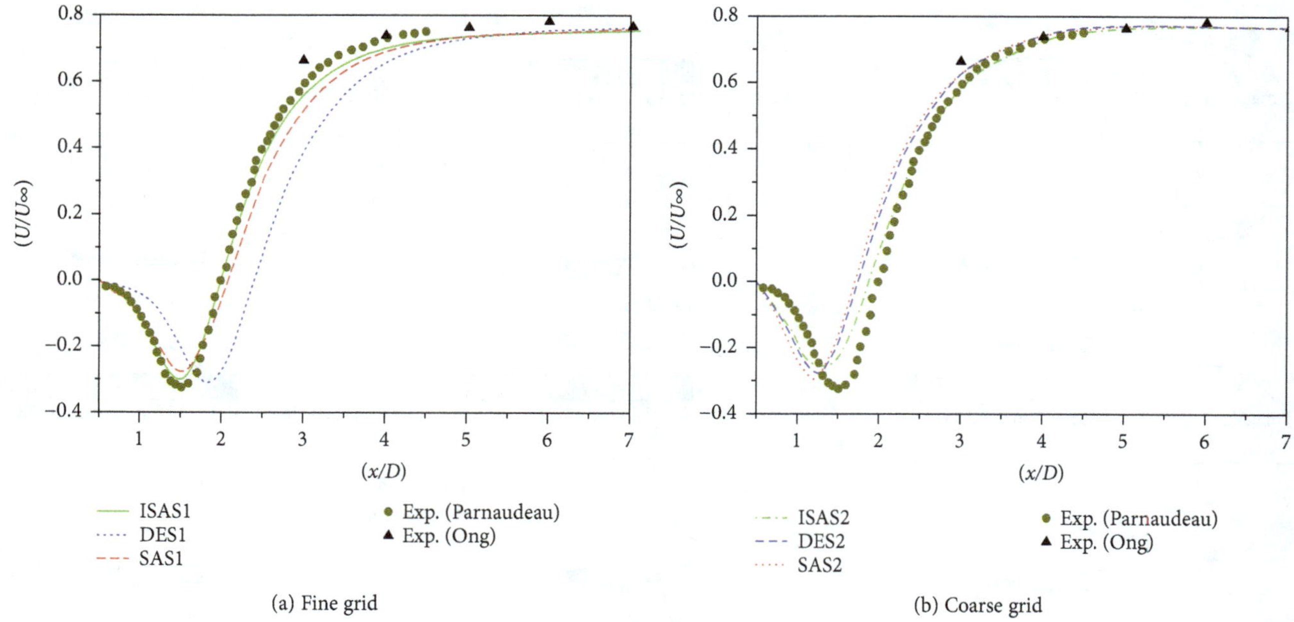

FIGURE 6: Distribution of mean streamwise velocity along the wake centerline.

length of shear layers, the DES model appears a delayed development of the recirculation region on fine grid, mainly due to the inherent grid-dependence feature. With the concept of scale-adaptive, both ISAS and SAS results do not appear exaggerated discrepancies considering different grids, and with a more accurate simulation of the shear layers, the recirculation profiles of ISAS simulation are accordingly closer to measurements. In addition, the specific results of the recirculation length (L_r/D) and the peak of back flow velocity (U_{min}/U_∞) are listed in Table 1.

From the above, we have simply discussed the relation between the simulation of boundary layer and grid using different turbulent models, while the ability of ISAS to simulate the separated flows is primitively validated. The further explanations for the LES-like resolution are based on the theoretical formulation and the physical mechanism of the three models. It is well known that the hybrid methods adjust the viscosity to different levels regarding the near-wall region and separated region in different ways. Considering the proposed ISAS model in this paper, it has a theoretical mechanism similar to DES but a physical mechanism like SAS.

Noting the different scales on the colour bar, the instantaneous ratio of the modelled turbulent length scale to the dominant length scale for different models is displayed in Figure 8. With a similar limiter formulation in the first k-scale equation as DES, the switch of the ISAS resolution from RANS to LES region is achieved by a comparison of the RANS scale (L) with the local von Karman scale (L_{vk}) rather than the grid scale used in DES. Both the L_{vk} and Δ scales are passive in the boundary layer and active in the detached region, which is distinguished by the dividing factor $L/L_{vk}=1.0$ or $L/\Delta=1.0$. Thus, as shown in Figures 8(a), 8(b), 8(d) and 8(e), the dominant turbulent length scales of ISAS and DES are nearly in the same level but visually differ from each other in the adjusting effort. For DES on fine grid, the grid scale is undesirably earlier activated and the switch from RANS to DES takes place somewhere inside the boundary layer, which will lead to the gray area issue and introduce uncertainty and inconsistent solution to the primary vertex, whereas this issue can be overcame in ISAS benefited from the L_{vk} scale, which is determined by the local flow physics and provides a finer adjusting effort for the flow unsteadiness. It should also be mentioned that the F_{ISAS} scale distribution (using fine mesh) fits the mechanism of hybrid approach very well with no very large value occurred, as plotted in Figure 9.

Both treating L_{vk} as the defining scale, the adjusting effort of ISAS is visually similar to SAS as shown in Figures 8(a), 8(c), 8(d) and 8(f). Both of them are independent to the grid spacing near wall and automatically adapt the resolvable flow scales down to the grid limit in separated region. Nevertheless, the adjusting mechanism of ISAS is not identical with SAS, and the main distinction between each other lies in the different adjusting formulation. In SAS, the switch from RANS to SAS mode is controlled by the additional source term (Q_{SAS}) in the second ω-scale equation, which therefore adjusts the flow by the change of the dissipation rate, and the dividing point from RANS to LES-like is near $L/L_{vk}=0$. By contrast, the ISAS adjusting formulation directly acting on the dissipative term of k equation is more efficient.

The mean turbulent viscosity (μ_t/μ) is presented in Figure 10. The viscosity level of ISAS is more consistent with that of DES results and lower than that of SAS on both the fine and coarse grids, which depicts that the SAS model is maybe more dissipative based on current grids. Meanwhile, concerning the DES results, it appears an extremely lower area near wall when using fine grid, which means the RANS mode is prematurely destroyed but the local grids are not dense enough for the true LES simulation, which is consistent with the analysis of the DES mechanism.

As the additional information to the first-order velocity magnitude, the time and space averaged Reynolds stresses

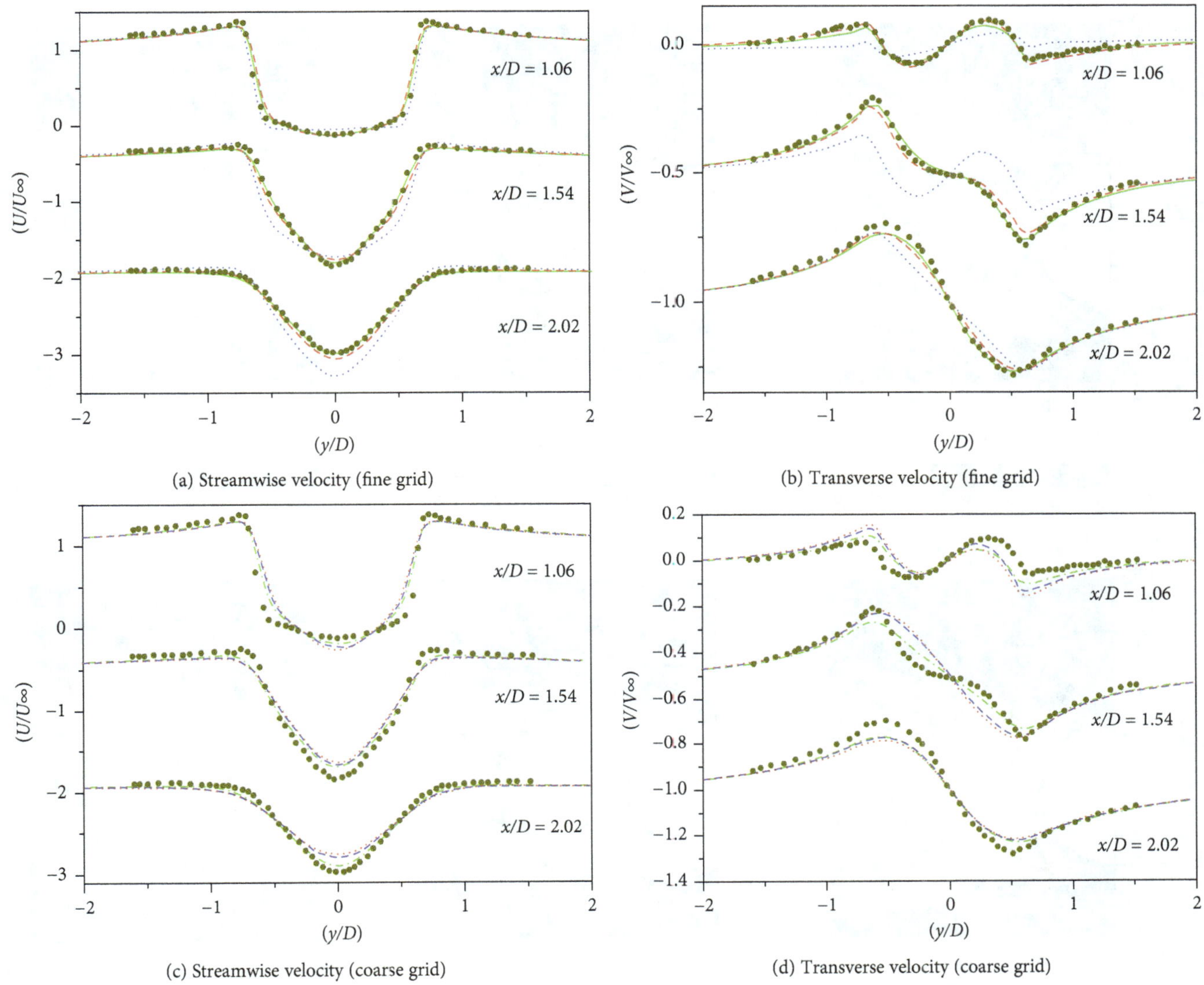

(a) Streamwise velocity (fine grid)

(b) Transverse velocity (fine grid)

(c) Streamwise velocity (coarse grid)

(d) Transverse velocity (coarse grid)

Figure 7: Mean velocity distributions at different sections (for details, see the caption for Figure 6).

are presented in Figure 11. As shown in Figures 11(a) and 11(b), the DES model arises terrible modeled stress depletion (MSD) issue, mainly due to the unacceptable lower viscosity level in the gray area, which leads to a drop of the RANS-modelled stress but without enough LES-resolved stress. At the same time, the MSD phenomenon is not evoked in the ISAS and SAS outputs. However, leaving aside the MSD issue appeared in DES, the agreement with measurements is not perfect for any model especially on the rough grid (Figures 11(c) and 11(d)), and this result is also observed for some other hybrid methods even the LES method, which is maybe associated with the inherent fault of the eddy viscosity models. It is also worth emphasizing that the ISAS result for the Reynolds stress is encouragingly closer to the experimental data than the original DES and SAS models, which is more obvious in near walls ($x/D = 1.06$, 1.54), and fairly reproduces the stress peaks in both streamwise and transverse direction.

Figure 12 demonstrates the mean wall parameters composed of the pressure coefficient (C_p) and the vorticity magnitude ($\mathrm{mag}|\Omega|$). It should be noted that the angle θ corresponds to the front stagnation point and the reference experimental data in this subsection are listed as follows:

(1) Pressure measurements conducted by Norberg [46], $\mathrm{Re} = 4020$

(2) Vorticity measurements conducted by Ma et al. [47], $\mathrm{Re} = 5000$

Concerning the mean pressure distribution first, as shown in Figures 11(a) and 11(c), when based on the fine grid, the ISAS result shows the best agreement with the measurement, while the DES and SAS results, respectively, underestimate and overestimate the pressure value, and all the three results overestimate the concerned value based on the coarse grid. Taking the vorticity magnitude into account, although the peak value on both fine and coarse grids is underestimated, all the calculated data are fairly consistent with the measurements, which is especially obvious on the fine grid. It should be mentioned that the simulated discrepancy related to the wall data among ISAS, DES, and SAS is

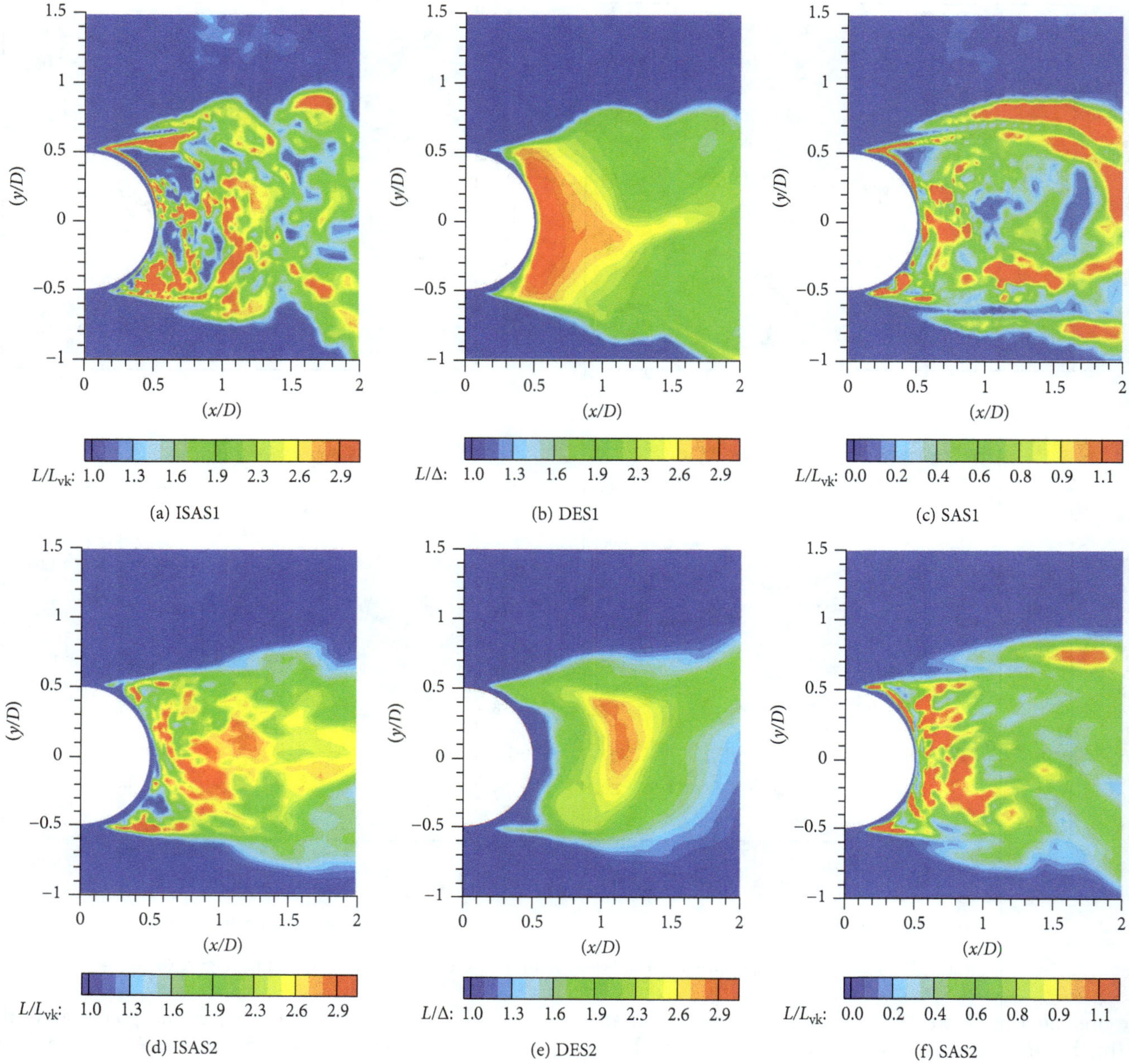

Figure 8: Contours of the dominant turbulent length scales.

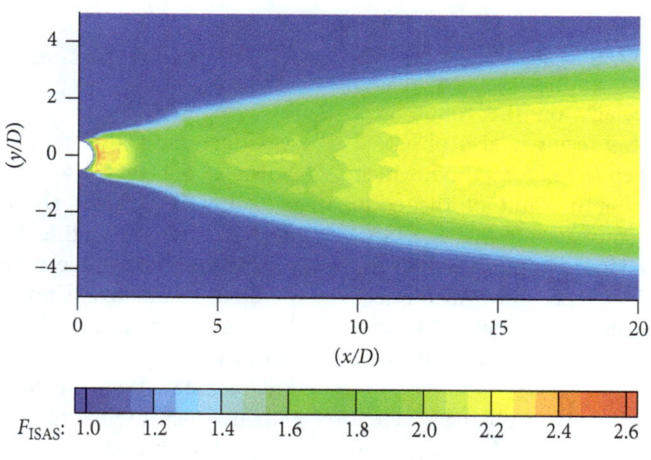

Figure 9: Time-averaged contours of the dominant turbulent length scales (ISAS1).

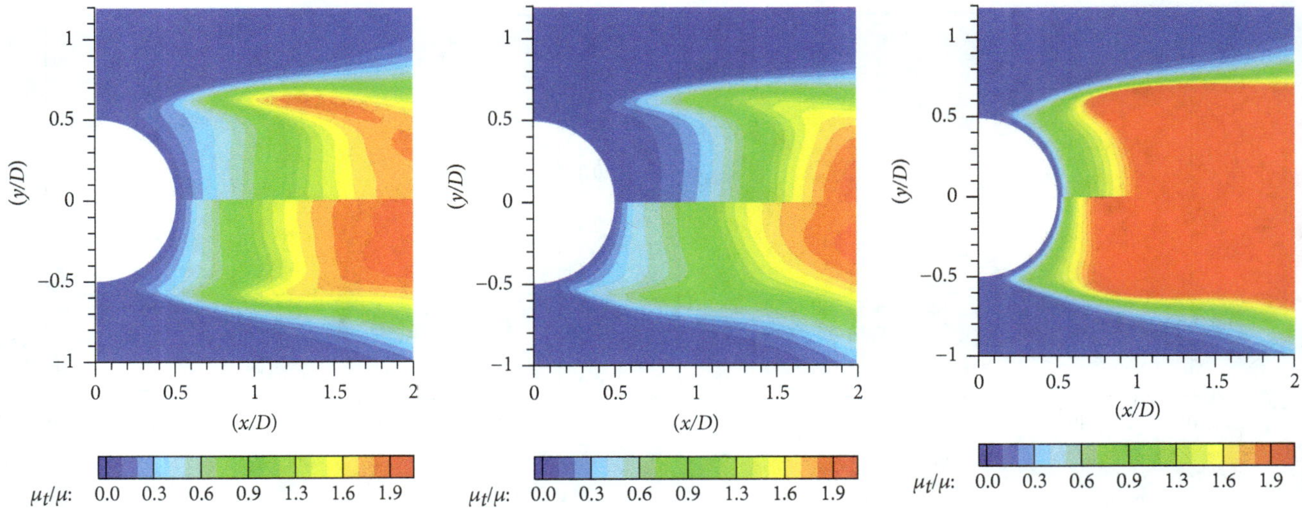

(a) ISAS viscosity

(b) DES viscosity

(c) SAS viscosity

FIGURE 10: Contours of mean turbulent viscosity (top, fine grid; bottom, coarse grid).

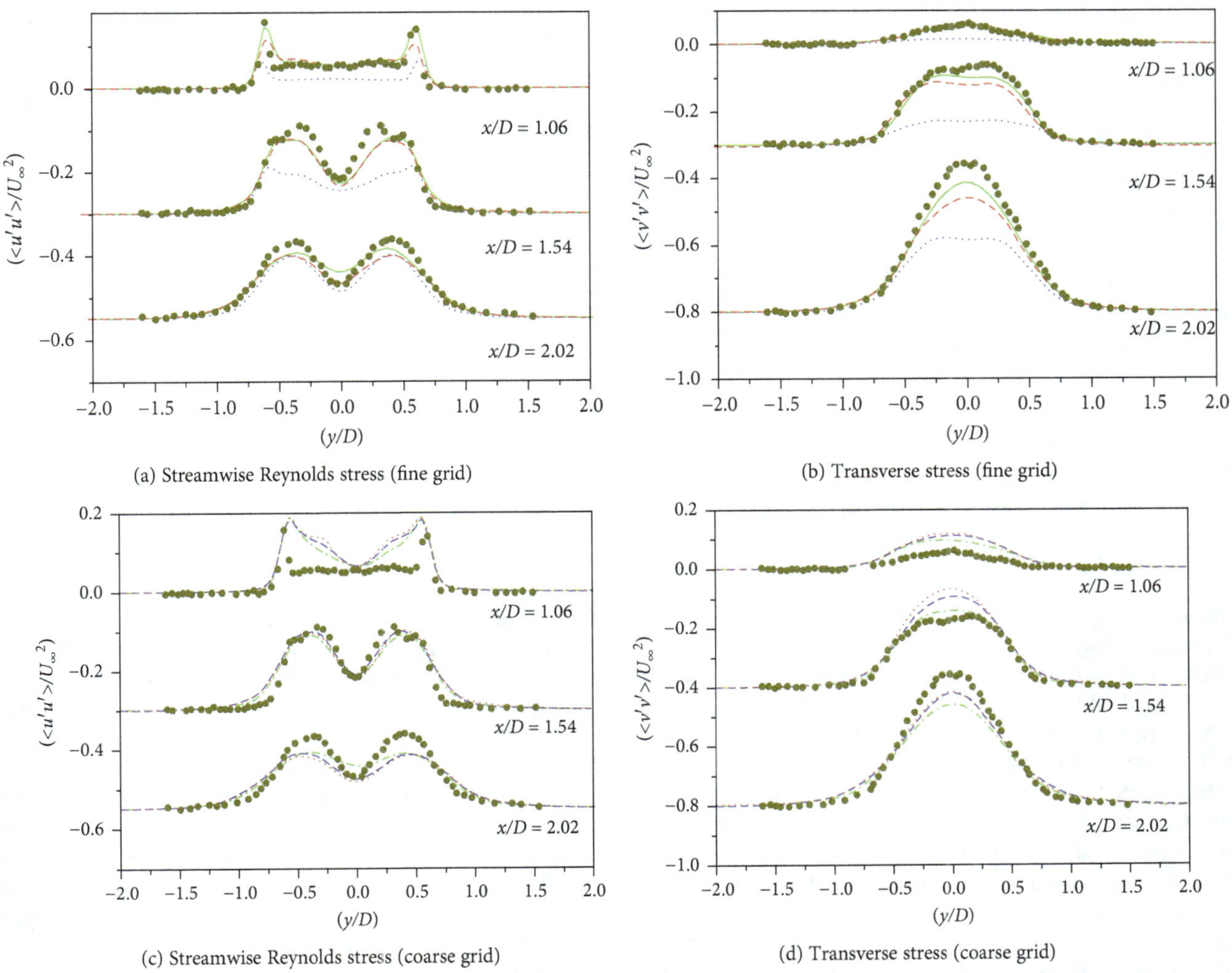

(a) Streamwise Reynolds stress (fine grid)

(b) Transverse stress (fine grid)

(c) Streamwise Reynolds stress (coarse grid)

(d) Transverse stress (coarse grid)

FIGURE 11: Mean Reynolds stress in different sections (for details, see the caption for Figure 6).

very minor and the numerical accuracy is also acceptable compared to measurements. The analysis shows the good ability of the three models in calculating the wall surface, which is greatly benefited from the SST model with strength of accurately and robustly integrating to the wall. Finally, for the sake of comparison, the calculated base pressure

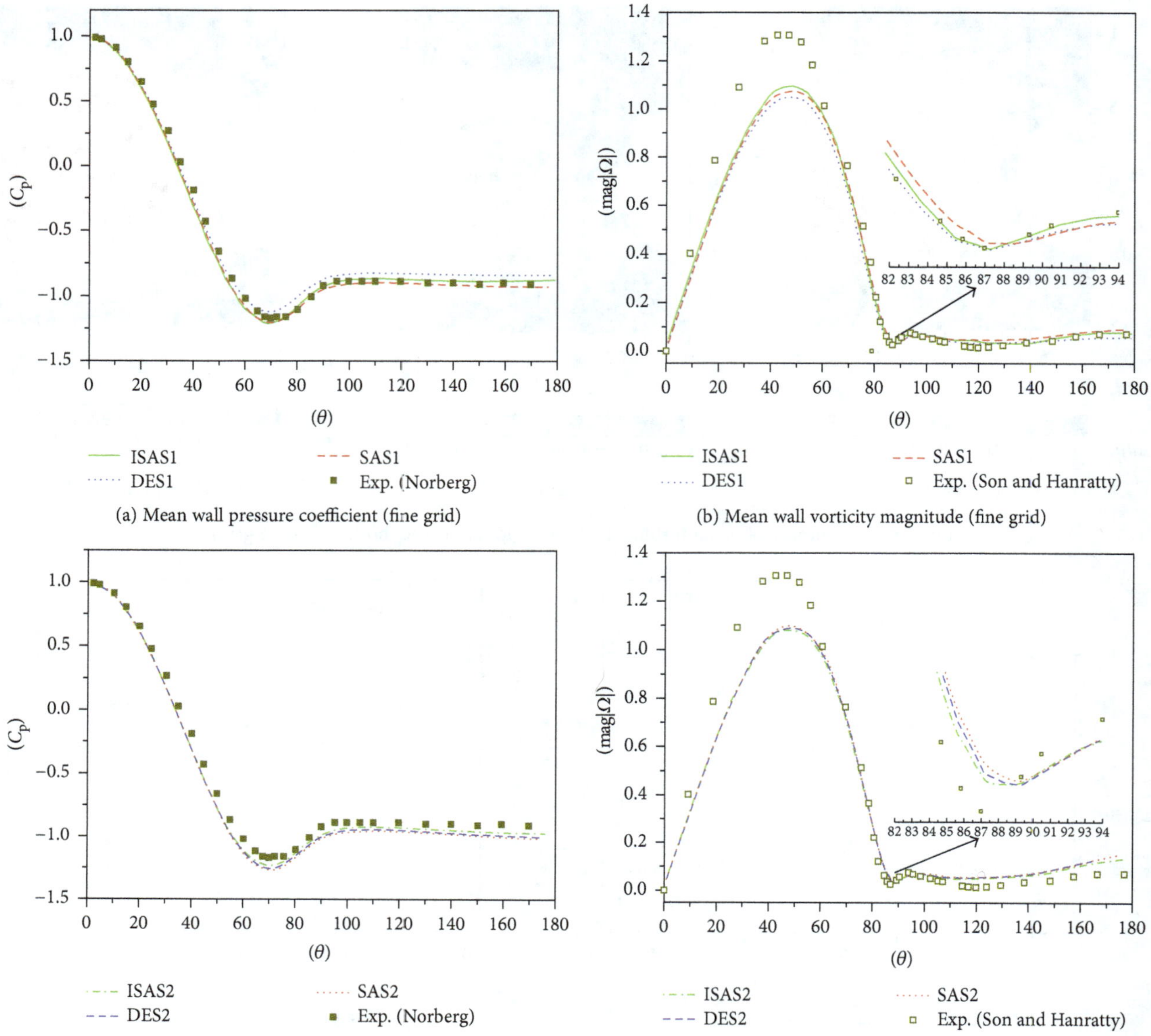

(a) Mean wall pressure coefficient (fine grid)

(b) Mean wall vorticity magnitude (fine grid)

(c) Mean wall pressure coefficient (coarse grid)

(d) Mean wall vorticity magnitude (coarse grid)

FIGURE 12: Comparison of the mean wall data.

coefficient (C_{pb}, C_p at $\theta = 180°$) and the separation angle (θ_{sep}) of the attached layer are also listed in Table 1.

Several important integral flow statistics are listed in Table 1, and meanwhile, the available experimental data [48] with relevant numerical results [49] are also presented here as reference. Concerning the parameters associated with the development of the shear layers such as the drag coefficient (C_d) and the recirculation length (L_r/D), the DES1 results based on the fine grid provide an underestimated C_d and excessively longer L_r/D with undesirable longer shear layers, and this result is consistent with that of [50], whereas both the ISAS and SAS results are in excellent agreement with the measurements, mainly benefiting from the grid-independent L_{vk} scale which will not introduce unpredictable efforts into the detachment mechanism of the shear layers. When using the coarse grid, the C_d and L_r/D obtained from ISAS2 are still closer to the measurements, while the results of DES and SAS are comparable with each other and the DES model gives a more accurate recirculation length. As for the minimum back flow velocity ($-U_{min}/U_{\infty}$), the ISAS results are not the best compared with DES and SAS results but still comparable with most of numerical references, while there is minor difference between the DES and SAS results. As for the base pressure coefficient ($-C_{pb}$), the ISAS results are better than the other two simulations, while between the DES and SAS results, it is still shown that each of them is better on the fine or coarse grids. Finally, the discrepancy from the three models is very small concerning the Strouhal number (St) and the separation angle (θ_{sep}), and all results are comparable with the other numerical results.

The capabilities of current models in reproducing turbulent unsteadiness are displayed in Figure 13, which shows the instantaneous turbulent structures identified by the Q-criterion iso-surface and coloured by velocity magnitude. As observed in Figures 13(a)–13(c), all the three models

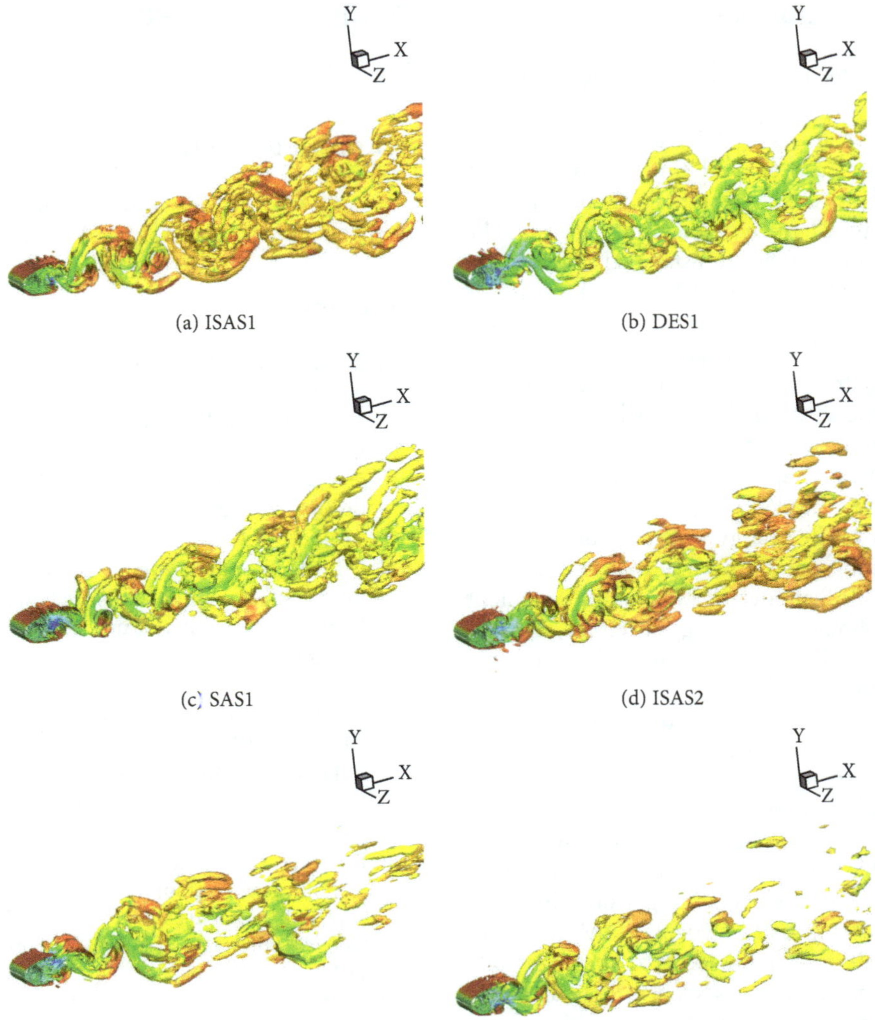

FIGURE 13: Instantaneous turbulent structures identified by the Q-criterion.

fairly represent the stably two-dimensional boundary layer and three-dimensional structures with LES-like modifications, while the coarse grid does not allow the same spatial resolution as the fine one, resulting in obviously rougher turbulent structures, shown in Figures 13(d) and 13(e). Through closer inspection, it shows that the ISAS model captures the dynamically smaller structures as well as DES model and better than SAS with higher viscosity level and excessive dissipation.

Finally, the one-dimensional energy spectrum is plotted in Figure 14, which is obtained through the analysis to the transverse velocity at $x/D = 3$ line of the wake. Noting that the frequency in x-axis is normalized by the characteristic frequency f_{vs} ($f_{vs} = \mathrm{St}\ U_\infty/D$), all the three models allow the formulation of turbulent spectra and capture the two peaks very well. Though the smaller turbulent structures with low energy and high frequency have been decayed under current grid limit, all the plots fairly match the −5/3 slope relevant to Kolmogorov's energy spectrum, enabling the models to simulate the larger structures with high energy and low frequency.

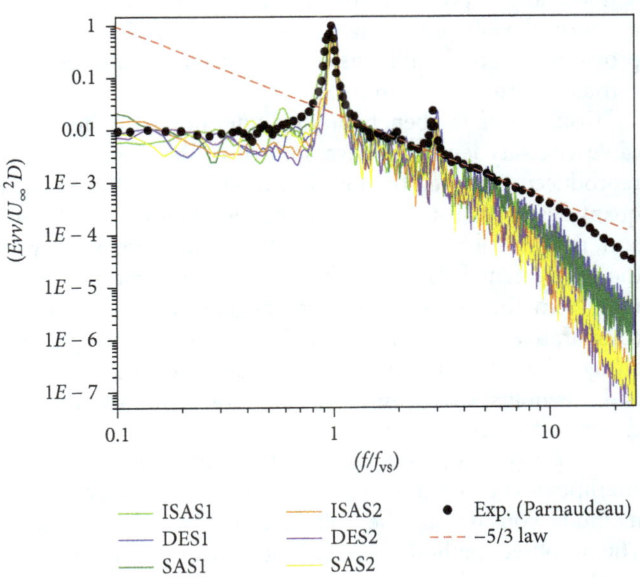

FIGURE 14: One-dimensional spectra of the transverse velocity at $x/D = 3$.

4. Summary and Conclusions

In this paper, by introducing the von Karman factor into the scale-determining equation, a new hybrid turbulent model combining the features of DES and SAS methods is proposed for massively separated flows, and with the concept of scale-adaptive, the underlying method is termed improved scale-adaptive simulation (ISAS). The derivation and calibration of the transformed formulation are presented, and the cylinder flow at $Re = 3900$ is selected as the benchmark case for the model validation. Detailed numerical comparisons with the available experimental data are conducted to study the properties of the proposed ISAS model. As expected, the ISAS model is capable to simulate turbulent details down to the grid limit in a simple and efficient way and, furthermore, appears some advantages over the original DES and SAS models.

From the analysis to the energy spectrum of the freely decaying isotropic homogeneous flow and the cylinder flow, it is shown that the ISAS model allows for the turbulent spectra close to the Kolmogorov's energy spectrum, demonstrating the capability of which in representing the resolvable contents with lower frequency.

By comparison with DES model, with similar formulation, the efficiency of ISAS in providing a convergent simulation is very close to that of DES. The L_{vk} length scale, according to the local flow physics, enables ISAS model to dynamically adjust the resolution to resolved spectra and appear comparable with DES in capturing the transient unsteadiness. Meanwhile, benefiting from the grid-insensitive character of L_{vk} scale, it enables ISAS model to operate a smooth switch from stable mode to unstable mode and avoid the grid-dependent issues such as gray area and modeled stress depletion appeared in DES simulations.

For the comparison with SAS model, with the reformulation in ISAS, the concept of scale-adaptive is made more intuitive and efficient than SAS. The resolving speed of SAS is noticeably improved through the simplified process. With a more flexible adjusting mechanism, the ISAS model provides a more feasible dissipative level and captures finer turbulent structures than SAS.

Concerning the benchmark cylinder flow, with an applicable viscosity level, the development of the shear layers is reproduced very well by the ISAS model. Thus, the focused quantities of industrial interest such as the drag coefficient and recirculation length are in the best agreement with the experimental data on fine grid. All the numerical accuracy from the coarse grid is reasonably decreased for the comparative ISAS, DES, and SAS models, though not always the best such as the minimum back flow velocity and Reynolds stress, most of the ISAS outputs are still closer to the measurements.

In general, the establishment of ISAS is a positive attempt to the combination and remedy of current hybrid methods considering the still unaffordable LES situation. The proposed method allows for a good compromise regarding the numerical accuracy, efficiency, and robustness, which is beneficial for the development of turbulence models with practical value for the engineering flows.

Conflicts of Interest

The authors declare that they have no conflicts of interest.

Acknowledgments

The authors wish to acknowledge the National Natural Science Foundation of China, Award no. 11202101.

References

[1] M. R. Malik and D. M. Bushnell, "Role of computational fluid dynamics and wind tunnels in aeronautics R&D," US Patent TP-2012-217602, 2012.

[2] J. Slotnick, A. Khodadoust, D. Darmofal, W. Gropp, E. Lurie, and D. Mavriplis, "CFD vision 2030 study: a path to revolutionary computational aerosciences," US Patent CR-2014-218178, 2014.

[3] J. Fröhlich and D. von Terzi, "Hybrid LES/RANS methods for the simulation of turbulent flows," *Progress in Aerospace Sciences*, vol. 44, no. 5, pp. 349–377, 2008.

[4] U. Piomelli, "Large eddy simulations in 2030 and beyond," *Philosophical Transactions of the Royal Society A: Mathematical, Phisical and Engineering Sciences*, vol. 372, no. 2022, article 20130320, 2014.

[5] C. D. Argyropoulos and N. C. Markatos, "Recent advances on the numerical modelling of turbulent flows," *Applied Mathematical Modelling*, vol. 39, no. 2, pp. 693–732, 2015.

[6] P. Spalart, M. Strelets, and S. Allmaras, "Comments on the feasibility of LES for wings, and on a hybrid RANS/LES approach," in *Paper Presented at the Meeting of the First AFOSR International Conference on DNS/LES*, Colombus, OH, USA, 1997.

[7] G. Kumar, S. K. Lakshmanan, H. Gopalan, and A. De, "Investigation of the sensitivity of turbulent closures and coupling of hybrid RANS-LES models for predicting flow fields with separation and reattachment," *International Journal for Numerical Methods in Fluids*, vol. 83, no. 12, pp. 917–939, 2017.

[8] D. Li, "Numerical simulation of thin airfoil stall by using a modified DES approach," *International Journal for Numerical Methods in Fluids*, vol. 54, no. 3, pp. 325–332, 2007.

[9] S. Tu, S. Aliabadi, R. Patel, and M. Watts, "An implementation of the Spalart-Allmaras DES model in an implicit unstructured hybrid finite volume/element solver for incompressible turbulent flow," *International Journal for Numerical Methods in Fluids*, vol. 59, no. 9, pp. 1051–1062, 2009.

[10] P. R. Spalart, "Detached-eddy simulation," *Annual Review of Fluid Mechanics*, vol. 41, no. 1, pp. 181–202, 2009.

[11] P. R. Spalart, S. Deck, M. L. Shur, K. D. Squires, M. K. Strelets, and A. Travin, "A new version of detached-eddy simulation, resistant to ambiguous grid densities," *Theoretical and Computational Fluid Dynamics*, vol. 20, no. 3, pp. 181–195, 2006.

[12] M. L. Shur, P. R. Spalart, M. K. Strelets, and A. K. Travin, "A hybrid RANS-LES approach with delayed-DES and wall-modelled LES capabilities," *International Journal of Heat and Fluid Flow*, vol. 29, no. 6, pp. 1638–1649, 2008.

[13] M. S. Gritskevich, A. V. Garbaruk, J. Schütze, and F. R. Menter, "Development of DDES and IDDES formulations for the k-ω shear stress transport model," *Flow, Turbulence and Combustion*, vol. 88, no. 3, pp. 431–449, 2012.

[14] Z. Xiao, J. Lu, K. Luo, J. Huang, and S. Fu, "Investigation of flows around a rudimentary landing gear with advanced detached-eddy-simulation approaches," *AIAA Journal*, vol. 51, no. 1, pp. 107–125, 2013.

[15] L. H. Xiao, Z. X. Xiao, Z. W. Duan, and S. Fu, "Improved-delayed-detached-eddy simulation of cavity-induced transition in hypersonic boundary layer," *International Journal of Heat and Fluid Flow*, vol. 51, pp. 138–150, 2015.

[16] W. L. Zheng, C. Yan, H. K. Liu, and D. H. Luo, "Comparative assessment of SAS and DES turbulence modeling for massively separated flows," *Acta Mechanica Sinica*, vol. 32, no. 1, pp. 12–21, 2016.

[17] F. R. Menter, M. Kuntz, and R. Langtry, "Ten years of industrial experience with the SST turbulence model," *Turbulence Heat and Mass Transfer*, vol. 4, 2003.

[18] F. R. Menter, M. Kuntz, and R. Bender, "A scale-adaptive simulation model for turbulent flow predictions," in *41st Aerospace Science Meeting and Exhibit*, Reno, NV, USA, 2003.

[19] F. R. Menter and Y. Egorov, "A scale adaptive simulation model using two-equation models," in *43rd AIAA Aerospace Sciences Meeting and Exhibit*, Reno, NV, USA, January 2005.

[20] F. R. Mentor and Y. Egorov, "The scale-adaptive simulation method for unsteady turbulent flow predictions. Part 1: theory and model description," *Flow, Turbulence and Combustion*, vol. 85, no. 1, pp. 113–138, 2005.

[21] Y. Egorov and F. R. Menter, "Development and application of SST-SAS turbulence model in the DESIDER project," in *Advances in Hybrid RANS-LES Modelling*, S. H. Peng and W. Haase, Eds., vol. 97 of Notes on Numerical Fluid Mechanics and Multidisciplinary Design, pp. 261–270, Springer, Berlin, Heidelberg, 2008.

[22] Y. Egorov, F. R. Mentor, R. Lechner, and D. Cokljat, "The scale-adaptive simulation method for unsteady turbulent flow predictions. Part 2: application to complex flows," *Flow, Turbulence and Combustion*, vol. 85, no. 1, pp. 139–165, 2010.

[23] L. Davidson, "Evaluation of the SST-SAS model: channel flow, asymmetrical diffuser and axisymmetric hill," in *European Conference on Computational Fluid Dynamics*, TU Delft, The Netherlands, 2006.

[24] M. Kaltenbacher, M. Escobar, S. Becker, and I. Ali, "Numerical simulation of flow-induced noise using LES/SAS and Lighthill's acoustic analogy," *International Journal for Numerical Methods in Fluids*, vol. 63, no. 9, pp. 1103–1122, 2010.

[25] P. Spalart, "Young-Person's guide to detached-eddy simulation grids," US Patent NASA/CR-2001-211032, 2001.

[26] S. R. Bruno N. Bertier, S. Deck, and F. Dupoirieus, "A DES method applied to a backward facing step reactive flow," *Comptes Rendus Mecanique*, vol. 337, no. 6-7, pp. 340–351, 2009.

[27] R. Z. Gao, J. L. Xu, R. Zhao, and C. Yan, "Evaluation of XY-SAS model for separated flows," *Journal of Beingjing University of Aernnautics and Astronauticss*, vol. 36, no. 4, pp. 415–419, 2010.

[28] C. Y. Xu, T. Zhou, C. L. Wang, and J. H. Sun, "Application of scale-adaptive simulation technique based on one-equation turbulence model," *Applied Mathematics and Mechanics*, vol. 36, no. 1, pp. 121–130, 2015.

[29] A. Travin, M. Shur, P. Spalart, and M. Strelets, "On URANS solutions with LES-like behaviour," in *Proceedings of ECCOMAS congress on Computational Methods in Applied Science and Engineering*, Jyväskylä, Finland, July 2004.

[30] F. R. Menter, "Two-equation eddy-viscosity turbulence models for engineering applications," *AIAA Journal*, vol. 32, no. 8, pp. 1598–1605, 1994.

[31] F. R. Menter, "Review of the shear-stress transport turbulence model experience from an industrial perspective," *International Journal of Computational Fluid Dynamics*, vol. 23, no. 4, pp. 305–316, 2009.

[32] M. Strelets, "Detached eddy simulation of massively separated flows," in *39th AIAA Aerospace Sciences Meeting and Exhibit*, Reno, NV, USA, 2001.

[33] F. R. Menter and Y. Egorov, "Revisiting the turbulent scale equation," in *IUTAM Symposium on One Hundred Years of Boundary Layer Research, 129*, Solid mechanics and its applications, G. E. A. Meier, K. R. Sreenivasan, and H. J. Heinemann, Eds., Springer, Dordrecht, Netherlands, 2006.

[34] L. Du and F. F. Ning, "Scale adaptive simulation of flows past an airfoil after stall," in *ASME 2012 Fluids Engineering Division Summer Meeting collocated with the ASME 2012 Heat Transfer Summer Conference and the ASME 2012 10th International Conference on Nanochannels, Microchannels, and Minichannels*, vol. 1 of *Symposia, Parts A and B*, pp. 1551–1560, Rio Grande, PR, USA, July 2012.

[35] G. Comtebellot and S. Corrsin, "Simple Eulerian time correlation of full-and narrow-band velocity signals in grid generated, 'isotropic' turbulence," *Journal of Fluid Mechanics*, vol. 48, no. 2, pp. 273–337, 1971.

[36] N. F. Ducros, "Subgrid-scale stress modelling based on the square of the velocity gradient tensor," *Flow, Turbulence and Combustion*, vol. 62, no. 3, pp. 183–200, 1999.

[37] D. H. Luo, C. Yan, H. K. Liu, and R. Zhao, "Comparative assessment of PANS and DES for simulation of flow past a circular cylinder," *Journal of Wind Engineering and Industrial Aerodynamics*, vol. 134, pp. 65–77, 2014.

[38] A. G. Kravchenko and P. Moin, "Numerical studies of flow over a circular cylinder at Re_D=3900," *Physics of Fluids*, vol. 12, no. 2, pp. 403–417, 2000.

[39] C. J. Doolan, "Large eddy simulation of the near wake of a circular cylinder at sub-critical Reynolds number," *Engineering Applications of Computational Fluid Mechanics*, vol. 4, no. 4, pp. 496–510, 2010.

[40] V. D. Alessandro, S. Montelpare, and R. Ricci, "Detached-eddy simulations of the flow over a cylinder at $Re = 3900$ using OpenFoam," *Computers & Fluids*, vol. 136, pp. 152–169, 2016.

[41] D. A. Lysenko, I. S. Ertesvag, and K. E. Rian, "Large-eddy simulation of the flow over a circular cylinder at Reynolds number 2×10^4," *Flow Turbulence and Combustion*, vol. 92, no. 3, pp. 673–698, 2014.

[42] H. Jasak, H. G. Weller, and A. D. Gosman, "High resolution NVD differencing scheme for arbitrarily unstructured meshes," *International Journal for Numerical Methods in Fluids*, vol. 31, no. 2, pp. 431–449, 1999.

[43] J. Franke and W. Frank, "Large eddy simulation of the flow past a circular cylinder at $Re_D = 3900$," *Journal of Wind Engineering and Industrial Aerodynamics*, vol. 90, no. 10, pp. 1191–1206, 2002.

[44] P. Parnaudeau, J. Carlier, D. Heitz, and E. Lamballais, "Experimental and numerical studies of the flow over a circular cylinder at Reynolds number 3900," *Physics of Fluids*, vol. 20, article 085101, 2008.

[45] L. Ong and J. Wallace, "The velocity field of the turbulent very near wake of a circular cylinder," *Experiments in Fluids*, vol. 20, no. 6, pp. 441–453, 1996.

[46] C. Norberg, *Effects of Reynolds Number and a Low-Intensity Free-Stream Turbulence on the Flow around a Circular Cylinder*, Department of Applied Thermodynamics and Fluid Mechanics: Publication No. 87/2, Chalmers University of Technology, Sweden, 1987.

[47] X. Ma, G. S. Karamanos, and G. E. Karniadakis, "Dynamics and low-dimensionality of a turbulent near wake," *Journal of Fluid Mechanics*, vol. 410, pp. 29–65, 2000.

[48] S. Dong, G. E. Karniadakis, A. Ekmekci, and D. Rockwell, "A combined direct numerical simulation–particle image velocimetry study of the turbulent near wake," *Journal of Fluid Mechanics*, vol. 569, pp. 185–207, 2006.

[49] S. Jee and K. Shariff, "Detached–eddy simulation based on the $v^2 - f$ model," in *Seventh International Conference on Computational Fluid Dynamics*, Big Island, Hawaii, July 2012.

[50] Y. Zhang, W. G. Habashi, and R. Khurram, "Zonal detached-eddy simulation of turbulent unsteady flow over iced airfoils," *Journal of Aircraft*, vol. 53, no. 1, pp. 1–14, 2016.

3

Prediction of Lift Coefficient for Tandem Wing Configuration or Multiple-Lifting-Surface System using Prandtl's Lifting-Line Theory

Hao Cheng and Hua Wang

School of Astronautics, Beihang University, Beijing 100191, China

Correspondence should be addressed to Hao Cheng; chenghao@buaa.edu.cn

Academic Editor: William W. Liou

In tandem airfoil configuration or multiple-lifting-surface layouts, due to the flow interaction among their lifting surfaces, the aerodynamic characteristics can be affected by each other. In accordance with Prandtl's classical lifting-line theory, a method to calculate the section lift coefficient for the tandem wing configuration or multiple-lifting-surface system is presented. In that method, the form of Fourier sine series is used to express the variation of the section circulation which changes continuously along the wingspan. The accuracy of the numerical solutions obtained by the method has been validated by the data obtained from computational fluid dynamics and tunnel experiment. By varying the design parameters, such as the gap, the stagger, the incidence angle, the wingspan, the taper ratio as well as the aspect ratio, a series of tandem wing configurations are tested to analyze the lift coefficient and the induced drag of each lifting surface. From the results, it can be seen that the bigger negative gap and stagger can produce better lift characteristic for tandem wing configuration. Besides, it will also be beneficial for the lift characteristic when the incidence angle and the wingspan of fore wing are appropriately declined or if the incidence angle and the wingspan of hind wing are appropriately increased.

1. Introduction

Compared to the traditional layout, the induced drag for tandem airfoil configuration is smaller and the wingspan can be reduced with the same stiffness, for the reason that the two wings in tandem airfoil configuration are all generating positive lift. Researches on tandem airfoil configuration with its unique aerodynamic advantage were conducted for decades. Bottomley [1, 2] summarized the history of aircrafts with tandem airfoil layouts and pointed out the advantages and disadvantages of that kind of design. To study the aerodynamic characteristics for tandem airfoil configuration, three different methods, including tunnel experiment, computational fluid dynamics (CFD) method, and theoretical calculation, were extensively used.

Through experiments, Feistal et al. [3] conducted experiments on a series of multiwing configuration with finite span wings at Reynolds number of 1.4*e*6, and the results indicated that the interaction between the two wings was favorable to the fore wing (the maximum lift coefficient increased), while the maximum lift coefficient of the hind wing did not decrease. Scharpf and Mueller [4] conducted wind tunnel experiments on different tandem wing configurations, to investigate the influence on the tandem wing aerodynamic characteristics by the horizontal distance, vertical distance, and the decalage between the two wings. Results obtained from tunnel experiments agree well with the real situation but with enormous financial costs.

Another high-fidelity method, CFD, has also been widely used to simulate the aerodynamic characteristics for different layouts. Alley et al. [5] presented a method, which used a correlation obtained from grid-resolved computational fluid dynamics solutions, to predict the maximum lift coefficient for wing of arbitrary planform including the effects of twist and sweep. Fanjoy and Dorney [6] performed a series of numerical experiments, in which a Navier-Stokes analysis adapted for external flow was used to calculate the flow field around the dual-wing geometry and the stagger was varied, to study the tandem-airfoil interaction in different flight regimes. By solving the incompressible Navier-Stokes equation on overlapping grids, Broering and Lian [7] investigated the effect of wing spacing between fore and hind wings on the aerodynamics of a tandem flapping wing configuration numerically at Reynolds number of 5000. Zhang and Yu [8] investigated the unsteady aerodynamics of an aircraft with tandem wing configuration in the morphing stage using CFD methods. Most recently, Patidar et al. [9] investigated the lift and drag characteristics of a Busemann biplane with staggered tandem airfoil configurations at various Mach numbers between 0.5 and 2.5 using computational fluid dynamics. Computational fluid dynamics methods can satisfy the required accuracy with little financial cost but low computational efficiency.

Compared with tunnel experiment and the CFD method, theoretical calculation with little financial cost and high efficiency also has been widely used during the aircraft conceptual design stage. Prandtl's classical lifting-line theory [10] can be used to predict the section lift coefficient by obtaining an infinite series solution for the spanwise distribution of vorticity generated on the wing, which can compromise between the crude approximations provided by analytic methods as well as low-fidelity analyses and the high level of geometric detail, time, and resources required for the high-fidelity tools, such as Reynolds-averaged Navier-Stokes computational fluid dynamics (CFD) methods, in the early aircraft conceptual design stage. With the assumption of the linear relationship between section lift and section angle of attack as well as the assumption of a straight lifting line, Prandtl's classical lifting-line theory can provide an analytical solution in the form of an infinite sine series for the circulation distribution to calculate the spanwise distribution of lift acting on a finite-lifting surface. For the evaluation of the coefficients in the sine series, Glauert [11] presented a very straightforward method. Moreover, a more rigorous and more rapidly converging method using Fourier series expansion was developed by Rasmussen and Smith [12]. Meanwhile, McCormick [13] proposed a purely numerical method to solve the lifting-line equation for a single-lifting surface with a straight lifting line, and Anderson and Corda [14] improved the numerical method by relaxing the linear assumption of the relationship between section lift and section angle of attack. However, all the aforementioned methods, to obtain the solution to the classical lifting-line equation, only applied for the particular configuration with a single-lifting surface with no sweep and no dihedral.

For the presence of the flow interaction between the fore wing and the hind wing, the wake generated by the fore wing will be hampered by the hind wing, which will also change the effective angle of attack of the hind wing, so the aerodynamic characteristics of the tandem airfoil configuration are different from the single-wing case. For the tandem wing configuration or the configuration with multiple-lifting surface, Phillips and Snyder [15] presented a numerical lifting-line method, which was based on the Prandtl's original model of a finite wing, to predict the forces and moments acting on a system of lifting surfaces. In this numerical lifting-line method, the continuous distribution of section circulation was equivalent to a finite number of discrete horseshoe vortices approximately, which was referred to what proposed by Katz and Plotkin [16]. Furthermore, Spall et al. [17] calculated the solutions obtained from a numerical method based on Prandtl's lifting-line theory, valid for multiple-lifting surface with arbitrary sweep, for a number of rigid wing and geometries. Most recently, Jacobs et al. [18] extended the numerical lifting-line method to transonic speeds and application to multiple-lifting-surface configurations, such as wing-canard configuration and three-lifting-surface business jet. Thus, the accuracy of the methods used to obtain the solution to the lifting-line equation of tandem wing configuration or multiple-lifting-surface configuration depends on the finite number of discrete horseshoe vortices, and the section lift coefficient is not continuous along the wingspan.

In this paper, the main objective is to predict the section lift coefficient for the tandem wing configuration or multiple-lifting-surface system. A lifting-line method based on Prandtl's classical lifting-line theory is proposed, in which the solution for the spanwise distribution of section circulation is expressed in the form of Fourier series truncated to a finite series. This approach can be used to obtain the lift acting on each lifting surface of tandem wing design and configuration with two lifting surfaces in arbitrary positions. Additionally, a series of tandem wing configuration with different parameters, including the gap, the stagger, the incidence angle, and the wingspan, are calculated and simulated using the proposed method to analyze the effect of the parameters on characteristics including lift and induced drag of each lifting surface. Results with this method are validated by experimental data and CFD results at small angle of attack. The accuracy from the presented method is shown to be comparable to that obtained from experiment and CFD with little experiment cost and computational cost.

2. Lifting-Line Theory for Tandem Wing Configuration

In this section, Prandtl's classical lifting-line theory for a single finite wing is presented firstly, and then, the lifting-line theory for tandem wing configuration is proposed on the foundation of Prandtl's classical lifting-line theory combined with the Biot-Savart law.

2.1. Prandtl's Classical Lifting-Line Theory. Traditionally, the distribution of section circulation along the span for a single finite wing, with no sweep or dihedral and having an arbitrary spanwise variation in chord length, can be obtained by solving Prandtl's classical lifting-line equation. Prandtl's lifting-line theory provides a simple and precise spanwise distribution of section circulation for the special case of an elliptic wing. Otherwise, to solve Prandtl's lifting-line equation for the wing with arbitrary shape, the change of variable is very necessary.

$$y = -\frac{1}{2}b\cos(\theta). \tag{1}$$

The variation in section circulation along the span of a finite wing, $\Gamma(\theta)$, as approximated by the infinite Fourier sine series solution to Prandtl's lifting-line equation, can be written as

$$\Gamma(\theta) = 2bV_\infty \sum_{n=1}^{\infty} A_n \sin(n\theta). \tag{2}$$

Based on the Kutta-Joukowski theorem and the definition of the lift coefficient, the following relationship between the section circulation, $\Gamma(\theta)$, and the lift coefficient, $C_L(\theta)$, is

$$C_L(\theta) = \frac{2\Gamma(\theta)}{V_\infty c(\theta)}. \tag{3}$$

Moreover, according to the assumption that the lift can be closely approximated as linear function of angle of attack for airfoil at small angle of attack, the section lift coefficient, $C_L(\theta)$, can be approximately calculated as

$$C_L(\theta) = C_{L,\alpha}(\alpha_e(\theta) - \alpha_{CL}), \tag{4}$$

where $C_{L,\alpha}$ is the airfoil section lift slope, α_{0L} is the zero-lift angle of attack, and $\alpha_e(\theta)$ is the effective local angle of attack.

Consequently, combining (3) and (4), the Prandtl's lifting-line equation applying on each lifting surface can be derived as

$$\frac{2\Gamma(\theta)}{V_\infty c(\theta) C_{L,\alpha}} - \alpha_e(\theta) = -\alpha_{0L}. \tag{5}$$

Prandtl's classical lifting-line equation can be solved by calculating the effective local angle of attack, $\alpha_e(\theta)$, which can be obtained from the induced velocity generated by its self-downwash effect at a series of control points along the wingspan.

2.2. Application on Tandem Wing Configuration. For the tandem wing configuration, the variation in section circulation along the span of the fore wing and the hind wing, $\Gamma_1(\theta_1)$ and $\Gamma_2(\theta_2)$, as approximated by the infinite Fourier sine series solution to Prandtl's lifting-line equation, can be, respectively, written as

$$\begin{aligned}\Gamma_1(\theta_1) &= 2b_1 V_\infty \sum_{n=1}^{\infty} A_n \sin(n\theta_1),\\ \Gamma_2(\theta_2) &= 2b_2 V_\infty \sum_{n=1}^{\infty} B_n \sin(n\theta_2).\end{aligned} \tag{6}$$

In order to distinguish the parameters, we defined some subscripts for the variables appeared in the flowing part, which can be explained in Table 1.

According to the classical Prandtl's lifting-line theory, the bound vortex is located along the quarter chord, and the trailing vortex is parallel to the x-axis along the freestream velocity, as shown in Figure 1.

TABLE 1: Definition of the subscripts of the parameters.

Subscripts	Definition
1	Parameters of the fore wing
2	Parameters of the hind wing
11	Parameters of the fore wing generated by itself
12	Parameters of the fore wing generated by the hind wing
21	Parameters of the hind wing generated by the fore wing
22	Parameters of the hind wing generated by itself
i	Parameters related to the induced effect, the index
x	The x-component of the parameters
y	The y-component of the parameters
z	The z-component of the parameters
-1	Parameters generated by the bound vortex
-2	Parameters generated by the trailing vortex

Considering the impact on aerodynamic between the two lifting surfaces and its self-downwash effect, the effective local angle of attack, $\alpha_{e1}(\theta_1)$ and $\alpha_{e2}(\theta_2)$, can be carried out as

$$\begin{aligned}\alpha_{e1}(\theta_1) &= \alpha + \delta_1 - \varepsilon_{11}(\theta_1) - \varepsilon_{12}(\theta_1),\\ \alpha_{e2}(\theta_2) &= \alpha + \delta_2 - \varepsilon_{22}(\theta_2) - \varepsilon_{21}(\theta_2),\end{aligned} \tag{7}$$

where α is the geometric angle of attack in an arbitrary spanwise section, δ_1 and δ_2 are the incidence angle of the fore wing and the hind wing, respectively, $\varepsilon_{11}(\theta_1)$ and $\varepsilon_{22}(\theta_2)$ are the induced angle of attack by its self-downwash effect, and $\varepsilon_{12}(\theta_1)$ and $\varepsilon_{21}(\theta_2)$ are the induced angle of attack by the upwash and downwash effect between the fore wing and the hind wing, respectively; the induced angle of attack is positive for the downwash effect.

Consequently, substituting (6) and (7) into (5), the Prandtl's lifting-line equation applying on each lifting surface can be derived as

$$\begin{aligned}\frac{2\Gamma_1(\theta_1)}{V_\infty c_1(\theta_1) C_{L1,\alpha}} + \varepsilon_{11}(\theta_1) + \varepsilon_{12}(\theta_1) &= \alpha + \delta_1 - \alpha_{0L_1},\\ \frac{2\Gamma_2(\theta_2)}{V_\infty c_2(\theta_2) C_{L2,\alpha}} + \varepsilon_{22}(\theta_2) + \varepsilon_{21}(\theta_2) &= \alpha + \delta_2 - \alpha_{0L_2}.\end{aligned} \tag{8}$$

With the traditional aerodynamic theory on a finite wing with no sweep, the induced velocity generated by its own bound vortex at an arbitrary location along the lifting line can be ignored. As a consequence, only the induced velocity generated by the trailing vortex should be taken into consideration to calculate the induced angle of attack by its self-downwash effect. From the Biot-Savart

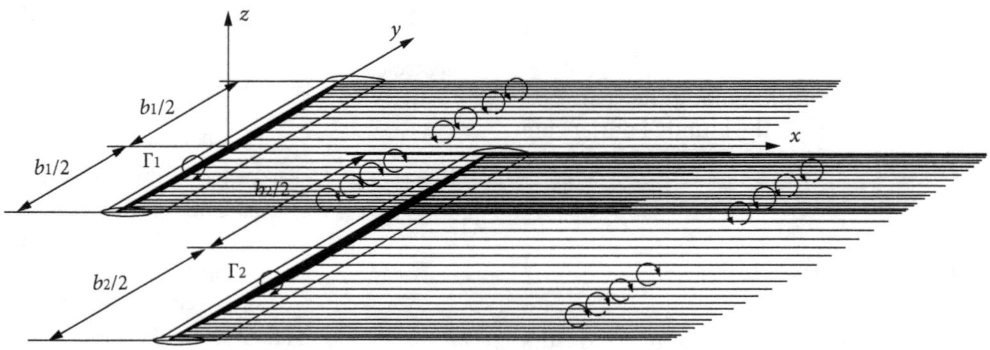

FIGURE 1: Prandtl's model for the bound vorticity and the trailing vortex parallel to the x-axis generated by tandem wings.

law, the z-velocity component induced by its own trailing vortex with the change of variable is

$$V_{iz11}(\theta_1) = V_\infty \sum_{n=1}^{\infty} \frac{nA_n \sin(n\theta_1)}{\sin(\theta_1)}, \qquad V_{iz22}(\theta_2) = V_\infty \sum_{n=1}^{\infty} \frac{nB_n \sin(n\theta_2)}{\sin(\theta_2)}. \tag{9}$$

Since the induced velocity is small compared to the freestream velocity, the induced angle of attack by its self-downwash effect, $\varepsilon_{11}(\theta_1)$ and $\varepsilon_{22}(\theta_2)$, would be approximated as the induced velocity divided by the freestream velocity, as

$$\varepsilon_{11}(\theta_1) = \arctan\left(\frac{V_{iz11}(\theta_1)}{V_\infty}\right) \approx \frac{V_{iz11}(\theta_1)}{V_\infty}, \qquad \varepsilon_{22}(\theta_2) = \arctan\left(\frac{V_{iz22}(\theta_2)}{V_\infty}\right) \approx \frac{V_{iz22}(\theta_2)}{V_\infty}. \tag{10}$$

In addition, the bound vortex as well as the trailing vortex generated by the other lifting surface can both produce an induced velocity at an arbitrary location along the lifting line. For the convenience of calculation, the z-velocity component induced by the horseshoe vortices generated by the other lifting surface would be separated into two parts, induced by the bound vortex and the trailing vortex. Applying the Biot-Savart law and the variable change, the z-velocity component induced by the bound vortex generated by the other lifting surface is

$$V_{iz12-1}(\theta_1) = -\cos(\varphi) \int_0^{\pi} \frac{\Gamma_2(\theta_2)\sin(\phi)}{4\pi r^2} \left[\frac{b_2}{2}\sin(\theta_2)\right] d\theta_2, \qquad V_{iz21-1}(\theta_2) = \cos(\varphi) \int_0^{\pi} \frac{\Gamma_1(\theta_1)\sin(\phi)}{4\pi r^2} \left[\frac{b_1}{2}\sin(\theta_1)\right] d\theta_1, \tag{11}$$

where φ is the angle between the induced velocity and the z-axis direction as shown in Figure 2, $\cos(\varphi) = S_t'/(S_t'^2 + G'^2)^{1/2}$, ϕ is the angle between the segment of the bound vortex and the line through the point and the vortex segment as shown in Figure 3, $\sin(\phi) = (S_t'^2 + G'^2)^{1/2}/r$,

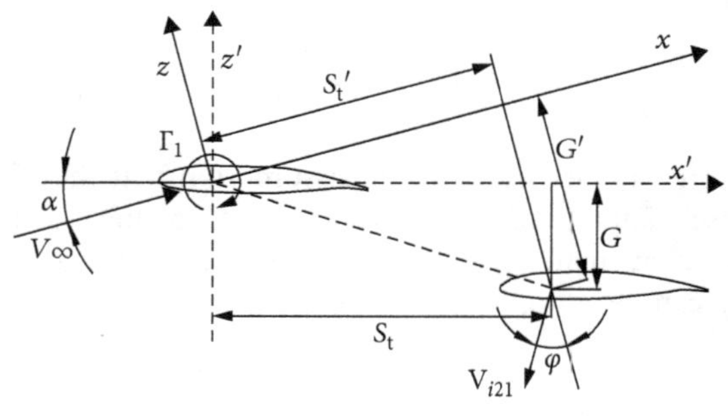

FIGURE 2: Relationship between the direction of the induced velocity and the z-axis.

r is the distance from the point to the vortex segment, and $r = ((y_2 - y_1)^2 + S_t'^2 + G'^2)^{1/2}$, S_t' is the longitudinal separation along the x-axis of the two wings and G' is the vertical distance along the z-axis between the two wings, as shown in Figure 2, and can be calculated with the coordinate transformation as

$$S_t' = S_t \cos(\alpha) + G \sin(\alpha), \qquad G' = -S_t \sin(\alpha) + G \cos(\alpha), \tag{12}$$

where S_t is the longitudinal separation of the two wings, stagger, and G is the vertical distance between the two wings, gap, as shown in Figure 2.

Meanwhile, applying the Biot-Savart law into the segment of the trailing vortex and integral along the spanwise, the z-velocity component induced by the trailing vortex generated by the other lifting surface is

$$V_{iz12-2}(\theta_1) = \frac{1}{4\pi} \int_0^{\pi} \frac{\delta y_{21}}{r'^2} \left(1 - \frac{S_t'}{r}\right) \frac{d\Gamma_2(\theta_2)}{d\theta_2} d\theta_2, \qquad V_{iz21-2}(\theta_2) = \frac{1}{4\pi} \int_0^{\pi} \frac{\delta y_{12}}{r'^2} \left(1 + \frac{S_t'}{r}\right) \frac{d\Gamma_1(\theta_1)}{d\theta_1} d\theta_1, \tag{13}$$

where δy_{21} and δy_{12} are the difference between the point and the trailing vortex segment in y direction, $\delta y_{21} = y_2 - y_1$ and $\delta y_{12} = y_1 - y_2$, and r' is the perpendicular distance from the point to the trailing vortex filament, $r' = (G'^2 + \delta y^2)^{1/2}$.

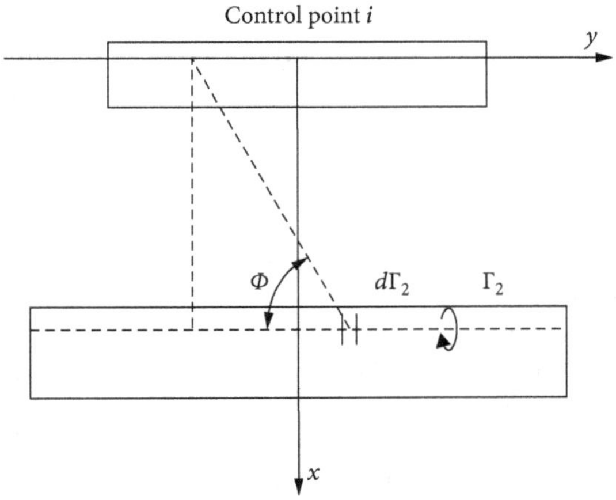

FIGURE 3: Definition of the angle describing the segment of the bound vortex and the vector.

Hence, with the same approximation in calculating the induced angle of attack by its self-downwash effect, the induced angle of attack by the other lifting surface would be

$$\varepsilon_{12}(\theta_1) = \arctan\left(\frac{V_{iz12-1}(\theta_1) + V_{iz12-2}(\theta_1)}{V_\infty}\right) \approx \frac{V_{iz12}(\theta_1)}{V_\infty},$$
$$\varepsilon_{21}(\theta_2) = \arctan\left(\frac{V_{iz21-1}(\theta_2) + V_{iz21-2}(\theta_2)}{V_\infty}\right) \approx \frac{V_{iz21}(\theta_2)}{V_\infty}. \quad (14)$$

Applying (9) to (14) into (8) at N_1 and N_2 points along the lifting line of the two lifting surfaces, respectively, the sequence $\{A_n\}$ and $\{B_n\}$ would be obtained by solving a system of N (where $N = N_1 + N_2$) linear equations and then the variation of section circulation along the span of the two lifting surfaces would be carried out.

Because of the wake-induced velocity, the freestream velocity vector will be rotated by the induced angle of attack, α_i, as shown in Figure 4 and the induced angles of attack for the two wings can be obtained as

$$\alpha_{i1}(\theta_1) = \varepsilon_{11}(\theta_1) + \varepsilon_{12}(\theta_1),$$
$$\alpha_{i2}(\theta_2) = \varepsilon_{22}(\theta_2) + \varepsilon_{21}(\theta_2). \quad (15)$$

As we assumed that the induced velocity is small compared with the freestream velocity, then $\cos\alpha_i = 1$ and $\sin\alpha_i = \alpha_i$, therefore the section lift coefficient would be calculated by using (3) easily, then the induced drag of the two wings, D_{i1} and D_{i2}, can be written as

$$D_{i1} = \rho V_\infty \int_0^\pi \Gamma_1(\theta_1)\alpha_{i1}(\theta_1)\left[\frac{b_1}{2}\sin(\theta_1)\right]d\theta_1,$$
$$D_{i2} = \rho V_\infty \int_0^\pi \Gamma_2(\theta_2)\alpha_{i2}(\theta_2)\left[\frac{b_2}{2}\sin(\theta_2)\right]d\theta_2. \quad (16)$$

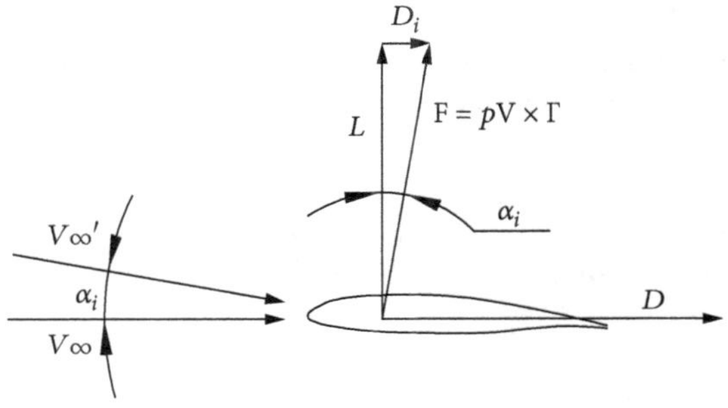

FIGURE 4: Schematic of the generation of the induced drag.

3. Correction of the Magnitude of Local Velocity

In the proceeding of calculating the section lift coefficients and the induced drag, the magnitudes of local velocity are both used in the Kutta-Joukowski theorem and the definition of the lift coefficient. To obtain the results closer to the real value, the magnitude of local velocity is corrected by that synthesized from the uniform freestream flow combined with the induced velocity at the corresponding control point.

3.1. Velocity Correction. Applying the Kutta-Joukowski theorem to a straight lifting surface, the force acting on a spanwise differential segment of the lifting surface related with the section circulation can be derived as

$$d\mathbf{F}(\theta) = \rho\Gamma(\theta)\mathbf{V}(\theta) \times d\mathbf{l}. \quad (17)$$

According to previous statements, the velocity used in calculating the lift coefficient in (3) has been set to equal to the freestream velocity. However, because the circulation distributing along the lifting line as well as the relevant trailing vortex of each lifting surface can also induce velocity at the control points, the velocity used in (17) should be synthesized from the uniform freestream flow combined with the induced velocity at the control points and can be shown as

$$\mathbf{V}(\theta) = V_\infty + \mathbf{V}_i(\theta), \quad (18)$$

where $\mathbf{V}(\theta)$ is the vector of the local velocity at the control points and $\mathbf{V}_i(\theta)$ is the vector of the induced velocity at the control points generated by the other lifting surface, $\mathbf{V}_i(\theta) = (V_{ix}(\theta), V_{iy}(\theta), V_{iz}(\theta))$.

Because the trailing vortex sheet parallel to the x-axis, so the x-velocity component induced by the trailing vortex equals to zero, the lift paralleled to the z-axis can be computed from (17) and (18) and the section lift coefficient of one lifting surface can be derived as

$$C_L(\theta) = \frac{|d\mathbf{F}(\theta)|}{1/2\left(\rho V_\infty^2 c(\theta)|d\mathbf{l}|\right)} = \frac{2V_x(\theta)\Gamma(\theta)}{V_\infty^2 c(\theta)}, \quad (19)$$

where $V_x(\theta)$ is the local x-velocity component of one lifting surface composed of the freestream velocity and the x-velocity component induced by the bound vorticity of the other lifting surface, $V_{x1}(\theta_1) = V_\infty + V_{ix12-1}(\theta_1)$, $V_{x2}(\theta_2) = V_\infty + V_{ix21-1}(\theta_2)$.

Meanwhile, by applying (16) and the related equations, the induced drag coefficient of the two wings, C_{Di1} and C_{Di2}, can be derived as

$$C_{D_{i1}} = \frac{2}{V_\infty^2 S}\int_0^\pi \Gamma_1(\theta_1)\alpha_{i1}(\theta_1)V_x(\theta_1)\left[\frac{b_1}{2}\sin(\theta_1)\right]d\theta_1,$$
$$C_{D_{i2}} = \frac{2}{V_\infty^2 S}\int_0^\pi \Gamma_2(\theta_2)\alpha_{i2}(\theta_2)V_x(\theta_2)\left[\frac{b_2}{2}\sin(\theta_2)\right]d\theta_2, \quad (20)$$

where S is the wing area.

Applying the Biot-Savart law into the segment of the bound vortex along the lifting line of the other lifting surface and integral along the spanwise, the x-velocity component induced by the bound vortex at an arbitrary point along the lifting line is

$$V_{ix12-1}(\theta_1) = -\sin(\varphi)\int_0^\pi \frac{\Gamma_2(\theta_2)\sin(\phi)}{4\pi r^2}\left[\frac{b_2}{2}\sin(\theta_2)\right]d\theta_2,$$
$$V_{ix21-1}(\theta_2) = \sin(\varphi)\int_0^\pi \frac{\Gamma_1(\theta_1)\sin(\phi)}{4\pi r^2}\left[\frac{b_1}{2}\sin(\theta_1)\right]d\theta_1, \quad (21)$$

where $\sin(\varphi) = G'/(S_t'^2 + G'^2)^{1/2}$, as shown in Figure 2.

Moreover, the magnitude of the local velocity used in (10) and (14) to calculate the induced angle of attack should also be corrected by the local x-velocity component of each lifting surface composed of the freestream velocity and the induced x-velocity component, as shown in (21).

3.2. Solving Prandtl's Lifting-Line Equation. After replacing the magnitude of the freestream velocity with the corrected magnitude of the local velocity in the achievement of the section lift coefficient and the induced angle of attack, the classical Prandtl's lifting-line equation would be derived as a system of N (where $N = N_1 + N_2$) nonlinear equation relating the N unknown coefficients (consists of $\{A_n\}$ and $\{B_n\}$) of the distribution of the section circulation to known properties of the lifting surface.

Traditionally, the Newton's method is commonly adopted to solve the system of nonlinear equation. However, because the unknown coefficients exist both in the numerator and the denominator in the equations, the partial derivatives cannot be obtained conveniently. So, the Prandtl's lifting-line equation shown as (8) can be changed as

$$\frac{2V_{x1}^2(\theta_1)\Gamma_1(\theta_1)}{V_\infty^2 c_1(\theta_1)C_{L1,\alpha}} + V_{iz11}(\theta_1) + V_{iz12}(\theta_1) = V_{x1}(\theta_1)\left(\alpha + \delta_1 - \alpha_{0L_1}\right),$$
$$\frac{2V_{x2}^2(\theta_2)\Gamma_2(\theta_2)}{V_\infty^2 c_2(\theta_2)C_{L2,\alpha}} + V_{iz22}(\theta_2) + V_{iz21}(\theta_2) = V_{x2}(\theta_2)\left(\alpha + \delta_2 - \alpha_{0L_2}\right). \quad (22)$$

To apply the Newton's method, writing the first equation of (22) at N_1 different points and writing the second equation of (22) at N_2 different points, respectively, combining the unknown sequence, $\{A_n\}$, $\{B_n\}$, as $\mathbf{X} = [A_n{}^T, B_n{}^T]$, then the system of N nonlinear equation can be written in the vector form as

$$T(\mathbf{X}) = \mathbf{R}, \quad (23)$$

where

$$T_i(\mathbf{X}) = \begin{cases} \dfrac{2V_{x1}^2(\theta_{1i})\Gamma_1(\theta_{1i})}{V_\infty^2 c_1(\theta_{1i})C_{L1,\alpha}} + V_{iz11}(\theta_{1i}) + V_{iz12}(\theta_{1i}) - V_{x1}(\theta_{1i})\left(\alpha + \delta_1 - \alpha_{0L_1}\right), & 1 \le i \le N_1, \\ \dfrac{2V_{x2}^2\left(\theta_{2(i-N_1)}\right)\Gamma_2\left(\theta_{2(i-N_1)}\right)}{V_\infty^2 c_2\left(\theta_{2(i-N_1)}\right)C_{L2,\alpha}} + V_{iz22}\left(\theta_{2(i-N_1)}\right) + V_{iz21}\left(\theta_{2(i-N_1)}\right) - V_{x2}\left(\theta_{2(i-N_1)}\right)\left(\alpha + \delta_2 - \alpha_{0L_2}\right), & N_1 + 1 \le i \le N. \end{cases} \quad (24)$$

In order to find the vector $\mathbf{X}$ of the N unknown coefficients which makes all components of the residual vector $\mathbf{R}$ equal to zero, an initial estimate for the vector $\mathbf{X}$ is selected and iteratively refined by applying the Newton's corrector equation:

$$[\mathbf{J}]\Delta\mathbf{X} = -\mathbf{R} = -T(\mathbf{X}), \quad (25)$$

where $[\mathbf{J}]$ is the $N \times N$ matrix of partial derivatives.

$$J_{ij}=\frac{\partial T_i}{\partial X_j}=\begin{cases}\dfrac{2V_{x1}^2(\theta_{1i})}{V_\infty^2 c_1(\theta_{1i})C_{L1,\alpha}}\dfrac{\partial\Gamma_1(\theta_{1i})}{\partial A_j}+\dfrac{\partial V_{iz11}(\theta_{1i})}{\partial A_j}, & 1\le i\le N_1, 1\le j\le N_1,\\ \dfrac{4\Gamma_1(\theta_{1i})V_{x1}(\theta_{1i})}{V_\infty^2 c_1(\theta_{1i})C_{L1,\alpha}}\dfrac{\partial V_{x1}(\theta_{1i})}{\partial B_{(j-N_1)}}+\dfrac{\partial V_{iz12}(\theta_{1i})}{\partial B_{(j-N_1)}}-\dfrac{\partial V_{x1}(\theta_{1i})}{\partial B_{(j-N_1)}}(\alpha+\delta_1-\alpha_{0L_1}), & 1\le i\le N_1, N_1<j\le N,\\ \dfrac{4\Gamma_2\left(\theta_{2(i-N_1)}\right)V_{x2}\left(\theta_{2(i-N_1)}\right)}{V_\infty^2 c_2\left(\theta_{2(j-N_1)}\right)C_{L2,\alpha}}\dfrac{\partial V_{x2}\left(\theta_{2(i-N_1)}\right)}{\partial A_j}+\dfrac{\partial V_{iz21}\left(\theta_{2(i-N_1)}\right)}{\partial A_j}-\dfrac{\partial V_{x2}\left(\theta_{2(i-N_1)}\right)}{\partial A_j}(\alpha+\delta_2-\alpha_{0L_2}), & N_1<i\le N, 1\le j\le N_1,\\ \dfrac{2V_{x2}^2\left(\theta_{2(i-N_1)}\right)}{V_\infty^2 c_2\left(\theta_{2(i-N_1)}\right)C_{L2,\alpha}}\dfrac{\partial\Gamma_2\left(\theta_{2(i-N_1)}\right)}{\partial B_{(j-N_1)}}+\dfrac{\partial V_{iz22}\left(\theta_{2(i-N_1)}\right)}{\partial B_{(j-N_1)}}, & N_1<i\le N, N_1<j\le N.\end{cases}\tag{26}$$

Furthermore, the correction vector can be calculated by applying (26) in (25), and the improved estimate of the vector $\mathbf{X}$ of the N unknown coefficients in the $(k+1)$th iteration based on the correction vector $\Delta\mathbf{X}$ and the vector $\mathbf{X}$ obtained in the kth iteration can be carried out as

$$\mathbf{X}^{(k+1)}=\mathbf{X}^{(k)}+\Omega\Delta\mathbf{X}^{(k)}.\tag{27}$$

This iterating process repeated until satisfying the convergence condition as

$$\frac{\left\|\Delta\mathbf{X}^{(k)}\right\|}{\left\|\mathbf{X}^{(k)}\right\|}\le\xi,\tag{28}$$

then the vector $\mathbf{X}$ of the N unknown coefficients equals to the vector $\mathbf{X}^{(k)}$.

For the purpose of fast convergence of the Newton's method, an accurate initial estimate of the vector $\mathbf{X}$ of the N unknown coefficients should be meticulously selected. So, it is appropriate to use the solution of the classical Prandtl's lifting-line equation regardless of the correction of the magnitude of local velocity as shown in (8).

After determining the circulation strength of each lifting surface, the aerodynamic force can be obtained by evaluating the integral of the aerodynamic force coefficient indicated as (19) along the wingspan of each lifting surface.

4. Methods to Generate Control Points

In order to calculate the Fourier coefficients for tandem wing configuration, the Prandtl's lifting-line equation, (8), must be applied at the control points of each lifting surface along its wingspan and then, the variation of the section circulation for the two lifting surface along the wingspan can be obtained.

Firstly, the simplest method to generate control points is that the lifting surface can be divided into elements of equal spanwise increment. In one manner, it would seem most reasonable to place the control point of each element along its bound vortex segment, midway between the two edges of each element, as shown in Figure 5.

Furthermore, for the pressure difference near the wing tip, the vorticity is shed from the wing more rapidly in the region near the wing tip, the control points should be clustered more tightly in this region to get best calculation efficiency. From [15], the conventional cosine clustering method could be used to generate element for control points of the two wings in tandem wing configuration. For the straight wing, the clustering method is only needed in the region near the wing tip. While for the wing with sweep and/or dihedral, the clustering method is needed both in the region near the wing tip and near the wing root since a step change in the slop of the quarter chord exists at the root for that wing, which caused the downwash to change very rapidly in the region near the wing root.

For the convenience that the method to generate control point can be appropriate for both the wing with/without sweep and/or dihedral, the clustering method should be applied independently over each semispan of each wing, as shown in Figure 6. Based on the variable change indicated in (1), the variables, θ_1 and θ_2, vary from 0 to π as y_1 varies from $-b_1/2$ to $b_1/2$, as well as y_2 varies from $-b_2/2$ to $b_2/2$ over each wing. Distributing the dividing line for the junctures of adjacent element uniformly in variable θ will provide the expected clustering in variable y.

After gathering the clustering elements over each wing, there are two different methods to get the location of the control point of each element. Firstly, it would seem most appropriately to set the control point at the midway between the two nodal of each element, on the bound segment of the vortex. Then, the position of the control point i can be easily obtained by applying the coordinates of the node $i-1$ and the node i. Secondly, another method has been used by Phillips and Snyder [15] to generate the control point based on the clustering. It may be more properly to achieve the control points by setting the coordinate midway in variable θ instead of midway in variable y.

Considering the interaction between the flow generated by the two lifting surfaces, control points in the region directly before or behind the wing tip region of the other lifting surface should also be clustered for the more accurate

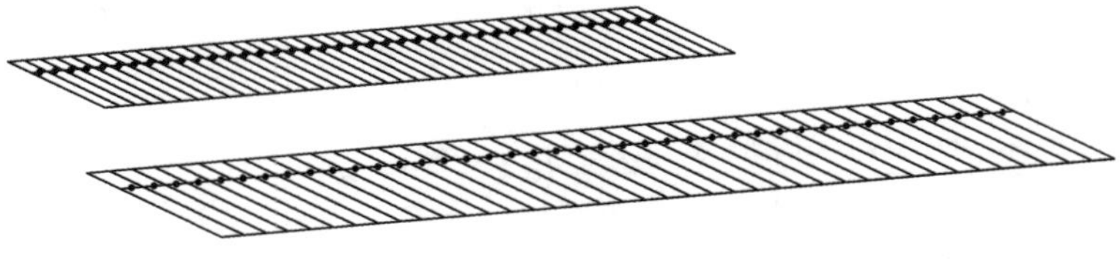

FIGURE 5: Grid generation for equal spanwise increment.

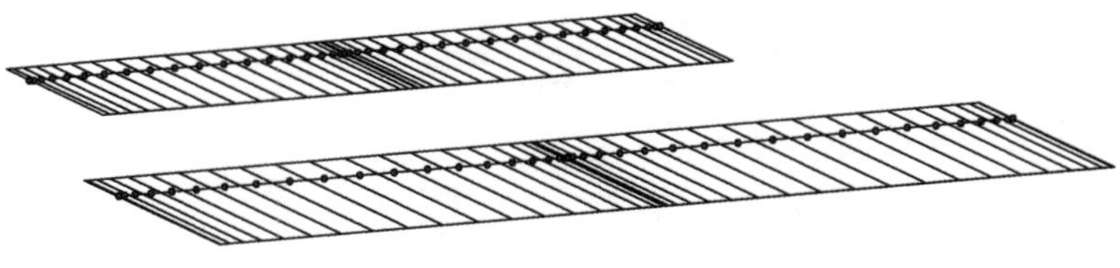

FIGURE 6: Grid generation with cosine clustering.

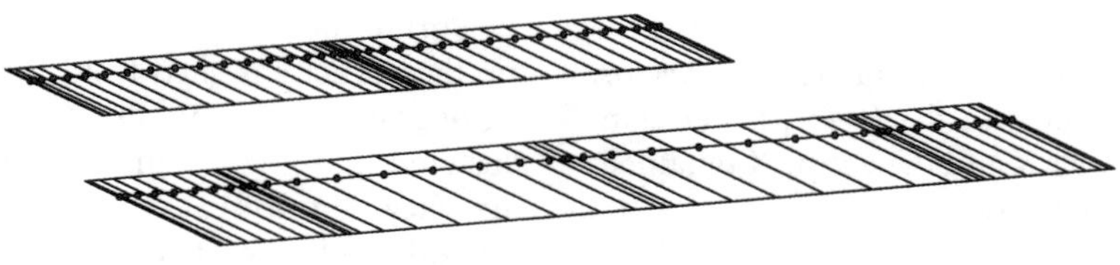

FIGURE 7: Grid generation with independent clustering.

section lift coefficients. For the tandem wing configuration with different wingspans of the two lifting surface, the cosine clustering method should be applied independently both in the region near the wing tip and wing root as well as the region aforementioned, as shown in Figure 7.

After validating and comparing, the clustering method with the midway in variable θ is selected to calculate the section lift coefficient for tandem wing configuration or multiple-lifting-surface system.

5. Results

5.1. Validation of the Prandtl's Lifting-Line Method. To validate the method for predicting the section lift coefficient of the tandem wing configuration or configuration with two lifting surfaces using the Prandtl's lifting-line theory and to analyze the grid sensitivity of the number of the control points, the section lift coefficient of tandem wing configuration by solving Prandtl's lifting-line equation at a series of different numbers of control points has been calculated and compared with the results obtained from an inviscid CFD solution and a limited experimental data.

For the experimental data comparison, the wing-canard configuration test cases presented by Feistel et al. [3] are selected. The experiment was conducted in the NASA Ames 7×10 foot wind tunnel in which the Reynolds number based on chord was 1.4×10^6 and the dynamic pressure was 60 lb/ft^2, in which the magnitude of the freestream velocity was about 69.15 m/s. The tested models were constructed from the NASA/LANGLEY LS(1)-0413 (GA(W)-2) airfoil section, and the canard had an incidence of 2° relative to the wing. The experiment data of the short-stagger, high-wing setup configuration was chosen for the significant interference effects between the wing and the canard presented in the wind tunnel.

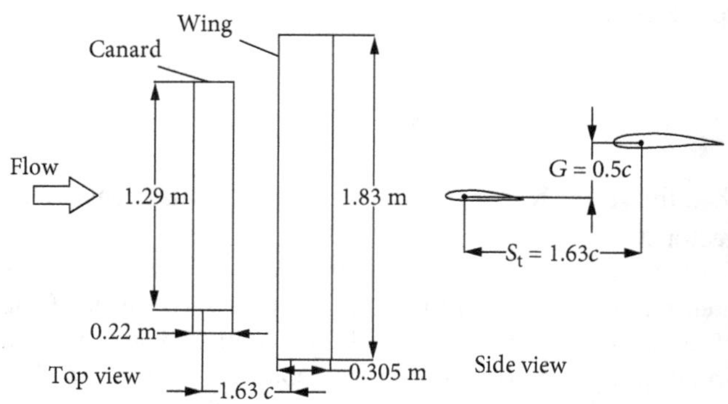

FIGURE 8: Wing-canard model top view and side view.

The CFD solutions are performed by applying a structured-grid, multizone code, which solves the three-dimensional, time-independent, Navier-Stokes equations using the finite volume method and SST (shear stress transport) k-ω turbulence model [19], for a rectangular tandem wing configuration, as shown in Figure 8. All aerodynamic simulations are calculated using O-grid generation about tandem wing configuration, which is clustered in the normal direction near the wing surface to capture the boundary layer conveniently. Velocity-in and pressure-out boundary conditions are defined on the far-field planes of the computational flow field. The flow condition used in CFD solutions is the same as that in the wind tunnel experimental conducted by Feistel et al. in [3].

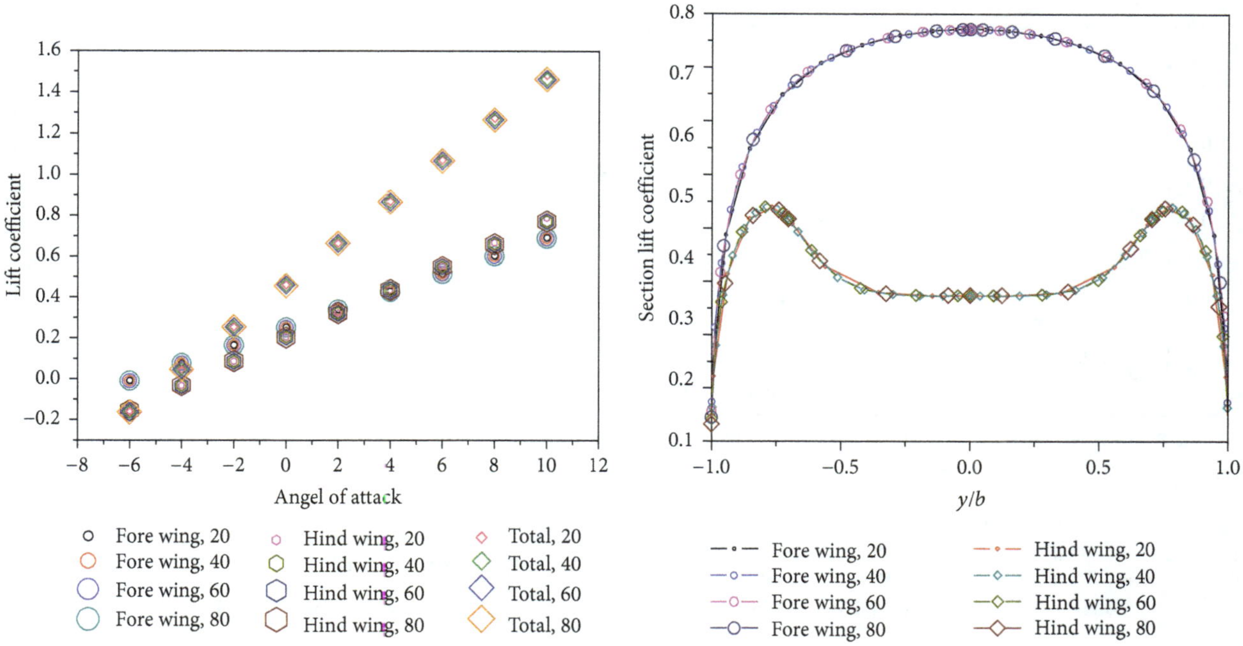

(a) Total lift coefficient

(b) Section lift coefficient

FIGURE 9: Grid sensitivity test.

In addition, the number of control points along the semispan of each lifting surface is set from 20 to 80 with an increment of 20 to analyze the grid sensitivity. The model used to calculate is constructed from the NASA/LANGLEY LS(1)-0413 (GA(W)-2) airfoil section and with the configuration as shown in Figure 8. Results of the section lift coefficient and the total lift coefficient of each lifting surface in tandem wing configuration are indicated in Figure 9.

As shown in Figure 9, the influence on the section lift coefficient and the wing lift coefficient by the number of control points per semispan can be ignored, then we set 40 control points per semispan for the best compromise between speed and accuracy.

The wing lift coefficient of the tandem wing configuration shown in Figure 8 is analyzed by comparing with the CFD solutions and the experiment data which were obtained with the same flow condition. Additional validations for the tandem wing configurations constructed from Wortmann FX63-137 airfoil by Scharpf and Mueller [4], and JED-EJ75 airfoil by Zhang and Yu [8], are conducted using the CFD method as well as the method presented in this paper with the same flow conditions in [4] and [8], respectively, and the results are indicated in Figure 10. As shown in Figure 11, the lift coefficient of the individual lifting surface is not predicted accurately by the lifting-line method compared with CFD and experiment data, while the total lift coefficient agrees well with the wind tunnel data and the CFD results over the majority of angles of attack in the range investigated. Furthermore, the lifting-line method proposed in this paper can predict the lift coefficient of the individual lifting surface as well as the total lift coefficient of the tandem wing configurations in [4, 8]. The deviation of the lift coefficients obtained by the lifting-line method and CFD method is acceptable, which is not more than 5% in the majority of angles of attack investigated, which indicated in Figure 10.

To verify whether the method on the foundation of Prandtl's classical lifting-line theory can estimate the lift coefficient of a whole configuration including the airframe, the three-dimensional model with the airframe is simulated by using the CFD method. As shown in Figure 12, the length of the model airframe, L, is 1.25 m, and the dimension of the cross airframe is 0.129 × 0.129 m. The NACA4309 airfoil is used as the fore wing and the hind wing. The main geometric characteristics of the two wings are a constant chord, c, of 0.129 m across the span, a reference wing area, S, of 0.544767 m^2, the fore wing's span, b_1, of 2.1805 m, and the hind wing's span, b_2, of 2.0425 m. The UAV model is simplified from a folding missile-carried vehicle, and the stagger, S_t, must be large enough to fit the entire wing span. Hence, S_t is chosen as 0.89 m and G is −0.101 m. The lift coefficient are evaluated at freestream $V = 30$ m/s, corresponding to a Reynolds number, Re, of $2.65e^5$ based on the mean aerodynamic chord. Velocity-in and pressure-out boundary conditions are defined on the far-field planes of the computational flow field.

As shown in Figure 13, when the angle of attack is below 6 degrees, the numerical results obtained by the lifting-line method are in excellent agreement with the CFD data for the tandem wing configuration with/without airframe, which indicates that the airframe has little inference on the lift coefficient. However, the method based on Prandtl's lifting-line theory cannot estimate the lift coefficient of the whole configuration with airframe accurately anymore when the angle of attack is above 6 degrees. Nevertheless, the lifting-line method can also be extensively applied during the

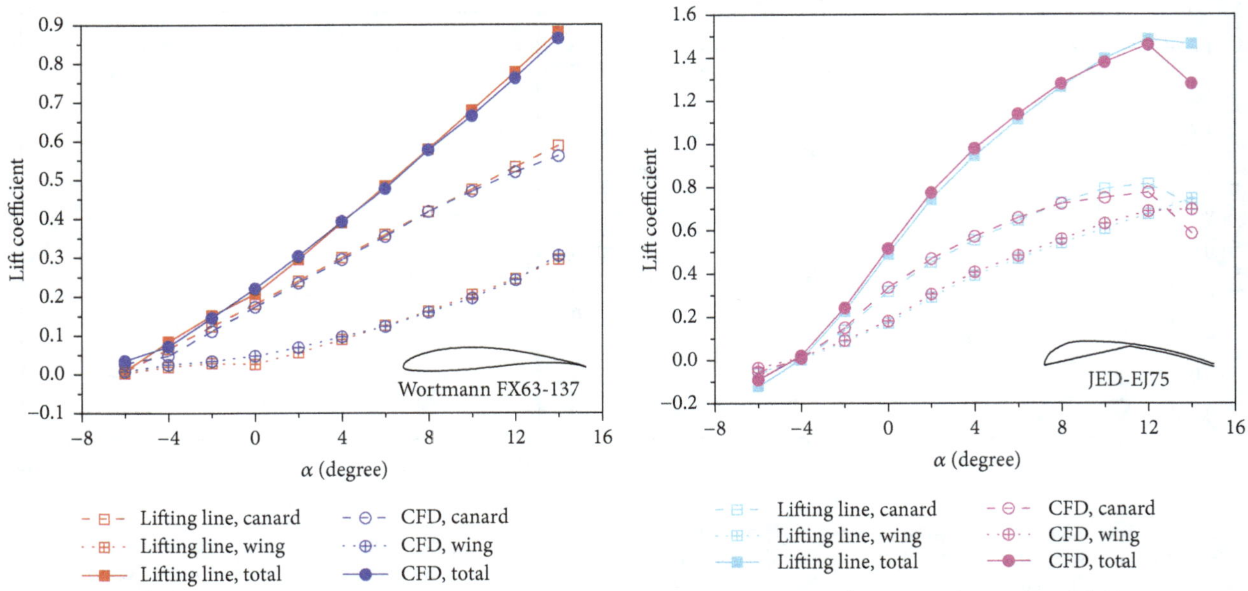

(a) Configuration with Wortmann FX63-137 airfoil

(b) Configuration with JED-EJ75 airfoil

FIGURE 10: Validation for different tandem wing configurations.

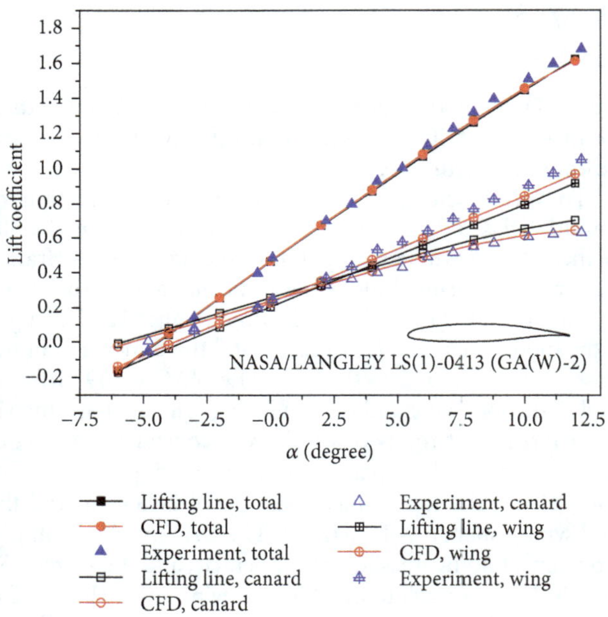

FIGURE 11: Lift curve comparison for tandem wing configuration.

aircraft conceptual design stage to obtain the aerodynamic characteristics of an aircraft with tandem wing configuration more efficiently.

5.2. Result Analysis. For tandem wing configuration, the gap represents the vertical distance between the aerodynamic centers of the two different lifting surfaces and the stagger represents the horizontal distance between the two different lifting surfaces' aerodynamic center. As the lifting-line method indicated, the induced velocity at the control points generated by the bound vortex and the trailing vortex of the other lifting surface would vary with the gap and the stagger. Meanwhile, the different incidence angle as well as the wingspan of one lifting surface would influence the region and the direction of its trailing vortex sheet, respectively, and further influence the induced velocity.

The section lift coefficient, the total lift coefficient, and the coefficient of the induced drag of each lifting surface of tandem wing configuration with different gap, stagger, incidence angle, and wingspan are calculated and analyzed by using the method proposed. All the simulations are calculated at freestream $V = 69.15$ m/s, corresponding to a Reynolds number of 1.4×10^6 based on chord, with the CFD method in the validation. The NASA/LANGLEY LS(1)-0413 (GA(W)-2) airfoil section is selected to construct all the tandem wing configurations, and the number of control points per semispan is set as 40 for the best compromise between speed and accuracy.

5.2.1. Effect of Gap. Effect of the gap can be analyzed from the section lift coefficient along the wingspan of each lifting surface for the tandem wing configuration, in which the stagger is $1.63c$, $S_t = 1.63c$, the incidence angle along its wingspan is constant, $\delta_1 = 2^\circ$, $\delta_2 = 0$, and the gap ranges from $-2.0c$ to $2.0c$. The results are calculated at angle of attack as 2 degrees, by using the Prandtl's classical method proposed above, and shown in Figure 14 and Table 2.

As shown in Figure 14, the section lift coefficient of the fore wing decreases along the entire wingspan with the vertical distance of the two lifting surfaces increasing, and it has a little bigger value with the positive gap, for the reason that the upwash effect by the hind wing becomes weaken. On the other hand, the section lift coefficient of the hind wing increases along the wingspan in the region directly behind the fore wing, and it is bigger with the negative gap, while it declines somewhat along the other wingspan region and has a smaller value with the negative gap, because the downwash effect by the fore wing weakens with the gap increasing. What is more, as Table 2 indicated, the wing lift coefficient

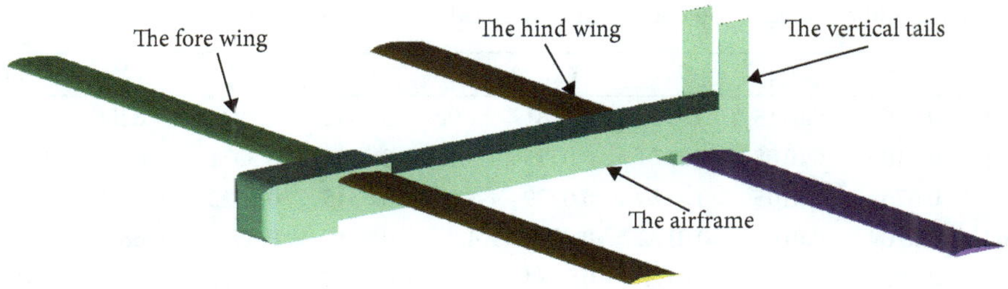

FIGURE 12: Model of tandem wing configuration with airframe.

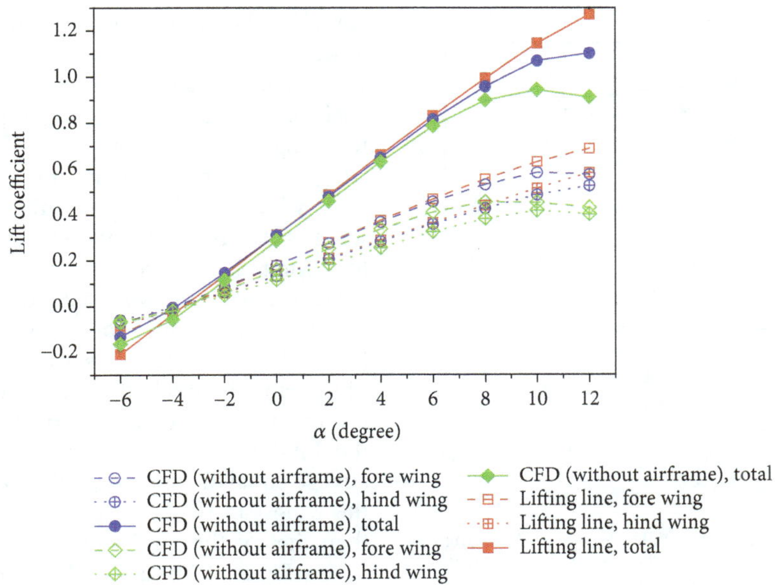

FIGURE 13: Lift curve comparison for tandem wing configuration with airframe.

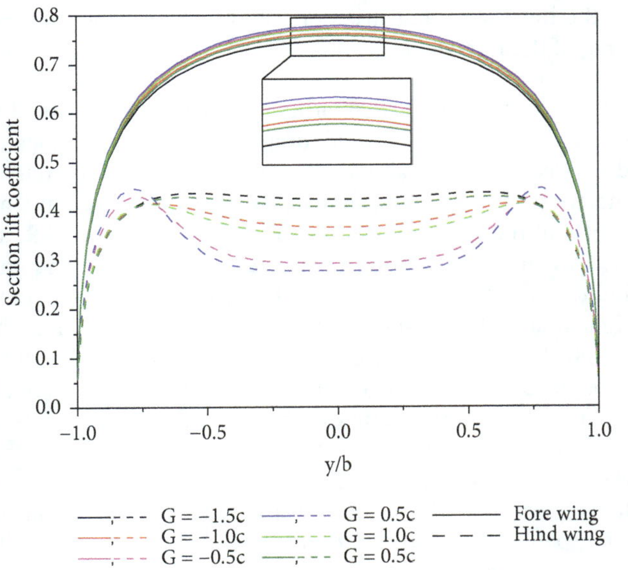

FIGURE 14: Effect of gap on section lift coefficient for tandem wing configuration.

has the biggest value for the fore wing with the small positive gap but for the hind wing as well as the tandem wing configuration with the big negative gap. Because the upwash effect generated by the hind wing can decrease the induced angle of attack of the fore wing, the induced drag of the fore wing is small when the two wings locate closer to each other in the vertical direction, as shown in Table 2. However, the induced drag of the hind wing and the whole configuration increase firstly and decrease subsequently with the gap increasing both in positive and negative directions. That is because the downwash effect produced by the fore wing on the hind wing is in the dominant situation for the whole tandem wing configuration.

5.2.2. Effect of Stagger. The range of stagger from 1.0c to 7.0c is calculated to analyze its effect on the section lift coefficient, the wing lift coefficient and the coefficient of the induced drag. As shown in Figure 15 and Table 2, the section lift coefficient as well as the wing lift coefficient of the fore wing decrease with the stagger increasing, but it is opposite for its induced drag and for that of the hind wing, and it tends to be constant gradually, for the reason that the aerodynamic interaction between the two lifting surfaces becomes weaken. Meanwhile, the total lift coefficient of the tandem wing configuration becomes larger with the stagger increasing as shown in Table 2.

5.2.3. Effect of Incidence Angle. For the effect of incidence angle, a series of tandem wing configurations with the range of incidence angle from −6 to 6 degrees of one lifting surface

TABLE 2: Lift coefficient and coefficient of induced drag with the effect of stagger and gap.

G	C_{L1}	C_{L2}	C_L	C_{Di1}	C_{Di2}	C_{Di}	S_t	C_{L1}	C_{L2}	C_L	C_{Di1}	C_{Di2}	C_{Di}
$-1.5c$	0.3325	0.4036	0.7361	0.0113	0.0168	0.0281	$1.0c$	0.3542	0.2944	0.6486	0.0099	0.0168	0.0267
$-1.0c$	0.3383	0.3723	0.7106	0.0108	0.0174	0.0283	$2.0c$	0.3408	0.3371	0.6779	0.0109	0.0173	0.0281
$-0.5c$	0.3433	0.3337	0.6769	0.0105	0.0174	0.0279	$3.0c$	0.3345	0.3522	0.6867	0.0113	0.0172	0.0286
$0.5c$	0.3447	0.3262	0.6709	0.0106	0.0172	0.0278	$4.0c$	0.3315	0.3583	0.6894	0.0115	0.0172	0.0287
$1.0c$	0.3415	0.3635	0.7050	0.0109	0.0175	0.0284	$5.0c$	0.3298	0.3612	0.6910	0.0117	0.0171	0.0288
$1.5c$	0.3364	0.3950	0.7315	0.0113	0.0169	0.0282	$6.0c$	0.3288	0.3629	0.6917	0.0117	0.0171	0.0288

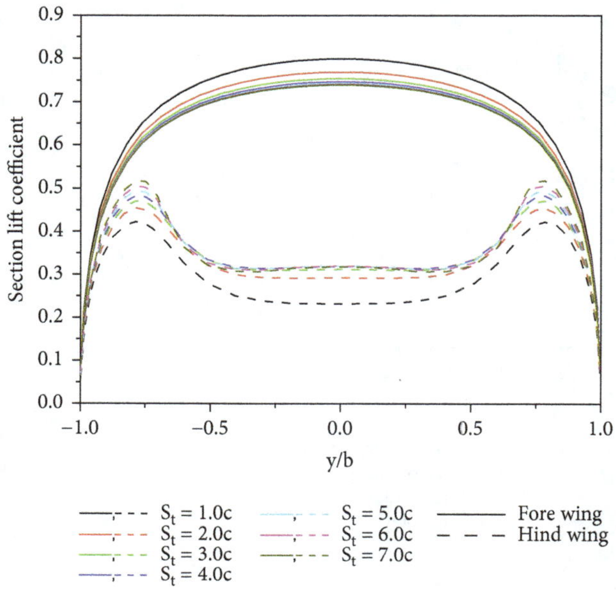

FIGURE 15: Effect of stagger on section lift coefficient for tandem wing configuration.

and the constant incidence angle of the other lifting surface are calculated, the section lift coefficients of the two lifting surfaces are shown in Figure 16, and the total lift coefficient and the coefficient of the induced drag of each lifting surface with the variation of the incidence angle of the other lifting surface are shown in Table 3.

The bigger incidence angle of the hind wing will produce the better section lift coefficient and smaller induced drag of the fore wing, for the reason that the upwash effect has been strengthened to increase the effective angle of attack of the fore wing section. However, the section lift coefficient of the hind wing decreases along the wingspan in the region directly behind the fore wing with the incidence angle of the fore wing increasing while it increases somewhat in the other wingspan region, because the direction of the wake stream is changed with the incidence angle of the fore wing increasing. The induced drag of the hind wing has the bigger value when the incidence angle of the fore wing is about zero, and it decreases with the magnitude of the incidence angle of the fore wing increasing.

5.2.4. Effect of Wingspan. When the wingspan of the fore wing, b_1, is bigger than the constant wingspan of the hind wing, b_2, the hind wing is in the region of the fore wing's wake stream entirely and is influenced along its whole wingspan. As shown in Figure 17(a), the section lift coefficient of the hind wing is positively correlated with the wingspan of the fore wing. If the wingspan of the fore wing, b_1, is smaller than the constant wingspan of the hind wing, b_2, the region influenced by the fore wing's wake is different with the change of the wingspan of the fore wing, so the section lift coefficient of the hind wing decreases with the wingspan of the fore wing increasing and the curve peak moves towards to the middle of the hind wing. As Table 3 shown, the lift coefficient of the hind wing is smallest when the wingspans of the two lifting surfaces are almost equal, but the induced drag of the hind wing is biggest when the two wings have the almost equal wingspan, and it continues to increase with the wingspan of the fore wing increasing, which is bigger than that of the hind wing.

As Figure 17(b) and Table 3 indicated, the wingspan of the hind wing has little influence on the section lift coefficient of the fore wing, and there is a positive correlation between the section lift coefficient of the fore wing and the wingspan of the hind wing, but it is opposite for that of the induced drag of the fore wing.

5.2.5. Effect of Wing Planform. Effects on the section lift coefficient and the induced drag by different parameters discussed above are all based on that the canard as well as the wing is straight rectangular wing. In this section, the effects of the wing platform, such as the taper ratio, AR, as well as the aspect ratio, λ, on the lift coefficient and the induced drag are analyzed with the wing-canard configuration indicated in Figure 8. A series of simulation cases, in which the taper ratio of one lifting surface varies from 0 to 1 based on its unchanged root chord with an unchanged taper ratio of another lifting surface, are calculated to analyze the effect of the taper ratio. What is more, the effect of the aspect ratio is discussed with the calculation cases, in which the aspect ratio of the one lifting surface varies while its wingspan remains constant and the aspect ratio of another lifting surface keeps unchanged.

As shown in Figure 18, the taper ratio of the fore wing has a significant effect on the lift coefficient of the hind wing, the lift coefficient of the hind wing decreases with the taper ratio of the fore wing increasing, but it is opposite for that of the fore wing, when the taper ratio is about 0.6, and the effects on the lift coefficient are similar to that by the elliptical wing. With the taper ratio of the hind wing increasing, the upwash

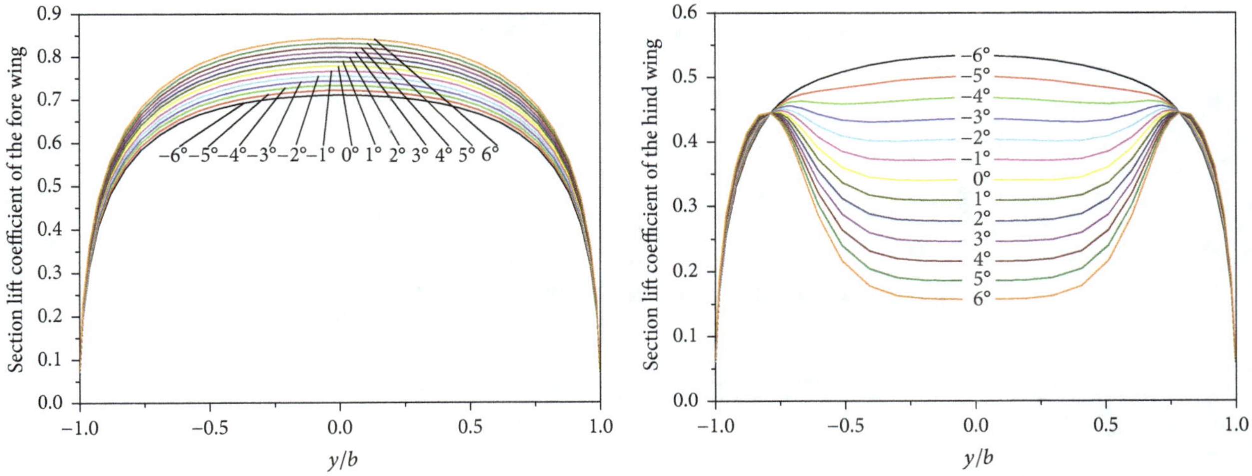

(a) Incidence angle of the hind wing

(b) Incidence angle of the fore wing

FIGURE 16: Effect of incidence angle on section lift coefficient for tandem wing configuration.

TABLE 3: Lift coefficient and coefficient of induced drag with the effect of incidence angle and wingspan.

δ_1, degree	C_{L2}	C_{Di2}	δ_2, degree	C_{L1}	C_{Di1}	b_2, m	C_{L2}	C_{Di2}	b_2, m	C_{L1}	C_{Di1}
−6	0.4739	0.0140	−6	0.3163	0.0125	0.99	0.3711	0.0166	0.53	0.6475	0.0795
−5	0.4550	0.0150	−5	0.3210	0.0122	1.09	0.3543	0.0168	0.63	0.6491	0.0665
−4	0.4361	0.0158	−4	0.3257	0.0119	1.19	0.3408	0.0170	0.73	0.6508	0.0571
−3	0.4173	0.0164	−3	0.3305	0.0116	1.29	0.3262	0.0172	0.83	0.6525	0.0499
−2	0.3990	0.0169	−2	0.3352	0.0113	1.39	0.3113	0.0173	0.93	0.6542	0.0443
−1	0.3809	0.0173	−1	0.3399	0.0109	1.49	0.2967	0.0174	1.03	0.6560	0.0398
0	0.3626	0.0174	0	0.3447	0.0106	1.59	0.2836	0.0201	1.43	0.6658	0.0276
1	0.3444	0.0174	1	0.3494	0.0102	2.09	0.2538	0.0155	1.53	0.6686	0.0257
2	0.3260	0.0172	2	0.3540	0.0098	2.19	0.2580	0.0164	1.63	0.6718	0.0239
3	0.3078	0.0169	3	0.3587	0.0094	2.29	0.2621	0.0165	1.73	0.6750	0.0223
4	0.2899	0.0164	4	0.3633	0.0090	2.39	0.2663	0.0166	1.83	0.6780	0.0208
5	0.2724	0.0157	5	0.3679	0.0086	2.49	0.2703	0.0167	1.93	0.6808	0.0195
6	0.2553	0.0150	6	0.3724	0.0082	2.59	0.2741	0.0168	2.03	0.6835	0.0184

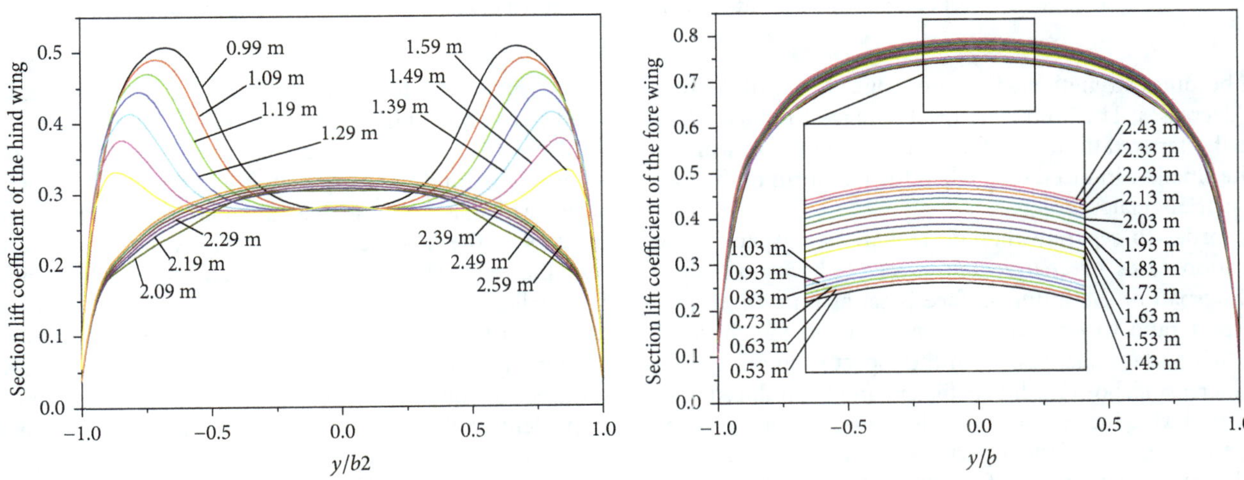

(a) Wingspan of the fore wing

(b) Wingspan of the hind wing

FIGURE 17: Effect of wingspan on section lift coefficient for tandem wing configuration.

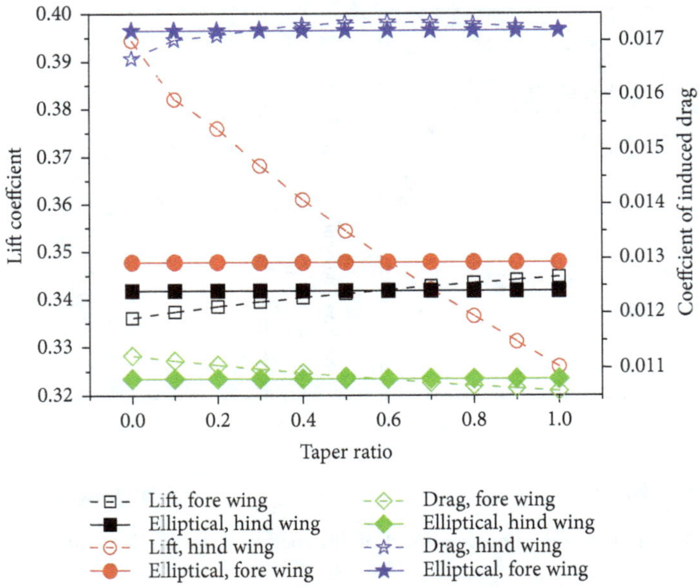

FIGURE 18: Effect of taper ratio on lift coefficient and induced drag for tandem wing configuration.

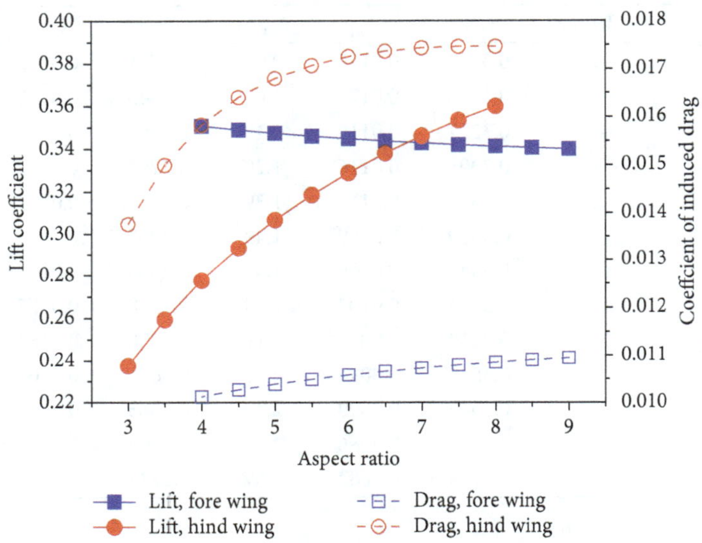

FIGURE 19: Effect of aspect ratio on lift coefficient and induced drag for tandem wing configuration.

effect becomes strengthened, so the induced drag of the fore wing decreases. The hind wing has the biggest induced drag when the taper ratio of the fore wing is about 0.6, while it has the smallest induced drag when the planform of the fore wing is triangular.

In order to separate the effect of the wingspan in the calculation cases to analyze the effect of the aspect ratio, the wingspan of the lifting surface is set as unchanged and the aspect ratio varies, which is equivalent to the chord of the lifting surface changes. With the aspect ratio of the fore wing increasing, both the lift coefficient and the induced drag of the hind wing increase as shown in Figure 19. On the other hand, the variation of the aspect ratio of the hind wing has slightly influenced on the lift coefficient and induced drag of the fore wing. When the aspect ratio of the hind wing increases (i.e., the chord of the hind wing decreases), the lift coefficient of the fore wing decreases slightly, whereas the coefficient of the induced drag increases slightly.

Note from all results in Figures 10–19, the lift coefficient of the hind wing is much lower than that of the fore wing, which is because that an interference phenomenon appears in the flow field around the two wings. Nevertheless, the induced drag for tandem airfoil configuration will be smaller with an appropriate location of the two wings compared to the traditional layout. Additionally, the wingspan of the wings in tandem wing configuration can be reduced to obtain enough stiffness, with the same lift coefficient of the present configuration for the reason that the two wings in tandem airfoil configuration are all generating positive lift. Meanwhile, the position of the aerodynamic center of tandem wing configuration moves backward compared with that of the traditional layout, so that it is more competitive to carry payloads. In conclusion, tandem wing configuration can be widely used in missile-carried vehicle as well as gun-launching vehicles with folding wings especially.

6. Conclusions

(1) In accordance with Prandtl's classical lifting-line theory, a method to calculate the section lift coefficient for the tandem wing configuration or multiple-lifting-surface system is presented and the form of Fourier sine series is used to express the variation of the section circulation which changes continuously along the wingspan.

(2) To obtain the results closer to the real value, the magnitude of local velocity is corrected by that synthesized from the uniform freestream flow combined with the induced velocity at the corresponding control point.

(3) For configuration with different wingspans of the two lifting surfaces, the cosine clustering method is applied independently in the region near the wing tip and wing root as well as the region directly before or behind the wing tip region of the other lifting surface.

(4) The accuracy of the numerical solutions obtained by the proposed method is validated by the data obtained from computation fluid dynamics and tunnel experiment.

(5) From the results of tandem wing configuration with different parameters, it indicates that the bigger negative gap and stagger can produce better lift characteristic for tandem wing configuration or multiple-lifting-surface system. What is more, it will also be beneficial for the lift characteristic when the incidence angle and the wingspan of fore wing are appropriately declined or if the incidence angle and the wingspan of hind wing are appropriately increased.

(6) The induced drag of tandem wing configuration is dominated by the downwash effect generated by the fore wing on the hind wing. Furthermore, the minimum value of induced drag can be obtained by optimizing the parameters which can influence the aerodynamic interaction between the two wings.

Conflicts of Interest

The authors declared that there is no conflict of interest regarding the publication of this paper.

References

[1] J. W. Bottomley, "Tandem wing aircraft," *Aerospace*, vol. 4, pp. 12–20, 1977.

[2] J. W. Bottomley, "The tandem wing concept applied to modern transports," *Aeronautical Journal*, vol. 78, pp. 523-524, 1974.

[3] T. Feistal, V. Corsiglia, and D. Levin, "Wind-tunnel measurements of wing-canard interference and a comparison with various theories," in *SAE Technical Paper Series*, pp. 1–14, Wichita, KS, USA, 1981, , Society of Automotive Engineers.

[4] D. F. Scharpf and T. J. Mueller, "Experimental study of a low Reynolds number tandem airfoil configuration," *Journal of Aircraft*, vol. 29, no. 2, pp. 231–236, 1992.

[5] N. R. Alley, W. F. Phillips, and R. E. Spall, "Predicting maximum lift coefficient for twisted wings using computational fluid dynamics," *Journal of Aircraft*, vol. 44, no. 3, pp. 911–917, 2007.

[6] D. Fanjoy and D. Dorney, "A study of tandem-airfoil interaction in different flight regimes," in *35th Aerospace Sciences Meeting and Exhibit*, pp. 1–18, Reno, NV, USA, January 1997, AIAA Inc.

[7] T. Broering and Y. Lian, "The effect of wing spacing on tandem wing aerodynamics," in *28th AIAA Applied Aerodynamics Conference*, pp. 1–20, Chicago, IL, USA, June 2010, AIAA Inc.

[8] G. Q. Zhang and S. C. M. Yu, "Unsteady aerodynamics of a morphing tandem-wing unmanned aerial vehicle," *Journal of Aircraft*, vol. 49, no. 5, pp. 1315–1323, 2012.

[9] V. K. Patidar, R. Yadav, and S. Joshi, "Numerical investigation of the effect of stagger on the aerodynamic characteristics of a Busemann biplane," *Aerospace Science and Technology*, vol. 55, pp. 252–263, 2016.

[10] L. Prandtl, "Applications of modern hydrodynamics to aeronautics," Report No.: NACA 116, Göttingen University, 1921.

[11] H. Glauert, *The Elements of Aerofoil and Airscrew Theory*, Cambridge University Press, Cambridge, UK, 2nd edition, 1983.

[12] M. L. Rasmussen and D. E. Smith, "Lifting-line theory for arbitrarily shaped wings," *Journal of Aircraft*, vol. 36, no. 2, pp. 340–348, 1999.

[13] B. W. McCormick, "The lifting line model," in *Aerodynamics, Aeronautics and Flight Mechanics*, pp. 340–348, Wiley, New York, NY, USA, 2nd edition, 1995.

[14] J. D. Anderson and S. Corda, "Numerical lifting-line theory applied to drooped leading-edge wings below and above stall," *Journal of Aircraft*, vol. 17, no. 12, pp. 898–904, 1980.

[15] W. F. Phillips and D. O. Snyder, "Modern adaptation of Prandtl's classic lifting-line theory," *Journal of Aircraft*, vol. 37, no. 4, pp. 662–670, 2000.

[16] J. Katz and A. Plotkin, "Lifting-line solution by horseshoe elements," in *Low-Speed Aerodynamics from Wing Theory to Panel Methods*, pp. 379–386, McGraw-Hill, New York, NY, USA, 1991.

[17] R. E. Spall, W. F. Phillips, and B. B. Pincock, "Numerical analysis of multiple, thin-sail geometries based on Prandtl's lifting-line theory," *Computers & Fluids*, vol. 82, pp. 29–37, 2013.

[18] R. Jacobs, H. Ran, M. Kirby, and D. Mavris, "Extension of a modern lifting-line method to transonic speeds and application to multiple-lifting-surface configurations," in *30th AIAA Applied Aerodynamics Conference*, pp. 787–799, New Orleans, LA, USA, June 2012, AIAA Inc.

[19] F. R. Menter, "Two-equation eddy-viscosity turbulence models for engineering applications," *AIAA Journal*, vol. 32, no. 8, pp. 1598–1605, 1994.

4

Aerodynamic Optimization based on Continuous Adjoint Method for a Flexible Wing

Zhaoke Xu and Jian Xia

College of Aerospace Engineering, Nanjing University of Aeronautics and Astronautics, Nanjing 210016, China

Correspondence should be addressed to Jian Xia; jxia@nuaa.edu.cn

Academic Editor: Pier Marzocca

Aerodynamic optimization based on continuous adjoint method for a flexible wing is developed using FORTRAN 90 in the present work. Aerostructural analysis is performed on the basis of high-fidelity models with Euler equations on the aerodynamic side and a linear quadrilateral shell element model on the structure side. This shell element can deal with both thin and thick shell problems with intersections, so this shell element is suitable for the wing structural model which consists of two spars, 20 ribs, and skin. The continuous adjoint formulations based on Euler equations and unstructured mesh are derived and used in the work. Sequential quadratic programming method is adopted to search for the optimal solution using the gradients from continuous adjoint method. The flow charts of rigid and flexible optimization are presented and compared. The objective is to minimize drag coefficient meanwhile maintaining lift coefficient for a rigid and flexible wing. A comparison between the results from aerostructural analysis of rigid optimization and flexible optimization is shown here to demonstrate that it is necessary to include the effect of aeroelasticity in the optimization design of a wing.

1. Introduction

Thanks to the rapid development of computer performance and Computational Fluid Dynamics (CFD), CFD methods have been increasingly used in aerodynamic analysis of aircrafts. In the meantime, aerodynamic shape optimization based on CFD technology and optimization strategies has been more and more interesting to aircraft designers. Continuous adjoint formulation method was developed by Jameson to design aircrafts in transonic speeds, which can significantly reduce computational cost of aerodynamic shape optimization [1–5]. Subsequently, continuous adjoint method was utilized by Anderson and Venkatakrishnan on unstructured grid [6]. And recently Economon et al. in Stanford University have developed an open-source CFD code named SU2 [7–12], which is an integrated computational environment consisting of multi-physics simulation and design. The main module of SU2 is aerodynamic shape optimization on the basis of continuous adjoint method and Reynolds-averaged Navier-Stokes (RANS) equations. Now SU2 has also developed aerodynamic shape optimization based on discrete adjoint method [13, 14]. Discrete adjoint method was derived from the discrete flow equations by Baysal et al. [15, 16]. Subsequently discrete adjoint method has been developed rapidly [17–22]. Discrete adjoint equations which come from discrete flow equations can provide a little more accurate gradients than continuous adjoint equations, but discrete adjoint equations are more difficult to be derived and implemented numerically. Continuous adjoint equations are directly derived from the governing equations, which can be solved with similar methods of the governing equations and implemented more easily. However, since the coupling between aerodynamics and structures is so tight that these two disciplines cannot be isolated from each other especially for flexible wings, aerodynamic shape optimization alone may lead to the performance that is not appropriate in realistic flight conditions. But successive discipline optimization may lead to suboptimal results of coupled systems, so MultiDisciplinary Optimization (MDO) is essential to aircraft design. Relatively low fidelity models have been used in relevant disciplines in conceptual design, but these models cannot accurately capture nonlinear phenomena such as wave drag. So MDO of CFD and CSM (Computational Structural Mechanics) has been more and more attractive to aircraft researchers. MultiDisciplinary design optimization based on Kriging and response surface methodology has been a UAV wing and supersonic fighter wing [23, 24]. Martins et al. [25–27] have developed systematic aerostructural optimization design codes, which use coupled high-fidelity sensitivity analysis based on discrete adjoint method, and have been applied to design full aircraft configurations with coupled adjoint method. Mavriplis et al. [28–30] have developed time-dependent aeroelastic shape optimization based on discrete adjoint method.

However, aerostructural optimization that has been performed so far mostly is grounded on discrete adjoint method. In present work, aerodynamic optimization on the basis of continuous adjoint method including the effect of aeroelasticity is developed using FORTRAN 90, which is for the purpose of the implementation of coupled aerostructural adjoint optimization in the future. The structural model of the wing is created consisting of two spars, 20 ribs, and skin. Aerostructural analysis is performed with high-fidelity models, with Euler equations on the aerodynamic side, and with a linear shell element model on the structure side. This shell element is suitable for thin and thick shell problems and is easy to implement, and in particular this shell element can deal with the shell problems with intersections which are common in the structural model of aircrafts. Thin-Plate Spline (TPS) method is applied to ensure the interpolation between aerodynamics and structures consistent and conservative [31]. sequential quadratic programming (SQP) method, a gradient-based optimizer, is used to search for the optimal point in a coupled system consisting of aerodynamics and structures.

The paper begins with a description of aerodynamic analysis in Section 2, where the flow governing equations and the solution method are described. Section 3 presents structural analysis method with the theory of the shell element and structural design of the wing. The interpolation method between aerodynamic and structural disciplines is introduced in Section 4. The derivation of the continuous adjoint method is shown in Section 5. Design strategies containing parameterization methods, mesh deformation, and optimization algorithm are presented in Section 6. In the end, Section 7 presents verification of sensitivities of ONERA M6 and the optimization results for rigid and flexible DPW-W1.

2. Aerodynamic Analysis

The governing equations of 3D steady inviscid compressible flow are

$$\frac{\partial}{\partial t}\int_{\Omega}\mathbf{U}\,d\Omega+\int_{S}\mathbf{F}\,dS=0, \tag{1}$$

where $\mathbf{U}$ is conservative variable vector and $\mathbf{F}$ is convective flux vector. They are defined as

$$\mathbf{U}=\begin{bmatrix}\rho\\ \rho u\\ \rho v\\ \rho w\\ \rho E\end{bmatrix},\qquad \mathbf{F}=\begin{bmatrix}\rho V_n\\ \rho u V_n+n_x P\\ \rho v V_n+n_y P\\ \rho w V_n+n_z P\\ \rho H V_n\end{bmatrix}, \tag{2}$$

where ρ is density, u, v, and w are Cartesian velocity components, E is total energy per unit mass, n_x, n_y, and n_z are Cartesian components of unit normal vector, P is static pressure, H is total enthalpy, and V_n is contravariant velocity, which is represented as

$$V_n=un_x+vn_y+wn_z. \tag{3}$$

The finite-volume method of HLLC (Harten, Lax, van Leer, Contact) scheme is applied to solve Euler equations, and in order to reduce the computational time, an efficient implicit parallel hybrid LU-SGS (Lower-Upper Symmetric Gauss-Seidel) method [32] is used here.

3. Structural Analysis

3.1. Shell Element. In the structural analysis of the present work, a linear quadrilateral shell element [33] is applied to model the wing structure. This four-node quadrilateral shell element with 5/6 nodal degrees of freedom is constructed based on Hellinger-Reissner variational principle with independent displacements and stress resultants. The element stiffness matrix is integrated analytically with the advantage of fast stiffness computation. The shell element can get rid of locking which many shell theories may encounter and it can deal with both thin and thick shell problems. And in essence this shell element can be implemented easily for the linear isotropic aeroelasticity.

With the stationary condition, the variational formulation based on Hellinger-Reissner principle is

$$\begin{aligned}\delta\Pi_{\mathrm{HR}}&(\mathbf{v},\boldsymbol{\sigma},\delta\mathbf{v},\delta\boldsymbol{\sigma})\\&=\int_{(\Omega_s)}\left[\delta\boldsymbol{\varepsilon}^T\boldsymbol{\sigma}+\delta\boldsymbol{\sigma}^T\left(\boldsymbol{\varepsilon}-\mathbf{C}^{-1}\boldsymbol{\sigma}\right)-\delta\mathbf{u}^T\overline{\mathbf{p}}\right]dA\\&\quad-\int_{(\Gamma_\sigma)}\delta\mathbf{u}^T\overline{\mathbf{t}}\,dS=0,\end{aligned} \tag{4}$$

where $\mathbf{v}$ and $\delta\mathbf{v}$ are displacement vector and virtual displacement vector, respectively, $\boldsymbol{\sigma}$ and $\delta\boldsymbol{\sigma}$ are independent stress resultants and virtual stress resultants, respectively, $\boldsymbol{\varepsilon}$ is a vector containing membrane strains, curvatures, and shear strains, $\delta\boldsymbol{\varepsilon}$ is corresponding virtual vector, $\mathbf{C}$ is the constitutive matrix, $\mathbf{u}$ is translational displacement vector,

$\overline{\mathbf{p}}$ is volume load in Ω_s, and $\overline{\mathbf{t}}$ is the boundary load on a part of boundary Γ_σ. The membrane strains $\varepsilon_{\alpha\beta}$, curvatures $\kappa_{\alpha\beta}$, and shear strains γ_α are evaluated, respectively, as follows:

$$\varepsilon_{\alpha\beta} = \frac{1}{2}\left(\mathbf{u}_{,\alpha}\cdot\mathbf{X}_{,\beta}+\mathbf{u}_{,\beta}\cdot\mathbf{X}_{,\alpha}\right),$$
$$\kappa_{\alpha\beta} = \frac{1}{2}\left(\mathbf{u}_{,\alpha}\cdot\mathbf{D}_{,\beta}+\mathbf{u}_{,\beta}\cdot\mathbf{D}_{,\alpha}+\mathbf{X}_{,\alpha}\cdot\Delta\mathbf{d}_{,\beta}+\mathbf{X}_{,\beta}\cdot\Delta\mathbf{d}_{,\alpha}\right), \quad (5)$$
$$\begin{bmatrix}\gamma_1\\ \gamma_2\end{bmatrix} = \mathbf{J}^{-1}\begin{bmatrix}\gamma_\xi\\ \gamma_\eta\end{bmatrix},$$

where

$$\gamma_\xi = \frac{1}{2}\left[(1-\eta)\gamma_\xi^B+(1+\eta)\gamma_\xi^D\right],$$
$$\gamma_\eta = \frac{1}{2}\left[(1-\xi)\gamma_\eta^A+(1+\xi)\gamma_\eta^C\right], \quad (6)$$

where $\mathbf{X}$ is the nodal coordinate vector, $\mathbf{D}$ is a unit vector normal to the shell midsurface, $\Delta\mathbf{d}$ is the deviation of $\mathbf{D}$, ξ and η are isoparametric coordinates with the value $\{\xi,\eta\}\in[-1,1]$, $\mathbf{J}$ is the Jacobian matrix with local coordinates to isoparametric coordinates and local coordinates are evaluated with the center of the elements, and A, B, C, and D are the middle nodes of relevant edges of the elements [33]. From (5) it is shown that the shear strains are evaluated with independent interpolation functions, while the others are interpolated from displacements. These interpolation functions are adopted to fulfill the bending patch test. The interpolation of the stress resultants is defined as follows:

$$\boldsymbol{\sigma}^h = \mathbf{S}\boldsymbol{\beta},$$
$$\mathbf{S} = \begin{bmatrix}\mathbf{1}_3 & \mathbf{0} & \mathbf{0} & \mathbf{S}^m & \mathbf{0} & \mathbf{0}\\ \mathbf{0} & \mathbf{1}_3 & \mathbf{0} & \mathbf{0} & \mathbf{S}^b & \mathbf{0}\\ \mathbf{0} & \mathbf{0} & \mathbf{1}_2 & \mathbf{0} & \mathbf{0} & \mathbf{S}^s\end{bmatrix},$$
$$\mathbf{S}^m = \mathbf{S}^b = \begin{bmatrix}J_{11}^0J_{11}^0(\eta-\overline{\eta}) & J_{21}^0J_{21}^0(\xi-\overline{\xi})\\ J_{12}^0J_{12}^0(\eta-\overline{\eta}) & J_{22}^0J_{22}^0(\xi-\overline{\xi})\\ J_{11}^0J_{12}^0(\eta-\overline{\eta}) & J_{21}^0J_{22}^0(\xi-\overline{\xi})\end{bmatrix}, \quad (7)$$
$$\mathbf{S}^s = \begin{bmatrix}J_{11}^0(\eta-\overline{\eta}) & J_{21}^0(\xi-\overline{\xi})\\ J_{12}^0(\eta-\overline{\eta}) & J_{22}^0(\xi-\overline{\xi})\end{bmatrix},$$

where $\mathbf{1}_2$ and $\mathbf{1}_3$ are the second and third order unit matrices, respectively, the vector $\boldsymbol{\beta}$ contains 14 parameters in total and 8 for the constant part and the others for the nonconstant part of the stress field, respectively, $\overline{\xi}$ and $\overline{\eta}$ are the coordinates of the center of gravity of the element, and $J_{\alpha\beta}^0$ is the parameter of the Jacobian matrix $\mathbf{J}$ with $\xi = 0$ and $\eta = 0$ [33]. After a series of transformation, (4) can be written as

$$\delta\Pi_{\mathrm{HR}} = \sum_{e=1}^{\mathrm{nele}}\begin{bmatrix}\delta\boldsymbol{\beta}\\ \delta\mathbf{v}_e\end{bmatrix}_e^T\left\{\begin{bmatrix}-\mathbf{H} & \mathbf{G}\\ \mathbf{G}^T & \mathbf{0}\end{bmatrix}\begin{bmatrix}\boldsymbol{\beta}\\ \mathbf{v}_e\end{bmatrix}-\begin{bmatrix}\mathbf{0}\\ \mathbf{p}_e\end{bmatrix}\right\}_e = 0, \quad (8)$$

where nele denotes the number of elements, $\mathbf{v}_e = [\mathbf{v}_1,\mathbf{v}_2,\mathbf{v}_3,\mathbf{v}_4]^T$ is the element displacement vector, and $\mathbf{p}_e = [\mathbf{p}_1,\mathbf{p}_2,\mathbf{p}_3,\mathbf{p}_4]^T$ is the element load vector which comes from the external virtual work. The matrices $\mathbf{H}$ and $\mathbf{G}$ are, respectively,

$$\mathbf{H} = \int_{(\Omega_s)}\mathbf{S}^T\mathbf{C}^{-1}\mathbf{S}\,dA,$$
$$\mathbf{G} = \int_{(\Omega_s)}\mathbf{S}^T\mathbf{B}\,dA. \quad (9)$$

In view of (9), we know that the matrix $\mathbf{H}$ only contains polynomials of ξ and η, so it can be evaluated analytically. While the matrix $\mathbf{G}$ can only be carried out analytically in the projected flat surface Ω_0^h [33]. As the matrix $\mathbf{G}$ is evaluated with the flat surface, then the corresponding transformation should be adopted to get the sound stiffness matrix.

In order to analyze the shell with intersections, a second transformation should be utilized [33]. The drilling degree of freedom is not available for the nodes on intersections. So the nodes on intersections have six degrees of freedom and the other nodes have five. The intersections can usually be seen in the internal structure of the wing, such as the intersection between the spars, the ribs, and the skin. Eventually the stiffness matrix of the shell element takes the form as follows [33]:

$$\mathbf{k}^e = \mathbf{G}^T\mathbf{H}^{-1}\mathbf{G} = \mathbf{k}_0+\mathbf{k}_{\mathrm{stab}},$$
$$\mathbf{k}_{IK} = \mathbf{G}_I^T\mathbf{H}^{-1}\mathbf{G}_K = \mathbf{A}_e\mathbf{B}_I^{0T}\mathbf{C}\mathbf{B}_K^0+\mathbf{g}_I^T\mathbf{h}^{-1}\mathbf{g}_K, \quad (10)$$

where $\mathbf{k}_{IK}$ denotes block parameters of the element stiffness matrix $\mathbf{k}^e$.

In the analysis of static aeroelasticity, structural equilibrium equation should be solved to obtain the structural deformation:

$$\mathbf{K}\mathbf{q} = \mathbf{f}, \quad (11)$$

where $\mathbf{K}$ is the global stiffness matrix which is assembled with the element stiffness matrix, $\mathbf{q}$ is the displacement vector of nodes, and $\mathbf{f}$ is nodal force vector which is interpolated from the aerodynamic forces. Equation (11) is a classical linear equation set, and here it is solved using GMRES [34] method to get the values of the nodal displacements. The global stiffness matrix is sparse so Compressed Sparse Row (CSR) which is popular for storing general sparse matrix is used here to store it.

3.2. Structural Design. The wing designed here is the wing-alone geometry called DPW-W1 which was developed for the Third AIAA Drag Prediction Workshop [35]. Figure 1 shows the overview of DPW-W1 (dimensions in m) and the geometric quantities are located in Table 1. The structural model is not necessary to represent exactly the internal components but is sufficient to capture the features of the wing. Furthermore since the realistic and precise structural model may have too many degrees of freedom, it is appropriate to simplify the model with the identical attributes. As the structural model of the DPW-W1 wing is not delivered, we create

TABLE 1: Geometric quantities for DPW-W1 wing.

Parameter	Value
Span	60 m
Aspect ratio	8.0
Sweep (leading edge)	17.2°
Sweep (quarter chord)	15.0°

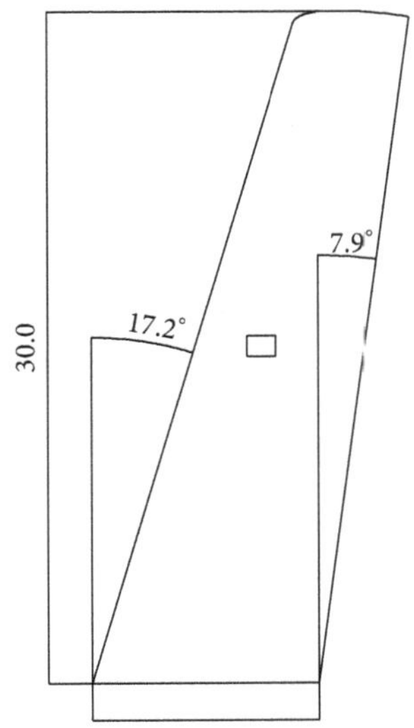

FIGURE 1: Overview of DPW-W1.

a structural model conforming to the wing which is a typical structure for a modern wing. The model used here consists of front and rear spars, ribs, and skin. There are 20 ribs distributed evenly along the span of the wing, not including the rib at the root for wing-body attachment. The front spar is located at 10% and the rear spar at 60% of the chord from the leading edge. In order to investigate appropriate effect of aerostructural deformation, the thickness of the two spars is adjusted to make sure that the displacement of the wing tip is about 3% of the semispan. Figure 2 shows the structural model of the wing. The skin is mirrored to the left to have a clear vision of the interior structure. The structural model of the wing is constructed with 1452 shell elements in total. The material used here for the structural model is assumed to be 7000 series of aluminium alloy with the properties shown in Table 2.

4. Aerostructural Analysis

In the aerostructural analysis, the aerodynamics provides loads to the structure and in turn the deflections are transferred from the structure to the CFD mesh. In general, the wall surface of the CFD mesh and the outer surface of the CSM mesh do not conform with each other. Thus a consistent and conservative data transfer method needs to be used. Here in the present work, TPS method which can deal with 3D problems with large deformation is applied.

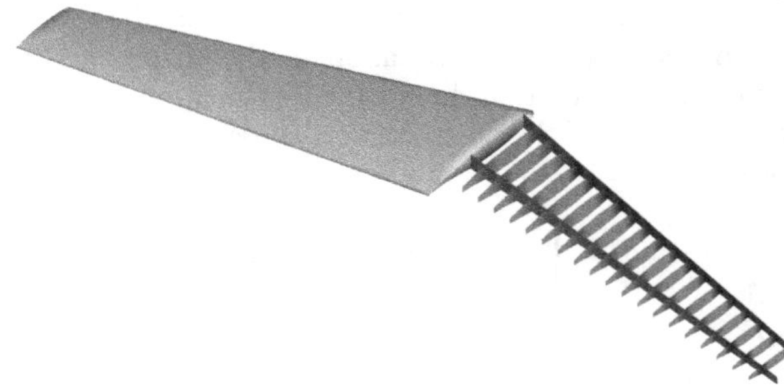

FIGURE 2: The wing structural model.

TABLE 2: Material properties.

Parameter	Value
Density	2780 kg/m^3
Modulus	73.1×10^9 Pa
Poisson ratio	0.33

The relational expression of displacements between the aerodynamics and structure is

$$\mathbf{d}_a = \mathbf{H}\mathbf{d}_s. \tag{12}$$

And the relational expression of forces is

$$\begin{aligned}\mathbf{F}_s &= \mathbf{H}^T\mathbf{F}_a,\\ \mathbf{H} &= \mathbf{B}_{as}\left[\mathbf{B}_s^{-1} - \mathbf{B}_s^{-1}\mathbf{R}_s\left(\mathbf{R}_s^T\mathbf{B}_s^{-1}\mathbf{R}_s\right)^{-1}\mathbf{R}_s^T\mathbf{B}_s^{-1}\right]\\ &\quad + \mathbf{R}_a\left[\left(\mathbf{R}_s^T\mathbf{B}_s^{-1}\mathbf{R}_s\right)^{-1}\mathbf{R}_s^T\mathbf{B}_s^{-1}\right],\end{aligned} \tag{13}$$

where the matrices $\mathbf{B}_{as}$ and $\mathbf{B}_s$ are defined as follows:

$$\begin{aligned}\mathbf{B}_{as,ij} &= r_{as,ij}^2 \ln\left(r_{as,ij}^2\right),\\ \mathbf{B}_{s,ij} &= r_{s,ij}^2 \ln\left(r_{s,ij}^2\right),\\ r_{as,ij}^2 &= \left(x_{a,i} - x_{s,j}\right)^2 + \left(y_{a,i} - y_{s,j}\right)^2 + \left(z_{a,i} - z_{s,j}\right)^2,\\ r_{s,ij}^2 &= \left(x_{s,i} - x_{s,j}\right)^2 + \left(y_{s,i} - y_{s,j}\right)^2 + \left(z_{s,i} - z_{s,j}\right)^2,\end{aligned} \tag{14}$$

where $\mathbf{d}_a$ and $\mathbf{F}_a$ are aerodynamic displacements and forces, respectively, $\mathbf{d}_s$ and $\mathbf{F}_s$ are structural displacements and forces, respectively, x_a, y_a, and z_a are Cartesian coordinates of CFD mesh points, and x_s, y_s, and z_s are Cartesian coordinates of CSM mesh nodes.

5. Sensitivity Analysis with Continuous Adjoint Method

Most cost functions in aerodynamic shape optimization can be represented as integrals of a function of flow variables over the solid surface:

$$J = \int_S f(V)\, dS, \tag{15}$$

where V denotes flow variables.

As the wall surface deforms, the change of the cost function comes from the variation of the shape and the alteration of the solution due to the deformation of shape. So the variation of the cost function contains two parts as follows:

$$\delta J = \int_{\delta S} f(V)\, dS + \int_S \frac{\partial f}{\partial V} \delta V\, dS, \tag{16}$$

where

$$\int_{\delta S} f(V)\, dS = \int_S \frac{\partial f}{\partial V} (\delta \mathbf{x} \cdot \nabla V)\, dS + \int_S \frac{\partial f}{\partial \mathbf{n}} \cdot \delta \mathbf{n}\, dS + \int_S f \delta\, dS, \tag{17}$$

where $\delta \mathbf{x}$ denotes the change of location of points on the wall surface and $\delta \mathbf{n}$ stands for the variation of unit normal vector of boundary. The cost function in the work is drag coefficient C_D and subject to that lift coefficient C_L is constant. As a result,

$$\begin{aligned} f(V) &= \mathbf{d} \cdot (P \mathbf{n}_S), \\ \mathbf{d} &= \frac{1}{C_\infty} (\cos\alpha \cos\beta, \sin\beta, \sin\alpha \cos\beta), \quad \text{for } C_D, \\ \mathbf{d} &= \frac{1}{C_\infty} (-\sin\beta, \cos\alpha, 0), \quad \text{for } C_L, \end{aligned} \tag{18}$$

where α is angle of attack, β is angle of sideslip, $C_\infty = (1/2)\mathbf{V}_\infty^2 \rho_\infty S_r$, and S_r is reference area. After a series of transformations [36], (16) turns into

$$\delta J = \int_S \mathbf{d} \cdot \delta P \mathbf{n}_S dS + \int_S (\mathbf{d} \cdot \nabla P)\, \delta S\, dS. \tag{19}$$

Consider the governing equations of 3D steady inviscid compressible flow as follows:

$$\nabla \cdot \mathbf{F} = \frac{\partial F_x}{\partial x} + \frac{\partial F_y}{\partial y} + \frac{\partial F_z}{\partial z} = 0, \tag{20}$$

where $\mathbf{F}$ is convective flux vector and F_x, F_y, and F_z are Cartesian components of $\mathbf{F}$. In order to circumvent the computation of δP in (19), the adjoint variables $\boldsymbol{\Psi}^T = (\psi_1, \psi_2, \psi_3, \psi_4, \psi_5)$ are introduced. Firstly linearizing Euler equations, calculating the inner product with the adjoint variables, and integrating the resulting equations over the entire domain by parts, then we obtain

$$-\int_\Omega \left(\nabla \boldsymbol{\Psi}^T\right) \mathbf{A} \delta \mathbf{U}\, d\Omega + \int_S \boldsymbol{\Psi}^T (\mathbf{n}_S \mathbf{A})\, \delta \mathbf{U}\, dS + \int_{\Gamma_\infty} \boldsymbol{\Psi}^T (\mathbf{n}_\infty \mathbf{A})\, \delta \mathbf{U}\, dS = 0, \tag{21}$$

where $\mathbf{A} = \partial \mathbf{F} / \partial \mathbf{U}$ is convective flux Jacobian matrix and $\mathbf{U}$ denotes conservative variables. As to the third part in (21) on the far-field boundary, we can get

$$\boldsymbol{\Psi}^T (\mathbf{n}_\infty \mathbf{A})\, \delta \mathbf{U} = \left(\boldsymbol{\Psi}^T \mathbf{T}\right) \boldsymbol{\Lambda} \left(\mathbf{T}^{-1} \delta \mathbf{U}\right), \tag{22}$$

where $\boldsymbol{\Lambda}$ is the diagonal matrix of eigenvalues and $\mathbf{T}$, $\mathbf{T}^{-1}$ are right and left eigenvectors, respectively. For shape optimization, the variation of characteristic variables of the flow is $\delta \mathbf{W} = \mathbf{T}^{-1} \delta \mathbf{U} = 0$, so the part on the far-field boundary of (21) can be eliminated with appropriate far-field boundary conditions for the adjoint variables [1, 6]. Applying wall boundary conditions of Euler equations, the integration over wall boundary turns into

$$\int_S \boldsymbol{\Psi}^T (\mathbf{n}_S \mathbf{A})\, \delta \mathbf{U}\, dS = \int_S (\mathbf{n} \cdot \delta \mathbf{v}) \cdot \left(\rho \psi_1 + \rho \left(u \psi_2 + v \psi_3 + w \psi_4\right) + \rho H \psi_5\right) dS + \int_S \left(n_x \psi_2 + n_y \psi_3 + n_z \psi_4\right) \delta P\, dS. \tag{23}$$

Using linearized nontranspiration wall boundary conditions $\mathbf{n} \cdot \delta \mathbf{v} = -(\delta S \partial_n \mathbf{v}) \cdot \mathbf{n}_S - \delta \mathbf{n}_S \cdot \mathbf{v}$, (23) gives

$$\int_S \boldsymbol{\Psi}^T (\mathbf{n}_S \mathbf{A})\, \delta \mathbf{U}\, dS = -\int_S \left(\left(\partial_n \mathbf{v} \cdot \mathbf{n}_s\right) \xi + \nabla_s (\mathbf{v} \xi)\right) \delta S\, dS + \int \left(n_x \psi_2 + n_y \psi_3 + n_z \psi_4\right) \delta P\, dS, \tag{24}$$

where $\xi = \rho \psi_1 + \rho(u \psi_2 + v \psi_3 + w \psi_4) + \rho H \psi_5$. Substituting (24) into (21) and subtracting (21) from (19), we can get the variation of the cost function as follows:

$$\delta J = \int_{\Omega_a} \left(\nabla \boldsymbol{\Psi}^T\right) \mathbf{A} \delta \mathbf{U}\, d\Omega + \int_S \left(\mathbf{d} \cdot \mathbf{n}_S - \left(n_x \psi_2 + n_y \psi_3 + n_z \psi_4\right)\right) \delta P\, dS + \int_S \left(\mathbf{d} \cdot \nabla P + \left(\partial_n \mathbf{v} \cdot \mathbf{n}_S\right) \xi + \nabla_S (\mathbf{v} \xi)\right) \delta S\, dS. \tag{25}$$

For eliminating $\delta \mathbf{U}$ and δP in (25), set the first two terms in (25) to zero, and then we can get the adjoint equations,

$$\mathbf{A}^T (\nabla \boldsymbol{\Psi}) = \left(\left(\nabla \boldsymbol{\Psi}^T\right) \mathbf{A}\right)^T = 0. \tag{26}$$

The wall boundary conditions of the adjoint equations are as follows:

$$\mathbf{d} \cdot \mathbf{n}_S = \left(n_x \psi_2 + n_y \psi_3 + n_z \psi_4\right). \tag{27}$$

In the end, the variation of the cost function is

$$\delta J = \int_S \left(\mathbf{d} \cdot \nabla P + \left(\partial_n \mathbf{v} \cdot \mathbf{n}_S\right) \xi + \nabla_S (\mathbf{v} \xi)\right) \delta S\, dS. \tag{28}$$

It can be seen that the right hand side terms of (28) contain flow variables (such as gradient of pressure ∇P and normal derivative of velocity to the surface S) and adjoint variables (such as ξ), so the geometry sensitivities can be acquired only after the governing and adjoint equations are solved to get flow and adjoint variables.

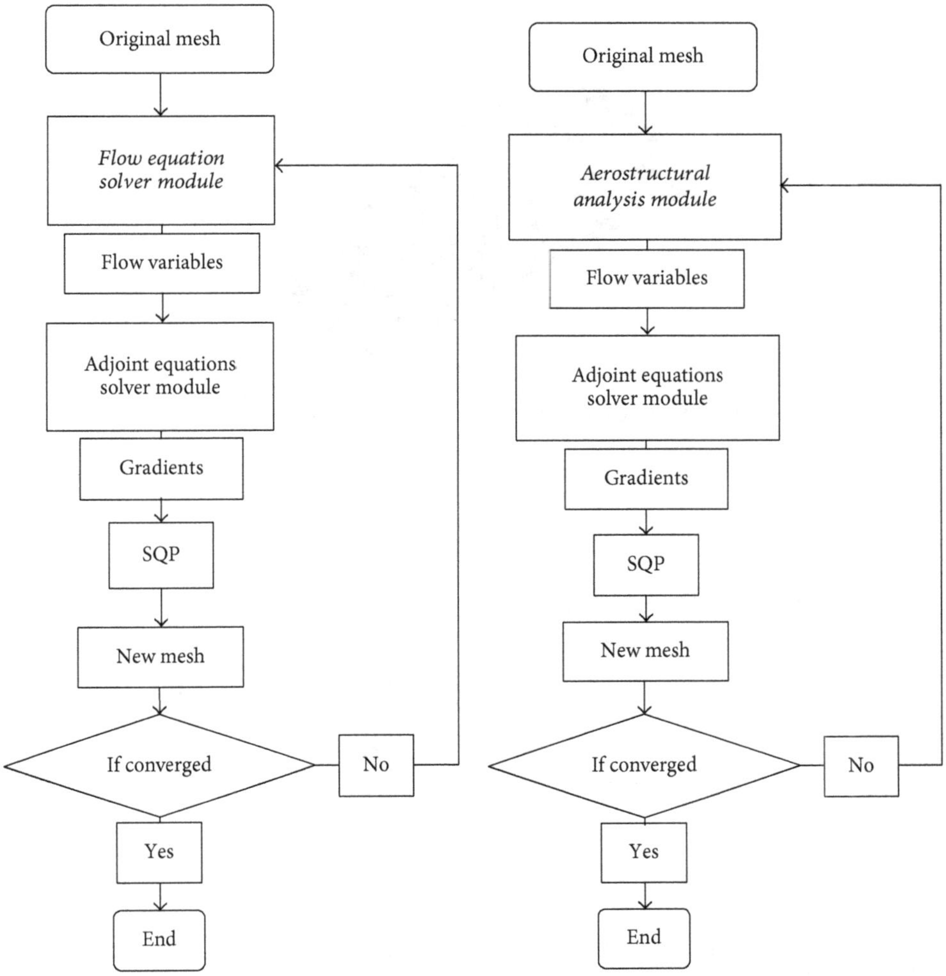

(a) Aerodynamic optimization

(b) Aerodynamic optimization including aeroelasticity

Figure 3: Optimization flow chart.

From (26), it can be known that the Jacobian matrix of the adjoint equations is transposed of the flow convective flux Jacobian matrix, so the adjoint equations have the same form with Euler equations. Then the adjoint equations can be solved using a finite-volume type of method similar to that used for Euler equations. As the cost of solving the adjoint equations is approximately equivalent to that of solving the Euler equations and the computational cost of the gradients does not depend on the number of design variables, the gradients with respect to all design variables for each objective function or constraint can be obtained with the cost roughly twice that of solving Euler equations. So the adjoint method is quite suitable for aerodynamic shape optimization with a large number of design variables.

6. Design Strategy

A Free-Form Deformation (FFD) method is used for shape parameterization. In FFD, the design variables are the control points of the control volume and as the design variables change, then the surface mesh inside the control volume makes corresponding changes. Then a simple classical edge spring method is adopted to deform the CFD mesh to conform to the new designed configuration. The optimization algorithm used here to drive the design to the optimal point is sequential quadratic programming (SQP) method which is gradient-based and can deal with linear or nonlinear problems with or without constraints.

The flow chart of aerodynamic optimization is shown as Figure 3(a), and Figure 3(b) is the flow chart of aerodynamic optimization including the effect of aeroelasticity. Comparing with Figure 3(a), the significant difference of Figure 3(b) is that in every design cycle aerostructural analysis is included. Because of the presence of the deflection of the static aeroelasticity, there are two shapes of the wing before and after aerostructural analysis, respectively. In the design, the deflection of the optimization is performed on the former, while the flow variables, adjoint variables, and gradients are evaluated on the latter.

7. Test Cases

The airfoil of DPW-W1 is representative of a supercritical section which is usually adopted by most modern commercial

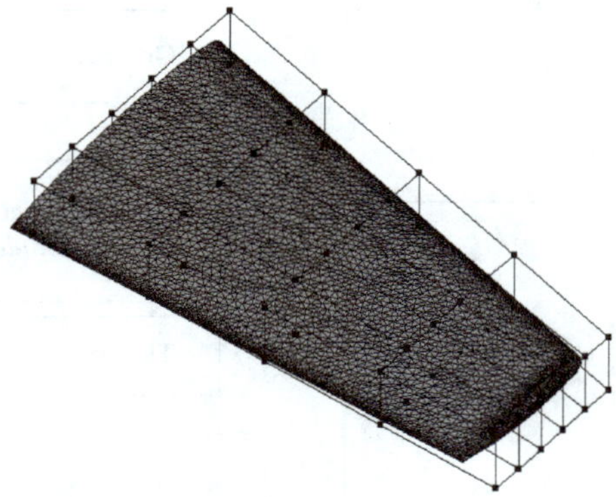

FIGURE 4: Surface mesh and FFD control volume of ONERA M6.

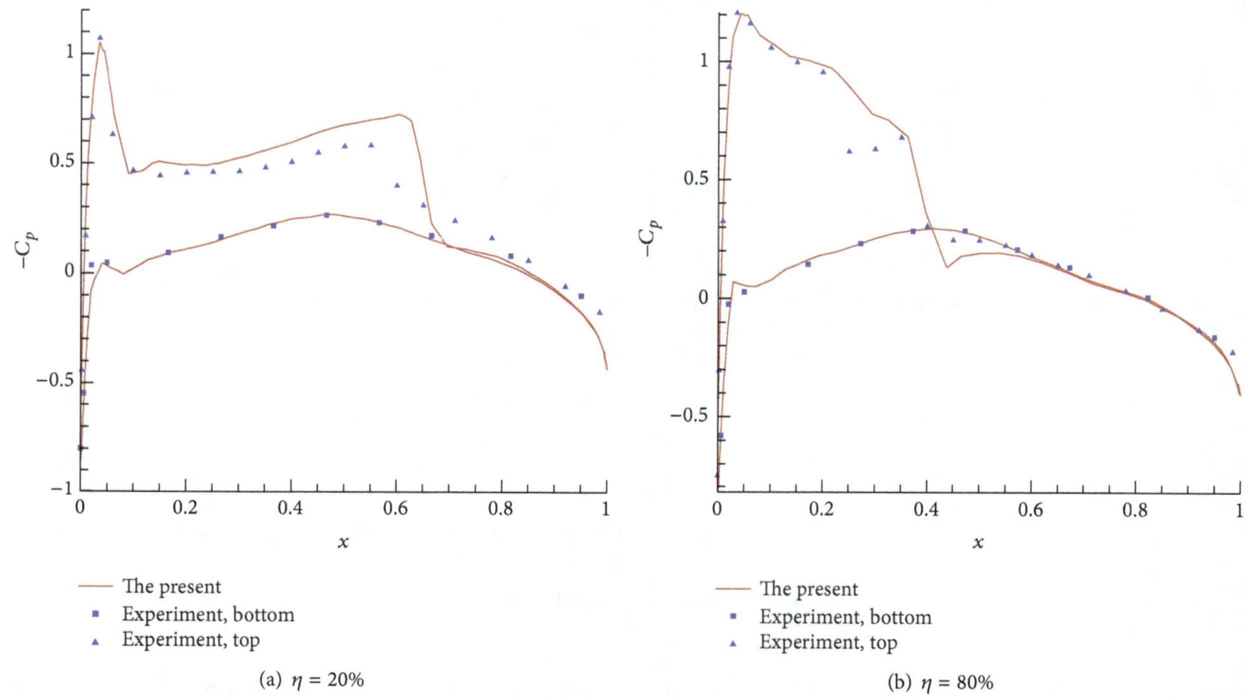

FIGURE 5: Comparison of section pressure coefficients of ONERA M6.

aircrafts. Here the wing has been modified with a sharp trailing edge in order to be compatible with inviscid flow solution. Firstly verification of sensitivities and aerodynamic analysis and aerodynamic optimization results for ONERA M6 wing are presented in Section 7.1. Then aerodynamic optimization for rigid DPW-W1 wing is shown in Section 7.2. As the aspect ratio of this wing is relatively large and as it deforms in realistic flight conditions, then the optimization results including the effect of static aeroelasticity are presented in Section 7.3. A comparison between the results from aerostructural analysis of rigid optimization and flexible optimization is shown. In both cases, drag coefficient is minimized with lift coefficient and the wing volume being fixed.

7.1. Verification. In this section, all the verifications are implemented using ONERA M6 wing. The flow field is computed using Euler equations with Mach 0.8395 and angle of attack $\alpha = 3.06^\circ$ on unstructured mesh with 310000 points. The surface mesh is shown in Figure 4.

The verification of the inviscid solver for ONERA M6 is presented. Figure 5 shows the comparison of pressure coefficients at different sections between aerodynamic analysis and experimental results. It can be seen that the pressure coefficient distributions match each other well.

The sensitivities of drag coefficient are verified. The sensitivities of the continuous adjoint method are compared with those of the finite difference method. The design variables are the control points on the upper surface of FFD control volume (shown as Figure 4), 5 of which are used to validate the sensitivities of the objective function C_D. The comparison of sensitivities of drag coefficient with respect to the design variables is shown in Table 3. It is seen that the sensitivities are accurate enough to be used for aerodynamic optimization.

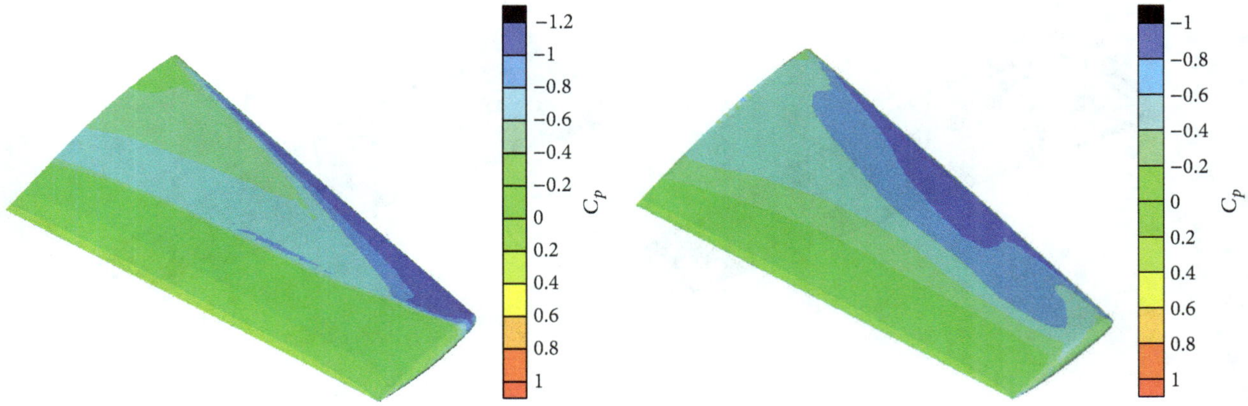

(a) Before design (b) After design

FIGURE 6: Contours of pressure coefficient of ONERA M6.

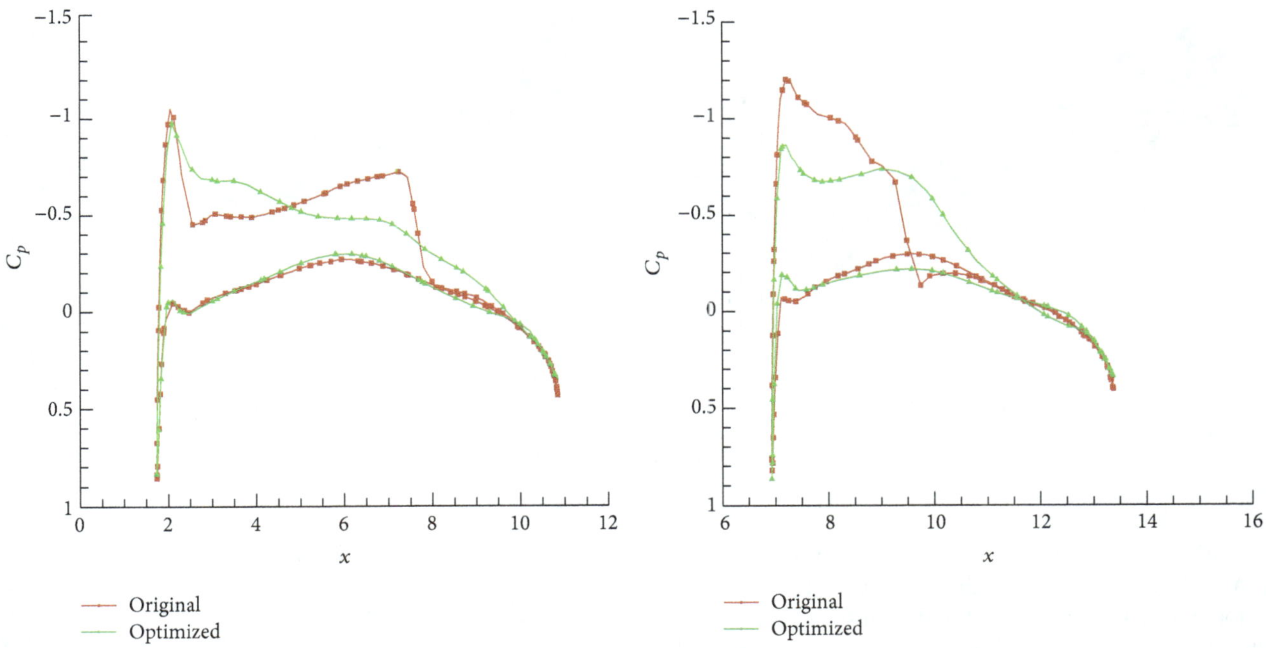

(a) η = 20% (b) η = 80%

FIGURE 7: Comparison of section pressure coefficients before and after design of ONERA M6.

TABLE 3: Comparison of the sensitivities of drag coefficient.

Design variables	Finite difference	Continuous adjoint	Δ%
1	3.4502E − 003	3.4982E − 003	1.39%
2	6.6485E − 003	6.3501E − 003	−4.49%
3	4.1414E − 003	4.2095E − 003	1.64%
4	4.4838E − 003	4.2258E − 003	−5.75%
5	−9.8593E − 004	−9.8141E − 004	−0.46%

TABLE 4: Optimization results of ONERA M6.

Parameter	Before optimization	After optimization	Δ%
C_L	0.2903	0.29	−0.10%
C_D	0.01417	0.01059	−25.26%
C_L/C_D	20.49	27.38	33.63%

The pressure coefficient of ONERA M6 before and after aerodynamic optimization is shown in Figure 6, respectively. Figure 7(a) is the pressure coefficient distributions at the section of 20% semispan while Figure 7(b) shows the pressure coefficient distributions at the section of 80% semispan. The optimization results of ONERA M6 are shown in Table 4. Figures 6 and 7 show that the shock wave on the upper surface has been eliminated. So the optimization results can show that the optimization method developed in the present work can be applied to aerodynamic optimization for a flexible wing.

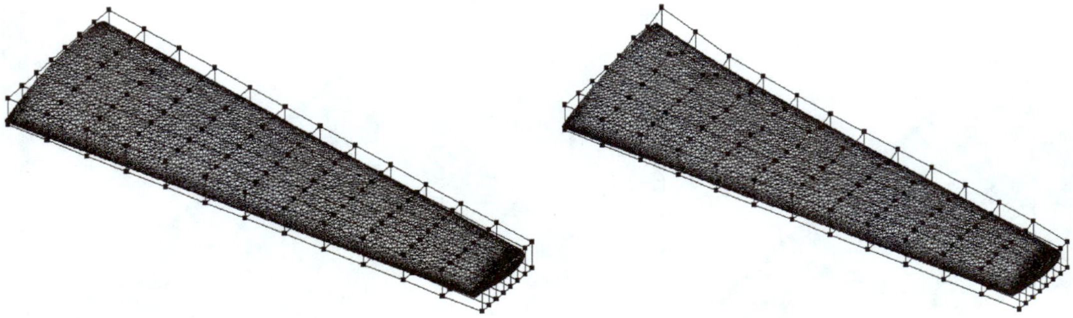

(a) Before design (b) After design

FIGURE 8: Surface mesh and FFD control volume of DPW-W1.

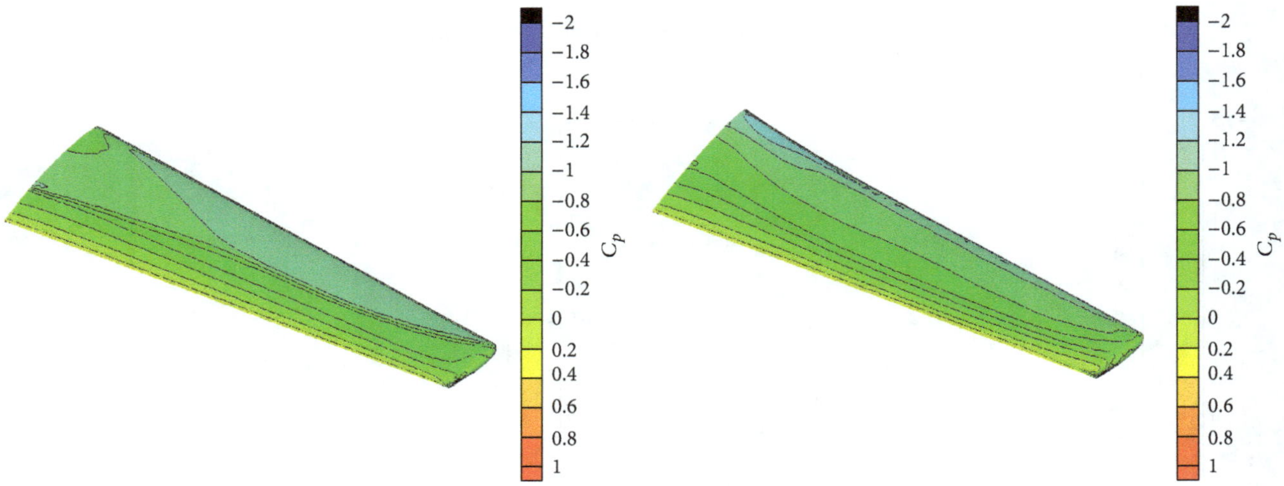

(a) Before design (b) After design

FIGURE 9: Contours of pressure coefficient of rigid wing.

7.2. Aerodynamic Optimization for Rigid DPW-W1 Wing. In this case, drag coefficient is minimized with lift coefficient fixed for rigid DPW-W1 wing. The design here is performed in typical transonic flow conditions with Mach 0.76 and angle of attack $\alpha = 0.5^\circ$. Aerodynamic analysis is implemented on unstructured mesh which has 270000 points. The surface meshes of DPW-W1 with the FFD control volume before and after the design are shown as Figure 8. There are 208 control points in total in the FFD control volume, while 91 of them are taken as design variables which are all located on the upper surface. Figure 9(a) is the original pressure coefficient contours, while Figure 9(b) is the pressure coefficient contours after optimization. Comparing the two figures, we can see that after optimization the shock wave on the upper surface is significantly weakened. Figure 10(a) is the pressure coefficient distributions at the section of 20% semispan while Figure 10(b) shows the pressure coefficient distributions at the section of 80% semispan. It is seen directly that the shock wave along the span is eliminated. The shapes of corresponding sections are shown in Figure 11, and it is clear that the incidences of the optimized sections are decreased, equivalently with decreased local angle of attack to reduce drag. Figure 12 shows the convergence history. In this case, the drag coefficient C_D reduces from 251 counts to 242 counts, namely, a reduction of 9 counts. While the lift coefficient C_L varies from 0.6211 to 0.6210 which almost keeps constant. The optimization results of rigid wing are presented in Table 5.

TABLE 5: Optimization results of rigid wing.

Parameter	Before design	After design	Δ%
C_L	0.6211	0.6210	−0.016%
C_D	0.02512	0.02415	−3.86%
C_L/C_D	24.7	25.7	4.04%

7.3. Aerodynamic Optimization for DPW-W1 Wing including Aerostructural Deflections. As shown in Figure 13, the aeroelasticity has apparent influences on the performance of the wing. Figure 13(a) shows the comparison of the rigid shape (black) and the deflected one (red), and in Figure 13(b) the section shape at 80% semispan is presented. It is seen that the wing has a deflection of negative twist angle. Due to the aerostructural deflection, drag and lift are affected by about 10% and 6%, respectively, with respect to the rigid wing. So it

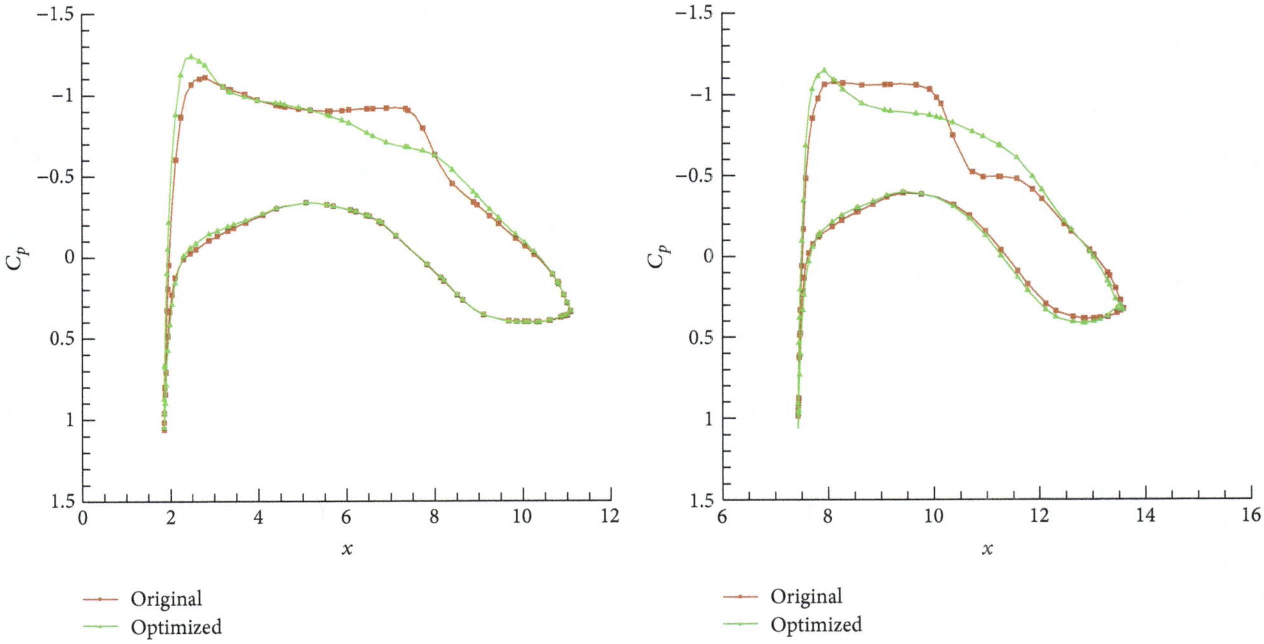

(a) $\eta = 20\%$ (b) $\eta = 80\%$

FIGURE 10: Comparison of section pressure coefficients before and after design of rigid wing.

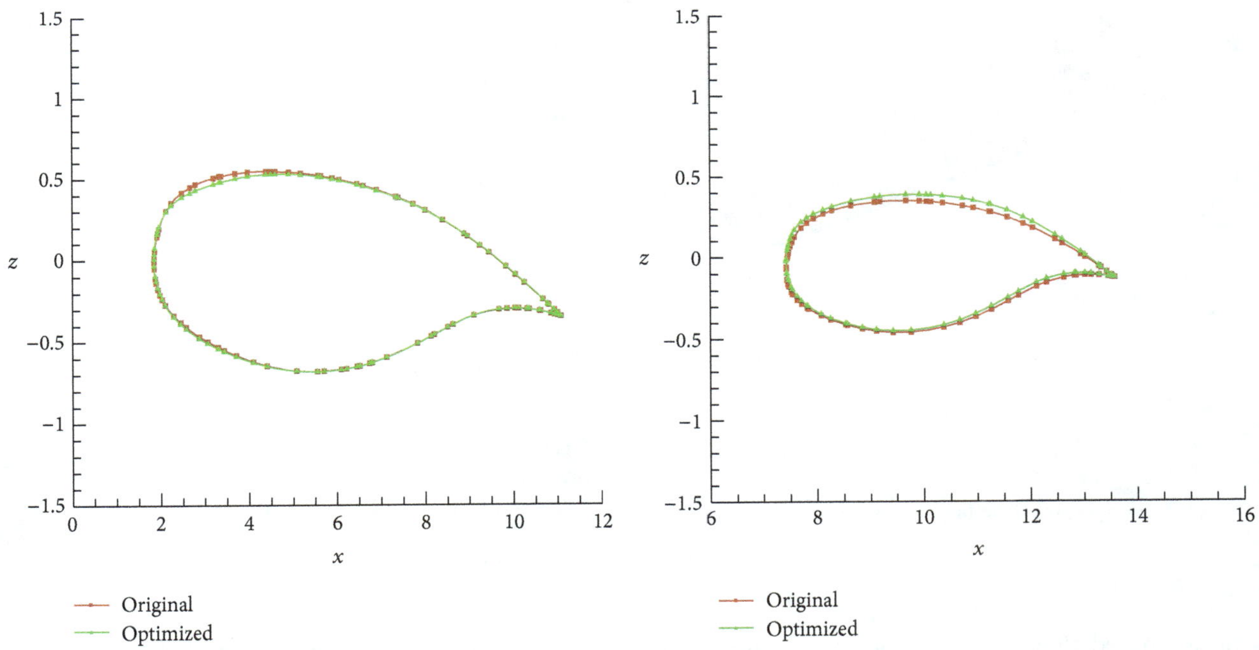

(a) $\eta = 20\%$ (b) $\eta = 80\%$

FIGURE 11: Comparison of section shapes before and after design of rigid wing.

is necessary to include the effect of aeroelasticity during the optimization design.

Here DPW-W1 wing is optimized to minimize drag coefficient with constant lift coefficient including aerostructural deflections. The flow conditions are the same as the rigid optimization case. As seen in Figures 14 and 15, after the optimization design of the flexible wing, the shock wave on the upper wing surface is eliminated. Since the effect of aeroelasticity is introduced, it is obvious that the pressure coefficient contours before optimization in Figure 14 are significantly different from those in Figure 9, as well as the pressure coefficient contours after optimization. The comparison of different section shapes before and after optimization is shown in Figure 16 and likewise the incidence of the

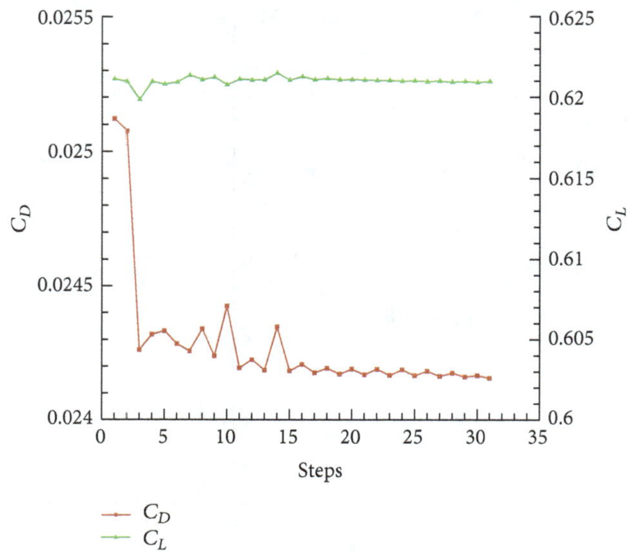

FIGURE 12: Convergence history for rigid wing design.

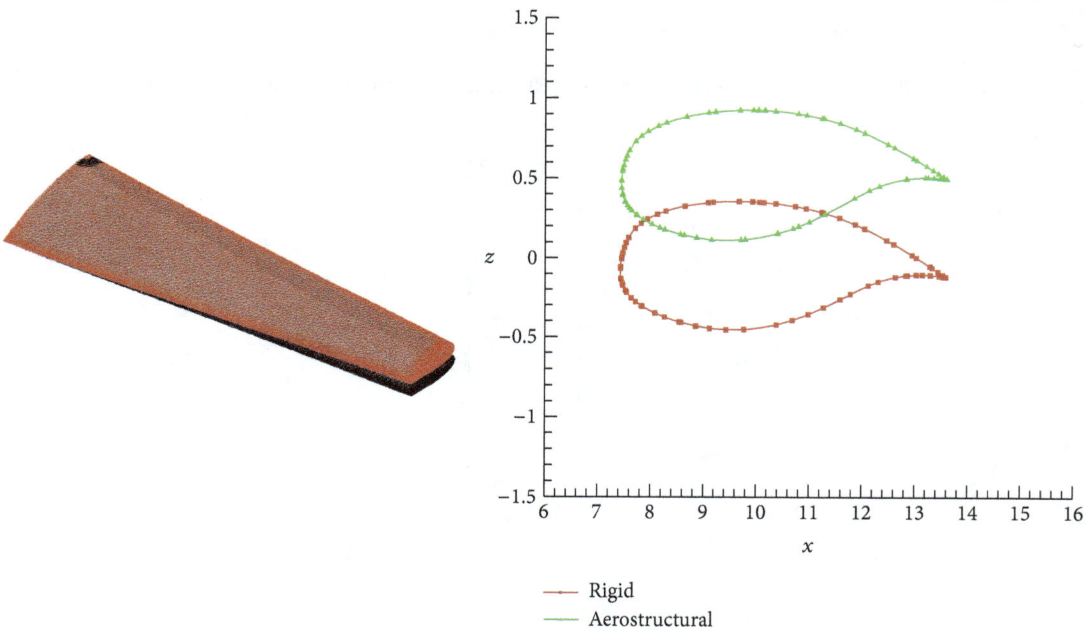

(a) Deflection of the wing (b) Section shape at η = 80%

FIGURE 13: Effect of static aeroelasticity.

wing tip is reduced after design. As shown in Figure 16(b) because of aeroelasticity the section at 80% semispan is deflected upward apparently. Convergence history for flexible wing design is shown in Figure 17. In the aerostructural analysis of the original DPW-W1 wing, the resulting values of drag coefficient and lift coefficient are 223.7 counts and 0.5815, while after the optimization introducing the effect of aeroelasticity, drag coefficient decreases to 218.6 counts with the reduction of 5.1 counts and lift coefficient is 0.5807 which nearly keeps constant. The optimization results of flexible wing are presented in Table 6.

TABLE 6: Optimization results of flexible wing.

Parameter	Before design	After design	Δ%
C_L	0.5815	0.5807	−0.19%
C_D	0.02237	0.02186	−2.28%
C_L/C_D	26.0	26.6	2.3%

A comparison of wing pressure distributions resulting from aerostructural analysis of rigid optimization and flexible optimization is shown in Figure 18, as well as the comparison of pressure coefficient distribution of the section at 80% semispan. In this comparison it is seen that the results from the two optimization cases are not the same, especially at the sections near the wing tip where the

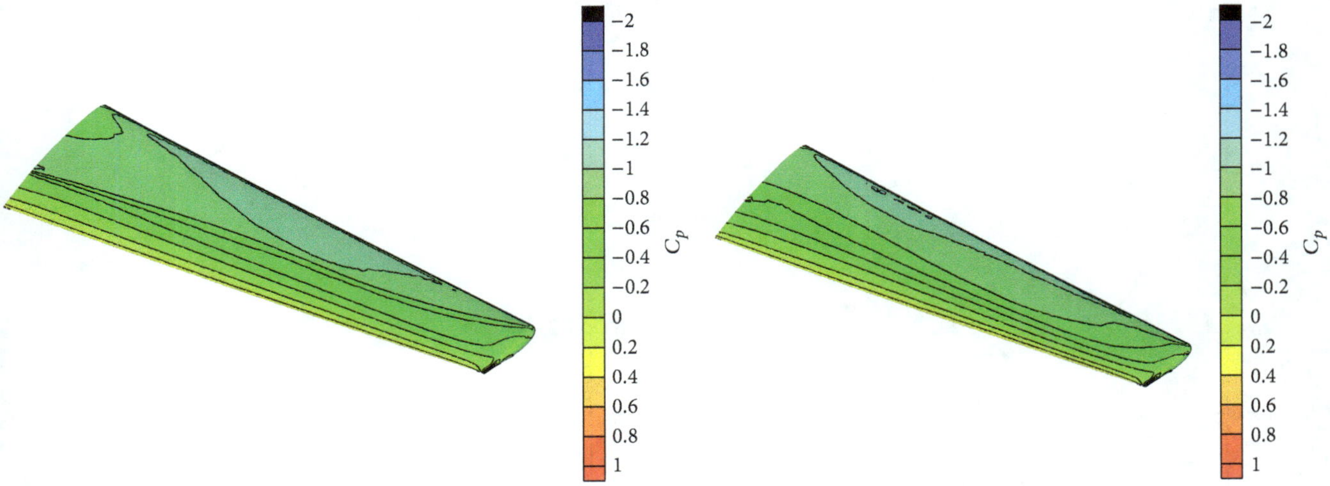

(a) Before design (b) After design

FIGURE 14: Contours of pressure coefficient of flexible wing.

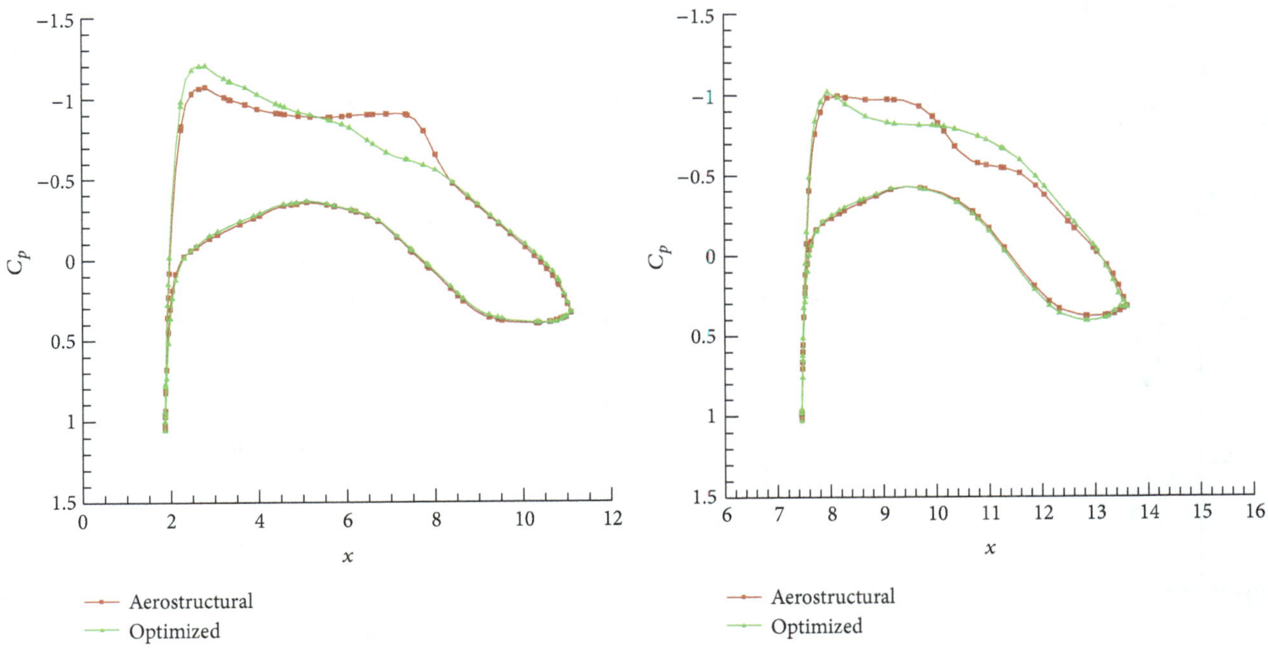

(a) $\eta = 20\%$ (b) $\eta = 80\%$

FIGURE 15: Comparison of section pressure coefficients before and after design of flexible wing.

aerostructural deflections are large. We can see that the wave on the upper surface of aerostructural analysis of rigid optimization is weakened less completely than that of flexible optimization. So it is proved that it is necessary to include the effect of aeroelasticity in the wing optimization design. Figure 19 shows the comparison of the corresponding resulting section shapes of the two optimization cases.

8. Conclusions

The main feature of this work is to develop an optimization strategy for a flexible wing based on the continuous adjoint method. The aerostructural analysis is implemented based on high-fidelity models, namely, the Euler equations on unstructured mesh and a linear quadrilateral shell element model which is suitable for both thin and thick shells but is relatively easily implemented. Sequential quadratic programming method is adopted to search for the optimal solution using the gradients from continuous adjoint method. The flow charts of rigid and flexible optimization are presented and compared. The continuous adjoint method is used to design the rigid wing and flexible wing and these two optimization cases are compared with each other to demonstrate that it is necessary to consider the effect of aeroelasticity in the wing optimization design.

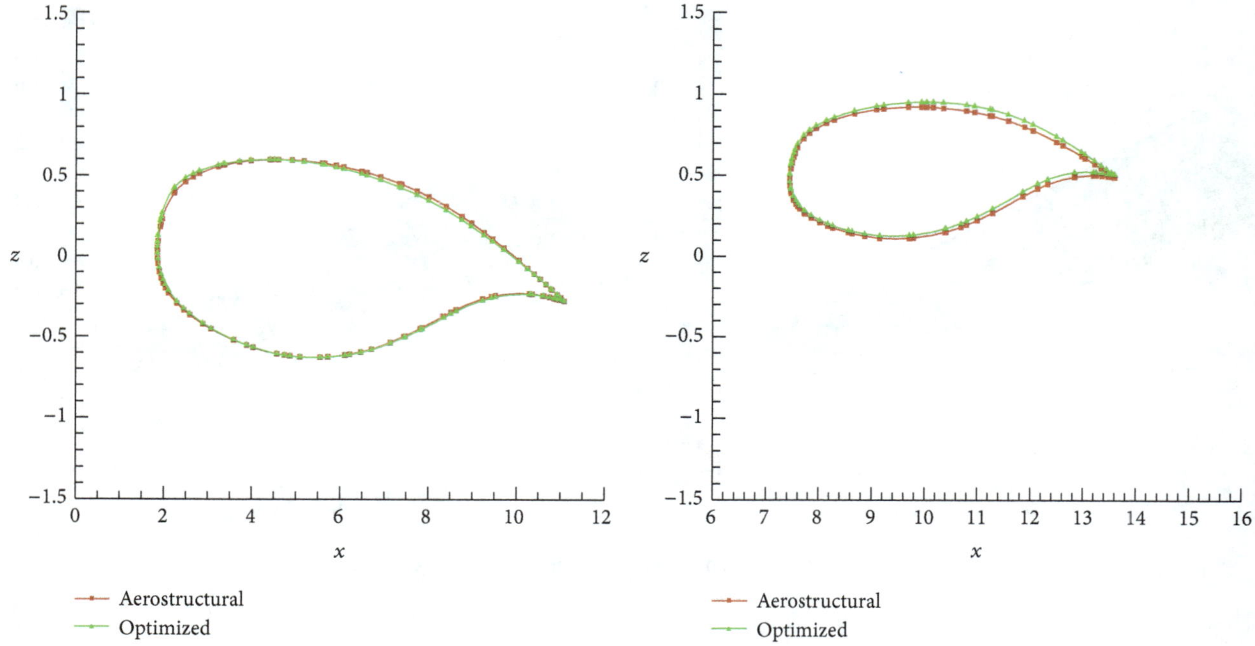

(a) η = 20%

(b) η = 80%

FIGURE 16: Comparison of section shapes before and after design of flexible wing.

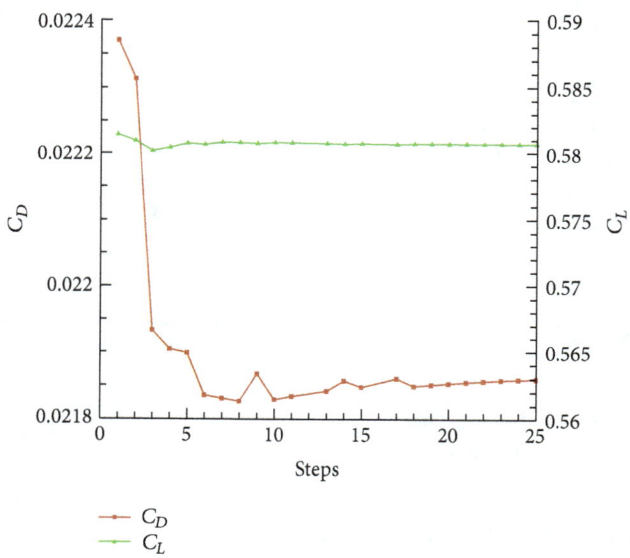

FIGURE 17: Convergence history for flexible wing design.

The future work is to implement the continuous adjoint methodology based on Navier-Stokes equations and to develop the optimization design system on the basis of aero-structural continuous adjoint method.

Nomenclature

A: Convective flux Jacobian matrix
A, B, C, D: Middle nodes of relevant edges of the elements
C: Constitutive matrix
D: Unit vector normal to the shell midsurface
$\mathbf{d}_a$: Aerodynamic displacements
$\mathbf{d}_s$: Structural displacements
E: Total energy per unit mass
F: Convective fluxes vector
f: Nodal force vector
$\mathbf{F}_a$: Aerodynamic forces
$\mathbf{F}_s$: Structural forces
F_x, F_y, F_z: Cartesian components of convective fluxes vector
H: Total enthalpy
J: Coordinate transformation Jacobian matrix
J: Objective or constraint function
K: Global stiffness matrix
$\mathbf{k}^e$: Element stiffness matrix
n_x, n_y, n_z: Cartesian components of unit normal vector
P: Static pressure
$\mathbf{p}_e$: Element load vector
q: Displacement vector of nodes
S: Solid wall domain boundary
S_r: Reference area
t: Time
T: Right eigenvectors
$\mathbf{T}^{-1}$: Left eigenvectors
U: Conservative variables vector
u: Nodal translational displacement vector
u, v, w: Cartesian velocity components
V: Flow variables
v: Structural nodal displacement vector
$\mathbf{v}_e$: Element displacement vector
V_n: Contravariant velocity

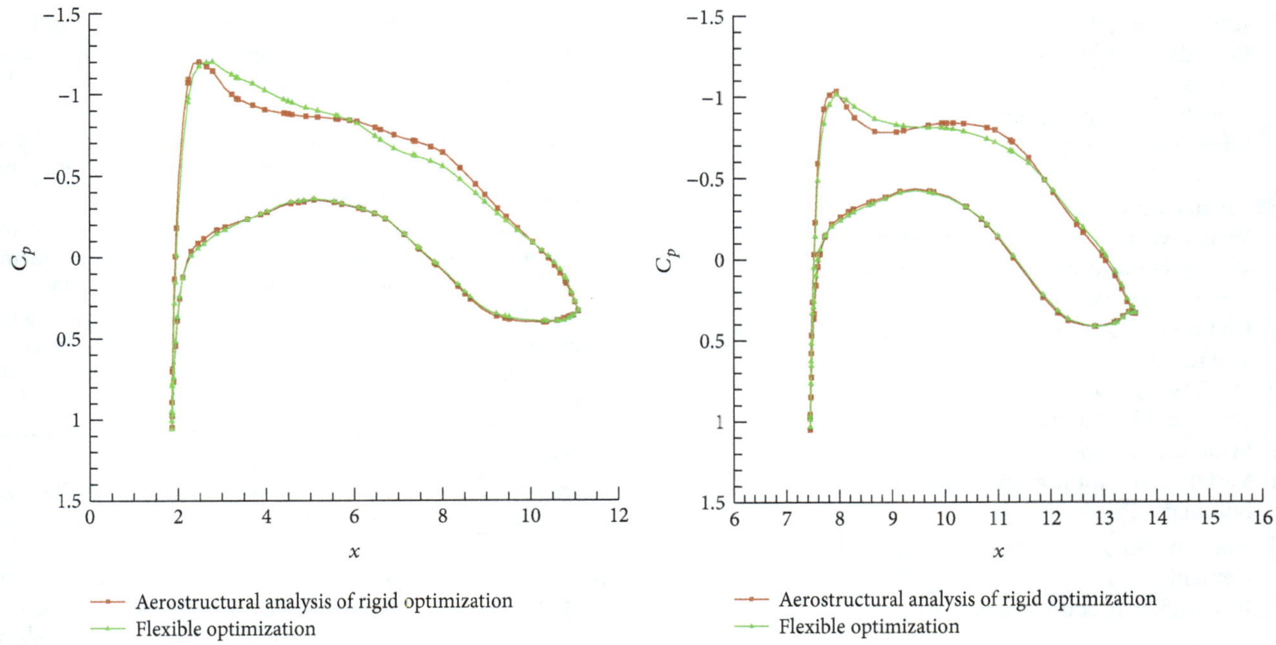

(a) $\eta = 20\%$ (b) $\eta = 80\%$

FIGURE 18: Comparison of section pressure distributions of aerostructural analysis of rigid optimization and flexible optimization.

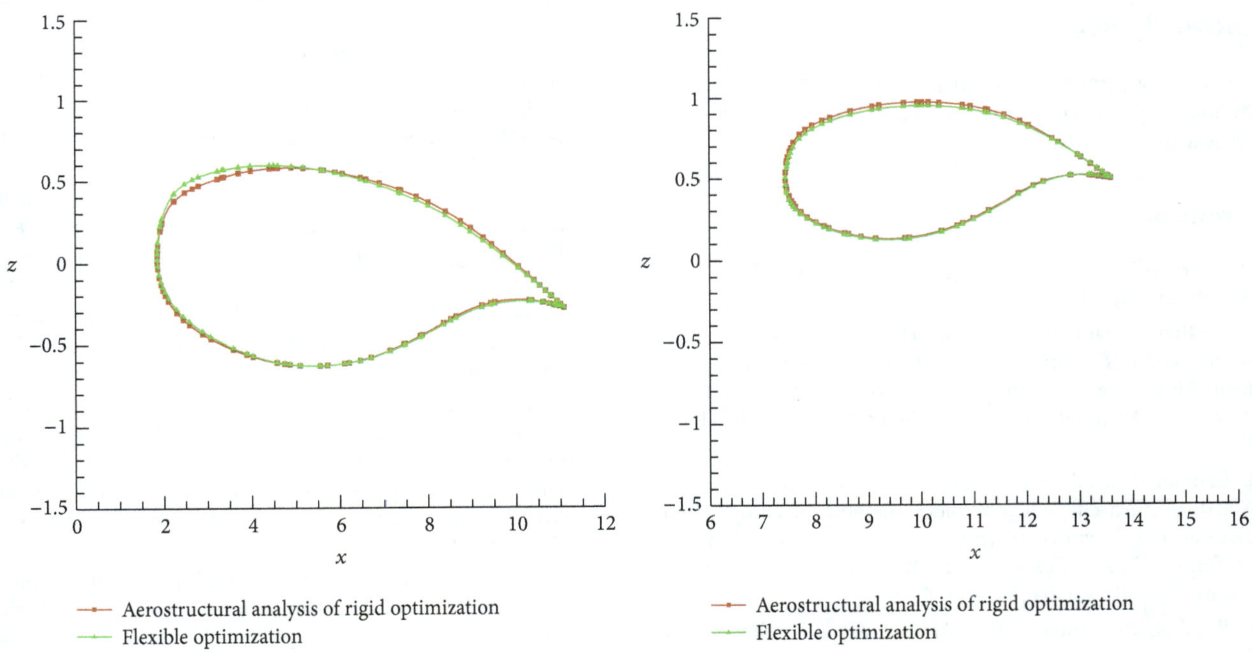

(a) $\eta = 20\%$ (b) $\eta = 80\%$

FIGURE 19: Comparison of section shapes of aerostructural analysis of rigid optimization and flexible optimization.

$\mathbf{X}$: Nodal coordinate vector
x_a, y_a, z_a: Cartesian coordinates of CFD mesh points
x_s, y_s, z_s: Cartesian coordinates of CSM mesh nodes
$\mathbf{\Lambda}$: Diagonal matrix of eigenvalues
ρ: Fluid density
$\boldsymbol{\sigma}$: Stress resultants
$\boldsymbol{\varepsilon}$: Vector containing membrane strains, curvatures, and shear strains
$\mathbf{\Psi}$: Adjoint variable vector
α: Angle of attack

β: Angle of sideslip
$\bar{\mathbf{t}}$: Boundary load vector on a part of boundary Γ_σ
Ω_a: Flow field computational domain
$\delta\mathbf{v}$: Structural nodal virtual displacement vector
$\delta\boldsymbol{\sigma}$: Virtual stress resultants
$\delta\boldsymbol{\varepsilon}$: Virtual vector containing membrane strains, curvatures, and shear strains
$\bar{\mathbf{p}}$: Volume load vector in Ω_s
$\delta\mathbf{x}$: Changes of location of points on the wall surface
$\Delta\mathbf{d}$: Deviation of $\mathbf{D}$
Γ_σ: Structural boundary
Ω_s: Structural computational domain
$\delta\mathbf{n}$: Variation of unit normal vector of boundary
$\bar{\xi}, \bar{\eta}$: Coordinates of the center of gravity of the element
ξ, η: Isoparametric coordinates.

Competing Interests

The authors declare no competing interests regarding the publication of this article.

Acknowledgments

This work is supported by a project funded by the Priority Academic Program Development (PAPD) of Jiangsu Higher Education Institutions.

References

[1] A. Jameson, "Aerodynamic design via control theory," *Journal of Scientific Computing*, vol. 3, no. 3, pp. 233–260, 1988.

[2] J. Reuther, A. Jameson, J. Farmer et al., "Aerodynamic shape optimization of complex aircraft configurations via an adjoint formulation," in *Proceedings of the 34th Aerospace Sciences Meeting and Exhibit Reno*, AIAA Paper 94, Reno, Nev, USA, 1996.

[3] J. Reuther, J. J. Alonso, J. C. Vassberg, A. Jameson, and L. Martinelli, "An efficient multiblock method for aerodynamic analysis and design on distributed memory systems," in *Proceedings of the 13th Computational Fluid Dynamics Conference*, Snowmass Village, Colo, USA, 1997.

[4] J. Reuther, A. Jameson, J. J. Alonso, M. J. Rimlinger, and D. Sauders, "Constrained multipoint aerodynamic shape optimization using an adjoint formulation and parallel computers," in *Proceedings of the AIAA Meeting Papers on Disc*, Reno, Nev, USA, January 1997.

[5] J. J. Reuther, A. Jameson, J. J. Alonso, M. J. Rimlinger, and D. Saunders, "Constrained multipoint aerodynamic shape optimization using an adjoint formulation and parallel computers, part 2," *Journal of Aircraft*, vol. 36, no. 1, pp. 61–74, 1999.

[6] W. K. Anderson and V. Venkatakrishnan, "Aerodynamic design optimization on unstructured grids with a continuous adjoint formulation," *Computers and Fluids*, vol. 28, no. 4-5, pp. 443–480, 1999.

[7] F. Palacios, T. D. Economon, A. C. Aranake et al., "Stanford University Unstructured (SU2): open-source analysis and design technology for turbulent flows," *AIAA Paper*, vol. 243, pp. 13–17, 2014.

[8] P. Francisco et al., "Large-scale aircraft design using SU2," in *Proceedings of the 53st AIAA Aerospace Sciences Meeting*, vol. 1946, Kissimmee, Fla, USA, 2015.

[9] T. D. Economon, F. Palacios, and J. J. Alonso, "Unsteady continuous adjoint approach for aerodynamic design on dynamic meshes," *AIAA Journal*, vol. 53, no. 9, pp. 2437–2453, 2015.

[10] T. D. Economon, F. Palacios, J. J. Alonso et al., "Towards high-performance optimizations of the unstructured open-source SU2 suite," in *Proceedings of the AIAA SciTech*, AIAA 2015-1949, Kissimmee, Fla, USA, 2015.

[11] H. L. Kline, F. Palacios, T. D. Economon, and J. J. Alonso, "Adjoint-based optimization of a hypersonic inlet," in *Proceedings of the 22nd AIAA Computational Fluid Dynamics Conference*, AIAA Paper 2015-3060, Dallas, Tex, USA, June 2015.

[12] H. L. Kline, T. D. Economon, and J. J. Alonso, "Mulit-objective optimization of a hypersonic inlet using generalized outflow boundary conditions in the continuous adjoint method," in *Proceedings of the 54th AIAA Aerospace Sciences Meeting*, San Diego, Calif, USA, 2016.

[13] B. Y. Zhou, T. A. Albring, N. R. Gauger et al., "A discrete adjoint framework for unsteady aerodynamic and aeroacoustic optimization," in *Proceedings of the 16th AIAA/ISSMO Multidisciplinary Analysis and Optimization Conference*, AIAA Aviation, June 2015.

[14] T. Albring, M. Sagebaum, and N. R. Gauger, "Development of a consistent discrete adjoint solver in an evolving aerodynamic design framework," in *Proceedings of the 16th AIAA/ISSMO Multidisciplinary Analysis and Optimization Conference*, AIAA 2015-3240, Dallas, Tex, USA, 2015.

[15] O. Baysal and M. E. Eleshaky, "Aerodynamic sensitivity analysis methods for the compressible Euler equations," *Journal of Fluids Engineering, Transactions of the ASME*, vol. 113, no. 4, pp. 681–688, 1991.

[16] G. W. Burgreen and O. Baysal, "Three-dimensional aerodynamic shape optimization of wings using sensitivity analysis," AIAA Paper 94-0094, 1994.

[17] D. J. Mavriplis, "Formulation and multigrid solution of the discrete adjoint problem on unstructured meshes," in *Computational Fluid Dynamics 2004: Proceedings of the Third International Conference on Computational Fluid Dynamics, ICCFD3, Toronto, 12–16 July 2004*, pp. 663–668, Springer, Berlin, Germany, 2004.

[18] D. J. Mavriplis, "Discrete adjoint-based approach for optimization problems on three-dimensional unstructured meshes," *AIAA Journal*, vol. 45, no. 4, pp. 740–750, 2007.

[19] F. Christakopoulos, D. Jones, and J.-D. Müller, "Pseudo-timestepping and verification for automatic differentiation derived CFD codes," *Computers & Fluids*, vol. 46, no. 1, pp. 174–179, 2011.

[20] Z. Lyu, G. K. W. Kenway, C. Paige, and J. R. R. A. Martins, "Automatic differentiation adjoint of the reynolds-averaged navier-stokes equations with a turbulence model," in *Proceedings of the 21st AIAA Computational Fluid Dynamics Conference*, San Diego, Calif, USA, June 2013.

[21] M. Biava, M. Woodgate, and G. N. Barakos, "Fully implicit discrete-adjoint methods for rotorcraft applications," *AIAA Journal*, vol. 54, no. 2, pp. 735–749, 2016.

[22] G. Carpentieri, *An Adjoint-Based Shape-optimization Method for Aerodynamic Design*, Delft University of Technology, Delft, The Netherlands, 2009.

[23] S. Rajagopal and R. Ganguli, "Multidisciplinary design optimization of a UAV wing using Kriging based multi-objective genetic algorithm," in *Proceedings of the 50th AIAA/ASME/ASCE/AHS/ASC Structures, Structural Dynamics, and Materials Conference*, AIAA Paper 2219, Palm Springs, Calif, USA, May 2009.

[24] Y. Kim, D.-H. Lee, Y. Kim, and K. Yee, "Multidisciplinary design optimization of supersonic fighter wing using response surface methodology," in *Proceedings of the 9th AIAA/ISSMO Symposium on Multidisciplinary Analysis and Optimization*, AIAA Paper 2002-5408, Atlanta, Ga, USA, September 2002.

[25] J. R. R. A. Martins and A. B. Lambe, "Multidisciplinary design optimization: a survey of architectures," *AIAA Journal*, vol. 51, no. 9, pp. 2049–2075, 2013.

[26] G. K. W. Kenway and J. R. R. A. Martins, "Multipoint high-fidelity aerostructural optimization of a transport aircraft configuration," *Journal of Aircraft*, vol. 51, no. 1, pp. 144–160, 2014.

[27] G. K. W. Kenway, G. J. Kennedy, and J. R. R. A. Martins, "Scalable parallel approach for high-fidelity steady-state aeroelastic analysis and adjoint derivative computations," *AIAA Journal*, vol. 52, no. 5, pp. 935–951, 2014.

[28] D. J. Mavriplis, E. Anderson, R. S. Fertig, M. Garnich, and Z. Yang, "Development of a high-fidelity time-dependent aerostructural capability for analysis and design," in *Proceedings of the 57th AIAA/ASCE/AHS/ASC Structures, Structural Dynamics, and Materials Conference*, San Diego, Calif, USA, January 2016.

[29] E. Fabiano and D. Mavriplis, "Adjoint-based aerodynamic design on unstructured meshes," in *Proceedings of the 54th AIAA Aerospace Sciences Meeting*, AIAA 2016-1295, San Diego, Calif, USA, 2016.

[30] E. Fabiano, A. Mishra, D. J. Mavriplis et al., "Time-dependent aero-acoustic adjoint-based shape optimization of helicopter rotors in forward flight," in *Proceedings of the 57th AIAA/ASCE/AHS/ASC Structures, Structural Dynamics, and Materials Conference*, vol. 3 of *AIAA 2016-1910*, San Diego, Calif, USA, 2016.

[31] J. Duchon, "Splines minimizing rotation-invariant semi-norms in Sobolev spaces," in *Constructive Theory of Functions of Several Variables*, W. Schempp and K. Zeller, Eds., vol. 571 of *Lecture Notes in Mathematics*, pp. 85–100, Springer, Berlin, Germany, 1977.

[32] M. J. Wright, G. V. Candler, and D. Bose, "Data-parallel line relaxation method for the Navier-Stokes equations," *AIAA Journal*, vol. 36, no. 9, pp. 1603–1609, 1998.

[33] F. Gruttmann and W. Wagner, "A linear quadrilateral shell element with fast stiffness computation," *Computer Methods in Applied Mechanics and Engineering*, vol. 194, no. 39–41, pp. 4279–4300, 2005.

[34] Y. Saad, *Iterative Methods for Sparse Linear Systems*, SIAM, 2003.

[35] J. C. Vassberg, E. Tinoco, M. Mani et al., "Summary of the third AIAA CFD drag prediction workshop," in *Proceedings of the 45th AIAA Aerospace Sciences Meeting and Exhibit*, Reno, Nev, USA, January 2007.

[36] A. Bueno-Orovio, C. Castro, F. Palacios, and E. Zuazua, "Continuous adjoint approach for the spalart-allmaras model in aerodynamic optimization," *AIAA Journal*, vol. 50, no. 3, pp. 631–646, 2012.

5

Efficient Method for Aeroelastic Tailoring of Composite Wing to Minimize Gust Response

Yang Yu,[1] Zhengjie Wang,[1] and Shijun Guo[2]

[1] *School of Mechatronical Engineering, Beijing Institute of Technology, Beijing 100081, China*
[2] *Aerospace Engineering, Cranfield University, Cranfield MK43 0BL, UK*

Correspondence should be addressed to Yang Yu; yangyu.bit2012@gmail.com

Academic Editor: Marine Mikilyan

Aeroelastic tailoring of laminated composite structure demands relatively high computational time especially for dynamic problem. This paper presents an efficient method for aeroelastic dynamic response analysis with significantly reduced computational time. In this method, a relationship is established between the maximum aeroelastic response and quasi-steady deflection of a wing subject to a dynamic loading. Based on this relationship, the time consuming dynamic response can be approximated by a quasi-steady deflection analysis in a large proportion of the optimization process. This method has been applied to the aeroelastic tailoring of a composite wing of a tailless aircraft for minimum gust response. The results have shown that 20%–36% gust response reduction has been achieved for this case. The computational time of the optimization process has been reduced by 90% at the cost of accuracy reduction of 2~4% comparing with the traditional dynamic response analysis.

1. Introduction

In aircraft design, the use of composite materials and increase of wing aspect ratio for higher aerodynamic efficiency becomes a trend [1]. As a result, the wing becomes more flexible and prone to not only static aeroelastic deformation and aileron reversal, but also aeroelastic stability [2]. As one of the approaches to solve the problem, aeroelastic tailoring techniques has been applied to the optimal design of composite wing structures [3, 4]. In addition, gust response is a critical concern and has attracted research attention especially for lightweight aircraft of high aspect ratio wing. In this particular subject, substantial research activities have been carried out to develop various active and passive control technologies for gust alleviation of aircraft since 1990s [5–10]. Therefore aeroelasticity should be considered in the early phase of aircraft design process [11]. For this purpose, an efficient method to predict the gust response of an aircraft with the aeroelastic coupling effect of the wing is required.

This current investigation is to develop an efficient method for gust response analysis applicable to optimization of a composite wing structure for minimum gust response. In aircraft design, there are two types of gust model and methods to predict the response according to the airworthiness regulation EASA CS-25 [12]. The discrete gust model is the simple one; while the continuous gust based on Von Kaman's gust power spectrum density model is usually solved by using random process theory. To focus our study on the efficient dynamic response analysis, the discrete gust model is chosen in the current investigation.

When the dynamic aeroelastic coupling effect of the wing is considered, the calculation for maximum gust response performed in either frequency-domain or time-domain is time consuming. For aeroelastic tailoring, this becomes more concerned since such calculation is usually conducted in the gradient calculation of each design variable and each iterative cycle in a traditional optimization process. Although efforts have been made to simplify the structural models, the time consuming analysis of gust response remains as a challenge in this field [13, 14]. In a previous study of aircraft structure optimization subject to gust response, the efficiency was improved in two ways [15], firstly by saving the modal analysis time for reduced structural model at each variation step, and secondly calculating the gust response only in limited iteration steps to check if the optimization was kept in track like the other work [13]. In the FFAST ("Future Fast Aeroelastic Simulation Technologies") project, modeling and optimization methods were studied to predict the worst-case gust load in fast and efficient manner [16]. Their work also considered changes made to aircraft structure as part of the evolving design process [17].

However, when more detailed structure optimization such as aeroelastic tailoring is required, dynamic response result is required in each iteration step or more frequently. Therefore an efficient method for calculating the gust response rather than cutting off the calculation is required. This paper presents a method for establishing a relationship of the maximum gust response with quasi-steady deflection of a wing. Based on this relationship, the time consuming dynamic response analysis can be replaced by a quasi-steady deflection analysis in a large proportion of the iteration cycles during the optimization process.

In the optimization of a wing structure subject to gust, a response peak value at a representative point is normally taken as a measurement of the gust response. The center of gravity of a rigid aircraft or the wingtip for a flexible wing is usually taken as the measurement point. This implies that it is unnecessary to perform the time consuming traditional dynamic response analysis in time history or frequency domain. However, the existing approach to evaluate the response peak value is to perform the traditional dynamic response analysis unless a more efficient alternative method is available. This current study presents a quasi-static (Q-S) method to predict the gust response peak value based on a mixture of static and dynamic response analysis in time-domain. This method increases the efficiency of gust response analysis significantly especially when applied to structure optimization where many iterations of the response calculation are performed. The investigation was conducted in three stages. In the first stage, an expression was established for the energy conversion between the work done by external dynamic force and structural deformation at the time when the response reaches its first maximum value. A proportional relationship was created between the elastic energy of a structure and work done by external force and resulted in a Q-S ratio. In the second stage, the relationship was extended to the optimization process in which the structural stiffness has a small variation in the iteration cycles. Based on the relationship, the dynamic response peak value of a structure at an iteration cycle can be predicted based on the Q-S ratio and the response obtained in previous cycle. In the third stage, the time period to reach the first response peak value was determined for the structure having a small stiffness variation during the optimization. By applying this method, the computational time for a structural optimization subject to dynamic response can be reduced significantly. To demonstrate the method, a small aircraft has been taken as example to minimize its gust response by tailoring the composite wing laminate layup. Different cases are studied including gust frequency and wing sweepback angles. The results show that this method is able to predict the gust response accurately with the computational time over the optimization reduced by over 85% at the cost of accuracy reduction by less than 5% comparing with the traditional method of traditional dynamic response analysis.

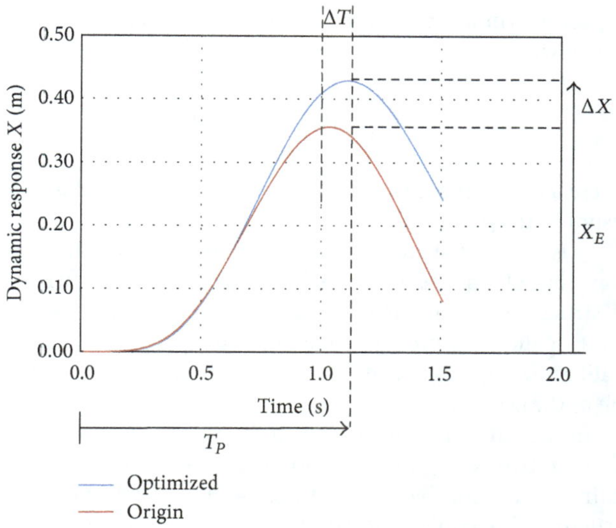

FIGURE 1: An illustration of T_P, X_E, and deviations between two optimization circles.

2. Theoretical Study of the Method

2.1. The Quasi-Static Model for Dynamic Response Analysis. The work done on an elastic body by external forces is absorbed and transformed to kinetic and elastic potential energy in a form of movement and deformation. For a linear elastic body subjected to a dynamic load, the total energy E includes both elastic potential and kinetic energy E_K written as [18]

$$E = \frac{1}{2}X^T KX + E_K, \tag{1}$$

where X is the elastic deformation vector and K is the stiffness matrix of a structure.

An estimation of the stiffness of a structure, K_E, can be obtained by the force applied against the deflection measured at representative points. In this article, an estimation of stiffness K_E of the wing structure is calculated from wingtip displacement and the total bending moment and torque on the wing obtained from static aeroelastic analysis. The estimated equivalent K_E is used in the following Q-Static method for gust response analysis.

Assuming T_P is time when the dynamic response reaches the first peak value as shown in Figure 1. The dynamic response such as the wingtip displacement reaches a peak value X_E at T_P. At this moment, the velocity and kinetic energy E_K of the structure becomes zero [19]. The system is in Q-Static status and total energy E can be simplified to an expression of elastic potential energy E_{EP}.

$$E = E_{EP} = \frac{1}{2}K_E X_E^2. \tag{2}$$

The work done by the external dynamic force $Q(t)$ through the structural deformation over a time period until a specified time such as the peak value of dynamic response can be written as

$$W_P = \int_0^{T_P} (Q(t) H(t))\, dt, \quad (3)$$

where $H(t)$ is the corresponding velocity of the structural response to $Q(t)$.

It is noted that the W_P gives an approximate total work done based on the estimated E_{EP}. For a small variation of structural K_E at each optimization cycle, it is a relative value against the previous one and will be updated using a "calibration process" as described later in Section 2.5 during the optimization.

In the current optimization study, the response variation between two contiguous iteration cycles is caused by a small stiffness variation by tailoring the laminate layup of a wing structure. The equivalent structure stiffness at iteration cycle N can be represented by the stiffness at previous cycle $N-1$:

$$K_{E|N} = \eta K_{E|N-1}, \quad (4)$$

in which η is the ratio of the stiffness variation between two contiguous iteration cycles. For linear elastic structural deformation, the response peaks at the two iteration steps are also in the same proportion as the stiffness:

$$X_{E|N} = \frac{1}{\eta} X_{E|N-1}. \quad (5)$$

This leads to the relationship of the E_{EP} between two iteration cycles:

$$E_{\text{EP}|N} = \frac{1}{2} K_{E|N} X_{E|N}^2 = \frac{1}{\eta} E_{\text{EP}|N-1}. \quad (6)$$

As structural stiffness is varied, the time to reach the response peak value is also changed from the previous iteration cycle:

$$T_{P|N} = T_{P|N-1} + \Delta T(\eta), \quad (7)$$

in which $\Delta T(\eta)$ represents the time difference of peak gust response between cycle N and $N-1$. The work done by the external force between two iteration cycles is related by

$$\begin{aligned} W_{\text{EP}|N} &= \int_0^{T_{P|N}} (Q(t) H_{E|N}(t))\, dt = A_N \left[\int_0^{T_{P|N-1}} \left(Q(t) \overline{H}_{E|N}(t)\right) dt + \int_{T_{P|N-1}}^{T_{P|N-1}+\Delta T} \left(Q(t) \overline{H}_{E|N}(t)\right) dt \right] \\ &= A_N \left[\left(\int_0^{T_{P|N-1}} \left(Q(t) \left(\overline{H}_{E|N-1}(t) + \overline{H}_{E|N}(t) - \overline{H}_{E|N-1}(t)\right)\right) dt \right) + \int_{T_{P|N-1}}^{T_{P|N-1}+\Delta T} \left(Q(t) \overline{H}_{E|N}(t)\right) dt \right] \\ &= \frac{A_N}{A_{N-1}} \left[A_{N-1} \int_0^{T_{P|N-1}} \left(Q(t) \overline{H}_{E|N-1}(t)\right) dt \right. \\ &\quad \left. + A_{N-1} \left(\int_0^{T_{P|N-1}} \left(Q(t) \left(\overline{H}_{E|N}(t) - \overline{H}_{E|N-1}(t)\right)\right) dt + \int_{T_{P|N-1}}^{T_{P|N-1}+\Delta T} \left(Q(t) \overline{H}_{E|N}(t)\right) dt \right) \right] = \frac{A_N}{A_{N-1}} \left(W_{\text{EP}|N-1} + \Delta W\right), \end{aligned} \quad (8)$$

in which $H(t) = A\overline{H}(t)$, and A_N/A_{N-1} is the Q-S ratio of response amplitude in two contiguous iteration cycles given by

$$\frac{A_N}{A_{N-1}} = \frac{X_{E|N}}{X_{E|N-1}} = \frac{1}{\eta}. \quad (9)$$

From (6) and (8), the following relation between E_{EP} and W_{EP} is obtained in two contiguous iteration cycles:

$$\begin{aligned} \frac{W_{\text{EP}|N}}{E_{\text{EP}|N}} &= \frac{\eta^{-1} (W_{\text{EP}|N-1} + \Delta W)}{\eta^{-1} E_{\text{EP}|N-1}} \\ &= \frac{W_{\text{EP}|N-1}}{E_{\text{EP}|N-1}} + \frac{\Delta W}{E_{\text{EP}|N-1}}, \end{aligned} \quad (10)$$

where $\Delta W / E_{\text{EP}|N-1}$ represents a deviation of the ratio $W_{\text{EP}}/E_{\text{EP}}$ between two contiguous iteration cycles.

In (10), ΔW is defined in (8) as

$$\begin{aligned} \Delta W = A_{N-1} &\left(\int_0^{T_{P|N-1}} (Q(t) \right. \\ &\cdot \left(\overline{H}_{E|N}(t) - \overline{H}_{E|N-1}(t)\right)) dt \\ &\left. + \int_{T_{P|N-1}}^{T_{P|N-1}+\Delta T} \left(Q(t) \overline{H}_{E|N}(t)\right) dt \right). \end{aligned} \quad (11)$$

The ΔW is expressed in two terms: the first term is due to the variation of deformation; the second term is due to the variation of peak response time all caused by tailoring the structure such as the laminate layup. For a small $\Delta W / E_{\text{EP}|N-1}$ below a value, it can be negligible as discussed in Section 2.5. In this case, a simplified ratio can be obtained from (10):

$$\frac{E_{\text{EP}|N}}{E_{\text{EP}|N-1}} = \frac{W_{\text{EP}|N}}{W_{\text{EP}|N-1}}. \quad (12)$$

When a ratio is obtained at iteration step $N - 1$, an approximate X_E at iteration cycle N can be obtained based on (12) instead of performing the traditional dynamic response analysis again. The more iteration cycles it performs, the more computational time to save.

In this current study, the (1 – cos) discrete gust model as defined in airworthiness regulation CS-25 is considered:

$$U_g = \frac{V_{ds}}{2}\left[1 - \cos\left(\omega_g t\right)\right], \tag{13}$$

where V_{ds} is a specified design gust velocity and $\omega_g = \pi s/tH$ is the gust frequency, in which $s = Vt$ is the penetrating distance of an aircraft at an equivalent flight speed V into a discrete gust at time t. H is gust gradient distance.

The symmetric aerodynamic force due to the gust can be calculated by

$$Q(t) = \frac{1}{2}\rho V^2 S a_1 F \frac{U_g}{V} = q\frac{V_{ds}}{2}\left[1 - \cos\left(\omega_g t\right)\right], \tag{14}$$

where ρ is the air density, S is the reference wing area, and a_1 is the lift curve slope of the wing. F is the gust alleviating factor expressed by

$$F = \frac{0.88\mu}{5.3 + \mu}, \qquad \mu = \frac{2m}{\rho S c a_1}. \tag{15}$$

It is noted that the above F is defined for a typical gust gradient distance $H = 12.5c$ where c is the mean chord. However the Q-static method is not related to the factor. It is only the accuracy of the gust response that will be affected when the factor is adapted for other H cases. Corresponding to the (1 – cos) gust, the responding wingtip deflection can also be expressed as a (1 – cos) function. Hence the response velocity $H_E(t)$ before approaching the peak time T_P can be expressed in a sinusoidal function by differentiating the wingtip deflection:

$$H_E(t) = A\pi T_P^{-1}\sin\left(\pi T_P^{-1} t\right), \tag{16}$$

where A represents the wingtip deflection amplitude at the response time T_P. Based on (3), (14), and (16), the W_{EP} can be rewritten as

$$W_{EP} = \int_0^{T_P}\left(Q(t)\times H_E(t)\right)dt = \int_0^{T_P}\left(q\times\frac{V_{de}}{2}\times\left[1-\cos\left(\omega_g t\right)\right]\times A\pi T_P^{-1}\sin\left(\pi T_P^{-1}t\right)\right)dt = A\int_0^{T_P}\left(Q(t)\times\overline{H}_E(t)\right)dt. \tag{17}$$

According to (2) and (17), (12) can be rewritten as

$$\frac{K_{E|N}X_{E|N}^2}{K_{E|N-1}X_{E|N-1}^2} = \frac{\int_0^{T_{P|N}}\left(q\times\left(V_{de}/2V\right)\left[1-\cos\left(\omega_g t\right)\right]\times A_N\pi T_{P|N}^{-1}\sin\left(\pi T_{P|N}^{-1}t\right)\right)dt}{\int_0^{T_{P|N-1}}\left(q\times\left(V_{de}/2V\right)\left[1-\cos\left(\omega_g t\right)\right]\times A_{N-1}\pi T_{P|N-1}^{-1}\sin\left(\pi T_{P|N-1}^{-1}t\right)\right)dt}. \tag{18}$$

By extracting the parameters q, π, T_P, A_N, and A_{N-1} out of the time integration and substituting (9) into (18), a simple recursive relation of $X_{E|N}$ at $T_{P|N}$ can be obtained from the $X_{E|N-1}$ as shown:

$$X_{E|N} = \left[\left(\frac{T_{P|N}^{-1}\int_0^{T_{P|N}}\left(\left[1-\cos\left(\omega_g t\right)\right]\times\sin\left(\pi T_{P|N}^{-1}t\right)\right)dt}{T_{P|N-1}^{-1}\int_0^{T_{P|N-1}}\left(\left[1-\cos\left(\omega_g t\right)\right]\times\sin\left(\pi T_{P|N-1}^{-1}t\right)\right)dt}\right)\times\left(\frac{K_{E|N-1}}{K_{E|N}}\right)\right]X_{E|N-1}. \tag{19}$$

In (19), since the $T_{P|N}$ in the iteration cycle N is different from previous iteration cycle $N - 1$, it will be predicted in each iteration cycle. The details of how to predict the $T_{P|N}$ are presented in the following section.

Although the deviation $\Delta W/E_{EP|N-1}$ as shown in (10) may be negligible, it will accumulate when (19) is used to predict the X_E in subsequent cycles beyond cycle N. After number of cycles, it is necessary to update the approximated ratio as shown in (12) by performing a full dynamic response analysis without approximation. This so-called "calibration process" is described in Section 2.5.

2.2. Prediction of the Peak Response Time. As an earlier description for (2), the wing structure was simplified to a second-order system with an equivalent stiffness. The system response time T_P to reach its peak value as illustrated in Figure 1 can be expressed in (19), in which ω_n is the structural natural frequency and ζ the damping ratio:

$$T_P = \frac{\pi}{\omega_n\sqrt{1-\zeta^2}}. \tag{20}$$

An expression of the angular value at peak response time T_P in general form can be written as

$$T_P\omega_n = f\left(\omega_g, \omega_n, \zeta\right), \tag{21}$$

where ω_g is the gust frequency. Since the constant damping ζ has little influence to the frequency of the system having a small variation of stiffness, the response frequency in the iteration cycle N has a small difference from that in cycle $N-1$ due to a small stiffness variation:

$$\omega_N = \omega_{N-1} + \Delta\omega. \tag{22}$$

Similarly the angular value of response function $T_{P|N}\omega_N$ has a small difference from $T_{P|N-1}\omega_{N-1}$:

$$T_{P|N}\omega_N = T_{P|N-1}\omega_{N-1} + e(\Delta\omega). \tag{23}$$

In this study, the first bending mode of the wing is taken as an approximate ω_N since it is in the same frequency range and hence dominates the response to the gust. This approximation may produce a small deviation $e(\Delta\omega)$ for the $T_{P|N}$ prediction. Equation (23) can be further simplified as below when this deviation is initially ignored for small $\Delta\omega$:

$$T_{P|N}\omega_N = T_{P|N-1}\omega_{N-1}. \tag{24}$$

During optimization process, the angle $T_{P|N-1}\omega_{N-1}$ in cycle $N-1$ is obtained from the traditional dynamic response analysis. For the cycle N and following circles, the time T_P at the peak gust response can be calculated based on (24). If $(T_{P|N}\omega_N)_R$ represents the correct value, the deviation of the above predicted result $(T_{P|N}\omega_N)_P$ due to the $e(\Delta\omega)$ can be evaluated by

$$e\left(T_{P|N}\omega_N\right)_P = 1 - \frac{\left(T_{P|N}\omega_N\right)_P}{\left(T_{P|N}\omega_N\right)_R}. \tag{25}$$

Since the $e(\Delta\omega)$ may increase with the variation of structure stiffness during the optimization, it is also necessary to obtain the correct gust response by performing a complete gust response analysis. The correct value will be used for calibrating the $T_{P|N+n}\omega_{N+n}$ to ensure the optimization converge. This is part of the "calibration process" similar to the $T_{P|N+n}$ prediction process in the Q-Static method.

If $X_{E|P}$ and $X_{E|R}$ represent the predicted and the correct gust response values, respectively, the deviation of the predicted result due to all the approximations can be evaluated by

$$e_P = \left|1 - \frac{X_{E|P}}{X_{E|R}}\right|. \tag{26}$$

2.3. Gust Response Analysis. The equations used in gust response analysis can be expressed in a simple form as [20]

$$\begin{gathered} M\ddot{X} + D\dot{X} + KX = F_{\text{steady}}(X) + F_{\text{unsteady}}\left(X, \dot{X}, \ddot{X}\right) \\ X(0) = X_{N-1} \\ \dot{X}(0) = \dot{X}_{N-1}, \end{gathered} \tag{27}$$

in which M is mass matrix and D is structural damping matrix. Unsteady aerodynamic forces are also included in simulation, which is calculated based on Theodorsen function. The structural analysis is based on thin-walled wing box beam model with the stiffness of each section depending upon the fiber orientations of the skin and spar web laminate. A program "AeroBeamSaw" for static aeroelastic analysis and model analysis is linked with the gust response analysis performed in MATLAB and an optimization program performed in FORTRAN environment [12, 21].

2.4. Optimization Method and Objective Function. A gradient-based deterministic optimization method was employed to tailor a composite wing structure for minimum gust loads with the objective function expressed below:

$$\begin{aligned} \text{Minimize} \quad & F_d(\alpha) \\ & = \left[\left(\frac{L_S(\alpha)}{L_S(\alpha_0)}\right)^{\eta} \times \left(\frac{X_E(\alpha)}{X_E(\alpha_0)}\right)^{\xi}\right]^2, \end{aligned} \tag{28}$$

in which L_S represents total shear force at wing root, X_E represents the peak value of wingtip bending deflection, and α_o and α represent a set of initial and tailored design variables respectively, that is, the fiber orientations in a range of 0~±90° of each layer of the laminated composite wing structure. In (28), η and ξ are weight coefficients, η is set at 5, and ξ is set at 0.15.

2.5. Calibration Process in the Optimization. In optimization procedure, there are two major parts, which are shown in Figure 2. The first major part is Q-S method, which is used to predict wingtip deflection in this article. The second major part is calibration process, which is used to eliminate the error accumulation due to the simplification of the Q-S method.

2.5.1. Q-S Method. Q-S method developed in the current investigation is based on a relationship between the dynamic response of a structure and the work done by an external force. For structural optimization, the relationship is extended to the responding deflections when the structure stiffness has a small variation between two iteration cycles as expressed in (12) and (24).

At the starting point of the optimization when the iteration cycle $N-1$ ($N-1=0$), a dynamic response analysis has to be performed for the initial structure design. Based on the results obtained in the cycle $N-1$, the deflection of the structure of a small stiffness variation in the iteration cycle N and subsequent cycles $N+n$ ($n = 1, 2, 3, \ldots$) can be predicted.

In this process, instead of performing a dynamic response analysis like the starting point, Q-S method should be used in each iteration cycle repeatedly by recalculating the parameters $K_{E|N}$, $T_{P|N}$ and the Q-S deflection $X_{E|N-1}$. It means that the results from one dynamic response analysis can be used for number of iteration cycles with the parameters in (12) and (24) kept updating.

2.5.2. Calibration Process. The calibration process in current study is designed on the purpose of improving prediction accuracy. There will be error accumulated in iteration cycles. The main cause of error comes from the variation of structure stiffness, which will affect (12) and (24), as error terms that have connection with small changes made to stiffness are omitted in order to get a simplified recursive relation. When calibration is launched, a traditional dynamic response analysis will be performed to update parameters used in recursive equations for the subsequent optimization process.

The launch condition of calibration process is vital for the performance of Q-S method in optimization. There are two major problems.

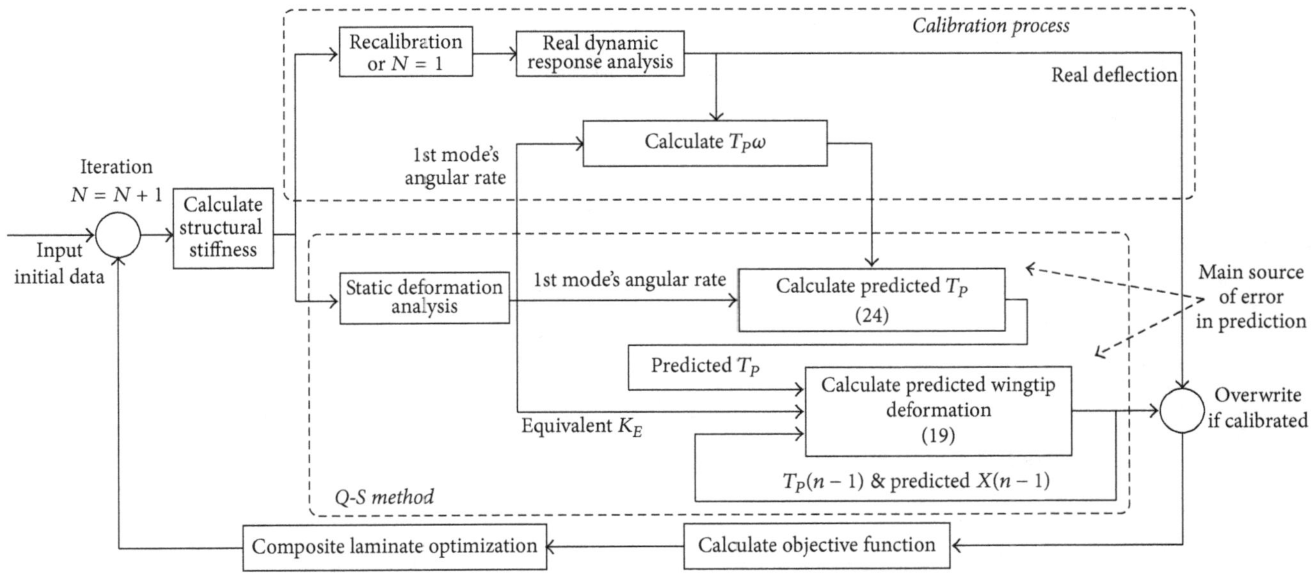

FIGURE 2: The flowchart of the Q-S method in structure optimization procedure.

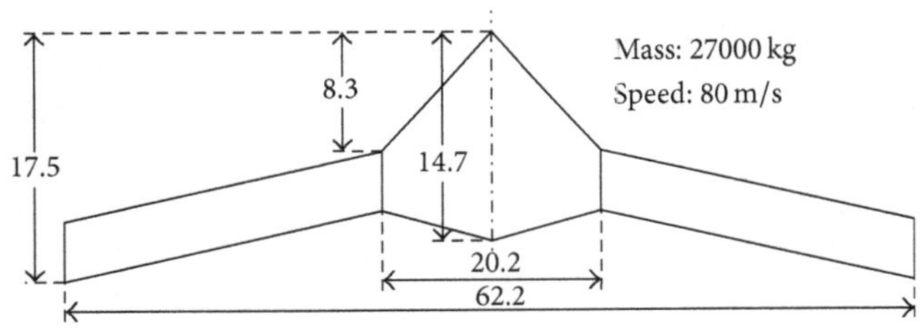

FIGURE 3: Planform of a flying wing aircraft model (unit: m).

The first problem is that, during optimization procedure, with only a few iteration cycles having "real" data, there is no reliable direct observation method for the assessment of prediction accuracy. Indirect measurement will be taken to set the launch condition. In the current study, equivalent stiffness $K_{E|N}$ obtained from static aeroelastic analysis is used as the indirect measurement to trigger calibration process.

The second problem is how much indirect measurement varies that can make it necessary to do the calibration. If the conditions are set too strict, it may lower down the efficiency of Q-S method. Otherwise, it may deteriorate Q-S method's accuracy.

In the current study, 4% stiffness variation is set as the criteria for performing the calibration process during optimization.

3. Example and Results

3.1. An Aircraft Planform and the Wing Structural Model. An aircraft of tailless flying-wing configuration as shown in Figure 3 is taken as an example to demonstrate the performance of the Q-S method. The planform and dimension of the aircraft outer wing of 20° swept angle are shown in Figure 4.

The primary structure of the wing made of laminated carbon/epoxy composite is modeled by assembly of eight segments of thin-walled wing-box beams enclosed between the spars. The leading and trailing edge cells of the wing were considered in mass, inertia, and aerodynamic force calculation. Each of the single-cell sections was further divided into four laminate panels representing the skins and spar webs of the wing-box shown in Figure 5.

Each of the panels is made of 8-layer carbon/epoxy laminate of symmetric layup. The composite material properties in the usual notation are $E_1 = 220.5$ GPa, $E_2 = 17.7$ GPa, $G_{12} = 7.05$ GPa, and density $\rho = 1900\,\text{Kg/m}^3$. Taking the fiber orientation of each ply as a design variable, the total number of independent design variables is 128 in the optimization of the outer wing structure. If the layups of eight span-wise segments were kept uniform, the design variables can be reduced to 16 in the optimization.

In this example, the outer wing as shown in Figure 4 is clamped at the inboard root section and subjected to the aerodynamic force. The wing box of original laminate layup was optimized to minimize the gust response. The specified gust velocity V_{ds} is 6 m/s and air speed is 80 m/s. The initial design and the optimized laminate layups and response results of the wing box structure are shown in Table 1. The shear force at wing root has been reduced 10.8% from original 142.7 KN to 127.2 KN and torque at wing root has been reduced from −87.7 KN·m to −78.2 KN·m. Meanwhile, the gust response in terms of wingtip deflection of the optimized wing has been reduced by 16.3% from original 0.43 m to 0.36 m as shown in Table 1.

The optimization process history can be divided into three stages as shown in Figure 6. In stage A, the gradient of objective function to each of the design variables is calculated. The variation of the objective function value was very small due to the small change of design variables and resulting $\Delta\omega$ at each step. In stage C, the variation of design

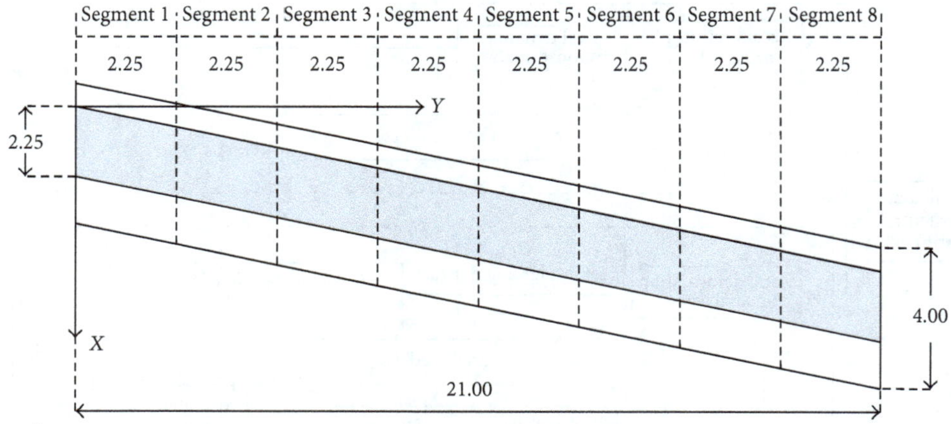

FIGURE 4: Planform of composite outer wing modeled in eight single-cell wing-box beams.

TABLE 1: Initial and optimized laminate layups, stiffness, and deflection.

Status	Panels	Laminate layup (deg)	Stiffness and deformation
Initial design	Upper skin	[0.0/45.0/−45.0/90.0] s	EI = 6.80 ·10^5 KN·m^2
	Lower skin	[0.0/45.0/−45.0/90.0] s	GJ = 3.77 · 10^5 KN·m^2
	Upper/lower skin thickness	2 mm for each layer	Wingtip deflection: 0.32 m
	Front spar web	[0.0/45.0/−45.0/90.0] s	Wingtip twist: −0.94 deg
	Rear spar web	[0.0/45.0/−45.0/90.0] s	Shear force at wing root: 142.7 KN
	Front/rear spar thickness	6 mm for each layer	Torque at wing root: −87.7 KN·m
Optimized design	Upper skin	[0.0/76.2/−84.1/90.0] s	EI = 5.49 · 10^5 KN·m^2
	Lower skin	[0.0/76.2/−84.1/90.0] s	GJ = 1.09 · 10^5 KN·m^2
	Upper/lower skin thickness	2 mm for each layer	Wingtip deflection: 0.36 m
	Front spar web	[0.0/89.9/−89.9/90.0] s	Wingtip twist: −1.70 deg
	Rear spar web	[0.0/89.9/−89.9/90.0] s	Shear force at wing root: 127.2 KN
	Front/rear spar thickness	6 mm for each layer	Torque at wing root: −78.2 KN·m

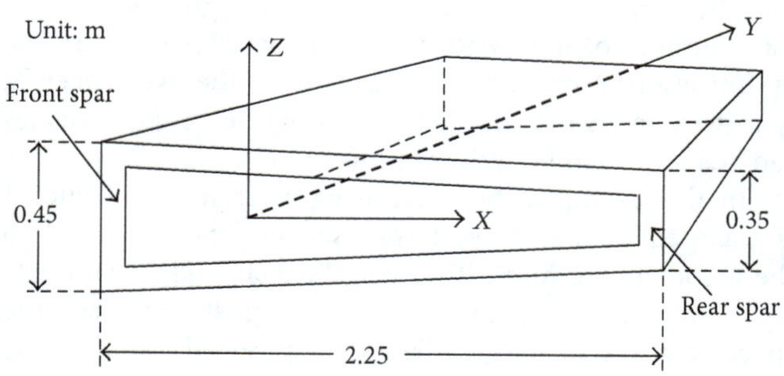

FIGURE 5: Cross section details of a wing-box beam element.

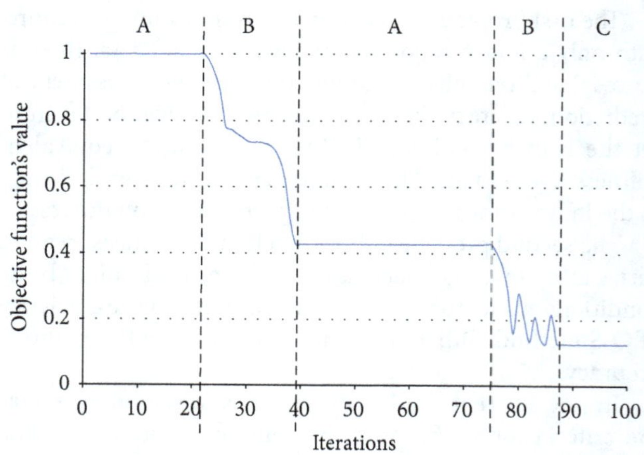

FIGURE 6: The optimization process using gradient-based deterministic method.

variables and resulting $\Delta\omega$ was also small when the variation converged to the final optimized results. In stage B however, the design variables and resulting $\Delta\omega$ varied in relatively large steps in the direction determined by the gradient. Therefore the necessary dynamic response analysis to update the parameters in (19) is normally required in stage B.

3.2. Reduced Computation Time in Optimization. Table 2 shows the computational time used for calculating the wing response to the gust by using the Q-S method and the traditional dynamic response analysis in the same optimization case. A relatively small percent of time was used by the "other calculation" including the gradient calculation and fiber orientation tailoring.

TABLE 2: CPU time distribution in optimization process.

Time distribution	Q-static method	Dynamic response	Other calculation
Time for a cycle	1.922 s	81.03 s	0.79 s
Total cycles	100	6	100
Total time used	192.2 s	486.2 s	79.0 s
Share of time	25.4%	64.2%	10.4%

In this case, the total number of iteration cycles in the optimization process is 100 with total 12.6 minutes of computational time. Although the traditional dynamic response analysis was performed only 6 times, it took 64.2% of the total computation time. If the traditional dynamic response analysis were performed in all 100 iteration cycles in the traditional approach, the total time to complete the optimization would be 135 minutes. The computation time saving is about 91% by employing the Q-S method.

4. Further Study of the Q-S Method Accuracy

Theoretically the accuracy and accuracy of the Q-S method largely depend upon two major factors when used in structural optimization. The first one is the accuracy of $T_{P|N}$ prediction or the amount of $e(\Delta\omega)$ as shown in (24). The second one is the amount of $\Delta W/E_{EP|N-1}$ given in (10). Since both factors are related to the stiffness variation step size, the Q-S method accuracy also depends on the step size.

In practice, the frequency difference between the first mode of the wing structure and gust load also affects the accuracy of the predicted $T_{P|N}$. For a higher gust frequency, the wing structure higher order modes have more influence on gust response. The aircraft free rigid-body modes have also influence on the prediction accuracy. As rigid-body mode is much lower than the gust frequency, the influence is negligible.

To evaluate the accuracy of the method, two parametric study cases were carried out and presented in the following subsections. The two major factors were also observed in both two groups, as the relative relation between the variation of prediction errors and the change of stiffness is the main object in the study.

4.1. Parametric Study of the Q-S Method. The first case is to quantify the influence of different gust frequency on the accuracy of $T_{P|N}$ prediction. Since the gust frequency varies with the flight speed, different flight speed from 75 m/s to 150 m/s corresponding to the gust frequency range from 0.75 Hz to 1.50 Hz as specified in CS-25 was considered. The first three modal frequencies of the wing structure 1.17 Hz, 2.29 Hz, and 5.61 Hz, respectively, were taken into account in the analysis. By using the Q-S method, the $T_{P|N}\omega_N$ $(N = 1, 2, 3, \ldots)$ results based on the predicted $T_{P|N}$ can be calculated.

In order to evaluate the accumulated deviation during the optimization, the calibration process in the Q-S method was locked in this case study. The deviation of predicted angle as defined in (25) from the correct value obtained from complete gust response analysis in the optimization process is shown in Figure 7. Only limited number of cycle steps was taken to calculate the deviation where a complete dynamic analysis was performed to obtain the correct value as reference. The results show that the deviation of predicted $T_{P|N}\omega_N$ is less than 4% for all four different gust frequencies in the CS25 specified range. For the higher gust frequencies, the prediction deviation increases because of the influence of higher order modes of the wing structure.

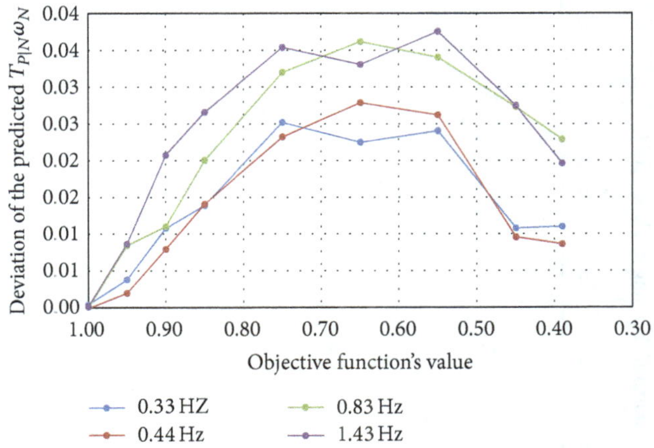

FIGURE 7: The deviation of predicted angles without calibration process for different gust frequencies.

When the proposed calibration process was performed, the key parameters affecting the accuracy are updated during the optimization. There are four cases investigated here, with different gust frequencies, the results are shown in Figure 8. The resulting deviation of the gust response as defined by (26) for the four different gust frequencies is shown in Figure 9. Again the correct gust response was obtained from performing complete gust response analysis at each optimization cycle, and the gust response was measured as wingtip deflection.

As shown in Figure 9, the accumulated maximum deviation is nearly 5%. When the calibration process was performed, the deviation was reduced to zero at two iteration cycle steps as circled in Figure 9 during the optimization. The results show the effectiveness of updating the predicted $T_{P|N}$ and $X_{E|P}$ by performing the calibration process in the Q-S method. The resulting deviation of the gust response at the end of optimization was reduced to about 1%. Table 3 lists the deviation of the predicted gust response against the computation time saving compared with the correct results obtained from performing the complete gust response analysis at each step of the optimization.

4.2. Case Studies of Aircraft Response Including Rigid-Body Motion. The second study case is to evaluate the influence of rigid-body mode on the prediction accuracy. In this case, a constant gust frequency corresponding to flight speed 150 m/s was taken. In this case, the gust response of the whole aircraft as shown in Figure 3 is taken in the gust response analysis. The

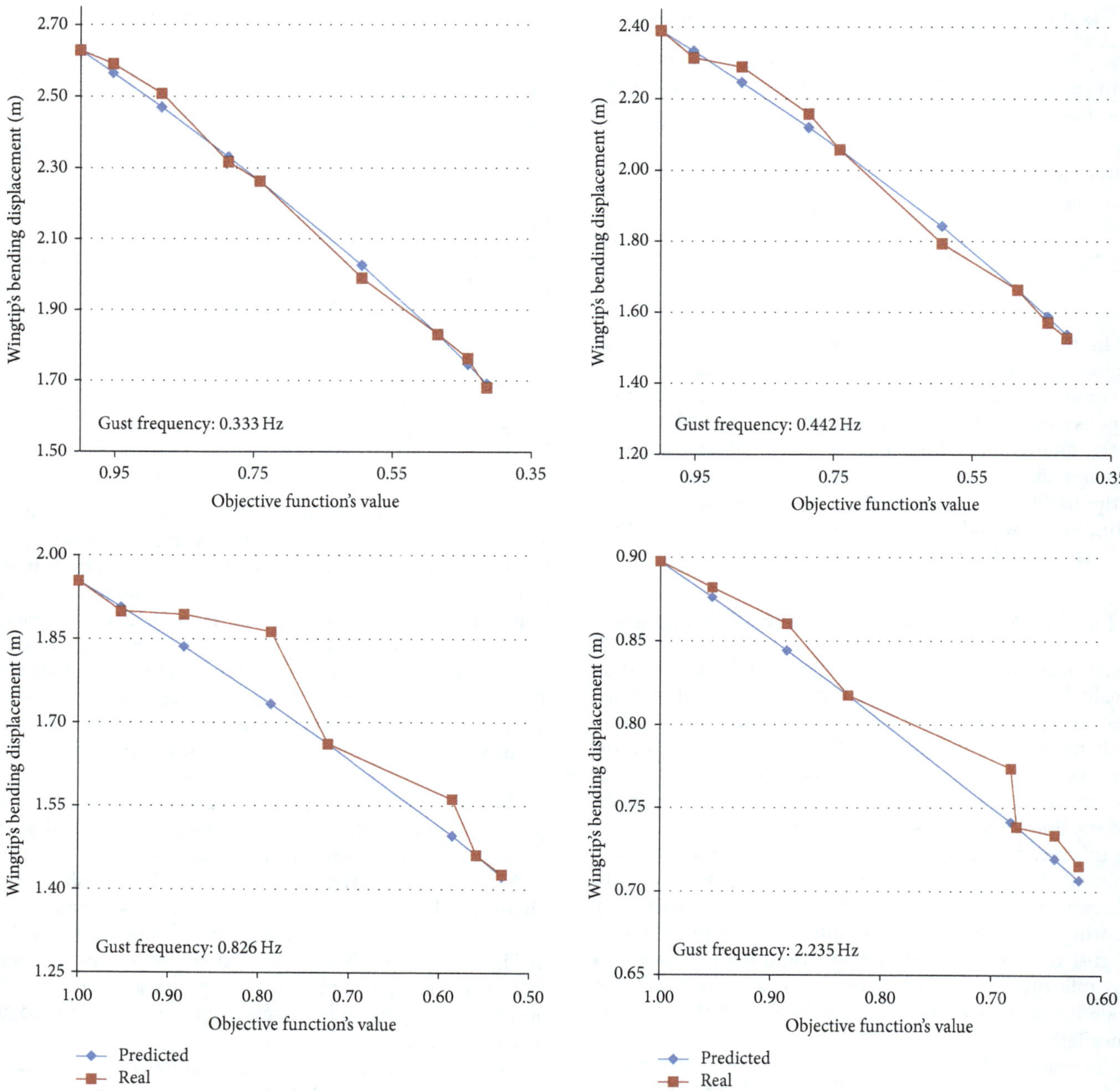

FIGURE 8: Predicted and real results under different gust frequencies with calibration process.

TABLE 3: Accuracy and computation time saving of the Q-static method in different gust frequencies.

Gust frequency	0.75 Hz	1.0 Hz	1.25 Hz	1.50 Hz
Accuracy reduction	1.1%	1.5%	2.1%	2.0%
Total cycles	100	100	100	100
Time reduced by	~90%	~90%	~90%	~90%

aircraft was set in a trimmed level flight condition with free motion in the symmetric plane before entering the (1 − cos) gust load uniformly in span-wise direction. The predicted and real deflection results at specific optimization points are shown in Figure 10, compared with results with clamped wing-root. Figure 11 shows the deviation of the predicted gust response in the optimization of the wing structure with and without the aircraft rigid-body mode. In the analysis, the gust response was measured as the wingtip deflection relative to the wing root. The results in Table 4 show that the rigid-body mode of the aircraft has little effect on the accuracy of the prediction using the Q-S method for the wing optimization.

5. Conclusions

This paper presents a quasi-static method to predict the peak value of gust response of a wing structure made of composite efficiently. Significant computational time can be saved when

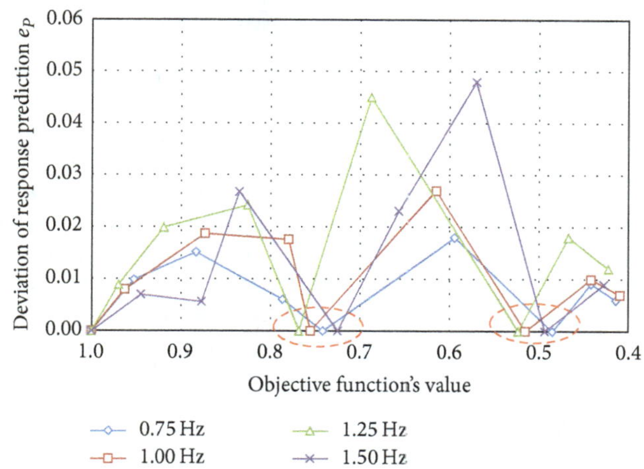

FIGURE 9: The deviation of predicted gust response with calibration process for different gust frequencies.

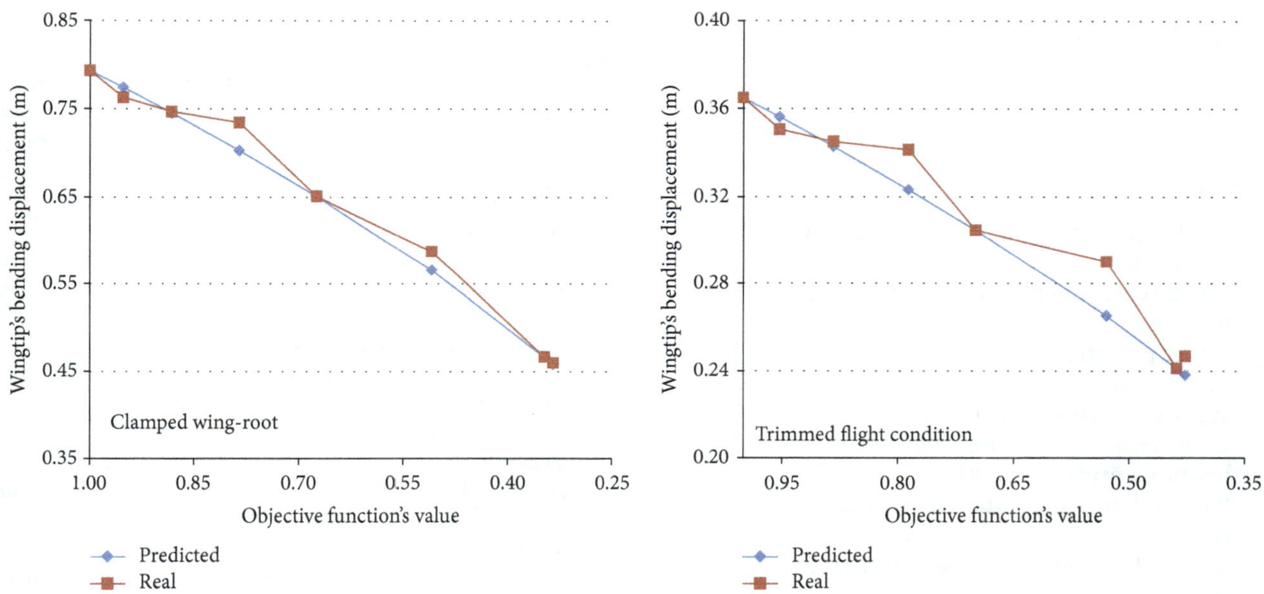

FIGURE 10: Predicted and real result for the cases with and without the aircraft rigid-body mode.

TABLE 4: Prediction method's performance in trimmed free-flight condition.

Gust model	(1 − cos) gust
Response reduction	36.7%
Accuracy reduction	2%~4%
Total cycles	80
Time saved	90%

applying this method to optimize the structural laminate layup subjected to gust load. An aircraft of flying-wing configuration has been taken as an example to demonstrate the effectiveness of the method when applied to minimizing response to a discrete (1 − cos) gust by optimizing the wing box structure. The validation of the example is shown by the results that the gust response has been reduced by 22% after optimizing the laminate layup of the composite wing structure. By employing the Q-static method, the computational time is reduced by over 90% comparing with the results from conducting the usual complete gust response analysis at each iterative cycle of the optimization. Although the prediction has to be calibrated by conducting the usual gust response analysis when the deviation is over a limit, the method is valid in most of the iterative cycles. In the total 80 times of iteration cycles of the example, only four times to perform the usual gust response analysis were required to update the result by the Q-S method. The parametric study has shown that the Q-static method can be applied to the whole gust frequency range as specified in CS-25. The deviation of the predicted results by the method is below 2.1% with the computation time saving nearly 90%. The rigid-body mode of the aircraft has little effect on the accuracy of the Q-S method.

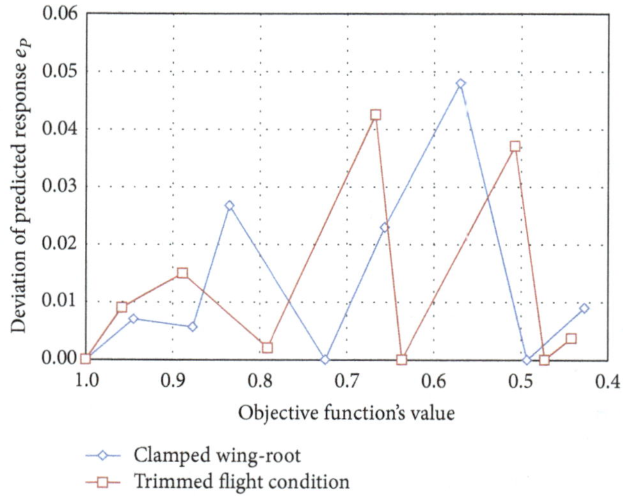

FIGURE 11: Deviation of the prediction gust response for the cases with and without the aircraft rigid-body mode.

Nomenclature

E: Total energy of a structure's deformation movement
E_K: Kinetic energy of a structure's deformation movement
E_{EP}: Elastic potential energy of a structure's deformation movement
K: Stiffness matrix of a structure
K_E: Estimation of a structure's stiffness
X: Elastic deformation vector
T_P: First peak time of a structure's dynamic response
X_E: A type of dynamic response of a structure at T_P
$Q(t)$: Time function of external dynamic force
$H(t)$: Time function of corresponding velocity of structural response to $Q(t)$
W_P: Work done by external force $Q(t)$ through structural deformation over a time period
A: Amplitude of structural deflection
ω_n: Structural natural frequency
L_S: Total shear force at wing root.

Competing Interests

The authors declare that there is no conflict of interests regarding the publication of this paper.

Acknowledgments

The authors acknowledge Chinese Natural Science Funding Council (NSFC 11302027, NSFC 11672033) for financial support of the project.

References

[1] W. W. Li and C.-G. Pak, "Aeroelastic optimization study based on the X-56A model," in *Proceedings of the AIAA Atmospheric Flight Mechanics Conference (AIAA AVIATION '14)*, AIAA, Atlanta, Ga, USA, June 2014.

[2] G. K. W. Kenway, G. J. Kennedy, and J. R. R. A. Martins, "Scalable parallel approach for high-fidelity steady-state aeroelastic analysis and adjoint derivative computations," *AIAA Journal*, vol. 52, no. 5, pp. 935–951, 2014.

[3] S. Guo, W. Cheng, and D. Cui, "Aeroelastic tailoring of composite wing structures by laminate layup optimization," *AIAA Journal*, vol. 44, no. 12, pp. 3146–3150, 2006.

[4] S. Guo, "Aeroelastic optimization of an aerobatic aircraft wing structure," *Aerospace Science and Technology*, vol. 11, no. 5, pp. 396–404, 2007.

[5] D. H. Baldelli, H. Ohta, and K. Nitta, "Gust load alleviation of an aeroelastic wing model," *Transactions of the Japan Society for Aeronautical and Space Sciences*, vol. 36, no. 113, pp. 125–142, 1993.

[6] H. Honlinger, H. Zimmermann, O. Sensburg, and J. Becker, "Structural aspects of active control technology," in *Proceedings of the AGARD Conference Proceeding, AGARD CP*, p. 18, 1995.

[7] S. Guo, W. Cheung, J. R. Banerjee, and R. Butlar, "Gust alleviation and flutter suppression of an optimized composite wing," in *Proceedings of the International Forum on Aeroelasticity and Structural Dynamics*, vol. 2426, pp. 41–49, 1995.

[8] B. Moulin and M. Karpel, "Gust loads alleviation using special control surfaces," *Journal of Aircraft*, vol. 44, no. 1, pp. 17–25, 2007.

[9] M. J. Dillsaver and C. E. S. Cesnik, "Gust load alleviation control for very flexible aircraft," in *Proceedings of the AIAA Atmospheric Flight Mechanics Conference*, vol. 6368, pp. 8–11, Portland, Oregon, August 2011.

[10] S. Guo, Q. Fu, and O. K. Sensburg, "Optimal design of a passive gust alleviation device for a flying wing aircraft," in *Proceedings of the 12th AIAA ATIO/14th AIAA/ISSMO MAO Conference, Session MAO-25*, vol. 1719, pp. 17–19, September 2012.

[11] E. Livne, "Future of airplane aeroelasticity," *Journal of Aircraft*, vol. 40, no. 6, pp. 1066–1092, 2003.

[12] European Aviation Safety Agency, CS-25 Certification Specification for Aircraft, 2012.

[13] S. Guo, D. Li, and Y. Liu, "Multi-objective optimization of a composite wing subject to strength and aeroelastic constraints," *Proceedings of the Institution of Mechanical Engineers, Part G: Journal of Aerospace Engineering*, vol. 226, no. 9, pp. 1095–1106, 2012.

[14] G. K. W. Kenway and J. R. R. A. Martins, "Multipoint high-fidelity aerostructural optimization of a transport aircraft configuration," *Journal of Aircraft*, vol. 51, no. 1, pp. 144–160, 2014.

[15] D. Amsallem and C. Farhat, "Interpolation method for adapting reduced-order models and application to aeroelasticity," *AIAA Journal*, vol. 46, no. 7, pp. 1803–1813, 2008.

[16] H. H. Khodaparast, G. Georgiou, and J. E. Cooper, "Rapid prediction of worst case gust loads," *Journal of Aeroelasticity and Structural Dyamics*, vol. 2, no. 3, pp. 33–54, 2012.

[17] H. Haddad Khodaparast and J. E. Cooper, "Rapid prediction of worst-case gust loads following structural modification," *AIAA Journal*, vol. 52, no. 2, pp. 242–254, 2014.

[18] S. Haghighat, J. R. R. A. Martins, and H. H. T. Liu, "Aeroservoelastic design optimization of a flexible wing," *Journal of Aircraft*, vol. 49, no. 2, pp. 432–443, 2012.

[19] M. R. Waszak and D. K. Schmidt, "Flight dynamics of aeroelastic vehicles," *Journal of Aircraft*, vol. 25, no. 6, pp. 563–571, 1988.

[20] N. Teunisse, L. Demasi, P. Tiso, and C. Rauno, "A computational method for structurally nonlinear joined wings based on modal derivatives," in *Proceedings of the 55th AIAA/ASMe/ASCE/AHS/SC Structures, Structural Dynamics, and Materials Conference*, p. 494, National Harbor, Md, USA, 2014.

[21] Z. Wang, Y. Sun, D. Yang, and S. Guo, "Active control using control allocation for UAVs with seamless morphing wing," in *SPIE Proceedings of the 3rd International Conference on Smart Materials and Nanotechnology in Engineering*, vol. 8409, Shenzhen, China, April 2012.

6

Study on Helicopter Antitorque Device based on Cross-Flow Fan Technology

Du Siliang, Tang Zhengfei, Xu Pei, and Ji Mengjiang

National Key Laboratory of Science and Technology on Rotorcraft Aeromechanics, Nanjing University of Aeronautics and Astronautics, Nanjing 210016, China

Correspondence should be addressed to Du Siliang; kjofchina@qq.com

Academic Editor: Maria G. De Giorgi

In order to improve low-altitude flight security of single-rotor helicopter, an experimental model of a helicopter antitorque device is developed for wind tunnel test. The model is based on the flow control technology of the cross-flow fan (CFF). Wind tunnel tests show that the model can produce side force. It is concluded that the influence of the CFF rotating speed, the rotor collective pitch, and the forward flight speed on the side force of the model is great. At the same time, the numerical simulation calculation method of the model has been established. Good agreement between experimental and numerical side force and power shows that results of numerical solution are reliable. Therefore, the results in actual helicopter obtained from Computational Fluid Dynamics (CFD) solution are acceptable. This proves that this antitorque device can be used for a helicopter.

1. Introduction

The basic advantage of the conventional tail rotors is that they require relatively little power, produce good yaw control, and contribute significantly to yaw damping and directional stability in forward flight [1]. Against these good features are the bad ones [2]: a tail rotor is dangerous to people on the ground and to itself if swung into trees, fence posts, or wires. Large helicopters circumvent these problems by incorporating tail rotors that are well above the ground, but the exposed whirling blades on smaller rotorcraft may be perceived as dangerous enough to prompt the consideration of another system. Some helicopter designs balance the main-rotor torque by using two or more rotors, while other designs eliminate the torque, applying a tip-jet drive concept. But the torque from a single shaft driven main rotor that is offset by a tail rotor is still the main direction of development of the helicopter. At present, the antitorque system is more mature with the tail rotor, the fenestron [3, 4], and no tail rotor (NOTAR) concept [5–10]. They all have some advantages or disadvantages. As with the tail rotor, the larger the diameter of the fenestron, the less the power it takes from the engine. However, in order to reduce the weight of the duct, the fenestron fan diameter can be as small as 50% of the tail rotor. Thus, it uses more power for antitorque than a typical tail rotor, taking power which could be used by the main rotor. To be effective, the depth of the duct should be at least 20% of the fan diameter, according to the theory. This means that it is difficult to streamline the thick fan for low drag in forward flight. This also means that the air path into and out of the fan is tortuous, so its efficiency is penalized. Another antitorque system to eliminate the tail rotor is the Hughes NOTAR. The system uses a variable pitch fan inside the tail boom to build a high volume of low-pressure air, which exits through two slots and creates a boundary layer flow of air along the tail boom utilizing the Coanda effect. The boundary layer changes the direction of airflow around the tail boom, creating thrust opposite the motion imparted to the fuselage by the torque effect of the main rotor. Benefits of the NOTAR system include increased safety (the tail rotor being vulnerable) and greatly reduced external noise. However, compared to the tail rotor, NOTAR is not widely used for reasons of low effective rate, weak maneuverability, and patent constraints. As for these bad factors, a new antitorque device, which can control flow with the assistance from the CFF, has been proposed. The CFF has been used in aircraft propulsion and flow control recently. The fan-wing aircraft [11–16] is one of them. Fan-wing aircraft makes use of a CFF to accelerate oncoming

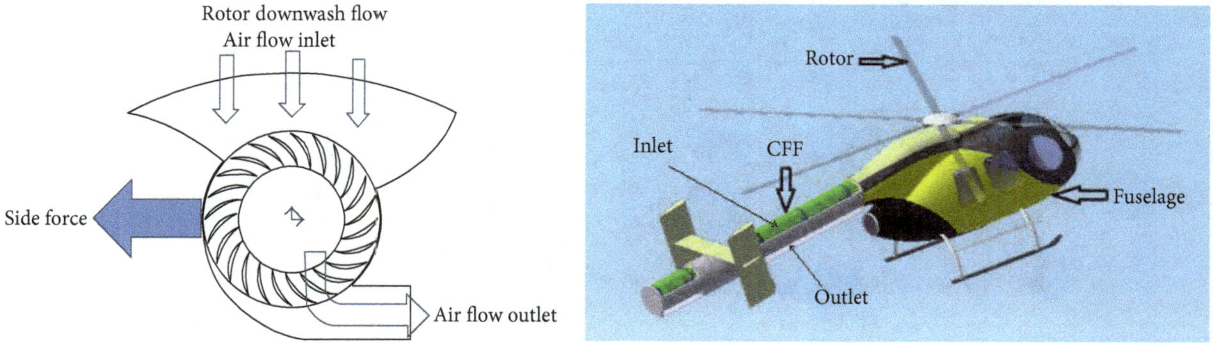

(a) Principle of side force generation

(b) The installation position of the antitorque device

FIGURE 1: Schematic diagram of CFF antitorque device.

airflow over a relatively thick aerofoil section, producing lift and thrust simultaneously. The advantages of fan-wing aircraft compared to the conventional aircrafts are short take-off and landing (STOL) at the low forward speed, no stall, and high power load. Another one using CFF is the propulsive wing [17, 18], which embeds a CFF into a thick wing for both lift enhancement and thrust production. It has the same advantages as fan-wing aircraft. There are also vertical take-off and landing (VTOL) aircraft using the CFF proposed by Gossett. It may be a good choice to use the CFF in helicopter antitorque system due to its advantages [19, 20].

A new antitorque system should have a little impact on helicopter aerodynamic characteristics to avoid complex dynamic phenomena such as the aerodynamic coupling with other components of the helicopter, reduce the aerodynamic noise, and avoid the aerodynamic force which is difficult to control. Finally, the antitorque system should be easy to operate and control. The aim of the present work is to build and test an antitorque device and to use CFF and verify that it has the ability to produce side force. It is also necessary to analyze the relationship with the side force and power by changing the rotating speed, the rotor collective pitch, and the forward flight speed. In order to simulate and analyze the aerodynamic characteristics of the device in a real helicopter, it is necessary to establish a CFD method, the results of which are validated with the experimental data presented in wind tunnel test.

2. Experimental Method

2.1. Wind Tunnel and Model Arrangements. The CFF consists of a drum-like rotor with forward curved blades, encased within housing walls. The inlet and outlet have rectangular cross sections. The advantage of CFF is its ability to extend lengthwise, producing a uniformly distributed inflow and outflow and approximately 90° flow turning from the inlet to the outlet. Therefore, the antitorque device we proposed is shown in Figure 1(a). Figure 1(b) illustrates the principle sketch of the antitorque device embedded in tail boom with MD600 fuselage.

This experiment is based on the low speed open return flow wind tunnel (Figure 2(a)), which is made by the National Key Laboratory of Rotorcraft Aeromechanics in Nanjing University of Aeronautics and Astronautics. The basic parameters of the experimental wind tunnel are shown in Table 1. Figure 2(b) shows the geometric parameters of the experimental antitorque device model. The experimental model is shown in Figure 2(c). The left interface and the right interface of the model can be separately removed, and the parameters are defined in Table 2. The whole experimental bench is placed in the relative position of the wind tunnel as shown in Figure 2(d). The basic parameters of the rotor used in the experiment are shown in Table 3. As the rotor thrust will change in different flows in wind tunnel, this article used rotor collective pitch to define the rotor downwash effects. Table 4 shows that when the rotor rotation speed is 900 r/min, rotor thrust changes with the collective pitch at different forward flight speed. The six-component balance was used in the experiment (Figure 2(e)).

TABLE 1: Parameters of wind tunnel.

Parameters	Value
Size of test area (m × m)	3.4 × 2.4
Maximum wind speed (m/s)	40
Minimum stable wind speed (m/s)	5
Shrinkage ratio	4

TABLE 2: Definitions of geometric parameters.

Definition	Value
Opening angle ψ (°)	90, 110, 130, 150
Outer radius of CFF $R1$ (mm)	100
Fan outlet height ζ (mm)	45
Blade mounting angle γ (°)	0, 10, 20, 30
Blade chord C (mm)	25
Number of blades (piece)	10, 12, 14, 16
Blade length (mm)	500
Clearance between CFF and shell σ (mm)	5

2.2. Experimental Method. The experiment includes three states, namely, static state, hover state, and forward flight state. Static state experiments are defined as the antitorque device has no rotor downwash flow and static forward flight

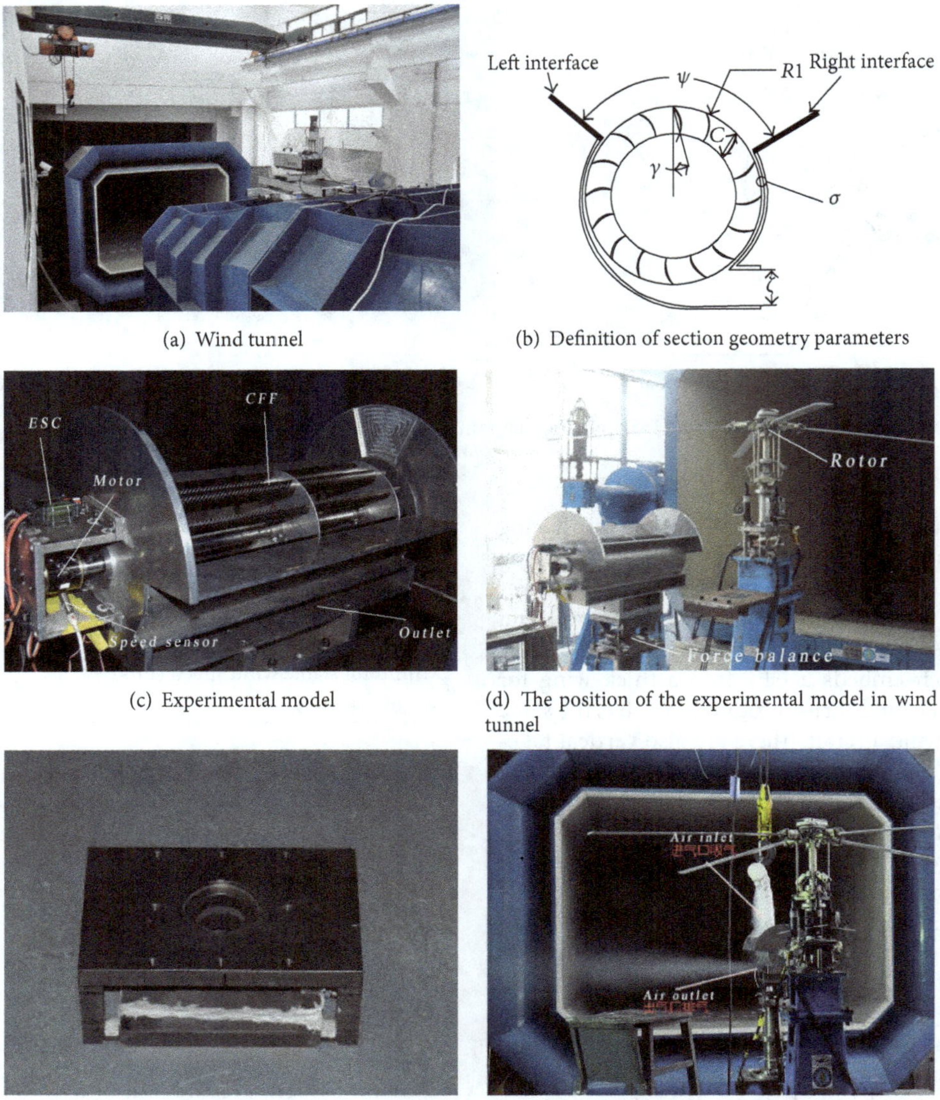

(a) Wind tunnel

(b) Definition of section geometry parameters

(c) Experimental model

(d) The position of the experimental model in wind tunnel

(e) Six-component balance

(f) Principle verification

FIGURE 2: Experimental model and equipment.

TABLE 3: Parameters of rotor test bench.

Parameters	Value
Number of blades (piece)	4
Blade radius (m)	1.25
Blade chord (m)	0.072
Geometric torsion (°)	0
Rotor solidity	0.077
Blade airfoil	NACA 0012

TABLE 4: Lift of rotor in wind tunnel test.

Free stream velocity (m/s)	0			5			10		
Collective pitch (°)	8	10	12	8	10	12	8	10	12
Rotor lift (kg)	35	45	59	37	49	62	41	53	65

inflow. The hover state experiment is defined as the antitorque device is in the rotor downwash flow, without the forward flight inflow. The forward flight state experiment is defined as the antitorque device is in the rotor downwash flow and forward flight inflow. First of all, the principle experiment of the side force generation has been carried out in static state. A smoke generator was placed on the top of the model. CFF rotational speed is set through the controller (Figure 3). Smoke billowed out from the model outlet (Figure 2(f)). This proved that the model can produce 90° flow turning from inlet to the outlet. So, it can generate a force in the opposite direction of the jet. The force data of the balance was collected and processed by a computer (data acquisition speed is 100 KHz).

2.3. Measurement Uncertainty. The experimental uncertainty is mostly expressed as fractional uncertainty of the measured value (Table 5). All measurement uncertainties were estimated from systematic and random sources of error that

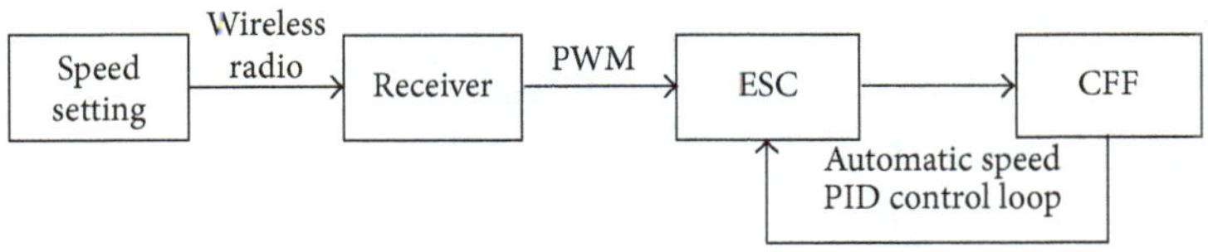

FIGURE 3: Control block diagram of the rotation speed of CFF.

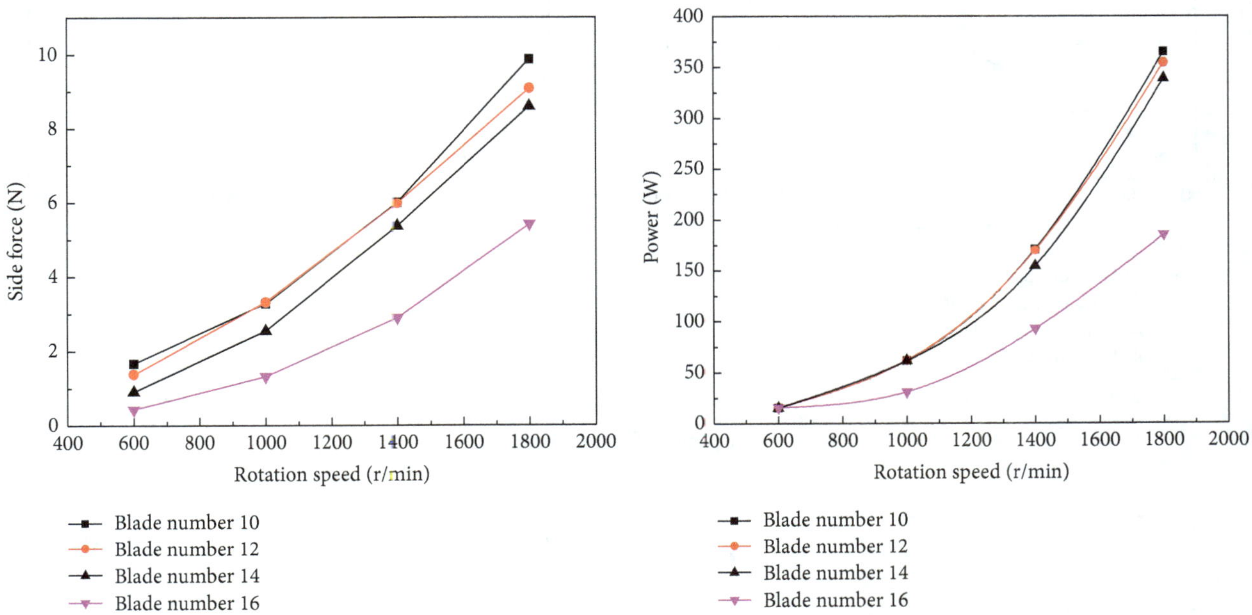

(a) Curves of side force with rotation speed

(b) Curves of power with rotation speed

FIGURE 4: Influence of blade number.

TABLE 5: Summary of the measurement uncertainties.

Parameters	Uncertainty
Force balance (kg)	±0.03%
Data acquisition resolution	±0.05%
Free stream velocity (m/s)	±0.5%
Free stream deflection angle (°)	±0.1%
Model installation angle (°)	±1%
Rotor collective pitch angle (°)	±1%

included the instrument calibration, instrument accuracy given by the manufacturer, and the data acquisition resolution.

3. Experimental Results and Analysis

3.1. Static State

3.1.1. Influence of Blade Number. Figure 4(a) shows the relationship between side force and rotation speed at different blade numbers when the blade installation angle is 10 degrees, with both interfaces, and the opening angle is 110 degrees in Figure 4(a). It can be figured out that the side force increases as the rotational speed increases and the increase rate is progressively larger with a rising trend. This means the rotation speed of CFF has a great influence on the side force. The higher the speed, the larger the side force. From the figure, we can find that the side force is greater with fewer blades, and when the number of the blades is less than 14 pieces, the increasing amount changes a little at different rotation speeds. Figure 4(b) shows the relationship between required power and rotational speed in different blade numbers. The faster the rotating speed of the CFF, the higher the power consumption. At the same rotational speed, the fewer the number of blades, the higher the power consumption. When the number of the blades is less than 14 pieces, the power consumption of the same speed is basically the same. When the blade number is 10 pieces, the maximum power load is 2.71 kg/kw (1800 r/min).

3.1.2. Influence of Blade Mounting Angle. Figure 5(a) shows the relationship between side force and rotational speed with different blade mounting angles when the blade number is 12 pieces, with both interfaces, and the opening angle is 110 degrees. It can be seen that the side force obviously increases as the rotational speed of the cross-flow fan increases. The higher the speed, the more obvious the side force increases. At high speed, the side force is different with the change of the angle of installation. From the figure, we can see that when the blade mounting angle is 20 degrees, it has a higher side force at the same speed compared with the other mounting

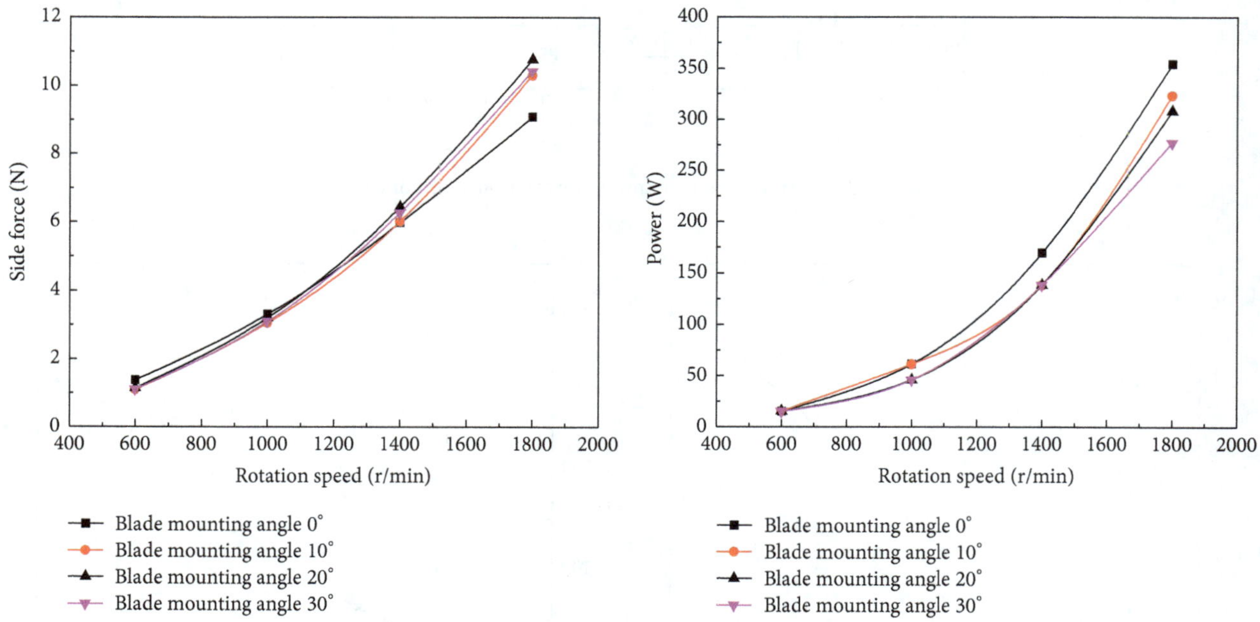

(a) Curves of lateral force with rotation speed

(b) Curves of power with rotation speed

FIGURE 5: Influence of blade mounting angle.

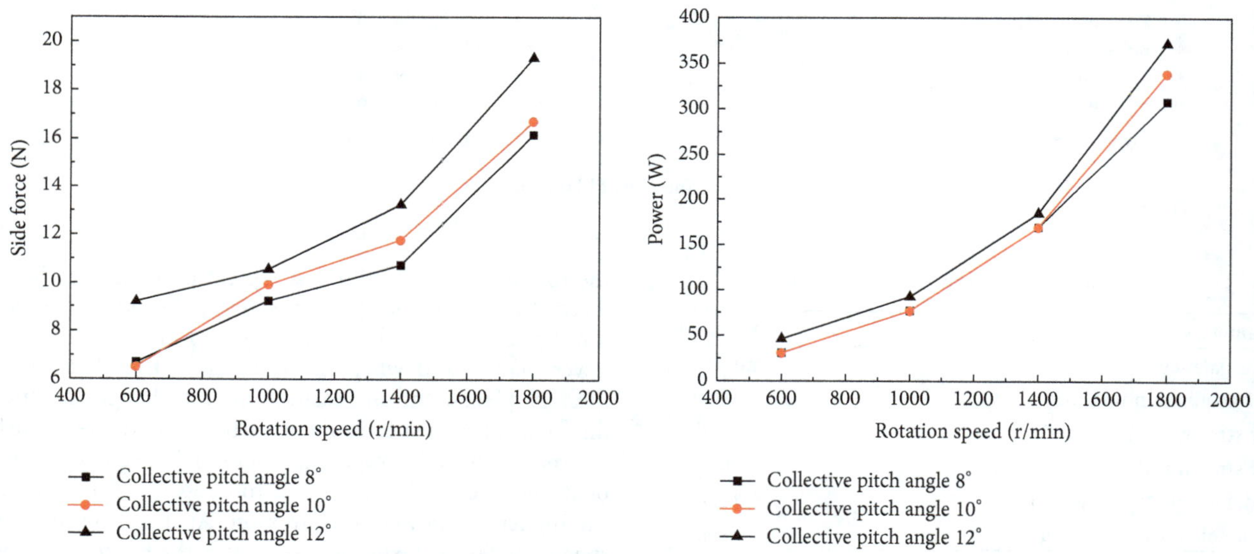

(a) Curves of side force with rotation speed

(b) Curves of power with rotation speed

FIGURE 6: Hover state.

angle. Figure 5(b) shows the relationship between power required and rotation speed with different blade mounting angles. Compared with Figure 5(a), it can be seen that when the rotation speed is 1800 r/min and the blade mounting angle is 20 degrees, the power load is max and is equal to 3.50 kg/kw.

3.2. Hover State. Figure 6(a) shows the relationship between side force and rotational speed with different rotor collective pitches when the blade number is 12 pieces, blade mounting angle is 20 degrees, with both interfaces, and the opening angle is 110 degrees. It can be seen from the chart that the greater the rotor collective pitch, the greater the side force, which means the rotor downwash flow has a significant impact on the side force of the antitorque device. Side force is large (equal to 19.32 N) when the CFF rotation speed is 1800 r/min and the rotor collective pitch is 12 degrees. The side force increased by about 80% compared with the maximum value of 10.76 N in Figure 4(a) (with no downwash flow). Figure 6(b) shows the relationship between required power of antitorque device and the rotation speed of CFF. Compared with Figure 5(b), when rotor downwash flow exists, with the same rotation speed of CFF, the required power of reactive torque device is higher than that of the

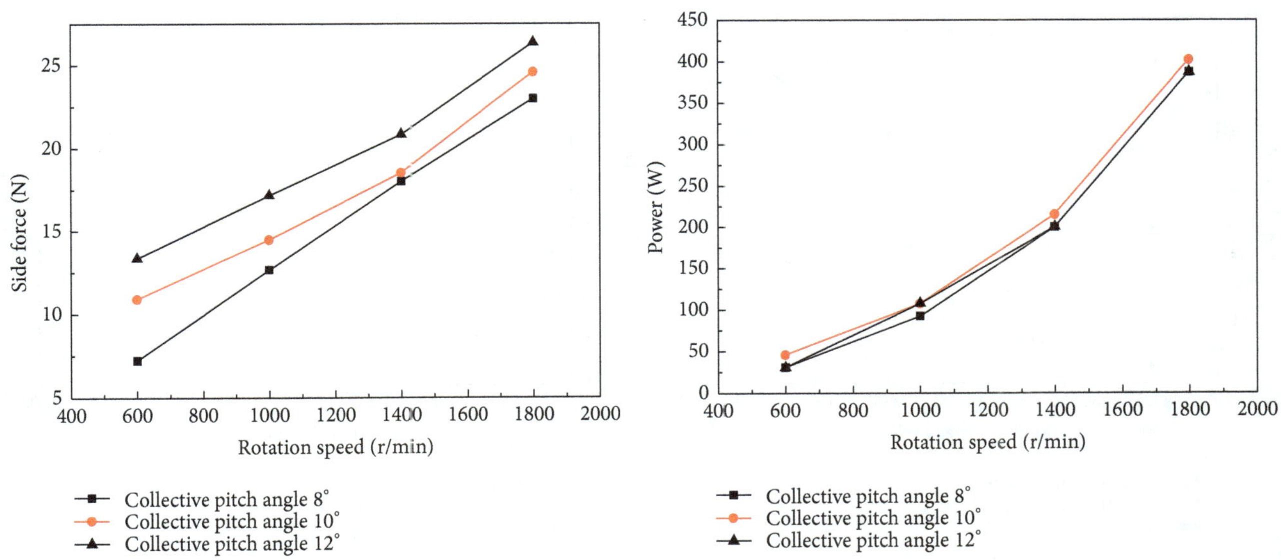

(a) Curves of side force with rotation speed

(b) Curves of power with rotation speed

FIGURE 7: Forward flight speed is 5 m/s.

static state. The maximum power load is 5.20 kg/kw, when the rotation speed is 1800 r/min.

3.3. Forward Flight State

3.3.1. Forward Flight Speed Is 5 m/s. Figure 7(a) shows the relationship between side force and rotational speed with different rotor collective pitches when the blade number is 12 pieces, blade mounting angle is 20 degrees, with both interfaces, the opening angle is 110 degrees, and forward speed is 5 m/s. As can be seen from the figure, the side force increases as the rotational speed increases and also increases with the rotor collective pitch. In contrast to Figure 6(a), it is shown that the side force produced by antitorque device when forward flight flow exists is larger than that without forward flight flow. Figure 7(b) shows the relationship between power and the rotation speed of CFF with different rotor collective pitches when the forward speed is 5 m/s. It can be found from the figure that the power consumption changes a little at the same rotation speed. It can be inferred that there is a part of the forward flight airflow under the effect of rotor downwash rushing into the cross-flow fan of the antitorque device, thereby increasing the side force.

3.3.2. Forward Flight Speed Is 10 m/s. Figure 8(a) shows the relationship between side force and rotational speed with different rotor collective pitches when the blade number is 12 pieces, blade mounting angle is 20 degrees, with an interface, the opening angle is 110 degrees, and forward speed is 5 m/s. Compared with Figure 15, we can find that when the rotor collective pitch and the rotation speed are the same, the side force increases more obviously when the forward speed is large. Figure 8(b) shows the power change curve. It shows that power changes a little at the same rotation speed and collective pitch and proves that the forward flight airflow under the effect of rotor downwash rushes into the CFF of the antitorque device. The maximum power load of 0 m/s, 5 m/s, and 10 m/s is 5.20 kg/kw, 6.94 kg/kw, and 7.45 kg/kw (rotor collective pitch 12 degrees, CFF rotation speed 1800 r/min). Therefore, the forward flow has the role of unloading on the antitorque device, which means when the helicopter needs certain reactive torque, the higher the flight speed, the lower the power consumption.

4. Numerical Methods and Comparison

4.1. Numerical Calculation Method. The numerical simulations are performed using the commercial general-purpose CFD code FLUENT 14.5 by Fluent Inc. The ANSYS ICEM software was used for grid division. Numerical calculation results will be compared with the experimental data. A two-domain model for the numerical calculation is shown in Figure 9. For this CFD analysis, the rotation speed of the fan varies between 400 and 1200 r/min, free stream velocity is 5 m/s, and angle of attack is zero. Free stream velocity and angle of attack are constant for all rotation velocities. Renormalization group (RNG) k-ε model was used for turbulence. The pressure-velocity coupling was calculated using the SIMPLEC algorithm. Second-order upwind discretization was considered for the convection terms. Finite volume method with rectangular elements was used for the whole solution domain. The rotating and stationary domains connected each other with a fluid-fluid interface, where the flow continuity is satisfied (Figure 9(a)). To simulate the fan rotation, the area surrounding the blades was designed as a sliding mesh region (Figure 9(b)). The dimensions of the computational box are 30 $R1 \times 20$ $R1$. A uniform velocity is imposed at the inlet, while a zero relative static pressure is prescribed at the outlet. Unsteady simulations require proper setting of both the time step size and the convergence criteria within each time step. For this simulation, a time step size equal to 1/20th the blade passing period captured the unsteady flow

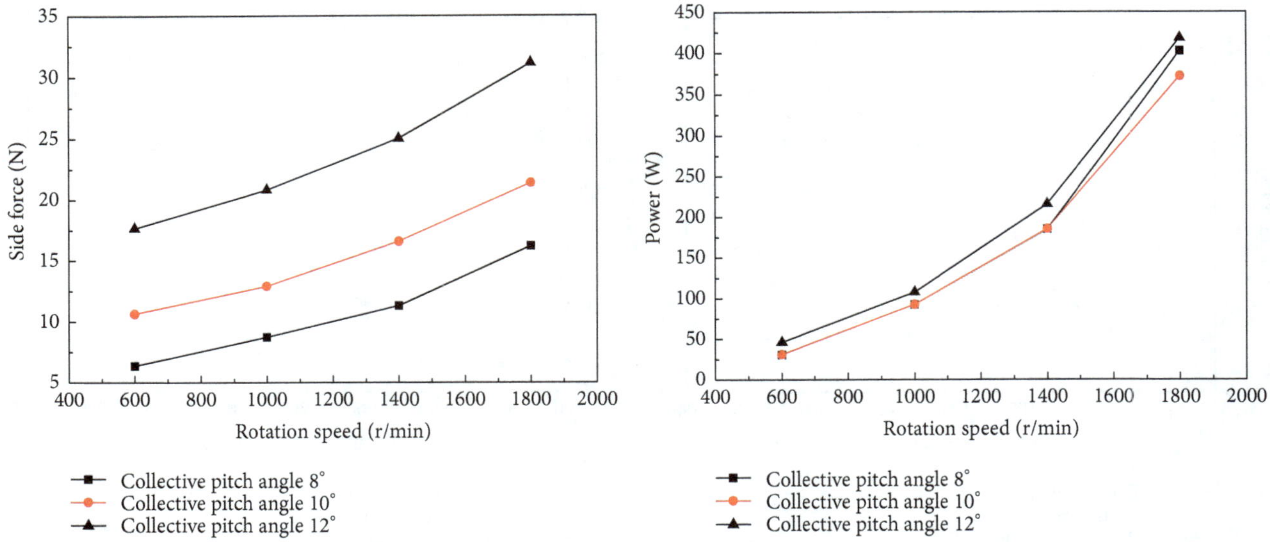

(a) Curves of side force with rotation speed

(b) Curves of power with rotation speed

FIGURE 8: Forward flight speed is 10 m/s.

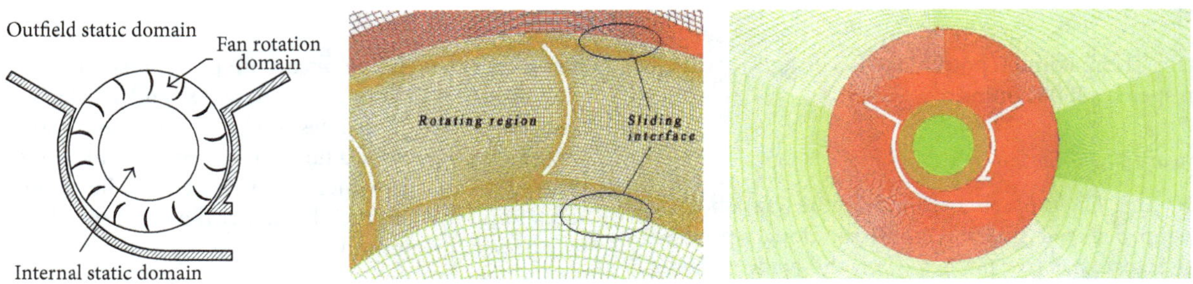

(a) Calculation domain definition

(b) Interface definition

(c) Mesh generation complete

FIGURE 9: Computational grid generation.

well. Within each time step, iterations were performed until the solution no longer changed. It was found to be necessary to reduce all residuals to at least 10^{-5}. Figure 9(c) shows the mesh generation complete.

4.2. Grid Convergence. There is measurement uncertainty in the experiments and numerical convergence in simulation: purely two-dimensional simulation in CFD and three-dimensionality of flow near the wing ends in the experiment, with inevitable mechanical losses in experiment but not in CFD. Domain size influence has been checked by extending the domain in each direction by an additional 50 per cent in length, respectively. Results (in terms of the side force) agree well with those from the baseline domain, confirming that the size of the current domain is large enough for an accurate prediction.

Grid independency studies were also conducted with three successive grids: coarse, medium, and fine meshes. The coarse mesh has 368,632 elements in total without any inflation layer near the walls. The medium mesh has 599,146 elements with a small element cluster around the blades and the central area. The inflation layer of 1 mm from the first element to the wall is added to both housing and shaft walls. The fine mesh has 1,163,316 elements with more strict mesh controls at the same locations as those used in the medium mesh. In particular, the inflation height is decreased considerably from 1 mm down to 0.1 mm. The study has shown that differences in results from the coarse and medium meshes are significant, while differences in results between the medium and fine meshes are almost negligible. Hence, it is decided to use the medium mesh as the baseline for further simulation.

4.3. Static State Comparison

4.3.1. Blade Mounting Angle Is 10 Degrees. Figure 10 shows numerical simulation calculation and test results contrast curve when blade number is 16, blade installation angle is 10 degrees, with both interfaces, and opening angle is 110 degrees. We can see that there is a little difference between the results of numerical simulation and the experimental results in the error range, and the tendency is consistent. This means the numerical calculation method for the side force and the required power of the reactive torque device is credible.

4.3.2. Blade Mounting Angle Is 20 Degrees. Figure 11 shows numerical simulation calculation and test results contrast

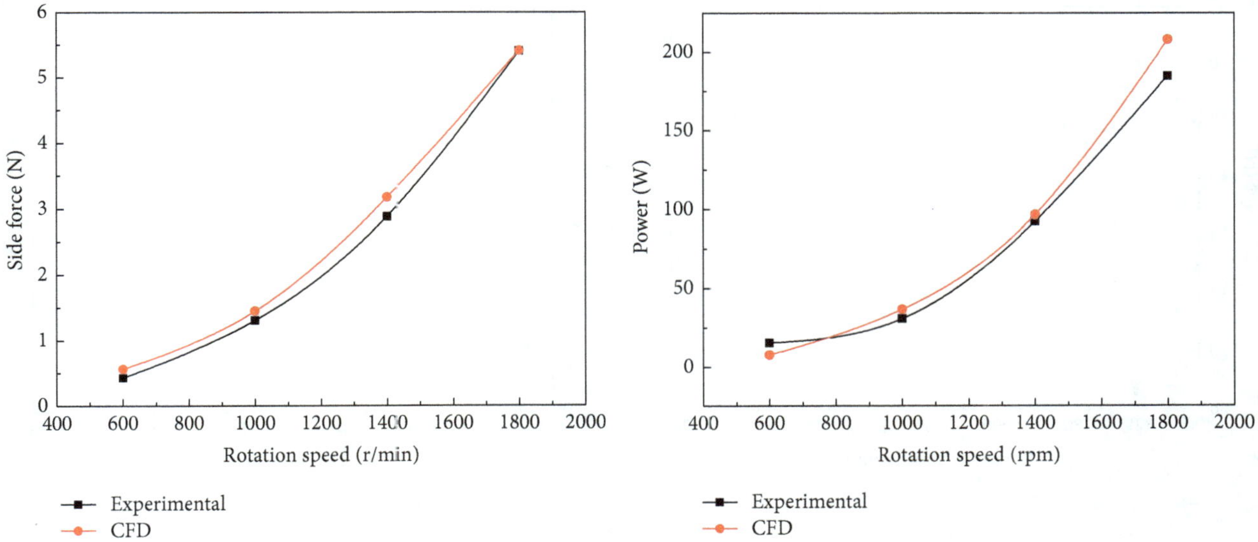

(a) Curves of side force with rotation speed
(b) Curves of power with rotation speed

FIGURE 10: Blade mounting angle is 10 degrees.

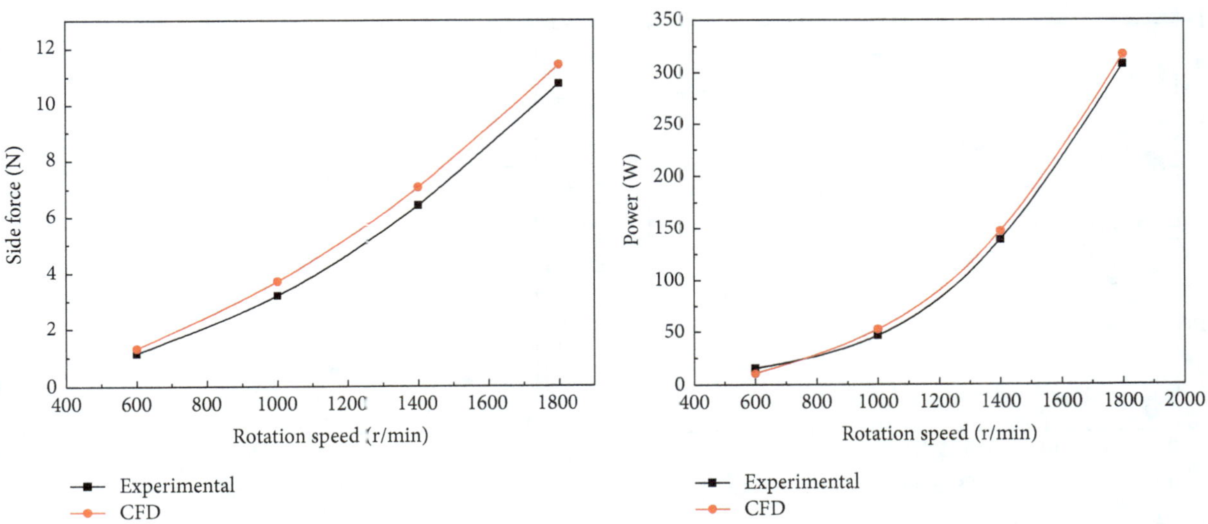

(a) Curves of side force with rotation speed
(b) Curves of power with rotation speed

FIGURE 11: Curves of power with rotation speed.

curve when blade number is 16, blade installation angle is 10 degrees, with interface, and opening angle is 110 degrees. We can see that there are few differences between the results of numerical simulation and the experimental results in the error range, and the tendency is consistent. This means the numerical calculation method for the side force and the required power of the reactive torque device keeps high accuracy even when blade mounting angle changes.

4.4. Hover State Comparison

4.4.1. Opening Angle Is 110 Degrees. Figure 12 shows the curves where the side force and the required power change with the rotation speed when the blade number is 12, blade mounting angle is 20 degrees, with interface, the opening angle is 110 degrees, and the collective pitch is 12 degrees. It can be seen from the figure that the numerical results are consistent with the experimental results, and the numerical method is reliable.

4.4.2. Opening Angle Changes. Test conditions are as follows: the CFF rotating speed is 1400 r/min, the blade number is 12, the blade installation angle is 20 degrees, and the rotor collective pitch is 10 degrees. Opening angle is 90 degrees, 110 degrees, 130 degrees, and 150 degrees. Figure 13 shows that the side force has an extreme point when the opening angle is between 110 degrees and 130 degrees. The power load calculated by the wind tunnel test data is 6.17 kg/kw, 6.83 kg/kw, 5.78 kg/kw, and 4.14 kg/kw with different opening angles. The antitorque device acceleration function of rotor

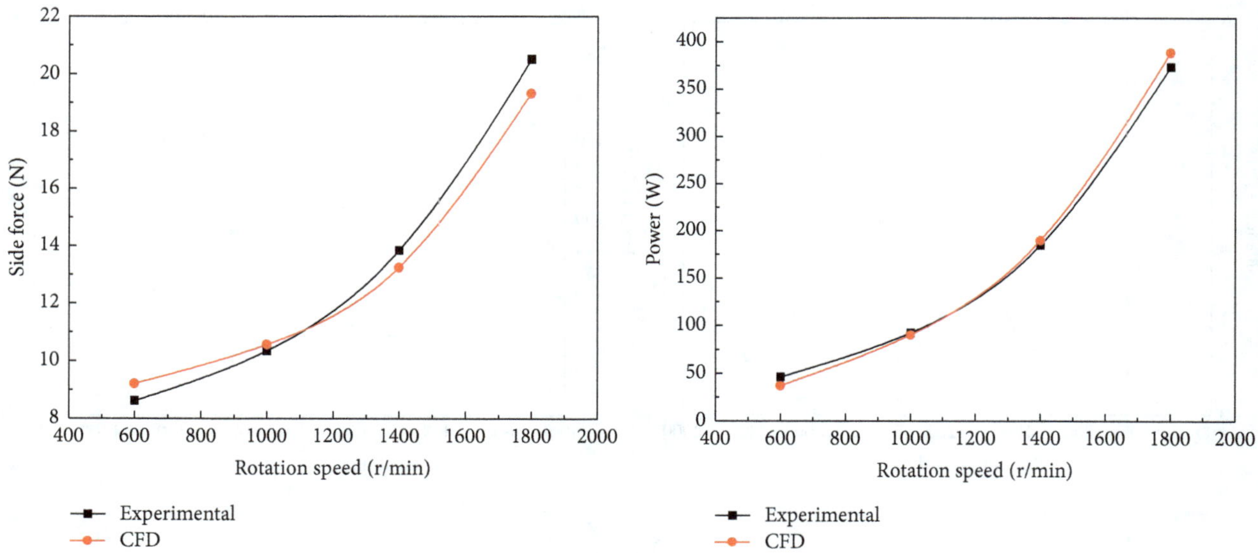

(a) Curves of side force with rotation speed

(b) Curves of power with rotation speed

FIGURE 12: Opening angle is 110 degrees.

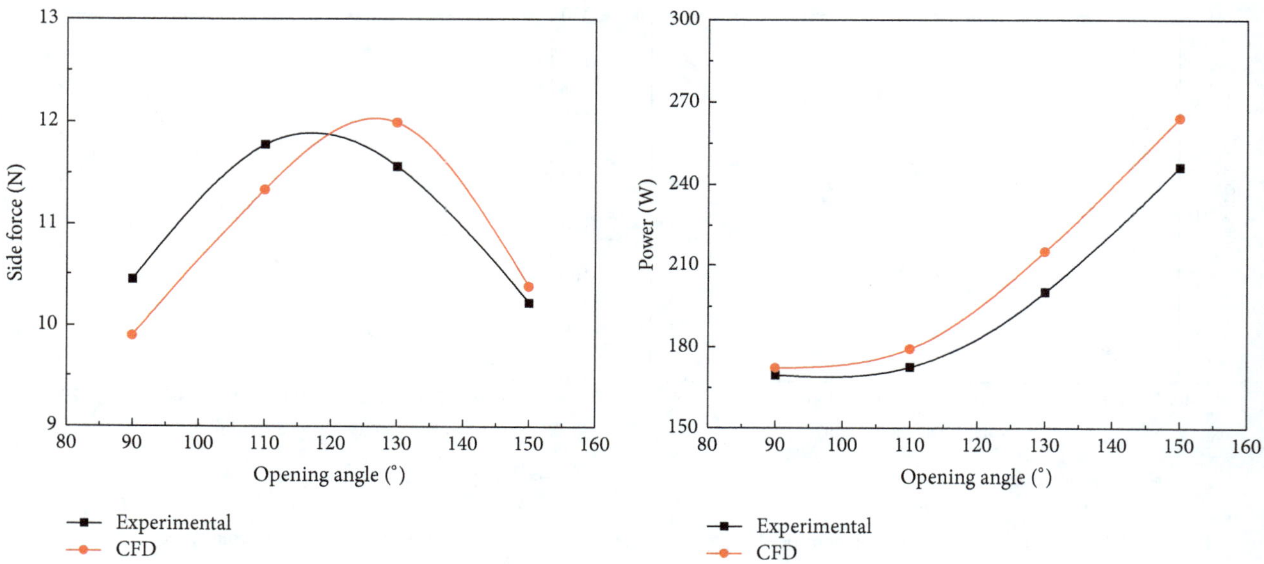

(a) Curves of side force with the opening angle

(b) Curves of power with the opening angle

FIGURE 13: Opening angle changes.

downwash flow becomes weaker when the opening angle is larger than 110 degrees. So, a proper opening angle is also an important parameter to the antitorque device. Figure 13(a) shows the experimental and numerical calculation of side force with the antitorque device at different opening angles. The error of numerical calculation is relatively large. That is because four grid computing models need to be, respectively, built when the model of the opening angle is 90°, 110°, 130°, and 150°. The grid connection and the grid number are different between the interface and the shell, which may cause errors. However, the deviation between the calculated results and the experimental values is also acceptable.

4.4.3. Left and Right Interface Changes. The experimental conditions are as follows: rotation speed of CFF is 1400 r/min, the blade number is 12, the blade installation angle is 20 degrees, the rotor collective pitch is 10 degrees, the flow speed is 0 m/s, and the opening angle is 110 degrees. It is shown in Figure 14(a) that when the antitorque device has an interface, the side force is smaller than the other two cases and side force is larger than the other two cases with half of the interface (no left interface). Compared with the experimental data of the wind tunnel, the side force generated by the configuration is higher than that of the other two cases, which is consistent with the numerical results of this paper. By numerical simulation, it is very convenient to see

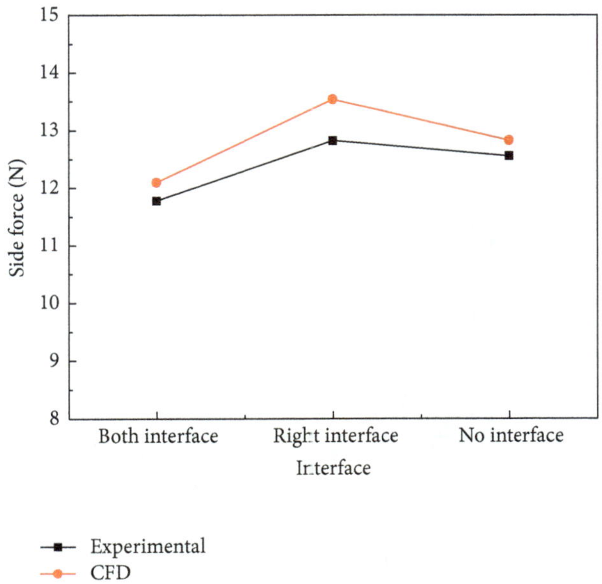

(a) The interface effect on side force curve

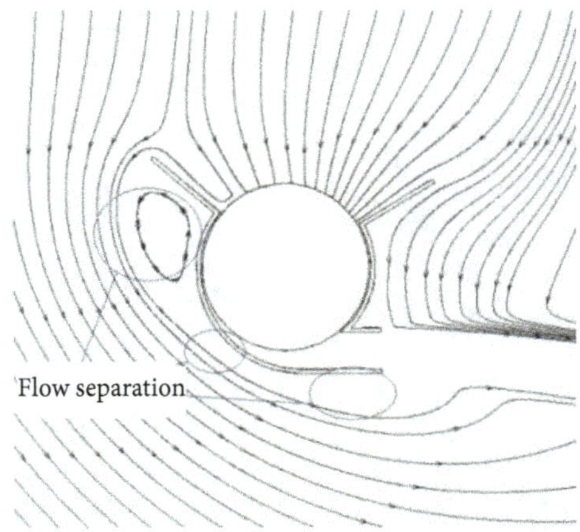

(b) Streamlines of both interfaces

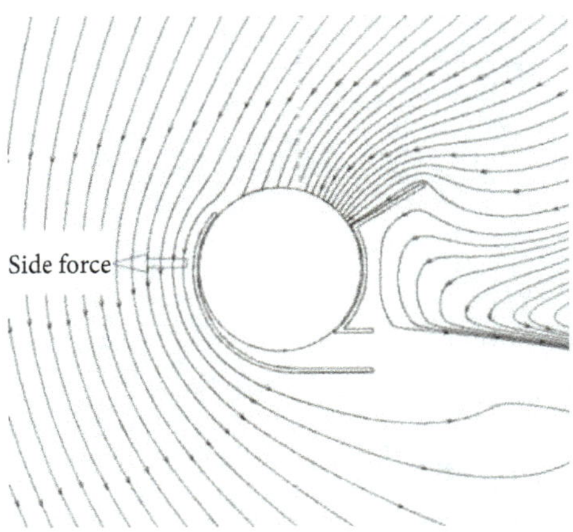

(c) Streamlines of right interface

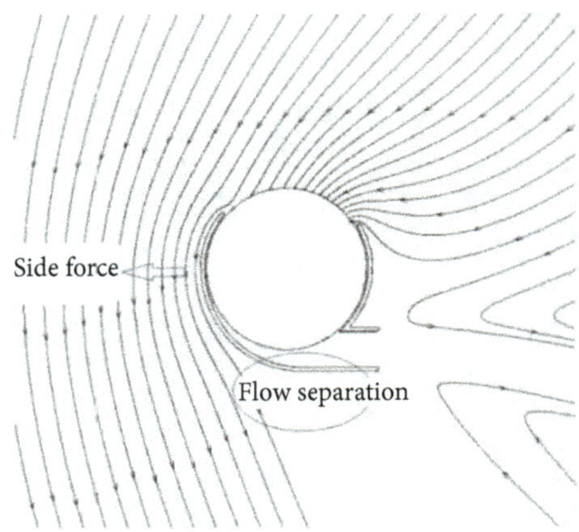

(d) Streamlines of no interface

Figure 14: Left and right interface changes.

how the air flows. Figure 14(b) is the external flow field streamlines of reactive torque device with interface. It can be figured out that the location of flow separation is in the left shell of outer surface. It formed vortex near surface part of the shell and the airflow does not go along the left shell to the bottom of the shell of reactive torque device exit, but it separates from left shell. Figure 14(c) is the flow chart when reactive torque device has half interface. Figure 14(d) is the external flow chart when reactive torque device has no interface. Comparing Figure 14(c) with 14(d), we can see that the antitorque device left shell is in the rotor downwash flow and CFF suction flow, so its streamline is denser and flow speed is higher. According to the Bernoulli theorem, pressure is lower than right shell and left side force is promoted. This is favorable for reactive torque. The left shell flow characteristics of the device with right interface are similar to that with no interface. The difference is that the airflow separates later when the device has right interface. So, it can be concluded that the antitorque device produces side force by two ways. A part is that the airflow is accelerated by CFF twice, when it goes through the fan and gets ejected to the outlet. According to Newton's third law, it obtains left side force. This part of the side force accounts for about 70%. Another part of the side force is the air going through the outer surface of the antitorque device, forming velocity circulation under the effect of rotor downwash flow and CFF suction flow and producing side force which is similar to the NOTAR helicopter.

4.5. Forward Flight State Comparison

4.5.1. Forward Flight Speed Is 5 m/s. Figures 15(a) and 15(b) show the curves where the side force and the required power change with the rotation speed when the blade number is 12, blade installation angle is 20 degrees, with both interfaces, the opening angle is 110 degrees, the collective pitch is 12 degrees, and the forward speed is 5 m/s. Figure 15(a) shows the side force at different rotation speeds when forward speed is 5 m/s. The errors between numerical calculation and experimental results at 600 r/min, 1000 r/min, 1400 r/min, and 1800 r/min

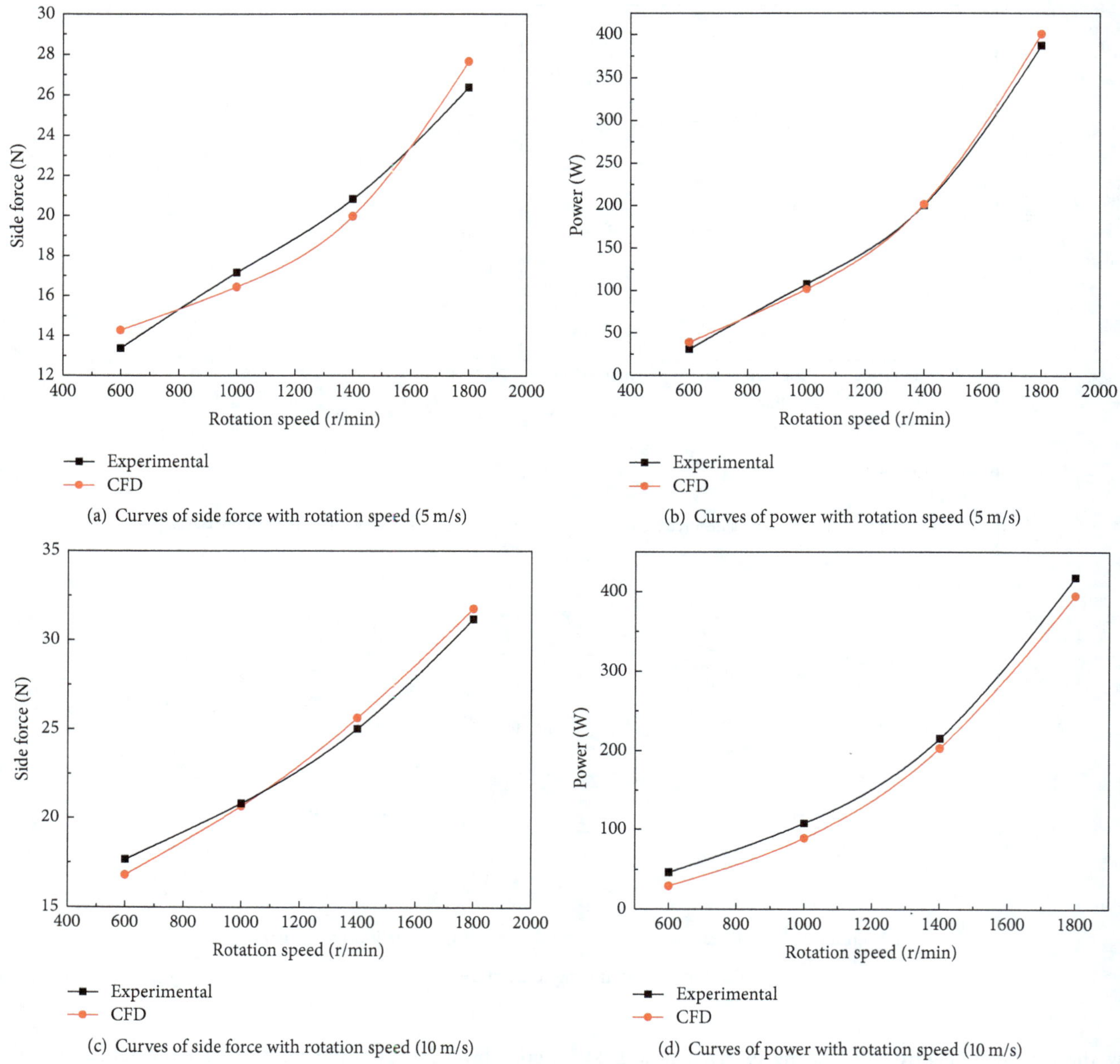

(a) Curves of side force with rotation speed (5 m/s)

(b) Curves of power with rotation speed (5 m/s)

(c) Curves of side force with rotation speed (10 m/s)

(d) Curves of power with rotation speed (10 m/s)

FIGURE 15: Forward flight state comparison.

are 6.7%, 4.2%, 4.1%, and 4.8%, respectively. The error is large and the experimental value is low at low speed. Probably, the eccentric vortex intensity inside the CFF is low and has not yet formed a stable pattern at low speed. And the calculated value is considered as the ideal situation; a stable flow field is formed, so the result is larger. It can be seen from the figure that the result of numerical calculation is credible.

Figures 15(c) and 15(d) show the curves where the lateral force and the required power change with the rotation speed when the blade number is 12, blade installation angle is 20 degrees, with both interfaces, the opening angle is 110 degrees, the collective pitch is 12 degrees, and the forward speed is 10 m/s. It can be seen from the figure that the results of numerical calculation are consistent with the experimental results.

5. Conclusion

The paper drew the following conclusions based on the wind tunnel test and numerical calculation method:

(1) Wind tunnel tests proved that the antitorque device can produce side force. CFF rotation speed, rotor downwash flow, and the forward flight flow have a great impact on side force. In this paper, the optimal geometric parameters for reactive torque device are as follows: the blade number is 10, with half interface, and the opening angle is 110 degrees. The test calculation power load of reactive torque device has a little difference with common helicopter and it can take place of a single-rotor helicopter tail rotor mechanism.

(2) We carried out numerical calculations at different blade installation angles, different blade numbers, different opening angles, with and without interface, different CFF rotation speeds, different rotor downwash flow speeds, different forward speeds, and compared them with the wind tunnel experiment result. Finally, this proves that the calculation result is credible.

(3) Numerical simulation flow pattern revealed initially the principle of producing side force. One part is that CFF accelerated the airflow, the other part is that the airflow bypassed the external surface of reactive torque device, formed velocity circulation, and generated the side force.

(4) According to the shape characteristic of the reactive torque device, the system can be arranged on the tail beam of the single-rotor helicopter to balance reactive torque and control course by controlling the rotation speed of the CFF.

Competing Interests

The authors declare that they have no competing interests.

Acknowledgments

The authors would like to acknowledge the financial support of the Priority Academic Program Development of Jiangsu Higher Education Institutions and the Nanjing University of Aeronautics and Astronautics Innovation Foundation (Grant no. 201501049).

References

[1] C. L. Zhang, "New developments of helicopter technology," *Journal of Nanjing University of Aeronautics and Astronautics*, vol. 29, no. 6, pp. 607–614, 1997.

[2] X. P. Ni, R. H. Cai, J. X. Cao, and J. Z. Yu, "Present situation and prospects of helicopter technology," *Acta aeronautica et Astronautica Sinica*, vol. 24, no. 1, pp. 16–20, 2003.

[3] Z.-W. Yu and Y.-H. Cao, "CFD simulation and validation of ducted tail rotor," *Journal of Aerospace Power*, vol. 21, no. 1, pp. 19–24, 2006.

[4] Y.-H. Cao, K. Chen, J. Wang, and W.-L. Yin, "CFD analysis of aerodynamic characteristics for the ducted tail rotor and its aerofoil," *Journal of Aerospace Power*, vol. 19, no. 6, pp. 792–797, 2004.

[5] S. Rao Carlo, "Tailboom design of the 520N NOTAR helicopter," in *Proceedings of the 46th Annual Forum of the American Helicopter Society, Part 1 (of 2) (AHS '90)*, vol. 1, pp. 549–556, 1990.

[6] I. Yavrucuk, H. M. Bakir, and O. Uzol, "Mathematical modeling of the NOTAR antitorque system for flight simulation," *Journal of the American Helicopter Society*, vol. 58, no. 2, Article ID 022002, pp. 407–415, 2013.

[7] J. Li and W. Yang, "An analysis of cross section of helicopter tail boom for NOTAR system," *Acta Aerodynamica Sinica*, vol. 33, no. 2, pp. 239–245, 2015.

[8] Z. X. Dong, Y. D. Gao, and H. M. Wang, "Influence of circulation control parameters on lateral thrust of helicopter tail boom in rotor downwash," *Helicopter Technique*, vol. 4, pp. 12–24, 2011.

[9] X. P. Luo, C. L. Zhang, and M. S. Shen, "No tail rotor and its experimental research," *Journal of Aerospace Power*, vol. 11, no. 1, pp. 15–18, 1996.

[10] Z. X. Dong, Y. D. Gao, and H. M. Wang, "Research on parameters of circulation control tail boom," *Aeronautical Science and Technology*, vol. 6, pp. 11–20, 2009.

[11] L. Meng, Y. Ye, and N. Li, "Research progress and application prospects of fan-wing aircraft," *Acta Aeronautica et Astronautica Sinica*, vol. 36, no. 8, pp. 2651–2661, 2015.

[12] T. Q. Dang and P. R. Bushnell, "Aerodynamics of cross-flow fans and their application to aircraft propulsion and flow control," *Progress in Aerospace Sciences*, vol. 45, no. 1–3, pp. 1–29, 2009.

[13] S. Askari and M. H. Shojaeefard, "Numerical simulation of flow over an airfoil with a cross flow fan as a lift generating member in a new aircraft model," *Aircraft Engineering and Aerospace Technology*, vol. 81, no. 1, pp. 59–64, 2009.

[14] S. Askari, M. H. Shojaeefard, and K. Goudarzi, "Experimental study of stall in an airfoil with forced airflow provided by an integrated cross-flow fan," *Proceedings of the Institution of Mechanical Engineers, Part G: Journal of Aerospace Engineering*, vol. 225, no. 1, pp. 97–104, 2011.

[15] S. Askari and M. H. Shojaeefard, "Experimental and numerical study of an airfoil in combination with a cross flow fan," *Journal of Aerospace Engineering*, vol. 227, no. 7, pp. 1173–1187, 2013.

[16] S. Askari and M. H. Shojaeefard, "Shape optimization of the airfoil comprising a cross flow fan," *Aircraft Engineering and Aerospace Technology*, vol. 81, no. 5, pp. 407–415, 2009.

[17] J. D. Kummer and T. Q. Dang, "High-lift propulsive airfoil with integrated crossflow fan," *Journal of Aircraft*, vol. 43, no. 4, pp. 1059–1068, 2006.

[18] N. Thouault, C. Breitsamter, and N. A. Adams, "Numerical and experimental analysis of a generic fan-in-wing configuration," *Journal of Aircraft*, vol. 46, no. 2, pp. 656–666, 2009.

[19] M. Tang, Z. F. Tang, and H. D. Wu, "The rotor anti-torque systems structure parameters on its aerodynamic influence," *Helicopter Technique*, vol. 2, pp. 15–19, 2013.

[20] Z. F. Tang, M. Tang, and H. D. Wu, "The analysis of the influence of rotor anti-torque system's structure parameters on its aerodynamic influence," in *Proceeding of the 2nd Asian/Australian Rotorcraft Forum and The 4th International Basic Research Conference on Rotorcraft Technology*, pp. 1–9, Tianjin, China, 2013.

7

Designing a Control System for an Airplane Wing Flutter Employing Gas Actuators

A. R. Ansari[1] and A. R. B. Novinzadeh[2]

[1] *Department of Mechanical and Aerospace Engineering, Science and Research Branch, Islamic Azad University, Tehran, Iran*
[2] *Faculty of Aerospace Engineering, K.N. Toosi University of Technology, Tehran, Iran*

Correspondence should be addressed to A. R. B. Novinzadeh; novinzadeh@kntu.ac.ir

Academic Editor: Enrico Cestino

The wing flutter is a dynamic instability of a flight vehicle associated with the interaction of aerodynamic, elastic, and inertial forces (aeroelastics phenomena). In this study, just the primary control is investigated. Also, in order to control the two-dimensional wing flutter, the force jet and pulse width pulse frequency (PWPF) are suggested. The PWPF modulator has the advantage of almost linear operation, low jet gas consumption, flexibility in addressing various needs, and good accuracy in presence of oscillations. This scheme makes use of quasi-steady dynamic premises and incompressible flow, as well as the thin airfoil theory. It should be noted that, to justify the application of the aerodynamic theory, we have speculated that the thruster jet ejected through a nozzle with a diameter smaller than several millimeters has a supersonic regime (with Mach number of the order of $M \approx 3.5$). Consequently, the interference of the thruster jet in the boundary layer, flow, and circulation around the airfoil which are characterized by low speed would be negligible. The operation of the jet as a thruster is handled by the PWPF modulator, and the process output is fed back to the system via a PD controller. In order to control the wing flutter oscillation, the location of installing the actuator on the airfoil is investigated.

1. Introduction

Flutter is a result of interaction of aerodynamic, inertial, and elastic forces as is shown in Figure 1; discussion and investigation of the phenomenon are related to the field of aeroelasticity [1, 2]. Such an event can occur in somewhat flexible structures, such as airplane wings, telegraph wires, bridges, and buildings [3]. It has not been long since controlling the wing flutter phenomenon has been investigated by aeroelastics researchers. Each has proposed a method to damp or evade flutter.

Zhang et al. [4, 5] suggested single- and multiple-input adaptive control solutions for unsteady airfoil while they considered mass unbalanced trailing- and leading-edge flaps in their model; that is, their model captures the dynamics of the attached control surfaces.

Utilizing piezoelectric materials and control surfaces is of the suggested methods. Conventional methods for linear and nonlinear control of wing flutter have deployed control surfaces and smart structures such as piezoelectric (PZT) and electrorheological fluids (ERFs) materials. For example, using two control surfaces and feedback linearization techniques, Ko et al. [6] proved global stability of an adaptive control law for an airfoil with nonlinear damping and stiffness operating in unsteady flow.

In the innovative scheme proposed in this paper, two-dimensional control of wing flutter is reached via the momentum resulting from the gas exhaled from the jet installed on the wing. The jet fluid used by the actuator can be supplied via the inlet vents in front of the airplane or through the airplanes compressor engine. In this study, ideal air is considered as the fluid operating the jet.

In practice, the PWPF modulator decides on the on- and off-time of the jet (thruster) by maintaining the merits of a Bang-Bang controller. Using two thrusters provides the opportunity of changing the direction of force and momentum (acted) on the airfoil. This has been made possible via a command block. Equations solutions and simulations are

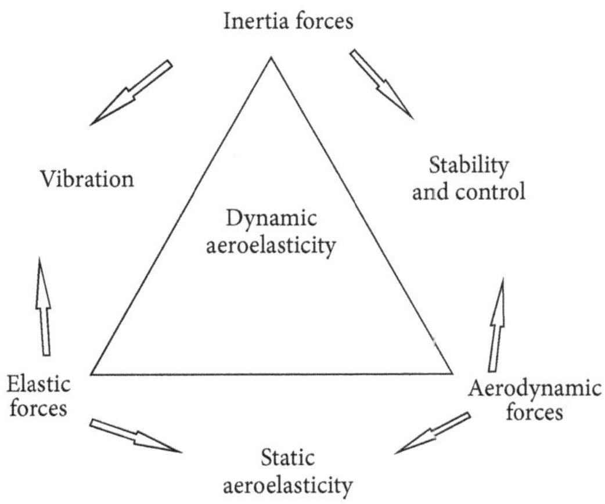

FIGURE 1: The summarized studies on an aeroelasticity triangle of collar [1].

carried out in MATLAB software. The model and equations governing this system are presented in the following sections.

2. Aeroelastic Model and Equations Governing the Two-Dimensional Wing Movement

In aeroelastics, the pitch and plunge oscillations of an airfoil are modeled via linear and torsion spring as illustrated in Figure 1. As mentioned before, due to using the thin airfoil theory, the angle of attack (α) is small. Another assumption employed in this scheme is quasi-steady aerodynamics. Dynamic equations of the system are obtained on the basis of Lagrange equations. The dynamic system has two degrees of freedom (i.e., the pitch and the plunge motions). The free stream velocity is V and the positive direction is downward. Accordingly, the upward lift force will be negative. All torque components are considered around the elasticity pivot. The airfoil chord is "c," and "b" is the mid-chord.

In Figure 2, X_f is the position of the elasticity pivot, X_c is the position of mass center, and K_h and K_α show the plunge and pitch spring stiffness coefficients, respectively. There e or a is dimensionless distance from the aerodynamic center to the pivot. α is the pitch angle.

For two-dimensional, incompressible flow aerodynamic center is at the airfoil quarter-chord; for supersonic flow it moves back to the half-chord [8].

I_α indicates the inertial moment around the elasticity pivot, and the moments are around the elasticity axis. In order to derive equations of aeroelastic airfoil, two sets of equations are coupled, namely, the equations of structural model of airfoil and the equations of the aerodynamic model of the airfoil.

In this study, a controlling model in order to indicate the effect of the actuators and monitoring elements is added to the total model. First, the structural equations are derived. The vertical displacement of the structure according to Figure 2 is described as in

$$z = h + \alpha x_f, \tag{1}$$

where x_f indicates the place of the elastic center from the leading edge:

$$T = \frac{1}{2} m \dot{h}^2 + s \dot{h} \dot{\alpha} + \frac{1}{2} I_\alpha \dot{\alpha}^2. \tag{2}$$

The overall kinetic energy of the structure is [2]

$$s = m\left(\frac{c}{2} - x_f\right), \tag{3}$$

where

$$I_\alpha = \frac{1}{3} m \left(c^2 - 3 c x_f + 3 x_f^2\right). \tag{4}$$

The potential energy for the above scheme is the sum of all energy saved in the spring:

$$V = \frac{1}{2} k_h h^2 + \frac{1}{2} k_\alpha \alpha^2. \tag{5}$$

The sum of kinetic and potential energies is the total energy of the system. Here, in order to find the motion equations, the total energy is placed in Lagrange equation [9]:

$$\frac{d}{dt}\left(\frac{\partial (T - V)}{\partial \dot{q}}\right) + \frac{\partial (T - V)}{\partial q} = Q, \quad q = [h, \alpha]^T. \tag{6}$$

The result of Lagrange equation is the following matrix equation:

$$M \ddot{q} + k q = Q, \tag{7}$$

where "Q" is the resultant vector of external forces (i.e., thruster and aerodynamic forces). The resultant equations of aerodynamic model of airfoil depend upon flow regimes. Here, since the fluid is incompressible, the Mach number is not included. The instantaneous aerodynamic forces on the airfoil depend upon both position and strength of the wake vertices (unsteady aerodynamics), which means the instantaneous aerodynamic forces rest not solely on the current motion of the airfoil, but also upon its past motion from the beginning of the movement. In quasi-steady aerodynamic forces, the simplest, possible modeling includes ignoring the effects of the wake. From the thin airfoil theory for the lift coefficient, the following equation is obtained [2]:

$$c_l = 2\pi\left(A_0 + \frac{A_1}{2}\right), \tag{8}$$

$$\frac{dz}{dx} = \left(x_f - x\right)\frac{\dot{\alpha}}{U}. \tag{9}$$

From the thin airfoil theory for moment and lift coefficients and (9), the following equations are obtained [2]:

$$\begin{aligned} c_l(t) &= 2\pi\left[\alpha + \frac{\dot{h}}{U} + \left(\frac{3}{4}c - x_f\right)\left(\frac{\dot{\alpha}}{U}\right)\right], \\ cm_{xf}(t) &= -\frac{c\pi}{8U}\dot{\alpha}(t) + \left(\frac{x_f}{c} - \frac{1}{4}\right)c_l(t). \end{aligned} \tag{10}$$

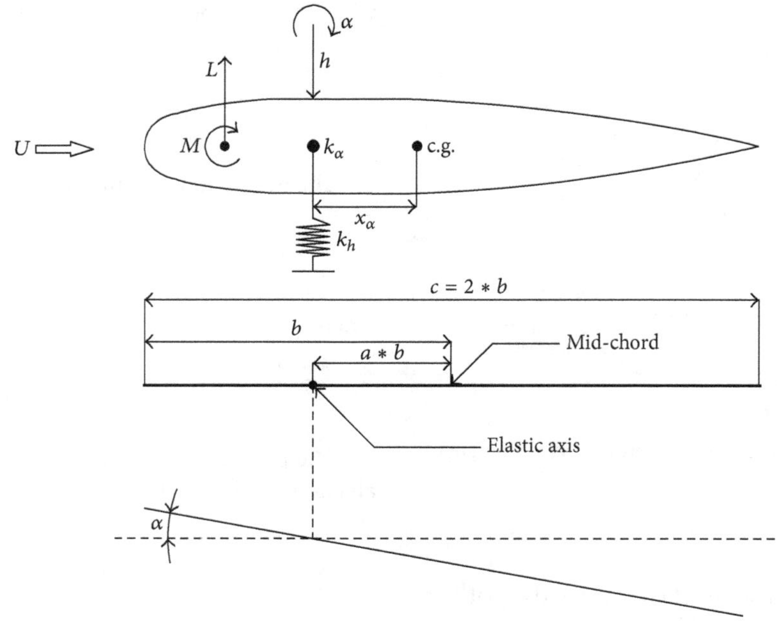

FIGURE 2: Two-dimensional model of the airplane wing.

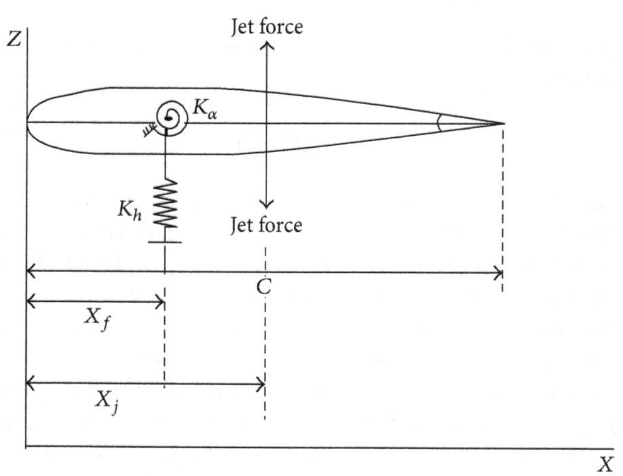

FIGURE 3: Two-dimensional model of an airplane wing, equipped with an actuator capable of on-off operation (positive direction is downward).

The moments are referenced around the elasticity pivot. Hence, the contribution of the lift to the distance from the leading edge (X_f) is added to moment around the flexural axis. As well as lift and moment forces, the force of the virtual mass due to the airfoil acceleration is also exerted to the airfoil. This force affects both the lift and the moment forces.

In case of diverging the wing oscillations at a certain speed, the activation of the actuator (thruster), force ($F(t)_{\text{jet}}$), and moment ($M(t)_{\text{jet}}$) generated by the thruster in two directions (Figure 3) leaves a damping impact on the force and moment. The total motion equations are presented in (11) and (12):

$$\begin{aligned} N(t) &= L(t) + F(t)_{\text{jet}} \\ &= \rho\pi b^2\left(\ddot{h} - \left(x_f - \frac{c}{2}\right)\ddot{\alpha}\right) + \rho\pi b^2 U\dot{\alpha} \\ &\quad + \rho\pi U^2 c\left(\alpha + \frac{\dot{h}}{U} + \left(\frac{3}{4}c - x_f\right)\frac{\dot{\alpha}}{U}\right) \\ &\quad + F(t)_{\text{jet}}. \end{aligned} \tag{11}$$

For the overall moment,

$$\begin{aligned} M(t) &= M_{xf}(t) + M(t)_{\text{jet}} \\ &= \left(x_f - \frac{c}{2}\right)\left[\rho\pi b^2\left(\ddot{h} - \left(x_f - \frac{c}{2}\right)\right)\ddot{\alpha}\right] \\ &\quad - \frac{\rho\pi b^2}{8}\ddot{\alpha} - \left(\frac{3}{4}c - x_f\right)\rho\pi b^2 U\dot{\alpha} \\ &\quad + \rho U^2 e c^2 \pi\left(\alpha + \frac{\dot{h}}{U} + \left(\frac{3}{4}c - x_f\right)\frac{\dot{\alpha}}{U}\right) \\ &\quad - \frac{1}{16}\rho U c^3 \pi\dot{\alpha} + M(t)_{\text{jet}}. \end{aligned} \tag{12}$$

In (11) and (12) equations, $L(t)$ and $M_{xf}(t)$ are, respectively, the lift and overall moment, created by aerodynamic forces and added mass, referenced around the elasticity pivot.

By substituting (2), (5), (11), and (12) in (7) and after simplification, the aeroelastic equations are derived:

$$\begin{pmatrix} m & s \\ s & I_\alpha \end{pmatrix}\begin{pmatrix} \ddot{h} \\ \ddot{\alpha} \end{pmatrix} + \rho\pi b^2 \begin{pmatrix} 1 & \left(\frac{c}{2} - x_f\right) \\ \left(\frac{c}{2} - x_f\right) & \left(\frac{c}{2} - x_f\right)^2 + \frac{b^2}{8} \end{pmatrix}\begin{pmatrix} \ddot{h} \\ \ddot{\alpha} \end{pmatrix} + \rho U c\pi \begin{pmatrix} 1 & \left(\frac{3}{4}c - x_f\right) + \frac{c}{4} \\ -ec & \left(\frac{c}{2} - x_f\right)^2 + \left(\frac{3}{4}c - x_f\right)\frac{c}{4} \end{pmatrix}\begin{pmatrix} \dot{h} \\ \dot{\alpha} \end{pmatrix} + \begin{pmatrix} k_h & 0 \\ 0 & k_\alpha \end{pmatrix}\begin{pmatrix} h \\ \alpha \end{pmatrix} + \rho U^2 c\pi \begin{pmatrix} 0 & 0 \\ 0 & -ec \end{pmatrix}\begin{pmatrix} h \\ \alpha \end{pmatrix} = \begin{pmatrix} F_{\text{jet}} \\ M_{\text{jet}} \end{pmatrix}. \tag{13}$$

Equations in (13) are the second-order linear differential equations. In these equations, there are the stiffness, damping, and mass matrices both structurally and aerodynamically. As denoted, the right side of the equations in (13) is nonzero. These equations are heterogeneous.

3. Gas Actuator

Controlling the wing oscillations is employed via a jet, as an actuator, to control flutter and damp oscillations. The momentum resulting from this jet affects wing oscillations amplitude. In the simulations, activating the thruster in both up and down directions is programmed. The flow rate and pressure of cold-gas system are provided by the vents in the front of the airplane or through the airplane's engine compressor. The actuator is provoked as the relative speed of the motion reaches the flutter speed. The influence of the thruster on the wing surface contains force and moment. In case the jet resides precisely at the elasticity pivot, it will not have a moment around the pivot. The distance from the jet to the leading edge is (x_j), Figure 4, which is a percentage of the chord (C).

The force acting upon the wing by the thruster is presented in (14), which is a function of speed, mass flow rate, and the difference between ambience and discharge pressures, as well as jet cross section. The amount of the thrust resulting from the jet equals [10]

$$F_{\text{jet}} = V_e \dot{m}_e + A_e \left[P_e - P_a\right]. \tag{14}$$

Here, V_e indicates emitted fluid speed, $\dot{m}_e$ is the mass flow rate, P_e is the emitted pressure, P_a denotes the ambient pressure, and A_e is the cross section of the jet nuzzle. The acting thruster fluid is the ideal air.

The kind of jet fluid in order to produce thrust is of importance. However, the second important characteristic parameter of the thrust source is the Specific Impulse (I_{sp}),

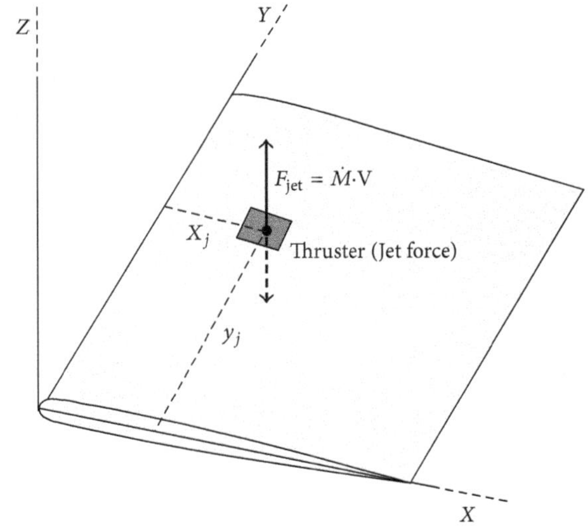

FIGURE 4: Three-dimensional perspective of the wing and the placement and operation of the thruster in up direction (symmetrically employed in down direction in appropriate intervals).

indicating the amount of fuel mass which can be converted to thrust stroke. According to definition, the Specific Impulse is

$$I_{\text{sp}} = \frac{F_{\text{jet}}}{\dot{m}g}, \qquad F_{\text{jet}} = I_{\text{sp}} \cdot \dot{m}g. \tag{15}$$

The moment produced by the thruster is expressed by (16). All moments are referenced around the elasticity pivot:

$$M_{\text{jet}} = x_{\text{jet}} \times F_{\text{jet}}. \tag{16}$$

4. Modulator

PWPF modulator is a pulse to pulse frequency-type thruster control system. This type of modulator consists of a two-variable Schmitt-trigger circuit, a first-order filter, and a feedback loop. In such a modulator, the filter precedes Schmitt-trigger, not in the control feedback (Figure 5). This modulator produces a sequence of pulses for the thruster through frequency- and width-modulated pulses [6]. The modulator has its merits over other thruster control systems. It also enjoys all the advantages of the Bang-Bang control system. Moreover, it is one of the popular systems of position control in the space.

In addition, the mentioned modulator is chosen as a candidate for being used in satellite system of European Student Moon Orbiter (ESMO) [11]. Static analysis indicates that the modulator, for a fixed input, has a quasi-linear (almost linear) performance. In order to analyze the stability, a dynamic analysis for a nonlinear input is required, and one can make use of the approach of fractional describing function [12]. In the present paper, the jet performance is controlled via the modulator; that is, the modulator, in appropriate intervals,

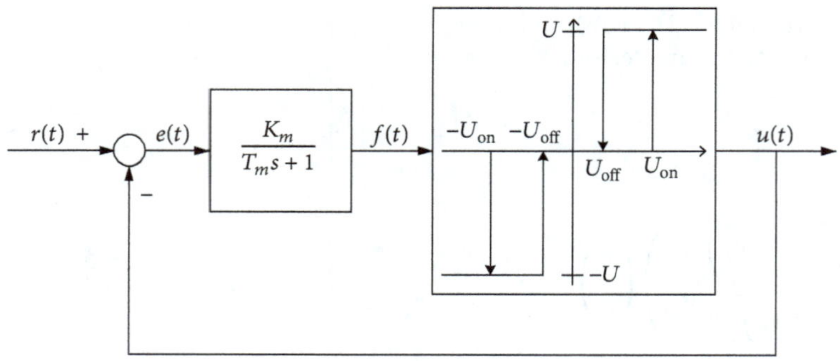

FIGURE 5: PWPF modulator circuit.

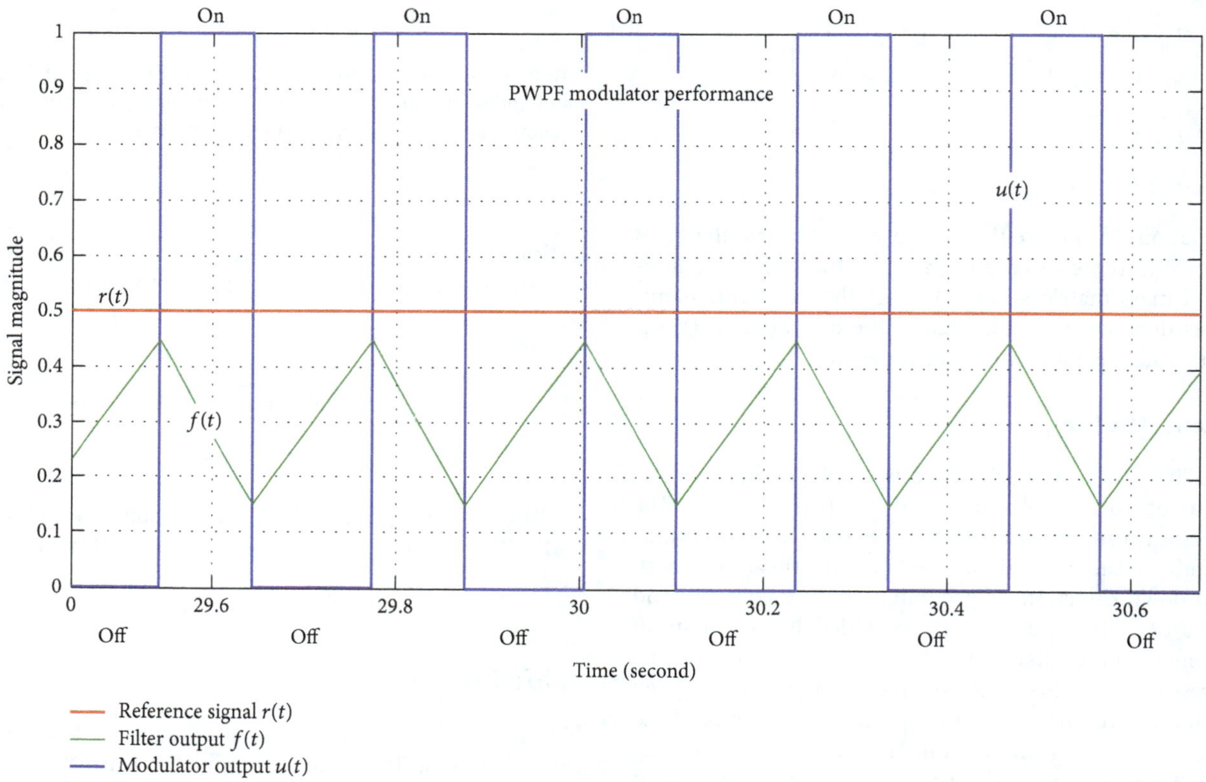

FIGURE 6: PWPF modulator response to t fixed input $C = 0.5$.

provides us with the situation to monitor the jet performance (on and off). This modulator runs like digital converters.

The jet (or, if needed, "jets") will operate in case the flight speed reaches the critical speed. In this speed, known as "flutter speed," the wing oscillations divergence is controlled via the thruster. By the operation of the thruster which is modeled as two-directional one (up and down), the wing oscillation amplitude is controlled in two directions.

According to Figure 5, the four parameters of this modulator are simply recognized, namely, k_m, U_{on}, U_{off}, and T_m. k_m, indicating the gain of the filter, and T_m is the time constant of the filter. U_{on} and U_{off} depict the turn-on and turn-off values of the Schmitt-trigger, respectively. In addition, h is the hysteresis band, which is derived by $U_{on} - U_{off} = h$.

The Schmitt-trigger, as illustrated in Figure 5, is an on-off relay composed of a dead-zone and a hysteresis band. When the input to the Schmitt-trigger "$f(t)$" is greater than U_{on}, the thruster output is U_m (i.e., the thruster switches on). If its input "$f(t)$" is smaller than U_{off}, the output is zero, and when input is in order between T_{on} and T_{off}, the thruster sustains its on-state and keeps on with its operation. The system error is the difference between the Schmitt-trigger output U_m and the system reference $r(t)$, Figure 6 [13].

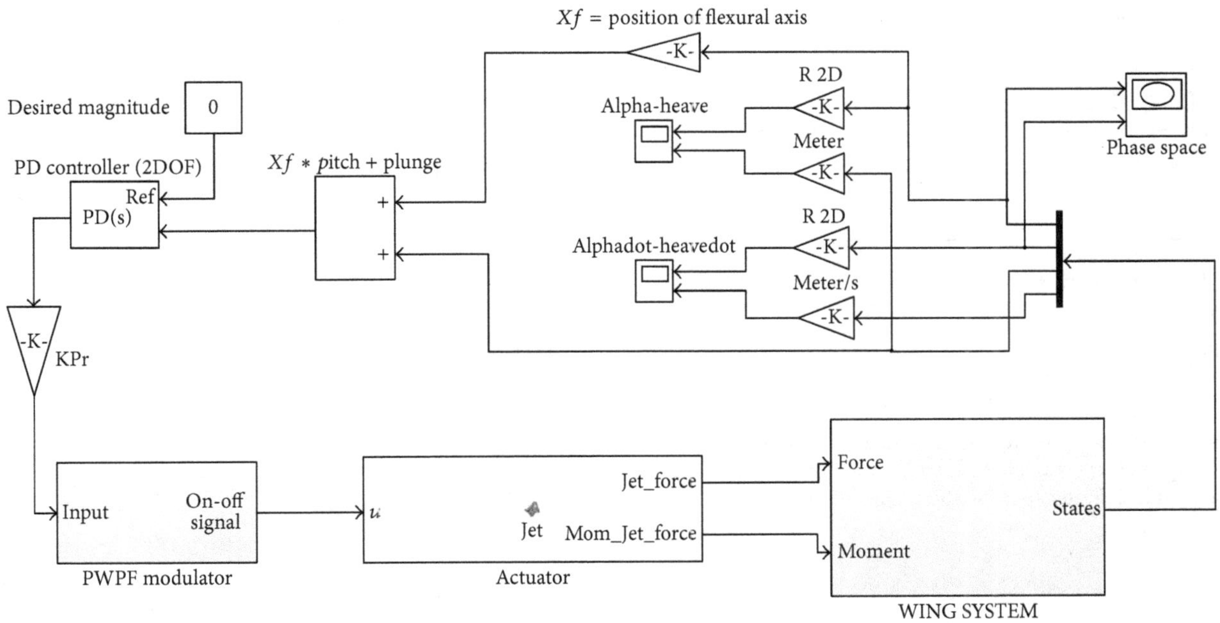

FIGURE 7: MATLAB block diagram of a two-dimensional wing controller with the modulator.

4.1. On- and Off-Time of the Modulator. For a fixed reference signal in Figure 5, the error $e(t)$ and the filter output $f(t)$ are correlated as

$$e(t) = r(t) - u(t), \tag{17}$$

$$f(t) = k_m e(t)\left(1 - \exp\left(-\frac{t}{T_m}\right)\right) + f(0)\exp\left(-\frac{t}{T_m}\right), \tag{18}$$

where $f(0)$ is the initial condition of the filter. When the reference signal is a step function with amplitude $r(t) = C$, the modulator output is also a step function $u(t) = U$ (Figure 6). Within the on-time period of the thruster, the filter output falls asymptotically. At the turn-on instant, the filter output reaches U_{on}; hence, the initial condition is $f(0) = U_{\text{on}}$.

After T_{on} passes, the filter output $f(t)$ will stop at U_{off}. Substituting these values in (18), one can find the thruster on-time:

$$T_{\text{on}} = -T_m \ln\left(1 - \frac{h}{U_{\text{on}} - k_m(C-U)}\right). \tag{19}$$

The modulator on-time is referred to as pulse width (PW) [14]. With analogous procedure, the modulator off-time is obtained as

$$T_{\text{off}} = -T_m \ln\left(1 - \frac{h}{k_m C - U_{\text{off}}}\right). \tag{20}$$

The modulator output frequency can also be expressed as

$$f = \frac{1}{T_{\text{on}} + T_{\text{off}}} \tag{21}$$

TABLE 1: Parameters for a thin symmetrical airfoil.

ρ	1.2254 kgs/m^3
I_α	6.1272 kgs·m^2
k_α	2563.83 N-m/Rad
k_h	1962.433 N/m
m	19.624 kgs
$C = 2b$	1.829 m
x_f	0.4 $*$ c m
a	−0.2

and the modulator duty cycle as

$$\text{DC} = \frac{T_{\text{on}}}{T_{\text{on}} + T_{\text{off}}}. \tag{22}$$

The duty cycle, also referred to as an adjustment factor, is a measure to determine how the modulator responds to an input; for example, to what extent the output tracks the reference input signal [13].

5. Results and Comparison

For a two-dimensional wing with the following features (Table 1), simulations are conducted in two ways: (1) without control at flutter speed; (2) with the thruster enabled by the modulator at the same speed. The simulated block diagram of this process is illustrated in Figure 7. The simulations are carried out by MATLAB software.

One of the important properties of the modulator is minimizing the acting thruster fluid and its appropriate

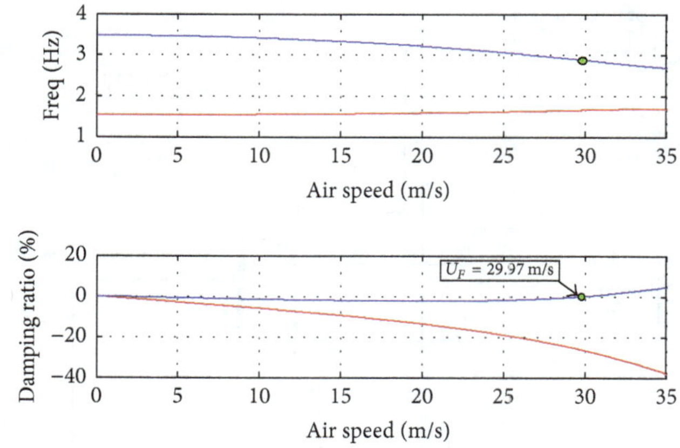

FIGURE 8: Variations in two damping coefficients and pitch and plunge frequencies versus speed derived from quasi-steady aerodynamic equations. Flutter speed is about $U_F = 29.97$ m/s.

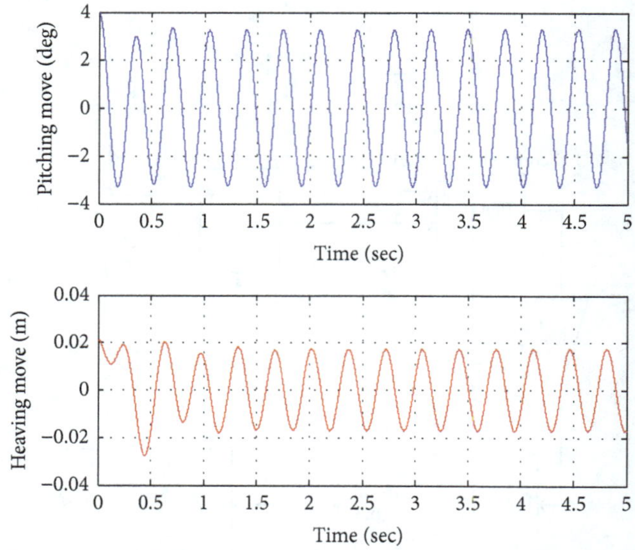

FIGURE 9: Open-loop responses (limit cycle) in pitch and plunge oscillation amplitude versus time for two-dimensional wing, in degree and meter, respectively.

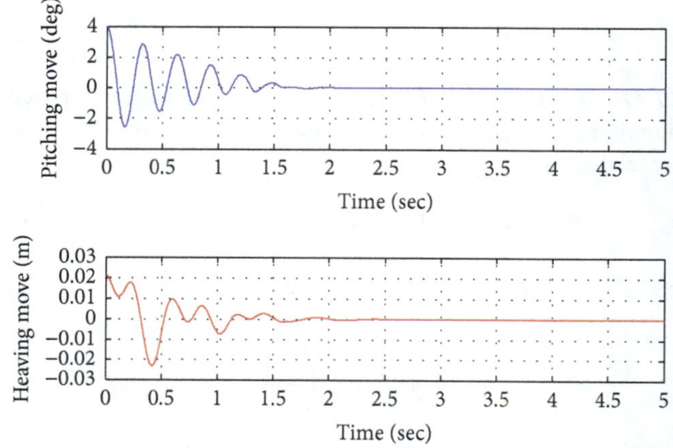

FIGURE 10: Closed-loop responses (controlled flutter) in pitch and plunge oscillation amplitude in two- dimensional wing, versus time (in degree and meter, resp.).

performance in the case of oscillations. In this paper, merely the control aspect is discussed.

In the first case, the thruster is disabled/off and there is no control over the wing. After the program is performed, the system condition is investigated. As mentioned earlier, the quasi-steady equations are applied and the results obtained from the same equations. However, the eigenvalues of the matrix describe the behavior of the system. Natural frequencies and damping coefficients can be calculated by them.

Figure 8 depicts the damping variations versus speed. At the moment when the damping coefficient gets positive, the flutter speed is measured, which is approximately $U_{\text{Flutter}} = 29.97$ m/s. The pitch and plunge oscillation amplitudes are displayed in Figure 9. Divergence in both cases occurs after the flutter speed.

The program data is offered in Table 1. The initial conditions for the wing are expressed in

$$x_0 = \left[\alpha, h, \dot{\alpha}, \dot{h}\right]^T = [0.07, 0.02, 0, 0]^T. \quad (23)$$

In the second case, the system is controlled in order to prevent the structure against damage. The process data is again derived from Table 1. Also, modulator parameters are $k_m = 16$, $T_m = 0.15$, $U_{\text{on}} = 0.45$, $h = 0.2$, and $U_m = 1$ in this study. In addition, (PD) controller gains are $k_p = 5000$ and $k_d = 3000$.

According to Figure 7, the jet and control blocks are enabled in this state. The time range of jet operation and fluid consumption rate are related to the modulator parameters.

Employing the modulator data and applying the force and moment to the right side of aeroelasticity equations (13), together with the help of the PD controller, the divergence of oscillation amplitude is controlled.

From (14), the amount of employed force from the thruster in this scheme is $F_{\text{jet}} = 37.6$ N and is placed at the distance of $x_{\text{jet}} = 0.98 * C$ from the chord. Due to the assumption of the two-dimensional wing (airfoil), the thruster can alter its location just on the x-axis.

Figure 10 shows controlling the divergence of the oscillation amplitude. In this picture, oscillations in both pitch and plunge are reduced and controlled. This procedure has been employed in a similar paper [7] in which the flutter occurrence of a two-dimensional wing has been investigated via applying a control surface.

The results are thoroughly satisfactory in order to control the flutter. As it can be noticed in Figure 11, the two-dimensional wing at the flutter speed of $U_{\text{Flutter}} = 15$ m/s displays periodic pitch and plunge oscillations amplitude.

Through activating the thruster at this speed, the oscillations amplitude tends to zero at $t = 5$ s, Figure 12. The demanded thrust is $F_{\text{jet}} = 12.45$ N and is employed at $X_{\text{jet}} = 0.10 * C$ from the leading edge. Also, modulator parameters are $k_m = 20$, $T_m = 0.2$, $U_{\text{on}} = 0.4$, $h = 0.2$, and $U_m = 1$ in this study. In addition, (PD) controller gains are $k_p = 800$ and $k_d = 350$.

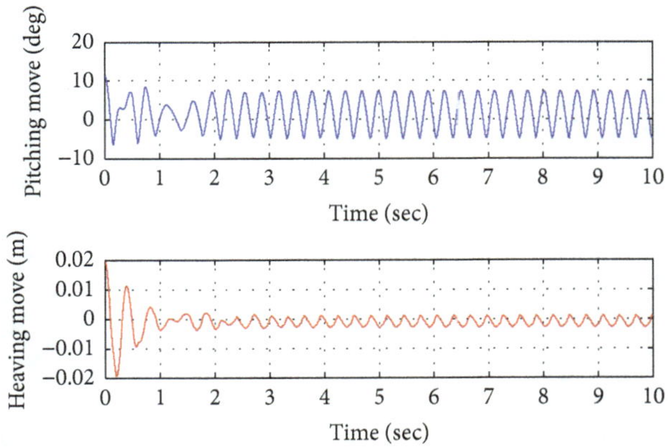

FIGURE 11: Results from [7] for a small two-dimensional wing with cycle-limited asymptotic oscillations.

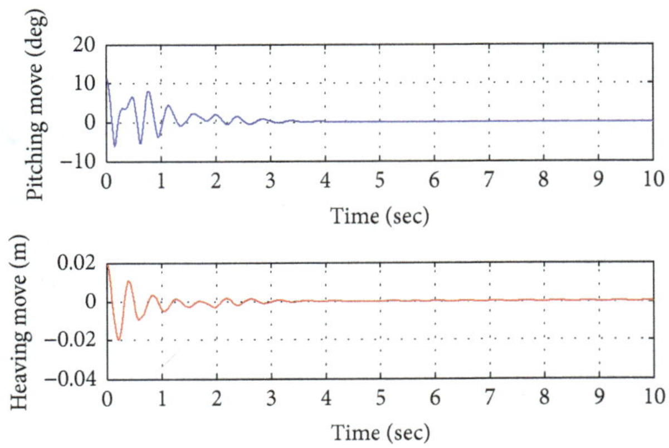

FIGURE 12: Oscillation control of a two-dimensional wing via PWPF modulator for a short time, from $t = 0$ s and $t = 10$ s (according to Table 2).

TABLE 2: Parameters for a thin symmetrical airfoil from [7].

ρ	1.2254 kgs/m^3
I_α	0.1065 kgs·m^2
k_α	Nonlinear N-m/Rad
k_h	2844.4 N/m
m	12.287 kgs
$C = 2b$	0.27 m
x_f	0.4 ∗ c m
a	−0.6

The characteristics of this wing are stated in Table 2. The spring stiffness coefficient is decided on by [15]

$$k_\alpha = 2.82 * \left(1 - 22.1 * \alpha + 1315.5 * \alpha^2 - 8580 * \alpha^3 + 17289.7 * \alpha^4\right). \tag{24}$$

The initial conditions for this wing are expressed in

$$x_0 = \left[\alpha, h, \dot{\alpha}, \dot{h}\right]^T = [0.2, 0.02, 0, 0]^T. \tag{25}$$

Additional Points

Another Merit of This Scheme. Employing this novel scheme not only controls and constrains the severe structural oscillation but also removes the need for piezoelectric materials and control surfaces. Furthermore, this method can be applied in case of weak operation of control surfaces, particularly at high flutter speed. *Future Implications and Suggestions*. (A) Conducting the study with a two-dimensional wing in a wind tunnel and comparing with analytical results and calibration of analytical results will cause the use of this design to be more practical. (B) Due to the higher density of the water than the air, fluid water can be used instead of air for the wings with lower flutter speed. (C) Another suggestion is designing a cold-gas piping system and making use of external current stagnation pressure instead of compressed air tanks thereby greatly reducing the system weight. The above-mentioned suggestions tend to encourage the use of the proposed method.

Conflicts of Interest

The authors declare that they have no conflicts of interest.

References

[1] J. R. Wright and J. E. Cooper, *Introduction to Aircraft Aeroelasticity and Loads*, John Wiley & Sons Ltd, University of Manchester and J2W Consulting Ltd and University of Liverpool, UK, 2nd edition, 2007.

[2] G. Dimitriadis, "Introduction to Aeroelasticity," in *Lecture Notes in Aeroelasticity*, Department of Aerospace and Mechanical Engineering University of Liege, 2007.

[3] C. Hebert, D. Cowan, A. Peter, and C. D. Weiseman, *Aerodynamic Flutter , Nextgen Aeronautics and ContrAfrl/Vaac and Nasa*, Langley Research Center.

[4] K. Zhang, Z. Wang, A. Behal, and P. Marzocca, "Novel nonlinear control design for a two-dimensional airfoil under unsteady flow," *Journal of Guidance, Control, and Dynamics*, vol. 36, no. 6, pp. 1681–1694, 2013.

[5] K. Zhang, P. Marzocca, and A. Behal, "Adaptive aeroelastic control of nonlinear airfoil with multiple flaps under unsteady flow," *Journal of Vibration and Control*, vol. 23, no. 10, pp. 1593–1606, 2017.

[6] J. Ko, T. W. Strganac, and A. J. Kurdila, "Stability and control of a structurally nonlinear aeroelastic system," *Journal of Guidance, Control, and Dynamics*, vol. 21, no. 5, pp. 718–725, 1998.

[7] S. A. Fazelzadeh, A. Rasti, and H. Sadat-Hoseini, "Optimal flutter suppression of nonlinear typical wing section using time-domain finite elements method," *Journal of Aerospace Engineering*, vol. 27, no. 5, 2014.

[8] E. H. Dowell, R. Clark, D. Cox et al., *A Modern Course in Aeroelasticity*, vol. 116 of *Solid Mechanics and Its Applications*, Kluwer Academic Publishers, Boston, MA, USA, 4th edition, 2004.

[9] Y. C. Fung, *An Introduction to the Theory of Aeroelasticity*, John Wiley & Sons, New York, NY, USA, 1955.

[10] J. S. Marcel, *Spacecraft Dynamic and Control*, Cambridge university press, 1997.

[11] G. Song and N. Ma, *Control of shape memory alloy actuators using pulse width pulse-Frequency (PWPF) modulation*, Department of Mechanical Engineering, University of Houston, Houston, TX, USA, 2003.

[12] X. Wang, D. Wang, S. Zhu, and E. K. Poh, "Fractional describing function analysis of PWPF modulator," *Mathematical Problems in Engineering*, vol. 2013, Article ID 287040, 5 pages, 2013.

[13] T. D. Krovel, "Optimal tuning of PWPF modulator for attitude control," Tech. Rep., Department of Engineering Cybernetics, Norwegian University of Science and Technology, 2005.

[14] G. Arantes Jr., L. S. Martins-Filho, and A. C. Santana, *Optimal on-off attitude control for the Brazilian multimission platform satellite*, vol. 2009, Article ID 750945, 17 pages, 2009.

[15] T. O'Neil, H. Gilliatt, and T. W. Strganac, *Investigations of aeroelastic response for a system with continuous structural nonlinearities*, Texas A & M Univ., College Station, 1996.

8

Receptivity of the Boundary Layer over a Blunt Wedge with Distributed Roughness at Mach 6

Zhenqing Wang,[1] **Mingfang Shi**,[1] **Xiaojun Tang**,[2] **Lidan Xu**,[1] **and Xiaokun Sun**[1]

[1]*College of Aerospace and Civil Engineering, Harbin Engineering University, Harbin 150001, China*
[2]*China Academy of Space Technology, Beijing 100094, China*

Correspondence should be addressed to Xiaojun Tang; xiaojuntang87@sina.com

Academic Editor: William W. Liou

A hypersonic flow field over a blunt wedge with or without roughness is simulated by a direct numerical simulation method. The effect of isolated and distributed roughnesses on the steady and unsteady hypersonic flow field and boundary layer is analyzed. The shape of roughness is controlled by cubic polynomial. The evolution of disturbance waves caused by slow acoustic wave in the boundary layer is investigated by fast Fourier spectrum analysis. The results show that there is a great influence of roughness on the evolution of disturbance waves in the hypersonic boundary layer. The disturbance waves are promoted in the upstream-half region of roughness while suppressed in the downstream-half region of roughness. There is always a mode competition among different modes both in the temporal domain and in the frequency domain in the boundary layer, and mode competition is affected by roughness. The location of the dominant mode which is changed to a second-order harmonic mode from the fundamental mode moves upstream. The vortices caused by roughness also impact the evolution of disturbance waves in the boundary layer. The fundamental mode is suppressed in the vortex region while other harmonic modes are promoted.

1. Introduction

Boundary-layer laminar-turbulent transition effectively affects the aerodynamic lift and drag of hypersonic vehicles and has a great effect on the design of hypersonic vehicles [1–3]. However, the process of transition is very complex and affected by many factors. In general, there are four phases of transition under freestream with small disturbances: (1) receptivity, (2) linear instability, (3) nonlinear stability and saturation, and (4) secondary instability and breakdown to turbulence. The receptivity is crucial in the process of transition because it defines the initial state of the disturbances in the boundary layer. Receptivity of the boundary layer is the response process of the boundary layer to external disturbances that include freestream disturbances, wall temperature, and roughness.

Investigating the effect of freestream disturbances is significant to understand the transition mechanism. The type and parameters of freestream disturbances both affect the process of transition, and it has been investigated by many scholars [4–11]. As the amplitude of freestream disturbances increases to finite amplitude, the second and third phases of transition may be short or even disappear. The transition mechanism is essentially changed by the change of amplitude of freestream disturbances. Ma and Zhong [4, 5] investigate the receptivity over a plate flat under the freestream with small disturbances by DNS and linear stability theory (LST). Balakumar and Kegerise [6] analyze the response of the boundary layer over straight and flared cones to small disturbances. Small disturbances and finite amplitude disturbances (10^{-4}–10^{-1}) in the boundary layer over the blunt wedge are both investigated to reveal the leading edge receptivity by Cerminara and Sandham [7], and the Fourier analysis method is used to analyze the evolution of disturbances in the boundary layer. D. Park and S. O. Park [8] and Olaf et al. [9] also investigate the evolution process under the wall roughness in nonlinear instability by Fourier spectrum analysis. As for the type of freestream disturbances, they can be decomposed into entropy, vortical, and acoustic waves which include slow acoustic waves and fast acoustic waves. Qin and Wu [10], Zhang et al. [11], and Ma and Zhong [5] research the receptivity characteristic of a supersonic/hypersonic boundary layer under small disturbances with different types of freestream disturbances and found that there is a great difference between the four kinds of disturbances on the transition of the boundary layer.

In view of the fact that the change of flow state in the hypersonic boundary layer directly affects the propulsion efficiency and maneuverability of the hypersonic vehicle, and results in aerodynamic drag and significantly increased wall heating, laminar flow control (LFC) technologies are addressed. Laminar flow control technologies can be divided into three categories [12]: (1) active techniques: such as reverse jet [13], local heating, or cooling; (2) passive techniques: such as smoothing and shaping the wall; and (3) reactive techniques: such as actuators and microelectromechanical systems. However, due to complicated working conditions and serious environment, active techniques and reactive techniques are difficult to be applied in engineering practice, and passive technology is widely used in hypersonic vehicle design [14].

Passive techniques include a variety of technologies such as shaping and passive coatings [15–17]. There is a very effective method to arrange roughness on the wall. Roughness can be classified by many methods, such as two-dimensional roughness and three-dimensional roughness and isolated roughness and distributed roughness [18].

The effects of different types of roughness on the receptivity and stability of the boundary layer vary greatly. A large number of investigations have been carried out to investigate the effects of roughness on the receptivity and stability of the hypersonic boundary layer, and some achievements have been made [19–28]. Wang and Zhong and Fong et al. [20, 21] analyzed the effect of the height of isolated roughness on the receptivity of the hypersonic boundary layer by direct numerical simulation. They found that the height of isolated roughness made the disturbance grow instantaneously, and the growth rate increased with the increase of the height of isolated roughness. Zhao et al. [22] found that the transition position moved forward with the increase of the height of isolated roughness. Holloway and Sterett and Fujii [23, 24] found that the position of the isolated roughness would also affect the transition. Duan et al. [25, 26] found that the change of the position of isolated roughness directly affects the propagation of mode S in the boundary layer by a direct numerical simulation method and then affected the receptivity mechanism of the hypersonic boundary layer.

In terms of isolated roughness, the research of distributed roughness is less [18, 27, 28]. Obviously, the effect of distributed roughness on the flow field is more complicated. Balakumar [18] used a numerical simulation method to analyze the influence of distributed roughness on the receptivity and stability of the hypersonic tip-cone flow boundary layer under slow acoustic disturbances, fast acoustic disturbances, and vortex wave. It was found that the second mode disturbance in the nose region was significantly suppressed. Desjouy et al. [27] discussed the effects of isolated and distributed roughnesses on the transition of the hypersonic boundary layer. They found that the transition of the boundary layer appeared more than that in the presence of a single roughness and the spacing between the two roughnesses has a great influence on the flow state in the boundary layer. Saric et al. [28] experimentally proved that the transition of the boundary layer can be effectively restrained by arranging the roughness array at the leading edge.

In this paper, a high-order accurate finite difference method is used to simulate the hypersonic flow field over the blunt wedge with a smooth wall and rough wall, and the effect of roughness on the flow state in the boundary layer is discussed. The shape of roughness is controlled by cubic polynomial, and the wall conditions are divided into a smooth wall and rough wall with isolated roughness and distributed roughness. The influence of roughness on the steady-state flow field is investigated firstly. On this basis, the influence of the roughness on the evolution process of disturbances in the boundary layer is analyzed by adding the slow acoustic disturbances to disturb the hypersonic flow field. The effect of roughness on evolution of different mode disturbances in the boundary layer is present by using the fast Fourier spectrum analysis (FFSA) method, and the influence mechanism of roughness on the receptivity of the hypersonic boundary layer is revealed.

2. Governing Equations and Numerical Methods

2.1. Governing Equations. A high-order finite difference method is employed to simulate the hypersonic flow field to analyze the evolution of the disturbance wave in the hypersonic boundary layer and to investigate the effect of roughness on receptivity of the boundary layer. The governing equations are two-dimensional conservation N-S equations which can be expressed as

$$\frac{\partial \mathbf{U}}{\partial t} + \frac{\partial (\mathbf{F} - \mathbf{F}_v)}{\partial x} + \frac{\partial (\mathbf{G} - \mathbf{G}_v)}{\partial y} = 0, \tag{1}$$

where t is the time and $\mathbf{U}$ is the vector of conservative variables, $\mathbf{F}$ and $\mathbf{G}$ are the convective fluxes corresponding to x and y directions, respectively, and $\mathbf{F}_v$ and $\mathbf{G}_v$ are the viscous fluxes corresponding to the x and y directions, respectively. Their specific expressions are written as

$$\begin{aligned}
\mathbf{U} &= (\rho, \rho u, \rho v, E)^T, \\
\mathbf{F} &= \left(\rho u, \rho u^2 + p, \rho u v, (E + p)u\right)^T, \\
\mathbf{G} &= \left(\rho v, \rho u v, \rho v^2 + p, (E + p)v\right)^T, \\
\mathbf{F}_v &= \left(0, \tau_{xx}, \tau_{xy}, K\frac{\partial T}{\partial x} + u\tau_{xx} + v\tau_{xy}\right)^T, \\
\mathbf{G}_v &= \left(0, \tau_{yx}, \tau_{yy}, K\frac{\partial T}{\partial y} + u\tau_{yx} + v\tau_{yy}\right)^T,
\end{aligned} \tag{2}$$

where ρ, u, v, p, T, and K denote density, velocity along the x-axis and y direction, pressure, temperature, and the heat transfer coefficient, respectively. τ and E is the shear stress and total energy, respectively. The state equation for ideal gas is written as

$$P = \rho RT, \quad (3)$$

where R is 286.94 J/(kg·K).

In order to be convenient for computation, the governing equations are transformed from the physical coordinates (x, y) to computational coordinates (ξ, η), and the coordinate transformation coefficient is the Jacobian matrix $\mathbf{J}$ that can be expressed as

$$\mathbf{J} = \frac{\partial(x, y)}{\partial(\xi, \eta)} = \begin{vmatrix} x_\xi & x_\eta \\ y_\xi & y_\eta \end{vmatrix}. \quad (4)$$

The governing equations in the computational coordinates are expressed as

$$\frac{\partial \hat{\mathbf{U}}}{\partial t} + \frac{\partial(\hat{\mathbf{F}} - \hat{\mathbf{F}}_x)}{\partial \xi} + \frac{\partial(\hat{\mathbf{G}} - \hat{\mathbf{G}}_y)}{\partial \eta} = 0, \quad (5)$$

where

$$\begin{aligned} \hat{\mathbf{U}} &= J\mathbf{U}, \\ \hat{\mathbf{F}} &= y_\eta \mathbf{F} - x_\eta \mathbf{G}, \\ \hat{\mathbf{G}} &= -y_\xi \mathbf{F} + x_\xi \mathbf{G}, \\ \hat{\mathbf{F}}_v &= y_\eta \mathbf{F}_v - x_\eta \mathbf{G}_v, \\ \hat{\mathbf{G}}_v &= -y_\xi \mathbf{F}_v + x_\xi \mathbf{G}_v. \end{aligned} \quad (6)$$

2.2. Numerical Methods. In order to accurately simulate the hypersonic flow field over a blunt wedge and correctly capture the changes of the hypersonic flow field and the evolution of the disturbance wave in the boundary layer, a high-order finite difference method is used to directly simulate the hypersonic flow field. The governing equation can be decomposed into three parts for discretization. The convective flux is split by the Steger-Warming (S-W) splitting method into positive and negative flux terms, and then the fifth weighted essential nonoscillatory (WENO) scheme [29–31] is used to discretize the positive and negative flux terms. The central [25] difference method not only introduces unnecessary information but also causes calculation error. It is worse that they are not robust enough in the shock region, and there are very serious, nonphysical numerical oscillations generated in the shock region. So the upwind scheme is proper to discretize the convective flux, and there is a great advantage in this respect. The WENO scheme can effectively suppress the numerical oscillation in a discontinuous or large gradient region, such as the shock region, and the information of the flow field can be acquired stably and accurately. The physical information of the flow field is spread around caused by viscosity, so the viscous fluxes are discretized by a six-order central difference scheme. This method only introduces phase errors but not dissipative errors, which can ensure the accuracy of the calculation. The three-step third-order Runge-Kutta method is employed to advance time. This method is now widely used in the direct numerical simulation of hypersonic compressible flow fields, and the results achieved are good.

3. Computational Condition

A computational model is a hypersonic flow over a blunt wedge. Freestream conditions are adopted at the inlet, and extrapolation boundary conditions are adopted at the outlet; symmetric boundary conditions are employed at $y = 0$ due to the symmetry of the model. The wall surface has no slip and no penetration and is isothermal. The influence of roughness on the receptivity of the hypersonic boundary layer is analyzed, and the wall surface conditions are divided into three types: wall surface without roughness, with isolated roughness, and with distributed roughness. The shape of roughness is controlled by a third-order polynomial, and the specific expression is as follows:

$$\begin{aligned} y_n(x) &= \sum_n^N \left[\frac{R_n + h}{\cos\theta} + x\tan\theta - \frac{3h}{\cos\theta}\left(\frac{x - x_n + h\sin\theta}{w\cos\theta + h\sin\theta}\right)^2 \right. \\ &\quad \left. + \frac{2h}{\cos\theta}\left(\frac{x - x_n + h\sin\theta}{w\cos\theta + h\sin\theta}\right)^3 \right], \\ x_n &= x_0 + nR_n\cos\theta, \end{aligned} \quad (7)$$

where R_n, w, h, and θ represent the nose radius, the half width of roughness, the height, and the half wedge angle, respectively, and x_n is the center coordinate of roughness. $x_0 = 0.4109$. N represents the number of roughness: when $N = 1$, the wall surface condition is the wall with isolated roughness, and when $N > 1$, it represents the wall with distributed roughness. For this paper, N is 6. The space between two roughnesses is $d = R_n \cdot \cos\theta = 0.9962$. The calculational conditions and grid are given in Figure 1, where α represents the angle of attack. The subscripts "∞" and "w" denote the freestream and wall condition, respectively. The geometric parameters of roughness are given in Table 1. The number of meshes is 600×150, and the exponential stretching method is used to close meshes near the wall surface and head area properly where a strong shear flow is in the area.

Slow acoustic wave with single frequency is introduced as the disturbance wave to analyze the influence of slow acoustic disturbance on hypersonic flow field over the blunt wedge and flow structure in the boundary layer and to investigate the effect mechanism of roughness on the receptivity of the hypersonic boundary layer.

The steady flow field is computed first, and then the plus slow acoustic disturbances which can be expressed as (8) are continuously impinged on the upstream boundary after $t = 100$ (dimensionless time), and the expression of slow acoustic is written as

$$\begin{bmatrix} u' \\ v' \\ p' \\ \rho' \end{bmatrix} = \begin{bmatrix} \varepsilon \\ 0 \\ -\dfrac{\varepsilon}{\mathrm{Ma}_\infty} \\ -\varepsilon \mathrm{Ma}_\infty \end{bmatrix} e^{i\left(kx - \left(F\cdot \mathrm{Re}/10^6\right)t + \pi/2\right)}, \quad (8)$$

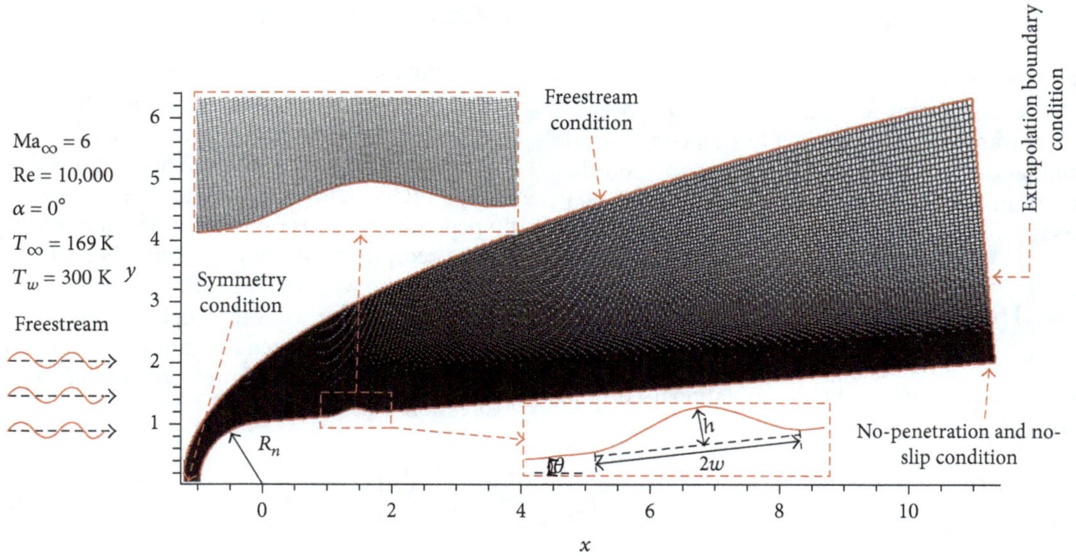

FIGURE 1: Calculational model and grid mesh.

TABLE 1: The condition of roughness.

R_n (mm)	θ (°)	w (mm)	h (mm)
1.0	5.0	0.40	0.10

where the variable ε is the amplitude ($\varepsilon = 0.06$), k is the wave number ($k = 3.1446 \times 10^{-4}$, and F is the generalized frequency ($F = 50\pi$) (dimensionless frequency $f = 0.25$). The incident wave angle of acoustic disturbances is 0. It should be noted that the superscript "$'$" represents the amount of disturbance in the flow field and the disturbance variable refers to the variable value of the unsteady flow field minus the variable value of the mean flow field at the same location. In this paper, each parameter is dimensionless, and flow parameter density ρ, velocity u, v, temperature T, and pressure p are dimensionless by corresponding values ρ_∞, U_∞, U_∞, T_∞, and $\rho_\infty U_\infty U_\infty$, respectively. The head radius R_n is chosen as the characteristic length; x, y, w, and h are dimensionless; and the time t is treated by R_n/U_∞. The correctness of the code and the grid sensitivity are verified previously, which are not given here [32].

4. Effect of Roughness on Mean Flow

In order to accurately analyze the effect of roughness on the receptivity of the hypersonic boundary layer, a high-order finite difference method is used to directly simulate the hypersonic mean flow field over the blunt wedge in this section. Pressure contour and streamline pattern of the undisturbed hypersonic flow field are given in Figure 2, and Figure 2(a, b, c) corresponds to smooth wall, isolated roughness, and distributed roughness, respectively. It can be seen that the roughness has a great influence on the hypersonic flow field and the flow structure in the boundary layer. Compression wave-expression wave and compression wave are formed caused by isolated roughness. The temperature of flow increases when the flow crosses over the compression wave and some local high-temperature regions are formed, as shown in the figure. The intensity of the waves caused by roughness decreases downstream and dissipates gradually. The distributed roughness results in more compression and expansion waves in the flow field.

Flow separation is formed in the boundary layer—a vortex is formed at the leading edges of the isolated roughness, and five more long narrow vortices are formed between two roughnesses of the distributed roughness. Note that the flow lines above the roughness are almost parallel to the wall surface, and it indicates that the shear structure of the region above the roughness is substantially unaffected and the flow smoothly "slips" over the roughness region. Bountin et al. [14] studied the effect of distributed roughness on the undisturbed hypersonic flat boundary layer by means of experimental and numerical simulations. They found similar results as shown in Figure 3 that the flow structure of the strong shear layer above the roughness is not significantly altered while flow separations are formed in the boundary layer. Balakumar and Egorov et al. [33, 34] pointed out that the second mode acoustic disturbance may make the disturbance wave grow exponentially at the upstream and downstream of the separation region but remain neutral in the separation zone, so it is reasonable to believe that the wall surface condition can restrain the disturbance growth in the boundary layer and maintain the stability of the boundary layer to some extent.

Flow separation in the boundary layer can significantly change the mechanical and thermodynamic mechanisms in the boundary layer; Hirschel [35] pointed out that the change of the mechanical and thermodynamic mechanisms in the boundary layer has a great influence on the stability of the boundary layer. The distribution of wall pressure, friction coefficient, and heat flux along the flow direction are given in Figures 4(a), 5, and 6, respectively, and Balakumar's results are given in Figures 4(b) and 4(c) where part of the calculation parameters is given. Figures 4(b) and 4(c) are the results corresponding to isolated roughness and disturbance

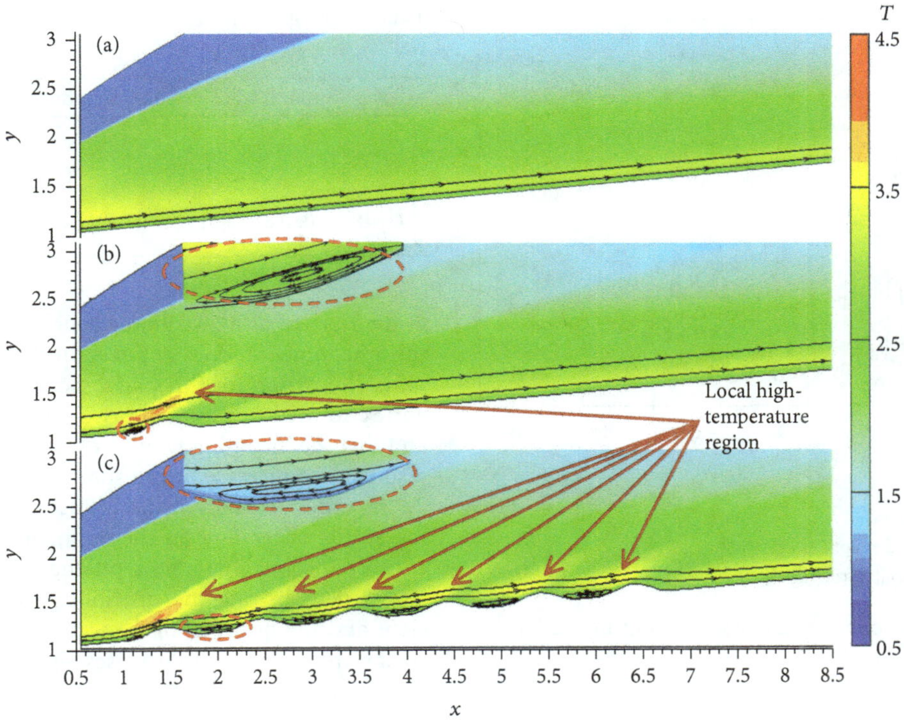

FIGURE 2: Pressure contour and streamline pattern of the undisturbed hypersonic flow field.

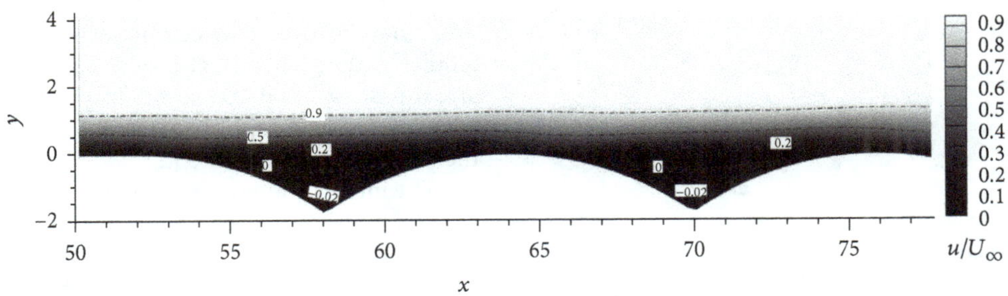

FIGURE 3: Velocity along the x-axis of steady flow in the boundary layer in literature [14].

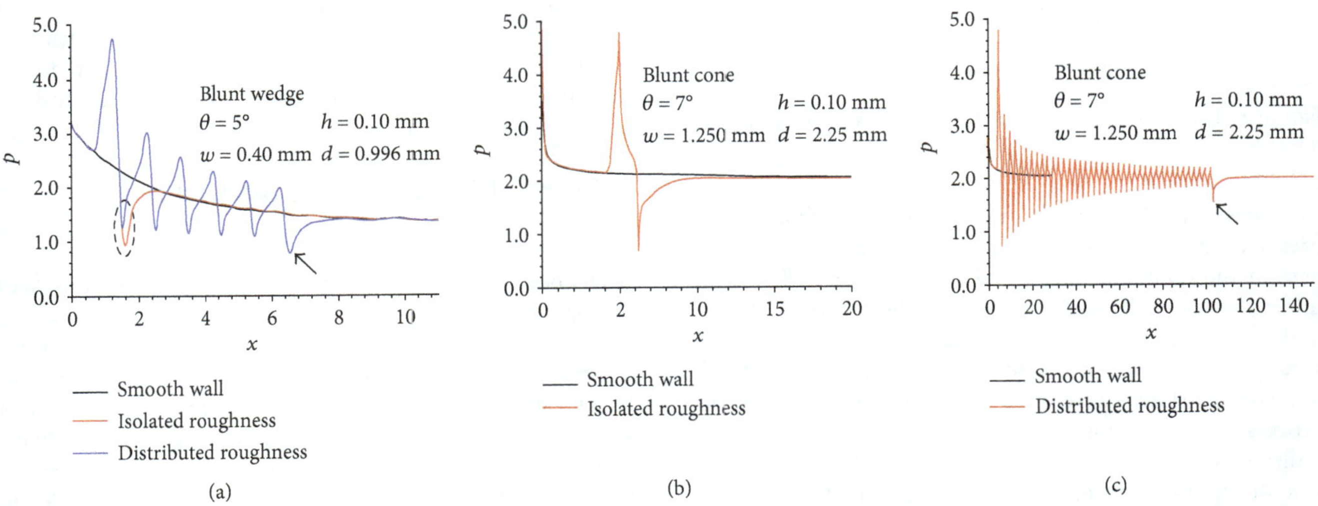

FIGURE 4: Comparison of distribution of wall pressure along the x-axis with the literature [36].

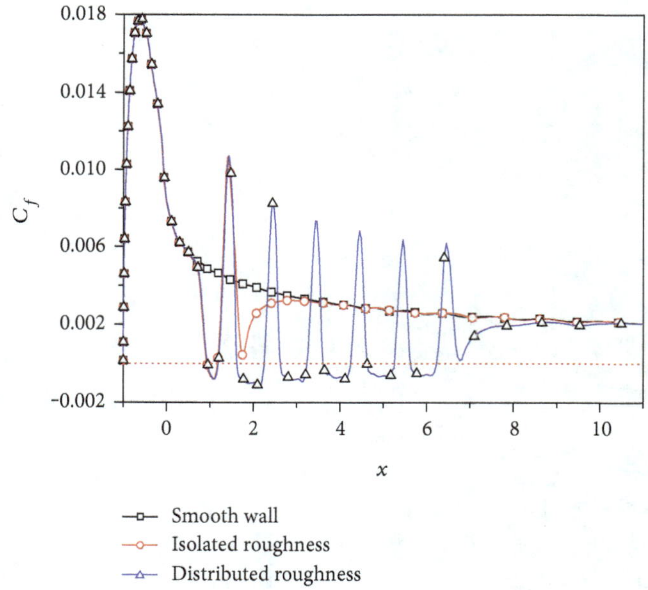

FIGURE 5: The distribution of friction coefficient of the wall along the flow distribution.

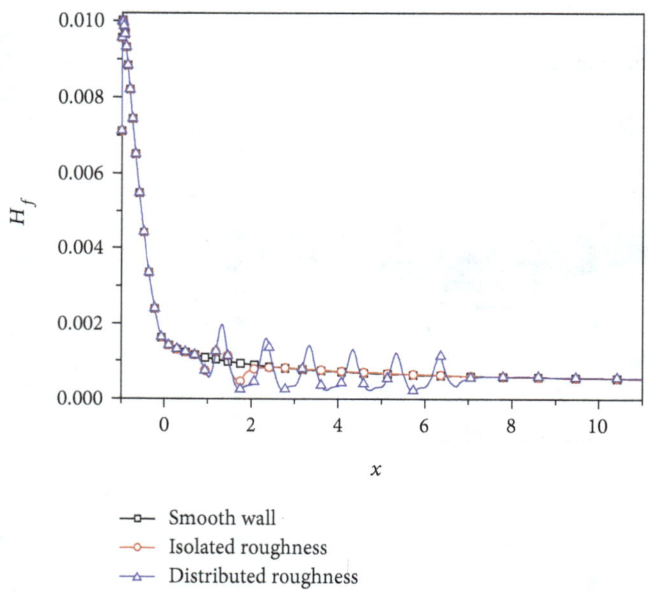

FIGURE 6: The distribution of heat flux of the wall along the flow direction.

roughness ($N = 40$), respectively. It can be seen from the figures that flow structure is significantly changed by the roughness in the boundary layer, and the trend of wall pressure under the two wall conditions is basically the same. As flow in the boundary layer evolves over each roughness, wall pressure increases, then decreases, and finally recovers to the undisturbed state. There is a strong expansion-compression process in the trailing edge region of the last roughness, as indicated by the arrows in the figures. It indicates that there is a strong adverse pressure gradient in this region, which may even result in flow separation. However, there also are some differences between the two kinds of roughness. In this paper, the position and magnitude of the minimum pressure of the isolated and distributed roughnesses are different, but this phenomenon does not exist in the results of Balakumar [33]. The reason for this phenomenon is a long and narrow vortex formed between the two roughnesses in this paper and that the incoming flow meeting the vortex in this region results in an increase in pressure.

TABLE 2: Integral value of friction coefficient and heat flux along the wall.

	Smooth wall	Isolated roughness	Distributed roughness
$\int C_f ds$	0.05488	0.05249	0.04639
$\int H_f ds$	0.01822	0.01817	0.01771

In order to analyze the influence of the roughness on the strong shear flow in the boundary layer and the influence of the incoming flow on the wall heating, Figures 5 and 6 demonstrate the distribution of the friction coefficient and the heat flux of the wall, respectively. The friction coefficient of the wall decreases significantly in the leading edge and the trailing edge of the roughness and reaches the maximum value in the center of the roughness, and the reason for this phenomenon is the change of boundary layer thickness. The friction coefficient is negative in some areas because of the existence of vortices in these areas. So friction drag may reduce to a certain extent caused by cortices. Similarly, the heat flux also decreases significantly in these areas. It shows that the vortex in the hypersonic boundary layer can reduce friction drag and restrain the heating of the wall to a certain extent.

The integral values of the friction coefficient and heat flux along the wall are given in Table 2. It can be seen that the friction drag of the wall and the total heat transfer of the distributed roughness are reduced by 15.47% and 2.80%, respectively, compared with those of the smooth wall. These values are similar to Bountin et al.'s [14] results, which show that distributed roughness reduces the friction drag of the hypersonic flat boundary layer by 17.64%. Through the above analysis, it can be found that arranging the distributed roughness on the wall can achieve the purpose of reducing drag and heat to a certain extent, while not significantly damaging the strong shear flow above the roughness.

5. Effect of Roughness on the Unsteady Flow Field

It has been pointed out above that in the steady flow field, arranging distributed roughness on the wall can reduce friction drag and total heat transfer to a certain extent. But the hypersonic vehicle inevitably encounters acoustic disturbance during the flight [37]. In this paper, a slow acoustic wave with single frequency is introduced as the disturbance and the effect mechanism of roughness on the receptivity of the boundary layer under the action of slow acoustic disturbance is revealed. The contours of velocity along the y-axis under different wall conditions are shown in Figure 7, and Figures 7(a)–7(c) correspond to smooth wall, isolated

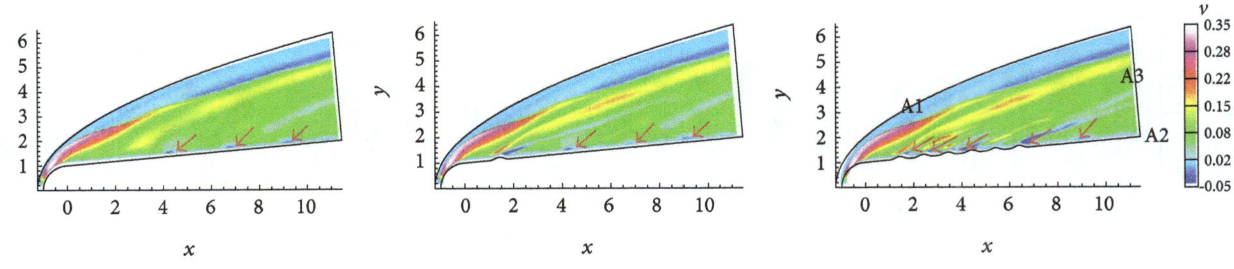

(a) Small wall (b) Isolated roughness (c) Distributed roughness

FIGURE 7: Contours of velocity along the y-axis under different walls.

roughness, and distributed roughness, respectively. It can be seen that the hypersonic flow field under slow acoustic disturbance is mainly divided into three regions: Region A1—the interaction between slow acoustic disturbance and shock wave leads to the shock wave deformation and the formation of a local high-velocity region that is basically unaffected by wall conditions. Region A2—the interaction between a slow acoustic disturbance and the boundary layer results in flow separation in the boundary layer, as shown in the figure. That more separation regions are formed in the boundary layer caused by distributed roughness may be inhibited by the heating of the wall by the flow. Region A3—the region between Regions A1 and A2, relatively speaking, is minimally affected by slow acoustic disturbance, but when distributed roughnesses are arranged on the wall, the region becomes more complex under the influence of slow acoustic disturbance. The hypersonic flow field of Region A3 is significantly changed by the compression and expansion waves induced by the distributed roughness, especially the first compression wave and the last two expansion-compression waves. The interaction between the acoustic disturbance and hypersonic flow field over the sharp cone is studied by Kara [36], and the author found that Region A3 is very quiet. So the author considered that acoustic waves are weakly transmitted through shock. There is a difference between the two results, and we believe that the reason for this difference is the amplitude of disturbance while the amplitude of acoustic waves in literature is 2×10^{-5}. It should be noticed that an inhomogeneity of perturbations is formed near the shock region. The inhomogeneity is caused by the interaction between the freestream disturbances and shock. The freestream disturbances go through the shock and result in the deformation of shock. The deformation of shock causes some tiny perturbations formed near the shock region. Thus, there is an inhomogeneity of perturbations in the oncoming flow. In addition, the great deformation of shock near the nose of the blunt wedge also affects the oncoming flow in the downstream region.

The variation of flow structure in the hypersonic flow field, especially in the boundary layer, has a great influence on the evolution of acoustic disturbance in the boundary layer [38]. Therefore, it should be believed that the wall conditions have a great influence on the receptivity mechanism of the hypersonic boundary layer. The instantaneous pressure disturbance in the boundary layer for acoustic disturbance at different stations is shown in Figure 8. It is worth emphasizing that the magnitude of acoustic disturbance is two to three orders higher than that of acoustic disturbance in freestream and is four to five orders higher than the results of Balakumar [18], even higher. This is because the magnitude of acoustic disturbance is magnified when acoustic waves are transmitted through the shock wave. In this paper, the influence of roughness on the flow in the boundary layer is more concerning, and the influence of isolated and distributed roughnesses on the receptivity is analyzed. Therefore, several stations along the flow direction are selected for analysis: (1) the stations around the isolated roughness or the first roughness of the distributed roughness ($x =$ 1.00427~1.793163, 5 stations), (2) the station between two roughnesses ($x = 1.897203$), (3) downstream region of the first roughness or around the second roughness ($x =$ 2.203272~2.606049, 3 stations), and (4) the downstream of the distributed roughness (x = 7.798937–9.719503, 3 points).

Obviously, the magnitudes of the disturbance wave at different stations under the action of a slow acoustic disturbance obviously change cyclically, and the boundary layer is disturbed by the slow acoustic wave more lately at the downstream station. The evolution of disturbance waves in the boundary layer can simply be divided into two stages: stages T1 and T2, as shown in Figure 8. At stage T1 is the response phase when the disturbance wave starts to affect the flow in the boundary layer and the pressure begins to change. The evolution of disturbance waves is not periodic, and the amplitude of the disturbance wave is small at this stage. At stage T2, the flow field is continuously and steadily affected by the slow acoustic disturbance, and the evolution of disturbance waves is periodic. The wall condition remarkably changes the evolution of the slow acoustic disturbance, and the magnitude of the acoustic disturbance becomes bigger at this stage.

After the acoustic disturbance enters the flow field through the bow shock wave, the disturbance propagates to the upstream-half region of roughness, ($x = 1.001427$~ 1.402937). The magnitude of disturbance waves increases, and the growth of disturbance is promoted by roughness in the boundary layer. Growth rate increases with the increase of wall curvature. For the downstream-half region of the roughness ($x = 1.402937$~1.793163), the magnitude of disturbance waves decreases. The growth of acoustic disturbance in the boundary layer is suppressed in this region and is suppressed with the increase of curvature. It should be noted that the magnitude of the disturbance

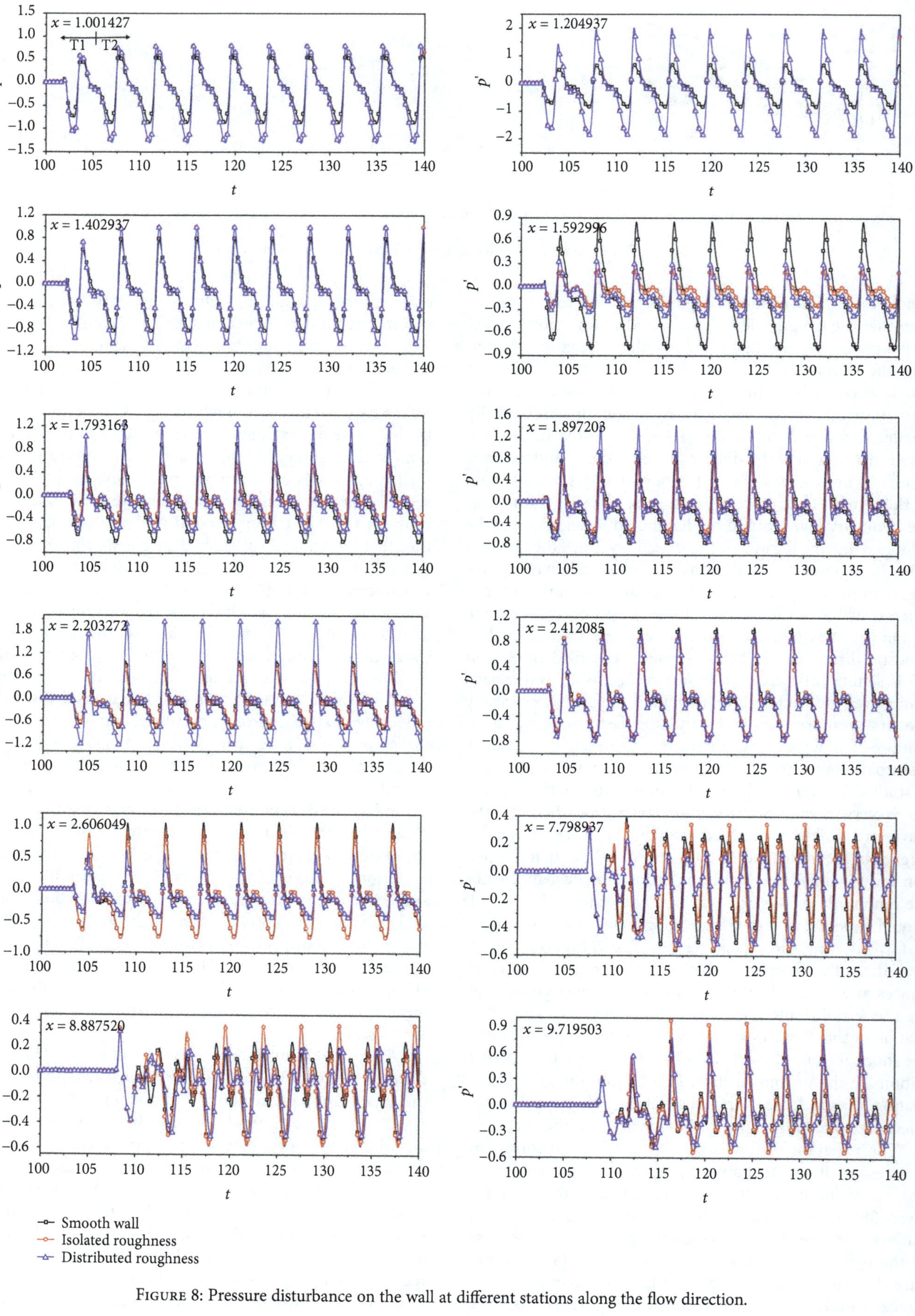

FIGURE 8: Pressure disturbance on the wall at different stations along the flow direction.

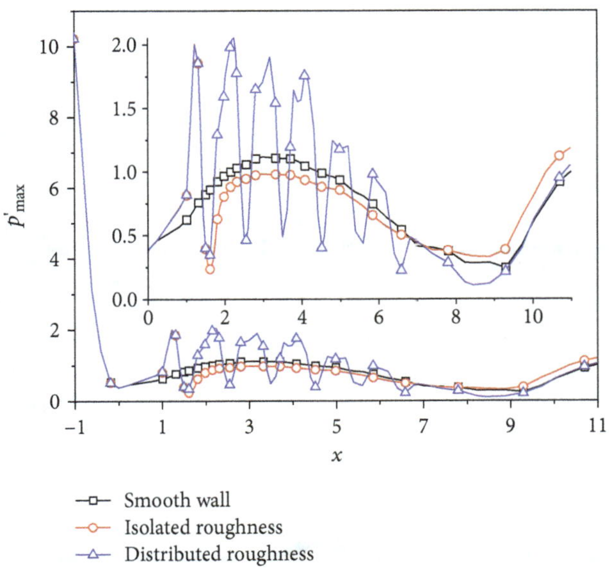

FIGURE 9: The distribution of maximum pressure disturbances along the flow direction.

wave corresponds to distributed roughness increases at the station ($x = 1.793163$). That indicates that the evolution of the acoustic disturbance at this station is affected by the downstream roughness or the vortex between two roughnesses. Flow is compressed at this position, and the magnitude of the acoustic disturbance increases. At $x = 1.897203$, the magnitude of disturbance waves corresponds to the isolated roughness which continues to decrease, but the magnitude of the disturbance wave corresponds to distributed roughness increases. It is worth noting that at this point, the pressure of the wall with distributed roughness or without roughness is the same in the undisturbed flow field. It indicates that the vortex amplifies the magnitude of the acoustic disturbance in the boundary layer. In the range of $x = 2.203272$ to 2.606049, the magnitude of the disturbance waves for isolated roughness decreases continuously. The magnitude variation trend of pressure disturbance of the distributed roughness is the same as that of the last upstream roughness which shows the process of enlargement-reduction during flow over the roughness. The magnitude of disturbance waves for distributed roughness is the same as that of the smooth wall at the center of roughness. The evolution of disturbance waves in the downstream region of the distributed roughness in the boundary layer is more concerning. In the range of $x = 7.798937$ to 9.719503, the acoustic disturbance becomes more complicated. It indicates that the modes of the disturbance waves in the boundary layer increase and there is the existence of competition between different mode disturbances in the boundary layer. The modal analysis of the acoustic disturbance in the boundary layer will be given in the next section. At $x = 7.798937$ and 8.887520, the variation of disturbance wave magnitude of the distributed roughness is small, and the growth of the disturbance wave is restrained. At $x = 9.719503$, the restraining effect weakens, and the magnitude of the disturbance wave for the distributed roughness approximates that of the smooth wall.

It is worth noting that in the range of $x = 7.798937$ to 9.719503, the magnitude of the disturbance waves for the isolated roughness is always high. That is to say, the disturbance waves for the isolated roughness are firstly restrained, then grow, and gradually exceed those for the smooth wall and distributed roughness.

Figure 9 shows the distribution of the maximum values of the magnitude of disturbance waves at different wall conditions along the flow direction. It can be seen that the acoustic disturbance is the magnitude in the upstream-half region of roughness and suppressed in the downstream-half region, and the intensity of amplification or inhibition is enhanced with the increase or decrease of the wall curvature.

In addition, the vortex also amplifies the acoustic disturbance in the boundary layer, which is the strongest at the center of the vortex.

6. Analysis of Disturbance Wave Modes in Boundary Layers

In order to reveal the influence of roughness on the receptivity of the boundary layer clearly, the evolution process of acoustic disturbances with different modes in the boundary layer is analyzed.

Fast Fourier spectrum analysis (FFSA) is used to decompose the time domain signal in the boundary layer and expand it in the frequency domain to get the evolution process of the acoustic disturbance with different frequencies. This method is widely used to analyze receptivity of the hypersonic boundary layer with roughness under freestream disturbances, and satisfactory results are attained [21, 39]. The Fourier transform formula is as follows:

$$p'(x, y, t) = \mathrm{Re}\left(\sum_{n=0}^{N} \left|p'_n(x, y)\right| e^{i[-n\omega t+\varphi n(x,y)]}\right). \tag{9}$$

Figure 10 shows the FFSA amplitude-frequency diagram at different stations, and it shows that under the slow acoustic wave with single frequency, the dominant modes in the boundary layer are fundamental frequency and harmonics while the other modes can be insignificant. In the region of $x = 1.402937$ to 1.793163, the fundamental frequency (frequency of the fundamental frequency mode f is 0.25, frequency of the second-order harmonic disturbance f is 0.50, frequency of the third-order harmonic disturbance f is 0.75, etc.) and each harmonic order mode are amplified in the upstream-half region of roughness. The dominant mode in the boundary layer is taken as the fundamental frequency mode under the three wall conditions, and the proportion of each order harmonic mode decreases with the increase of frequency. The evolution of fundamental and each order harmonic mode is restrained in the downstream-half region of the roughness. For the smooth wall, the fundamental mode is still the dominant mode, but for the rough wall, the dominant mode becomes the second-order harmonic mode. That is to say, in this region, the roughness can restrain the fundamental mode better than the second-order harmonic.

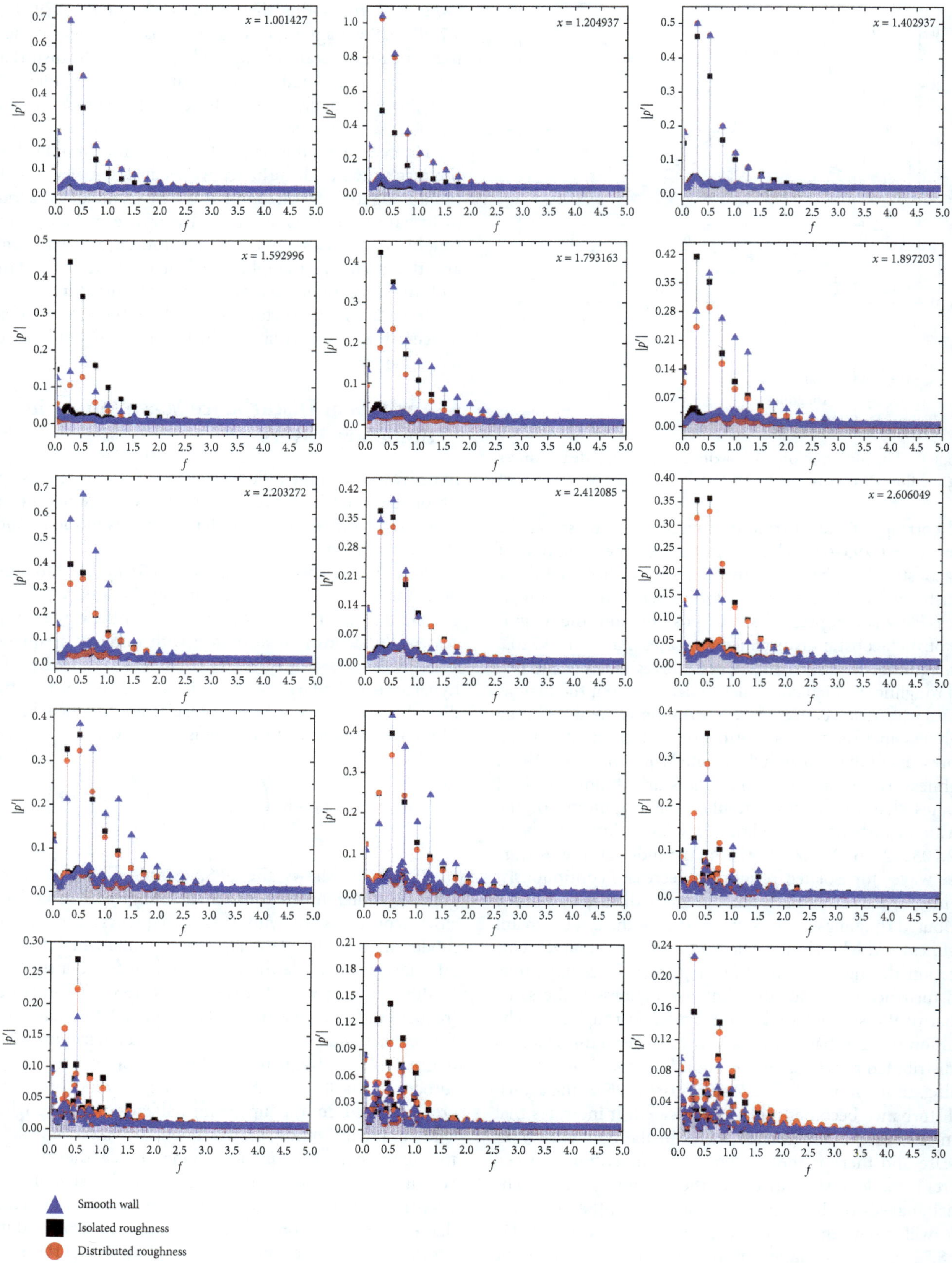

FIGURE 10: Comparison of Fourier frequency spectral analysis of pressure disturbance in the boundary layer for different wall conditions.

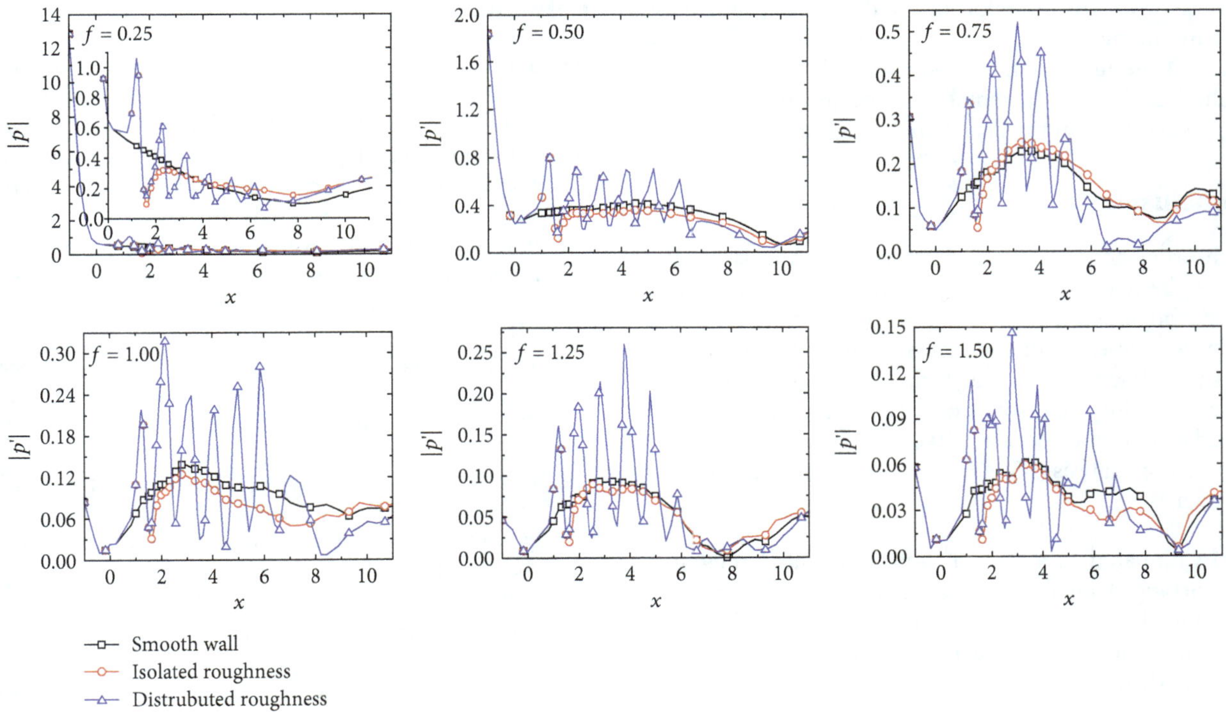

FIGURE 11: Comparison of evolution of different modes along the wall for different wall conditions.

At $x = 1.897203$, the FFSA magnitude of the fundamental mode and each order harmonic mode of the isolated roughness are smaller than those of the smooth wall and distributed roughness wall, and the FFSA magnitude of each order harmonic disturbance for the distributed roughness is larger than that of the smooth wall except the fundamental mode. It implies that for the distributed roughness, the fundamental mode is suppressed in the vortex region, but the evolution of each order harmonic mode is promoted. The same phenomenon occurs at the next two vortices ($x = 2.860089$ and 3.891759). In the region of $x = 2.203272$ to 2.606049, the trends of the fundamental mode and harmonic mode of each order are similar to those of the smooth wall but smaller except for the third-order harmonic mode at $x = 2.606049$. It is noteworthy that at $x = 2.606049$, the dominant mode for the smooth wall transforms from the fundamental mode to the second-order harmonic mode. The fundamental mode and harmonic mode for the distributed roughness are suppressed and then amplified in this region. The dominant mode is always the second-order harmonic frequency mode, but its proportion is decreasing gradually, while the proportion of the harmonic mode of each order is increasing gradually.

For the downstream region of distributed roughness, the second-order harmonic mode is still the dominant mode in the range of $x = 6.817512$ to 7.798937 under different wall conditions, but the magnitude of the distributed roughness is significantly smaller than that of the other two types of wall conditions; even the third-order and fifth-order harmonic modes tend to disappear. That is to say, the evolution of the acoustic disturbances in the boundary layer is obviously inhibited in the downstream region of the distributed roughness, which does not last for a long distance. At the station of $x = 8.888520$ and 9.719503, the FFSA magnitude of the acoustic disturbances of the distributed roughness gradually increases, especially the fundamental mode, which rapidly increases and becomes the dominant mode in the boundary layer.

Figure 11 shows the distribution of the fundamental mode and harmonic modes from the second order to the sixth order in the boundary layer along the flow direction. It can be seen that the fundamental frequency and each order harmonic mode decay rapidly in the head region. The degree of attenuation of different modes becomes greater with the frequency decrease.

The variation trend for different wall conditions on the fundamental mode and harmonic mode is basically the same, but there are still some differences. In general, different modes are amplified in the upstream-half region of the roughness and suppressed in the downstream-half region of the roughness, and this suppression will last for a distance. The different modes are suppressed for different distances; some disturbance modes are suppressed for a long distance, while others are short.

For the isolated roughness, the second-order harmonic, fourth-order harmonic, and sixth-order harmonic modes in the boundary layer are suppressed for a long distance while the fundamental mode and the third-order harmonic mode are suppressed by isolated roughness for a short distance in the boundary layer. Especially for the fundamental mode, the fundamental mode corresponding to the isolated roughness is always higher than that for the smooth wall after the inhibition effect of roughness is not effective.

In the downstream region of the distributed roughness, the harmonic modes of each order are suppressed, but the increase of the fundamental mode in the boundary layer is not inhibited. The fundamental mode begins to grow rapidly after passing through the distributed roughness.

7. Conclusions

A high-order accurate finite difference method is used to simulate the hypersonic flow field over a blunt wedge with isolated roughness and distributed roughness or without roughness. The effect of roughness on steady and unsteady hypersonic flow fields is analyzed. The effect of roughness on the evolution of disturbance waves caused by a slow acoustic wave in the boundary layer is discussed by fast Fourier spectrum analysis (FFSA) method, and some conclusions are drawn.

(1) Compared to a smooth wall, distributed roughness arranged on the wall can reduce friction drag and total heat transfer to a certain extent, while the strong shear flow above the roughnesses does not significantly damage.

(2) The amplitude of disturbance waves under the fundamental mode and the harmonic mode of each order increases markedly in the upstream-half region of roughness while decreases markedly in the downstream-half region of roughness. Disturbance waves are suppressed in the downstream region of isolated and distributed roughnesses for a distance. Some disturbance modes are suppressed for a long distance, while others are short.

(3) The mode competition always exits among different modes both in the temporal domain and in the frequency domain in the boundary layer. The dominant mode is changed by roughness: the fundamental mode is the dominant mode until $x = 2.606049$ for a smooth wall in the boundary layer while the dominant mode is changed to the second-order harmonic mode at $x = 1.592996$ for a rough wall.

(4) The evolution of disturbance waves in the boundary layer is affected by the vortices. The vortex amplifies the magnitude of the acoustic disturbance in the boundary layer: the fundamental mode is suppressed in the vortex region, but the evolution of each order harmonic mode is promoted.

Conflicts of Interest

The authors declare that there is no conflict of interests regarding the publication of this paper.

Acknowledgments

The authors gratefully acknowledge the financial support of the National Natural Science Foundation of China (Grant nos. 11472086 and 11532013).

References

[1] S. P. Schneider, "Flight data for boundary-layer transition at hypersonic and supersonic speeds," *Journal of Spacecrafts and Rockets*, vol. 36, no. 1, pp. 8–20, 1999.

[2] S. P. Schneider, "Hypersonic laminar–turbulent transition on circular cones and scramjet forebodies," *Progress in Aerospace Sciences*, vol. 40, no. 1-2, pp. 1–50, 2004.

[3] P. J. P. Jeroen, V. D. Eynde, and N. D. Sandham, "Numerical simulations of transition due to isolated roughness elements at Mach 6," *AIAA Journal*, vol. 54, no. 1, pp. 53–65, 2016.

[4] Y. B. Ma and X. L. Zhong, "Receptivity of a supersonic boundary layer over a flat plate. Part 2. Receptivity to free-stream sound," *Journal of Fluid Mechanics*, vol. 488, pp. 79–121, 2003.

[5] Y. B. Ma and X. L. Zhong, "Receptivity of a supersonic boundary layer over a flat plate. Part 3. Effects of different types of free-stream disturbances," *Journal of Fluid Mechanics*, vol. 532, pp. 63–109, 2005.

[6] P. Balakumar and M. A. Kegerise, "Receptivity of hypersonic boundary layers over straight and flared cones," *AIAA Journal*, vol. 53, no. 8, pp. 2097–2109, 2015.

[7] A. Cerminara and N. D. Sandham, "Acoustic leading-edge receptivity for supersonic/hypersonic flows over a blunt wedge," *AIAA Journal*, vol. 55, no. 12, pp. 4234–4244, 2017.

[8] D. Park and S. O. Park, "Influence of two-dimensional smooth humps on linear and non-linear instability of a supersonic boundary layer," *Computers & Fluids*, vol. 79, pp. 140–149, 2013.

[9] M. Olaf, I. Gianluca, and E. S. G. Shaqfeh, "Nonlinear instability of a supersonic boundary layer with two-dimensional roughness," *Journal of Fluid Mechanics*, vol. 752, pp. 497–520, 2014.

[10] F. Qin and X. Wu, "Response and receptivity of the hypersonic boundary layer past a wedge to free-stream acoustic, vortical and entropy disturbances," *Journal of Fluid Mechanics*, vol. 797, pp. 874–915, 2016.

[11] Y. D. Zhang, D. X. Fu, Y. W. Ma, and X. L. Li, "Receptivity to free-stream disturbance waves for hypersonic flow over a blunt cone," *Science in China Series G: Physics, Mechanics and Astronomy*, vol. 51, no. 11, pp. 1682–1690, 2008.

[12] R. L. Kimmel, "Aspects of hypersonic boundary layer transition control," in *41st Aerospace Sciences Meeting and Exhibit*, Reno, NV, USA, January 2003.

[13] L. W. Chen, G. L. Wang, and X. Y. Lu, "Numerical investigation of a jet from a blunt body opposing a supersonic flow," *Journal of Fluid Mechanics*, vol. 684, no. 11, pp. 85–110, 2011.

[14] D. Bountin, T. Chimitov, A. Maslov, A. Novikov, I. Egorov, and F. Alexander, "Stabilization of a hypersonic boundary layer using a wavy surface," *AIAA Journal*, vol. 51, no. 5, pp. 1203–1210, 2013.

[15] A. V. Fedorov, N. D. Malmuth, A. Rasheed, and H. G. Hornung, "Stabilization of hypersonic boundary layers by porous coatings," *AIAA Journal*, vol. 39, no. 4, pp. 605–610, 2001.

[16] A. Rasheed, H. G. Hornung, A. V. Fedorov, and N. D. Malmuth, "Experiments on passive hypervelocity boundary layer control using an ultrasonically absorptive surface," *AIAA Journal*, vol. 40, no. 3, pp. 481–489, 2002.

[17] N. D. Sandham and H. Lüdeke, "Numerical study of Mach 6 boundary-layer stabilization by means of a porous surface," *AIAA Journal*, vol. 47, no. 9, pp. 2243–2252, 2009.

[18] P. Balakumar, "Receptivity of hypersonic boundary layers to distributed roughness and acoustic disturbances," in *51st AIAA Aerospace Sciences Meeting including the New Horizons Forum and Aerospace Exposition*, Dallas, TX, USA, January 2012.

[19] A. Sescu and D. Thompson, "On the excitation of Görtler vortices by distributed roughness elements," *Theoretical and Computational Fluid Dynamics*, vol. 29, no. 1-2, pp. 67–92, 2015.

[20] X. Wang and X. Zhong, "Receptivity of a hypersonic flat-plate boundary layer to three-dimensional surface roughness," *Journal of Spacecraft and Rockets*, vol. 45, no. 6, pp. 1165–1175, 2008.

[21] K. D. Fong, X. Wang, and X. Zhong, "Numerical simulation of roughness effect on the stability of a hypersonic boundary layer," *Computers & Fluids*, vol. 96, no. 96, pp. 350–367, 2014.

[22] Y. Zhao, W. Liu, D. Xu, D. Gang, and S. Yi, "A combined experimental and numerical investigation of roughness induced supersonic boundary layer transition," *Acta Astronautica*, vol. 118, pp. 199–209, 2016.

[23] P. Holloway and J. Sterett, "Effect of controlled surface roughness on boundary-layer transition and heat transfer at Mach number of 4.8 and 6.0," Tech. rep. D-2054. NASA Technical Note, Hampton, VA, USA, 1964.

[24] K. Fujii, "Experiment of the two-dimensional roughness effect on hypersonic boundary-layer transition," *Journal of Spacecraft and Rockets*, vol. 43, no. 4, pp. 731–738, 2006.

[25] L. Duan, X. Wang, and X. Zhong, "A high-order cut-cell method for numerical simulation of hypersonic boundary-layer instability with surface roughness," *Journal of Computational Physics*, vol. 229, no. 19, pp. 7207–7237, 2010.

[26] L. Duan, X. Wang, and X. Zhong, "Stabilization of a Mach 5.92 boundary layer by two-dimensional finite-height roughness," *AIAA Journal*, vol. 51, no. 1, pp. 266–270, 2013.

[27] C. Desjouy, S. Ollivier, O. Marsden, M. Karzova, and P. Blanc-Benon, "Irregular reflection of weak acoustic shock pulses on rigid boundaries: Schlieren experiments and direct numerical simulation based on a Navier-Stokes solver," *Physics of Fluids*, vol. 28, no. 2, article 027102, 2016.

[28] W. S. Saric, R. B. Carrillo Jr, and M. S. Reibert, "Leading-edge roughness as a transition control mechanism," in *36th AIAA Aerospace Sciences Meeting and Exhibit*, Reno, NV, USA, 1998.

[29] G. S. Jiang and C. W. Shu, "Efficient implementation of weighted ENO schemes," *Journal of Computational Physics*, vol. 126, no. 1, pp. 202–228, 1996.

[30] C. W. ShuA. Quarteroni, "Essentially non-oscillatory and weighted essentially non-oscillatory schemes for hyperbolic conservation laws," in *Advanced Numerical Approximation of Nonlinear Hyperbolic Equations*, vol. 1697 of Lecture Notes in Mathematics, pp. 325–432, Springer, Berlin, Heidelberg, 1998.

[31] D. S. Balsara, S. Garain, and C. W. Shu, "An efficient class of WENO schemes with adaptive order," *Journal of Computational Physics*, vol. 326, pp. 780–804, 2016.

[32] J. Shi, X. Tang, Z. Wang, M. Shi, and W. Zhao, "Receptivity of boundary layer over a blunt wedge due to freestream pulse disturbances at Mach 6," *International Journal of Aerospace Engineering*, vol. 2016, Article ID 3196057, 14 pages, 2016.

[33] P. Balakumar, "Stability of hypersonic boundary-layers over a compression corner," in *32nd AIAA Fluid Dynamics Conference and Exhibit*, St. Louis, MO, USA, June 2002.

[34] I. V. Egorov, A. V. Fedorov, and A. V. Novikov, "Numerical modeling of the disturbances of the separated flow in a rounded compression corner," *Fluid Dynamics*, vol. 41, no. 4, pp. 521–530, 2006.

[35] E. H. Hirschel, *Basics of Aerothermodynamics*, Springer, Berlin Heidelberg, 2005.

[36] K. Kara, "Receptivity of hypersonic boundary layers due to acoustic disturbances over blunt cone," in *45th AIAA Aerospace Sciences Meeting and Exhibit*, Reno, NV, USA, January 2007.

[37] Y. Bandadi and A. Sbaiba, *Localized Boundary Layer Receptivity To a Free Stream Acoustic Disturbance*, 23ème Congrès Français de Mécanique, Lille, France, 2017.

[38] X. Liang, X. L. Li, D. X. Fu, and Y. Ma, "Effects of wall temperature on boundary layer stability over a blunt cone at Mach 7.99," *Computers & Fluids*, vol. 39, no. 2, pp. 359–371, 2010.

[39] D. Fong, X. Wang, and X. Zhong, "Finite roughness effect on modal growth of a hypersonic boundary layer," in *50th AIAA Aerospace Sciences Meeting including the New Horizons Forum and Aerospace Exposition*, Nashville, TN, USA, January 2012.

9

Free Vibration Analysis of Rotating Pretwisted Functionally Graded Sandwich Blades

L. T. Liu,[1] Y. X. Hao,[1] W. Zhang,[2] and J. Chen[3]

[1]*College of Mechanical Engineering, Beijing Information Science and Technology University, Beijing 100192, China*
[2]*College of Mechanical Engineering, Beijing University of Technology, Beijing 100124, China*
[3]*Architecture and Civil Engineering Research Center, Shenzhen Research Institute, City University of Hong Kong, Shenzhen, China*

Correspondence should be addressed to Y. X. Hao; bimhao@163.com

Academic Editor: Ratneshwar Jha

A new structural dynamic model for the free vibration characteristic analysis of rotating pretwisted functionally graded (FG) sandwich blades is developed. The sandwich blade is made up of two functionally graded skins and a homogeneous material core. The thick shell theory is applied to derive the basic equations of motion of the rotating FG sandwich blade by considering the effects of centrifugal and Coriolis forces. The mode shapes are expanded in terms of two-dimensional algebraic polynomials in the Rayleigh–Ritz method, and the static and dynamic natural frequencies of the blade are obtained. The convergence analysis is studied, and the accuracy of the proposed model is verified by comparing with the literature results and ANSYS data. The effects of frequency parameters such as the twist angle, the thickness ratio, the aspect ratio, the layer thickness ratio, the scalar parameter of volume fraction, the stagger angle, and the rotation velocity on the vibration characteristics for pretwist FG sandwich blade are investigated in detail. In addition, the phenomena of frequency locus veering and mode shape exchanging occur in the static and dynamic states. Frequency locus veering is essentially caused by the coupling between different modes.

1. Introduction

The aero gas turbine engine is the key technology of the aircraft; blade is one of the most important parts of the system. Vibration, especially resonance of the blade, will produce larger stress, which will lead to fatigue failure. Thus, the dynamic behavior of rotating blades has been studied by numerous researchers.

Some simplified blade models such as beams, plates, and shells are used to investigate the vibration characteristics of the blades in the published literatures. Leissa and Jacob [1] investigated the free vibration of pretwisted, cantilevered beams and plates by using the Ritz method. The work was the first three-dimensional study of the problem. Yoo et al. [2] derived the equations of motion with a concentrated mass by using a modeling method with hybrid deformation variables and investigated the effects of the nondimensional parameters on the vibration characteristics of the pretwisted rotating blade through numerical analysis. Chandiramani et al. [3] simplified the pretwisted rotating blade as a laminated composite, hollow uniform box-beam model, which considers the centrifugal and Coriolis effects, transverse shear flexibility, and restrained warping, and studied the free and forced vibration by using HSDT and Galerkin method. Carrera et al. [4] used the Carrera unified formulation and FEM to study the free vibration analysis of rotating blades; the Coriolis and centrifugal force fields were included in their work.

The beam models are quite suitable for blades with large aspect ratio and low width thickness ratio. However, flexible rotating structures such as blades with low aspect ratios are also widely used in actual engineering applications. Therefore, more and more literatures were found to study the vibration of blade with plate and shell models.

Qatu and Leissa [5] were the first to study the effect of plate parameters such as twist angle, stacking sequence, and lamination angle on the natural frequency and mode shapes of laminated composite pretwisted cantilever plates by using the shallow shell theory and Ritz method. Nabi and Ganesan [6] analyzed the vibration characteristics of metal matrix composite pretwisted blades by using beam and plate theories, respectively, and summed up the quantitative comparison of natural frequency. Yoo and Chung [7] developed a linear dynamic modeling method for plates by using two in-plane stretch variables and analyzed and studied the transient characteristics of rotating plates. Hu et al. [8] proposed a numerical procedure for the free vibration analysis of pretwisted thin plates based on the thin shell theory and studied the vibration characteristics considering different twist rates and aspect ratios. Hashemi et al. [9] used the Mindlin plate theory and Kane dynamic method to develop a finite element formulation for vibration analysis of rotating thick plates. The coupling between in-plane and out-of-plane deformations and Coriolis effect was considered in the study. Sinha and Turner [10] derived the governing partial differential equation of motion for the rotating pretwisted plate by using the thin shell theory and studied the free vibration of a turbomachinery cantilevered airfoil blade with the Rayleigh–Ritz technique by considering the centrifugal force filed as a quasi-static load. Sun et al. [11] presented a dynamic model based on CLPT, investigated the vibration behavior of a rotating blade by using Hamilton's principle, and studied the point and distribution forced response using a proportional damping model.

Rao and Gupta [12] studied the vibration of natural frequencies with various parameters of a rotating pretwisted blade with a small aspect ratio by using the classical bending theory of thin shells. Sun et al. [13] developed a novel dynamic model for pretwisted rotating compressor blades by using the general shell theory and studied the eigenfrequencies and damping properties of the pretwisted rotating blade. Sinha and Zylka [14] derived the governing partial differential equation of motion based on the thin shell theory and formulated the free vibration of a turbomachinery cantilevered airfoil by considering it as an anisotropic shell in a centrifugal force field. Kielb et al. [15] gave the complete theoretical and experimental results of a joint research study on the vibration characteristics of pretwisted cantilever plates.

Claassen and Thorne [16] were the first to find out and discuss the curve veering in the plate vibration based on CLPT. Afolabi and Mehmed [17] studied the curve veering and flutter of rotating blades based on the theory of algebraic curves and catastrophe theory. They found that the frequency loci of undamped rotating blades do not cross, but must either repel each other or attract each other. Yoo and Pierre [18, 19] studied the vibration characteristics of rotating cantilever rectangular plates and composite plates, respectively, and found out and discussed the natural frequency locus veering, crossing, and associated mode shape variations in detail.

Most of the early studies were focused on homogeneous materials, and in recent years, some researchers have extended the range to composite materials because of their excellent properties. The structure was developed from single layer to multilayer and composite layer, and more and more literature has been published.

Oh et al. [20] studied the vibration of turbomachinery rotating blades made up of functionally graded materials and investigated the influence of parameters on the frequency of rotating blades. A refined dynamic theory involving the coupling between flapping, lagging, and transverse shear was used in their work. Ramesh and Rao [21] studied the natural frequencies of a pretwisted rotating FG cantilever beam by using Lagrange's equation and the Rayleigh–Ritz method. The influences of different parameters and coupling between chordwise and flapwise bending modes on the natural frequency were investigated. Oh and Yoo [22] presented a new structural dynamic model to study the vibration characteristics of rotating blades. The blade was modeled as a functionally graded material pretwisted beam by using the Rayleigh–Ritz and Kane method. Mantari and Granados [23] studied the free vibration of FG plates by using a new novel FSDT with 4 unknowns. Li and Zhang [24] developed a dynamic model for FG plates undergoing large overall motions, studied the free vibration of rotating cantilever FGM rectangular plates, and found frequency locus veering and associated mode shift phenomena. Sun et al. [25] developed a novel dynamic model for multilayer blades by using the quadratic layerwise theory and studied the structural dynamics, particularly the damping properties of the rotating blade. Frequency locus veering with rotation speed was also found. Cao et al. [26] developed a rotating cantilever pretwisted sandwich plate model with a prestagger angle and investigated the vibrational behavior of a turbine blade with thermal barrier coating layers by using FSDT, von Karman plate theory, and Chebyshev–Ritz method.

In the present paper, a new structural dynamic model is developed to study the free vibration characteristics of pretwisted rotating FG sandwich blades. The thick shell theory is applied to derive the basic equations of motion by considering the effects of centrifugal and Coriolis forces. The accuracy of the proposed model is verified by comparing with the literature and ANSYS. The effects of frequency parameters such as the twist angle, the thickness ratio, the aspect ratio, the layer thickness ratio, the scalar parameter of volume fraction, the stagger angle, and the rotation velocity on the vibration characteristics are investigated in detail. Furthermore, frequency locus veering and mode shape exchanging phenomena are found both in both static and dynamic states.

2. Theoretical Analysis

2.1. Basic Equations. As shown in Figure 1, the functionally graded sandwich blade is modeled as a cantilever pretwisted thick plate with the twist angle θ at the free end, which is clamped to the rigid disk with a radius R, mounted with a stagger angle φ. The geometric parameters of the blade are the span (length) dimension L, the chord (width) dimension b, the total thickness h, and the angular rotating velocity about the rigid disk axis Ω.

In this paper, three coordinate systems are established for dynamic modeling. One is the XYZ-coordinate system (with unit vectors $(\mathbf{i}_X, \mathbf{i}_Y, \mathbf{i}_Z)$), where the x-axis is along the spanwise direction of the blade, Y is the rotation axis, and the Z-axis is perpendicular to the XY-plane following the right-hand rule. The other is the xyz-coordinate system (with unit vectors $(\mathbf{i}_x, \mathbf{i}_y, \mathbf{i}_z)$), where the origin lies in the root of the mid-surface of the blade and the x-axis is parallel to the X-axis; the y-axis can be obtained by rotating around the x-axis with the stagger angle φ starting from the Y-axis

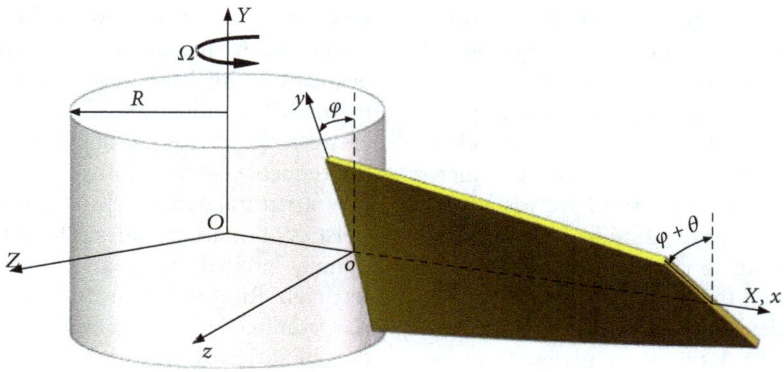

FIGURE 1: A typical pretwisted rotating blade model.

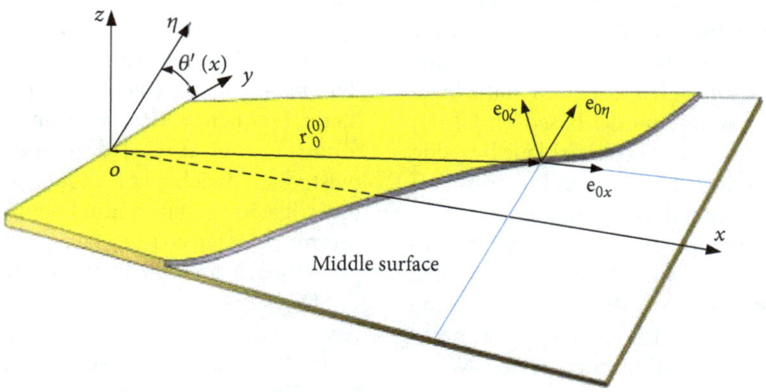

FIGURE 2: The dynamic coordinate system of the pretwisted rotating blade.

direction. Figure 2 shows the last $x\eta$ coordinate, a dynamic surface coordinate where the η-axis is perpendicular to the x-axis and equal to the y-axis at $x = 0$, rotating at a uniform twist rate $k = \theta/L$. Thus, the rotating angle at x can be expressed as

$$\theta'(x) = kx, \quad x \in [0, L]. \tag{1}$$

The position vector $\mathbf{r}_0^{(0)}(x, \eta)$ of a typical point on the mid-surface of the pretwisted blade in the $x\eta$-coordinate can be written as

$$\mathbf{r}_0^{(0)}(x, \eta) = \begin{bmatrix} x \\ \eta \cos\theta' \\ \eta \sin\theta' \end{bmatrix}^T \begin{pmatrix} \mathbf{i}_x \\ \mathbf{i}_y \\ \mathbf{i}_z \end{pmatrix}. \tag{2}$$

Based on the Frenet–Serret formula [28], the corresponding unit base vectors can be expressed as

$$\begin{pmatrix} \mathbf{e}_{0x} \\ \mathbf{e}_{0\eta} \\ \mathbf{e}_{0\zeta} \end{pmatrix} = \frac{1}{A} \begin{bmatrix} 1 & -k\eta \sin\theta' & k\eta\cos\theta' \\ 0 & A\cos\theta' & A\sin\theta' \\ -k\eta & -\sin\theta' & \cos\theta' \end{bmatrix} \begin{pmatrix} \mathbf{i}_x \\ \mathbf{i}_y \\ \mathbf{i}_z \end{pmatrix}, \tag{3}$$

where A is the Lame parameter of the middle surface in the x-direction, given by

$$A = \sqrt{1 + k^2\eta^2}. \tag{4}$$

Then, an arbitrary point of the blade before deformation can be expressed by a position vector $\mathbf{r}^{(0)}$ as

$$\mathbf{r}^{(0)} = \mathbf{r}_0^{(0)} + \zeta \mathbf{e}_{0\zeta} = \begin{bmatrix} x - k\eta \dfrac{\zeta}{A} \\ \eta\cos\theta' - \zeta \sin\dfrac{\theta'}{A} \\ \eta \sin\theta' + \zeta\cos\dfrac{\theta'}{A} \end{bmatrix}^T \begin{pmatrix} \mathbf{i}_x \\ \mathbf{i}_y \\ \mathbf{i}_z \end{pmatrix}. \tag{5}$$

As illustrated in Figure 3, the core of the blade ($h_1 \le \zeta \le h_2$) is fully metal (isotropic) and two skins are composed of functionally graded material (FGM) in the thickness direction. The top skin varies from ceramic-rich surface ($\zeta = h_3$) to metal-rich surface ($\zeta = h_2$), while the bottom skin varies from ceramic-rich surface ($\zeta = h_0$) to metal-rich surface ($\zeta = h_1$). There are no interfaces between the core and skins of the functionally graded sandwich blade, and two skins are symmetrical about the middle

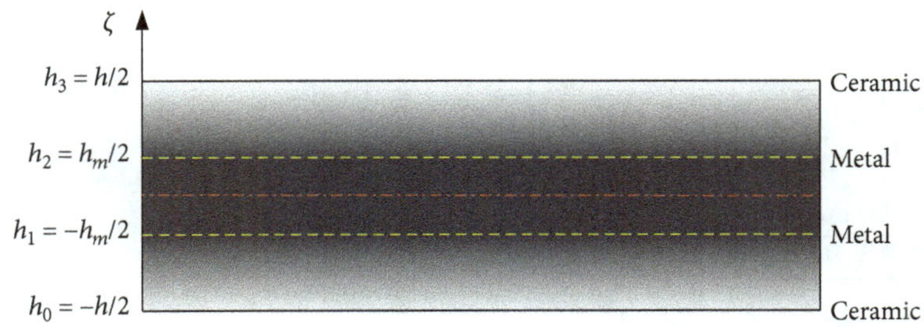

FIGURE 3: The section of the FG sandwich blade.

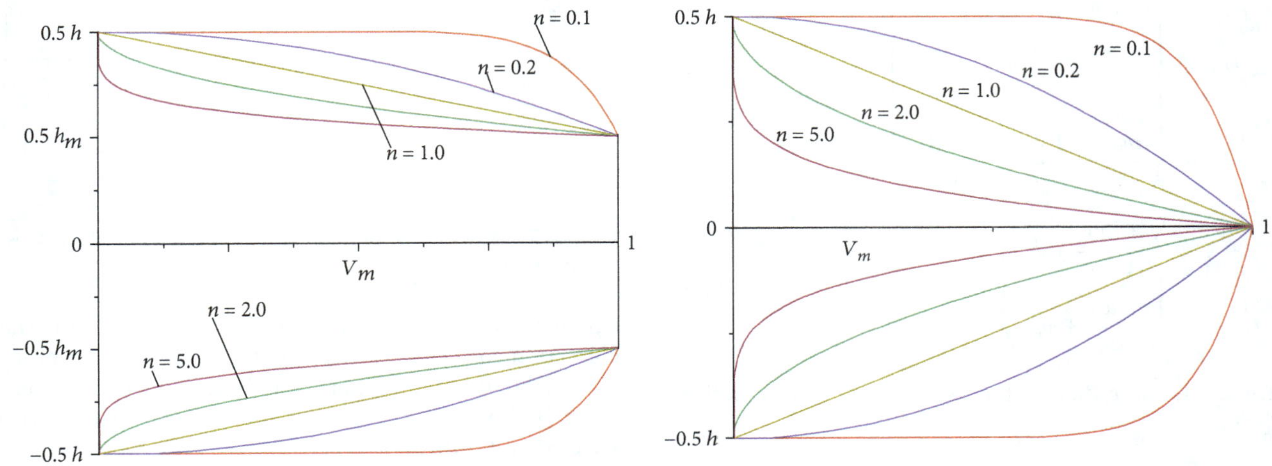

(a) $h_m : h = 0.5$

(b) $h_m = 0$

FIGURE 4: Effect of the scalar parameter of volume fraction in the FG sandwich blade.

surface in this paper. The volume fraction of the metal phase is obtained as [27]

$$V_m(\zeta) = \begin{cases} \left(\dfrac{h+2\zeta}{h-h_m}\right)^n & \zeta \in [h_0, h_1], \\ 1 & \zeta \in [h_1, h_2], \\ \left(\dfrac{h-2\zeta}{h-h_m}\right)^n & \zeta \in [h_2, h_3]. \end{cases} \tag{6}$$

As shown in Figure 4, n is the scalar parameter that allows users to define the gradation of material properties in the thickness direction. The $n = 0$ case corresponds to a fully metal blade. The volume fraction for the ceramic phase is given as

$$V_c(\zeta) = 1 - V_m(\zeta). \tag{7}$$

In the present work, the homogenization procedure and the law of mixtures are used. The variation of the parameters in the thickness direction is given by

$$P(\zeta) = P_m V_m(\zeta) + P_c V_c(\zeta), \tag{8}$$

where $P(\zeta)$ is the physical parameter, which refers to density ρ, elastic modulus E, and Poisson's ratio μ. In this paper, P_m and P_c are the physical parameters of metal and ceramic phase.

2.2. The Strain Energy. According to the first-order shear deformation theory, the displacements can be written as

$$\begin{aligned} u(x,\eta,\zeta) &= u_0(x,\eta) + \zeta\phi_x(x,\eta), \\ v(x,\eta,\zeta) &= v_0(x,\eta) + \zeta\phi_\eta(x,\eta), \\ w(x,\eta,\zeta) &= w_0(x,\eta), \end{aligned} \tag{9}$$

where ϕ_x and ϕ_η are mid-surface rotations and u_0, v_0, and w_0 are mid-surface displacements of the blade along the x, η, and ζ directions.

According to the thick shell theory [29] and the improved version of the Novozhilov nonlinear shell theory [30], the strains at any point in the blade can be written as

$$\begin{pmatrix} \varepsilon_x \\ \varepsilon_\eta \\ \gamma_{x\eta} \end{pmatrix} = \begin{pmatrix} \varepsilon_x^{(0)} \\ \varepsilon_\eta^{(0)} \\ \gamma_{x\eta}^{(0)} \end{pmatrix} + \frac{2}{R_{x\eta}} \begin{pmatrix} 0 \\ 0 \\ w_0 \end{pmatrix} + \zeta \begin{pmatrix} \kappa_x^{(1)} \\ \kappa_\eta^{(1)} \\ \kappa_{x\eta}^{(1)} \end{pmatrix}, \tag{10a}$$

$$\begin{pmatrix} \gamma_{x\zeta} \\ \gamma_{\eta\zeta} \end{pmatrix} = \begin{pmatrix} \gamma_{x\zeta}^{(0)} \\ \gamma_{\eta\zeta}^{(0)} \end{pmatrix} - \frac{1}{R_{x\eta}} \begin{pmatrix} v_0 \\ u_0 \end{pmatrix}, \tag{10b}$$

where

$$\begin{pmatrix} \varepsilon_x^{(0)} \\ \varepsilon_\eta^{(0)} \\ \gamma_{x\eta}^{(0)} \end{pmatrix} = \begin{pmatrix} \dfrac{1}{A}\dfrac{\partial u_0}{\partial x} + \dfrac{v_0}{A}\dfrac{\partial A}{\partial \eta} + \underline{\dfrac{1}{2}\left(\dfrac{1}{A}\dfrac{\partial w_0}{\partial x}\right)^2} \\ \dfrac{\partial v_0}{\partial \eta} + \underline{\dfrac{1}{2}\left(\dfrac{\partial w_0}{\partial \eta}\right)^2} \\ \dfrac{\partial u_0}{\partial \eta} + \dfrac{1}{A}\dfrac{\partial v_0}{\partial x} - \dfrac{u_0}{A}\dfrac{\partial A}{\partial \eta} + \underline{\dfrac{1}{A}\dfrac{\partial w_0}{\partial x}\dfrac{\partial w_0}{\partial \eta}} \end{pmatrix}, \quad (11a)$$

$$\begin{pmatrix} \kappa_x^{(1)} \\ \kappa_\eta^{(1)} \\ \kappa_{x\eta}^{(1)} \end{pmatrix} = \begin{pmatrix} \dfrac{1}{A}\dfrac{\partial \phi_x}{\partial x} + \dfrac{\phi_\eta}{A}\dfrac{\partial A}{\partial \eta} \\ \dfrac{\partial \phi_\eta}{\partial \eta} \\ \dfrac{\partial \phi_x}{\partial \eta} + \dfrac{1}{A}\dfrac{\partial \phi_\eta}{\partial x} - \dfrac{\phi_x}{A}\dfrac{\partial A}{\partial \eta} \end{pmatrix}, \quad (11b)$$

$$\begin{pmatrix} \gamma_{x\zeta}^{(0)} \\ \gamma_{\eta\zeta}^{(0)} \end{pmatrix} = \begin{pmatrix} \dfrac{1}{A}\dfrac{\partial w_0}{\partial x} + \phi_x \\ \dfrac{\partial w_0}{\partial \eta} + \phi_\eta \end{pmatrix}, \quad (11c)$$

where $R_{x\eta}$ is the radius of the twist, which can be derived as [31]

$$\frac{1}{R_{x\eta}} = -\frac{k}{A^2}. \quad (12)$$

The nonlinear parts, which are underlined in (11), will be ignored in the free vibration analysis of the blade.

The stress–strain equations can be written as

$$\begin{pmatrix} \sigma_x \\ \sigma_\eta \\ \sigma_{x\zeta} \\ \sigma_{\eta\zeta} \\ \sigma_{x\eta} \end{pmatrix} = \begin{bmatrix} Q_{11}(\zeta) & Q_{12}(\zeta) & 0 & 0 & 0 \\ Q_{21}(\zeta) & Q_{22}(\zeta) & 0 & 0 & 0 \\ 0 & 0 & Q_{44}(\zeta) & 0 & 0 \\ 0 & 0 & 0 & Q_{55}(\zeta) & 0 \\ 0 & 0 & 0 & 0 & Q_{66}(\zeta) \end{bmatrix} \begin{pmatrix} \varepsilon_x \\ \varepsilon_\eta \\ \gamma_{x\zeta} \\ \gamma_{\eta\zeta} \\ \gamma_{x\eta} \end{pmatrix}, \quad (13)$$

where the elastic constants $Q_{ij}(\zeta)$ are functions of thickness ζ, which is defined as

$$Q_{11}(\zeta) = Q_{22}(\zeta) = \frac{E(\zeta)}{1-\mu^2(\zeta)}, \quad (14a)$$

$$Q_{12}(\zeta) = Q_{21}(\zeta) = \frac{\mu(\zeta)E(\zeta)}{1-\mu^2(\zeta)}, \quad (14b)$$

$$Q_{44}(\zeta) = Q_{55}(\zeta) = \frac{K_s E(\zeta)}{2(1+\mu(\zeta))}, \quad (14c)$$

$$Q_{66}(\zeta) = \frac{E(\zeta)}{2(1+\mu(\zeta))}, \quad (14d)$$

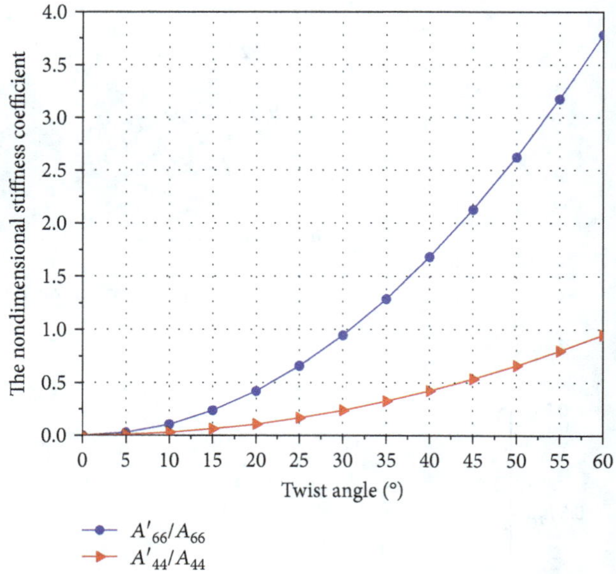

FIGURE 5: Variations of nondimensional stiffness coefficients with the initial twist angle ($\eta = 0$).

where K_s is the shear correction coefficient. Since the shear correction coefficient depends on the blade material properties and boundary conditions, it is difficult to obtain the exact value. In this paper, the shear correction coefficient K_s will be selected uniformly as 5/6 [32].

The strain energy of the blade is defined as

$$\begin{aligned} U_S &= \frac{1}{2}\int_0^L\int_{-b/2}^{b/2}\int_{-h/2}^{h/2}\left(\sigma_x\varepsilon_x + \sigma_\eta\varepsilon_\eta + \sigma_{x\zeta}\gamma_{x\zeta} + \sigma_{\eta\zeta}\gamma_{\eta\zeta} + \sigma_{x\eta}\gamma_{x\eta}\right)A\,d\zeta\,d\eta\,dx \\ &= \int_0^L\int_{-b/2}^{b/2}\left[\frac{1}{2}\left(D_{11}\kappa_x^{(1)2} + D_{22}\kappa_\eta^{(1)2} + D_{66}\kappa_{x\eta}^{(1)2} + 2D_{12}\kappa_x^{(1)}\kappa_\eta^{(1)}\right)\right. \\ &\quad + B_{11}\varepsilon_x^{(0)}\kappa_x^{(1)} + B_{22}\varepsilon_\eta^{(0)}\kappa_\eta^{(1)} + B_{66}\gamma_{x\eta}^{(0)}\kappa_{x\eta}^{(1)} + B_{12}\varepsilon_x^{(0)}\kappa_\eta^{(1)} + B_{21}\varepsilon_\eta^{(0)}\kappa_x^{(1)} \\ &\quad + \frac{1}{2}\left(A_{11}\varepsilon_x^{(0)2} + A_{22}\varepsilon_\eta^{(0)2} + A_{66}\gamma_{x\eta}^{(0)2} + 2A_{12}\varepsilon_x^{(0)}\varepsilon_\eta^{(0)} + A_{44}\gamma_{x\zeta}^{(0)2}\right. \\ &\quad \left. + A_{55}\gamma_{\eta\zeta}^{(0)2}\right) + B'_{66}w_0\gamma_{x\eta}^{(0)} + B'_{44}v_0\gamma_{x\zeta}^{(0)} + B'_{55}u_0\gamma_{\eta\zeta}^{(0)} + A'_{66}w_0^2 \\ &\quad \left. + A'_{44}v_0^2 + A'_{55}u_0^2\right]A\,d\eta\,dx, \end{aligned} \quad (15)$$

where A_{ij}, B_{ij}, and D_{ij} are the extensional, bending-extensional coupling, and bending stiffness [32], A'_{ij} and B'_{ij} are the stiffness due to the initial twist of the blade. They are defined as

$$\begin{aligned} \begin{pmatrix} A_{ij} \\ B_{ij} \\ D_{ij} \end{pmatrix} &= \int_{-h/2}^{h/2} Q_{ij}(\zeta)\begin{pmatrix} 1 \\ \zeta \\ \zeta^2 \end{pmatrix} d\zeta \\ &= \sum_{m=1}^{3}\int_{h_{m-1}}^{h_m} Q_{ij}(\zeta)\begin{pmatrix} 1 \\ \zeta \\ \zeta^2 \end{pmatrix} d\zeta \quad (i, j = 1, 2, 6), \end{aligned} \quad (16a)$$

TABLE 1: Convergence of nondimensional frequency parameters for the first six modes of the homogeneous material blade ($\bar{l}=3$, $\bar{h}=20$, and $\theta=15^\circ$).

Mode	ANSYS	5×5×5		5×6×6		5×7×7		5×8×8		5×9×9	
		Value	Error	Value	Error	Value	Error	Value	Error	Value	Error
1st	3.4225	3.4217	−0.02%	3.4102	−0.36%	3.4077	−0.43%	3.4194	−0.09%	3.4124	−0.30%
2nd	20.7177	20.7606	0.21%	20.7156	−0.01%	20.7222	0.02%	20.7540	0.18%	20.7270	0.04%
3rd	21.9608	22.2650	1.39%	21.9718	0.05%	22.1278	0.76%	22.0861	0.57%	22.0340	0.33%
4th	58.8524	59.6322	1.33%	54.3477	−7.65%	58.1334	−1.22%	58.1917	−1.12%	58.0295	−1.40%
5th	64.5829	66.5017	2.97%	64.0481	−0.83%	65.6567	1.66%	65.4872	1.40%	65.6406	1.64%
6th	68.0114	68.9640	1.40%	68.1982	0.27%	68.6168	0.89%	68.4144	0.59%	68.3678	0.52%

TABLE 2: Convergence of nondimensional frequency parameters for the first six modes of the FG sandwich blade ($\bar{l}=3$, $\bar{h}=20$, $n=2$, $\bar{h}_l=1/2$, and $\theta=15^\circ$).

Mode	ANSYS	5×5×5		5×6×6		5×7×7		5×8×8		5×9×9	
		Value	Error	Value	Error	Value	Error	Value	Error	Value	Error
1st	4.5272	4.5320	0.11%	4.5209	−0.14%	4.5228	−0.10%	4.5211	−0.13%	4.5215	−0.13%
2nd	27.3966	27.5019	0.38%	27.4741	0.28%	27.4639	0.25%	27.4684	0.26%	27.4665	0.26%
3rd	29.0406	29.5393	1.72%	29.3030	0.90%	29.3432	1.04%	29.2158	0.60%	29.1542	0.39%
4th	76.2376	75.8763	−0.47%	70.8471	−7.07%	74.8356	−1.84%	74.8922	−1.76%	75.0109	−1.61%
5th	80.3331	84.3843	5.04%	78.8139	−1.89%	82.1073	2.21%	82.2715	2.41%	82.2392	2.37%
6th	89.9847	91.5581	1.75%	90.9359	1.06%	91.0424	1.18%	90.7128	0.81%	90.5917	0.67%

TABLE 3: Comparison of nondimensional frequency parameters for the first six modes of homogeneous material blade ($\bar{h}=20$, $\mu=0.3$).

θ	Mode	$\bar{l}=3$						$\bar{l}=1$					
		1st	2nd	3rd	4th	5th	6th	1st	2nd	3rd	4th	5th	6th
	Present	3.4205	20.8483	21.3316	59.6783	62.2910	65.4353	3.4757	8.3462	20.9625	26.6650	30.1067	43.5409
0°	ANSYS	3.4223	20.8230	21.3590	59.8793	62.2508	65.3690	3.4814	8.3793	21.0993	26.7267	30.2352	43.5819
	Ref [15]	3.4140	21.2200	21.3600	60.1200	61.8200	66.6800	3.4620	8.4870	21.2900	27.4700	30.9800	43.5200
	Present	3.4194	20.7540	22.0861	58.1917	65.4872	68.4144	3.4479	10.4233	20.4526	26.6975	31.5336	44.6423
15°	ANSYS	3.4225	20.7177	21.9608	58.8524	64.5829	68.0114	3.4540	10.3394	20.3668	26.6264	31.3218	44.4679
	Ref [15]	3.4150	20.6700	22.2800	58.1500	65.2900	69.3200	3.4450	10.4000	20.5700	27.3500	32.1100	44.5800
	Present	3.4035	19.2320	25.4184	54.4781	70.6496	76.3746	3.4030	14.8939	18.9948	27.1437	35.2177	46.9877
30°	ANSYS	3.4153	18.9950	25.0621	55.7643	71.6994	75.9291	3.3794	14.3509	18.4374	26.7434	34.0572	45.9808
	Ref [15]	3.4170	19.0000	25.1500	55.9500	71.4000	76.5700	3.3970	14.3300	18.6900	27.3900	34.9000	46.5200
	Present	3.4163	17.4172	29.7580	54.0239	80.8963	87.6963	3.3387	20.0628	17.1876	28.2967	40.1453	48.6332
45°	ANSYS	3.4176	16.9599	29.4234	53.3063	79.0586	87.0233	3.2668	18.4291	16.0421	27.1718	37.3254	46.1459
	Ref [15]	3.4200	16.9900	29.2000	53.5800	78.8200	87.0500	3.3230	18.3200	16.3900	27.7100	38.1700	47.3000

$$A_{ij} = \int_{-h/2}^{h/2} Q_{ij}(\zeta)\,d\zeta = \sum_{m=1}^{3}\int_{h_{m-1}}^{h_m} Q_{ij}(\zeta)\,d\zeta \quad (i,j=4,5), \tag{16b}$$

$$B'_{ij} = \begin{cases} -\dfrac{1}{R_{x\eta}} B_{ij} & (i=j=4,5), \\ \dfrac{2}{R_{x\eta}} B_{ij} & (i=j=6), \end{cases} \tag{16c}$$

$$A'_{ij} = \begin{cases} \dfrac{1}{2R_{x\eta}^2} A_{ij} & (i=j=4,5), \\ \dfrac{2}{R_{x\eta}^2} A_{ij} & (i=j=6). \end{cases} \tag{16d}$$

It should be noted that the stiffness B_{ij} and B'_{ij} will be zero while the structure is symmetrical about the middle surface in the thickness direction. Figure 5 shows the variations of the nondimensional stiffness coefficients (A'_{44}/A_{44} and A'_{66}/A_{66} at $\eta = 0$) with the initial twist angle. It can be found that both the nondimensional stiffness coefficients increase with increase in the initial twist angle.

2.3. The Potential Energy of Centrifugal Force. The coordinate system transformation relationship between the xyz-coordinate and XYZ-coordinate is given by

$$\begin{pmatrix} \mathbf{i}_x \\ \mathbf{i}_y \\ \mathbf{i}_z \end{pmatrix} = \begin{bmatrix} 1 & 0 & 0 \\ 0 & \cos\varphi & \sin\varphi \\ 0 & -\sin\varphi & \cos\varphi \end{bmatrix} \begin{pmatrix} \mathbf{i}_X \\ \mathbf{i}_Y \\ \mathbf{i}_Z \end{pmatrix}. \quad (17)$$

Then, the angular velocity $\boldsymbol{\omega}$ in the xyz-coordinate of the pretwisted plate is expressed as

$$\boldsymbol{\omega} = \begin{bmatrix} 0 \\ \Omega\cos\varphi \\ -\Omega\sin\varphi \end{bmatrix}^T \begin{pmatrix} \mathbf{i}_x \\ \mathbf{i}_y \\ \mathbf{i}_z \end{pmatrix}. \quad (18)$$

The centrifugal force $\mathbf{F}_C$ per unit volume in the blade is given by

$$\mathbf{F}_C = \rho\boldsymbol{\omega} \times \left(\mathbf{r}^{(0)} + R\mathbf{i}_x\right) \times \boldsymbol{\omega} = \rho\Omega^2 \begin{bmatrix} R + x - k\eta\dfrac{\zeta}{A} \\ \eta\sin\Phi\sin\varphi + \zeta\cos\Phi\sin\dfrac{\varphi}{A} \\ \eta\sin\Phi\cos\varphi + \zeta\cos\Phi\cos\dfrac{\varphi}{A} \end{bmatrix}^T \begin{pmatrix} \mathbf{i}_x \\ \mathbf{i}_y \\ \mathbf{i}_z \end{pmatrix}, \quad (19)$$

where Φ is the sum of the stagger angle and the twist angle of the point, given by

$$\Phi = \varphi + \theta'. \quad (20)$$

The rotation of the blade about the turbine axial direction produces a significant amount of membrane stresses, which are essentially acting in the radial direction (the spanwise and chordwise directions of the blade). The component of centrifugal force along the spanwise direction σ_{cs} and the chordwise direction σ_{cc} can be written as [10, 14]

$$\begin{aligned}\sigma_{cs} &= \int_X^L \mathbf{F}_C \cdot \mathbf{e}_{0x} A dx = \rho\Omega^2\left\{\left(R - \frac{1}{2A}k\eta\zeta\right)(L-x)\right. \\ &\quad + \frac{1}{2}(L^2 - x^2) + \frac{1}{4}\eta^2(\cos 2\Phi - \cos 2\Theta) \\ &\quad \left. + \frac{1}{4A}\eta\zeta(\sin 2\Phi - \sin 2\Theta)\right\},\end{aligned} \quad (21a)$$

TABLE 4: The materials and properties of the FG sandwich blade.

Materials	Material properties		
Metal (alloy C-276)	$E_m = 2.08 \times 10^{11}$ Pa,	$\mu_m = 0.32$,	$\rho_m = 8900\,\text{kg/m}^3$
Ceramic (Al_2O_3)	$E_c = 3.50 \times 10^{11}$ Pa,	$\mu_c = 0.26$,	$\rho_c = 3970\,\text{kg/m}^3$

$$\begin{aligned}\sigma_{cc} &= \int_{b/2}^0 \mathbf{F}_C \cdot \mathbf{e}_{0\eta} d\eta \\ &= \rho\Omega^2\left\{\frac{1}{2k}\zeta\sin 2\Phi\sinh^{-1}\left(\frac{1}{2}kb\right) - \frac{1}{8}b^2\sin^2\Phi\right\},\end{aligned} \quad (21b)$$

where Θ is the sum of the stagger angle and the total twist angle, given by

$$\Theta = \varphi + \theta. \quad (22)$$

The centrifugal potential energy can be derived as

$$U_{CF} = \frac{1}{2}\int_0^L\int_{-b/2}^{b/2}\int_{-h/2}^{h/2}(\sigma_{cs}\varepsilon_{cs} + \sigma_{cc}\varepsilon_{cc})A d\zeta d\eta dx. \quad (23)$$

According to [26], ε_{cs} and ε_{cc} are given by

$$\begin{aligned}\varepsilon_{cs} &= \left(\frac{\partial v_0}{\partial x}\right)^2 + \left(\frac{\partial w_0}{\partial x}\right)^2, \\ \varepsilon_{cc} &= \left(\frac{\partial u_0}{\partial \eta}\right)^2 + \left(\frac{\partial w_0}{\partial \eta}\right)^2.\end{aligned} \quad (24)$$

Substituting (21) and (24) into (23), the potential energy can be written as

$$\begin{aligned}U_{CF} = \int_0^L\int_{-b/2}^{b/2} &\left\{\left(\frac{I_1}{8}b\sin 2\Phi - \frac{I_0}{16}b^2\sin^2 2\Phi\right)\right. \\ &\cdot\left[\left(\frac{\partial u_0}{\partial \eta}\right)^2 + \left(\frac{\partial w_0}{\partial \eta}\right)^2\right] \\ &+ I_1\left[\frac{\eta}{8A}(\sin 2\Phi - \sin 2\Theta) - \frac{k\eta(L-x)}{4A}\right] \\ &\cdot\left[\left(\frac{\partial v_0}{\partial x}\right)^2 + \left(\frac{\partial w_0}{\partial x}\right)^2\right] \\ &+ I_0\left[\frac{1}{2}R(L-x) + \frac{1}{4}(L^2 - x^2) + \frac{\eta^2}{8}(\cos 2\Phi - \cos 2\Theta)\right] \\ &\left.\cdot\left[\left(\frac{\partial v_0}{\partial x}\right)^2 + \left(\frac{\partial w_0}{\partial x}\right)^2\right]\right\}A d\eta dx.\end{aligned} \quad (25)$$

The total potential energy can be written as

$$U = U_S + U_{CF}. \quad (26)$$

TABLE 5: Comparisons of nondimensional frequency parameters for the first six modes of the FG sandwich blade ($\bar{l}=3$, $\bar{h}=20$, $n=2$, and $\bar{h}_l=1/2$).

Mode	0°				15°				30°			
	Present	ANSYS	Error	Mode shape	Present	ANSYS	Error	Mode shape	Present	ANSYS	Error	Mode shape
1st	4.5233	4.5270	−0.08%	1B	4.5206	4.5272	−0.15%	1B	4.4729	4.5290	−1.24%	1B
2nd	27.9633	27.7729	0.69%	1 T	27.4621	27.3966	0.24%	2B	25.4167	25.0992	1.26%	2B
3rd	28.3106	28.2605	0.18%	2B	29.1565	29.0406	0.40%	1T	33.0633	32.5549	1.56%	1T
4th	75.9386	75.8281	0.15%	1 EB	74.8841	76.2376	−1.78%	3B	73.0937	72.9145	0.25%	3B
5th	79.3035	79.1454	0.20%	3B	82.0284	80.3331	2.11%	1EB	89.2792	88.7755	0.57%	1EB
6th	87.5852	87.0515	0.61%	2 T	90.5827	89.9847	0.66%	2T	99.5237	98.9166	0.61%	2T

2.4. The Kinetic Energy. The displacement vector $\mathbf{u}$ can be written as

$$\mathbf{u}=[u\,v\,w]\begin{pmatrix}\mathbf{e}_{0x}\\ \mathbf{e}_{0\eta}\\ \mathbf{e}_{0\zeta}\end{pmatrix}=\frac{1}{A}\begin{bmatrix}u-k\eta w\\ Av\cos\theta'-(k\eta u+w)\sin\theta'\\ Av\sin\theta'+(k\eta u+w)\cos\theta'\end{bmatrix}^T\begin{pmatrix}\mathbf{i}_x\\ \mathbf{i}_y\\ \mathbf{i}_z\end{pmatrix}. \tag{27}$$

An arbitrary point of the blade after deformation can be expressed by the position vector $\mathbf{r}$ as

$$\mathbf{r}=\mathbf{r}^{(0)}+\mathbf{u}=\frac{1}{A}\begin{bmatrix}u+Ax-k\eta(w+\zeta)\\ A(v+\eta)\cos\theta'-(k\eta u+w+\zeta)\sin\theta'\\ A(v+\eta)\sin\theta'+(k\eta u+w+\zeta)\cos\theta'\end{bmatrix}^T\begin{pmatrix}\mathbf{i}_x\\ \mathbf{i}_y\\ \mathbf{i}_z\end{pmatrix}. \tag{28}$$

The velocity $\mathbf{V}_r$ of the point due to the rotation can be derived as

$$\mathbf{V}_r=\boldsymbol{\omega}\times(\mathbf{r}+R\mathbf{i}_x)=\frac{\Omega}{A}\begin{bmatrix}A(v+\eta)\sin\Phi+(k\eta u+w+\zeta)\cos\Phi\\ [k\eta(w+\zeta)-A(R+x)-u]\sin\varphi\\ [k\eta(w+\zeta)-A(R+x)-u]\cos\varphi\end{bmatrix}^T\begin{pmatrix}\mathbf{i}_x\\ \mathbf{i}_y\\ \mathbf{i}_z\end{pmatrix}. \tag{29}$$

The kinetic energy reads

$$T=\frac{1}{2}\int_0^L\int_{-b/2}^{b/2}\int_{-h/2}^{h/2}\rho\Big[(\dot{u}+\mathbf{V}_r\cdot\mathbf{e}_{0x})^2+(\dot{v}+\mathbf{V}_r\cdot\mathbf{e}_{0\eta})^2+(\dot{w}+\mathbf{V}_r\cdot\mathbf{e}_{0\zeta})^2\Big]Ad\zeta d\eta dx=T_{in}+T_{gy}+T_{ro}, \tag{30a}$$

where

$$T_{in}=\int_0^L\int_{-b/2}^{b/2}\Big[\frac{1}{2}I_2\big(\dot{\phi}_x^2+\dot{\phi}_\eta^2\big)+I_1\big(\dot{u}_0\dot{\phi}_x+\dot{v}_0\dot{\phi}_\eta\big)+\frac{1}{2}I_0\big(\dot{u}_0^2+\dot{v}_0^2+\dot{w}_0^2\big)\Big]Ad\eta dx, \tag{30b}$$

$$\begin{aligned}T_{gy}=\Omega\int_0^L\int_{-b/2}^{b/2}&\Big\{\big(\phi_\eta\sin\Phi+A\cos\Phi\big)\big(I_2\dot{\phi}_x+I_1\dot{u}_0\big)\\&+(k\eta-\phi_x)\big(I_2\dot{\phi}_\eta+I_1\dot{v}_0\big)\sin\Phi+[(\eta+v_0)\sin\Phi\\&+(Aw_0-k\eta(R+x))\cos\Phi]\big(I_1\dot{\phi}_x+I_0\dot{u}_0\big)\\&+(k\eta w_0-AR-Ax-u_0)\big(I_1\dot{\phi}_\eta+I_0\dot{v}_0\big)\sin\Phi\\&-I_0\big[(k\eta^2+k\eta v_0)\sin\Phi+(Au_0+R+x)\cos\Phi\big]\dot{w}_0\\&-I_1\big(A\varphi_x\cos\Phi+k\eta\varphi_\eta\sin\Phi\big)\dot{w}_0\Big\}d\eta dx,\end{aligned} \tag{30c}$$

$$\begin{aligned}T_{ro}=\Omega^2\int_0^L\int_{-b/2}^{b/2}&\Big\{\frac{1}{2}(I_2+I_0)\Big(\phi_\eta\sin\Phi+\frac{1}{A}(1+k\eta\phi_x)\cos\Phi\Big)^2\\&+\frac{1}{2A^2}I_2(\phi_x-k\eta)^2+\frac{1}{2}I_0\Big(R+x+\frac{1}{A}u_0-\frac{1}{A}k\eta w_0\Big)^2\\&+I_1\Big[(\eta+v_0)\sin\Phi+\frac{1}{A}(w_0+k\eta u_0)\cos\Phi\Big]\\&\cdot\Big(\phi_\eta\sin\Phi+\frac{1}{A}(1+k\eta\phi_x)\cos\Phi\Big)\\&+\frac{1}{A^2}I_1[A(R+x)+u_0-k\eta w_0](\phi_x-k\eta)\Big\}Ad\eta dx,\end{aligned} \tag{30d}$$

where "." is the first-order derivation of time; I_0, I_1, and I_2 in (25) and (30) are defined in terms of the density $\rho(\zeta)$ as

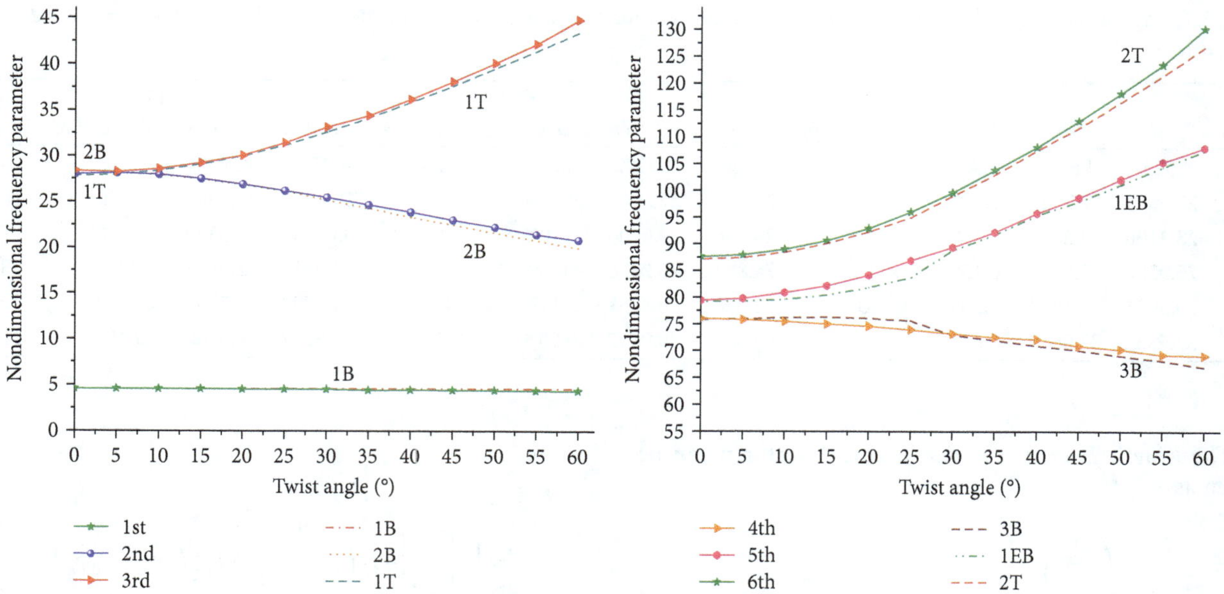

(a) 1st to 3rd modes (b) 4th to 6th modes

FIGURE 6: Variations of nondimensional frequency parameters with the twist angle ($\bar{l} = 3$, $\bar{h} = 20$, $n = 2$, and $\bar{h}_l = 1/2$).

$$\begin{pmatrix} I_0 \\ I_1 \\ I_2 \end{pmatrix} = \int_{-h/2}^{h/2} \rho(\zeta) \begin{pmatrix} 1 \\ \zeta \\ \zeta^2 \end{pmatrix} d\zeta = \sum_{m=1}^{3} \int_{h_{m-1}}^{h_m} \rho(\zeta) \begin{pmatrix} 1 \\ \zeta \\ \zeta^2 \end{pmatrix} d\zeta. \tag{31}$$

3. Frequency Solving

The Rayleigh–Ritz method is used to solve the natural frequency of the blade; the expressions of the displacements and rotations are assumed

$$u_0(x, \eta, t) = \sum_{N=1}^{\infty} U_N(x, \eta) e^{i\omega_N t}, \tag{32a}$$

$$v_0(x, \eta, t) = \sum_{N=1}^{\infty} V_N(x, \eta) e^{i\omega_N t}, \tag{32b}$$

$$w_0(x, \eta, t) = \sum_{N=1}^{\infty} W_N(x, \eta) e^{i\omega_N t}, \tag{32c}$$

$$\phi_x(x, \eta, t) = \sum_{N=1}^{\infty} \Phi_{xN}(x, \eta) e^{i\omega_N t}, \tag{32d}$$

$$\phi_\eta(x, \eta, t) = \sum_{N=1}^{\infty} \Phi_{\eta N}(x, \eta) e^{i\omega_N t}. \tag{32e}$$

The functions $U_N(x, \eta)$, $V_N(x, \eta)$, $W_N(x, \eta)$, $\Phi_{xN}(x, \eta)$, and $\Phi_{\eta N}(x, \eta)$ represent the N order mode shape, and ω_N is the corresponding circular frequency. The algebraic polynomials can form a mathematically complete set of functions, which guarantees convergence to the exact solution as the number of terms taken increases [5]. Thus, in this paper, the mode shapes are expanded in terms of two-dimensional algebraic polynomials in terms of the nondimensional coordinates χ and ξ as

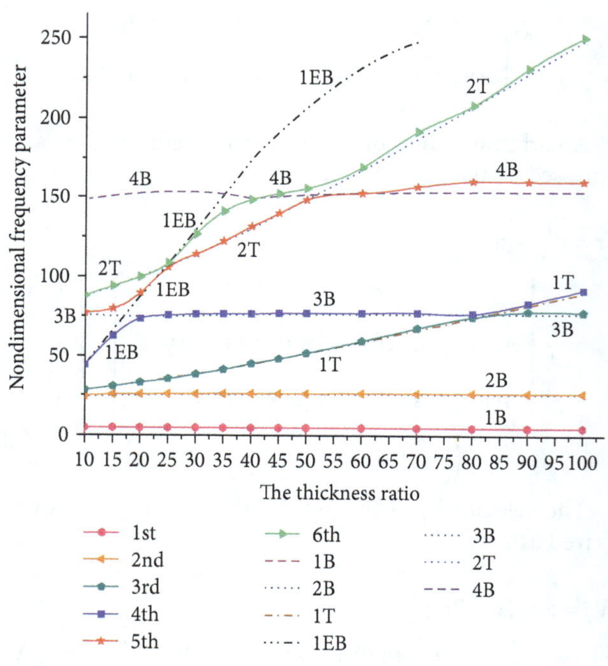

FIGURE 7: Variations of nondimensional frequency parameters with the thickness ratio ($\bar{l} = 3$, $n = 2$, $\bar{h}_l = 1/2$, and $\theta = 30°$).

$$U_N(\chi, \xi) = \sum_{i=1}^{I} \sum_{j=0}^{J} U_{ij} \chi^i \xi^j, \tag{33a}$$

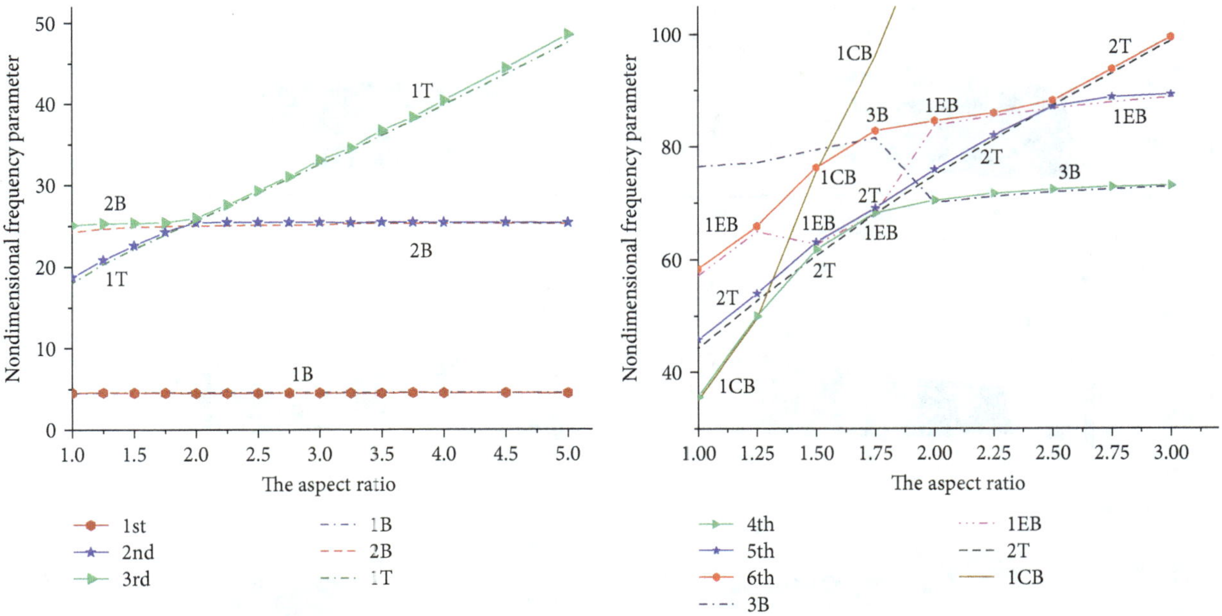

(a) 1st to 3rd modes

(b) 4th to 6th modes

FIGURE 8: Variations of nondimensional frequency parameters with the aspect ratio ($\bar{h} = 20$, $n = 2$, $\bar{h}_l = 1/2$, $\theta = 30^\circ$).

$$V_N(\chi, \xi) = \sum_{k=1}^{K} \sum_{l=0}^{L} V_{kl} \chi^k \xi^l, \tag{33b}$$

$$W_N(\chi, \xi) = \sum_{m=2}^{M} \sum_{n=0}^{N} W_{mn} \chi^m \xi^n, \tag{33c}$$

$$\Phi_{xN}(\chi, \xi) = \sum_{i=1}^{I} \sum_{j=0}^{J} \Phi_{xij} \chi^i \xi^j, \tag{33d}$$

$$\Phi_{\eta N}(\chi, \xi) = \sum_{k=1}^{K} \sum_{l=0}^{L} \Phi_{\eta kl} \chi^k \xi^l, \tag{33e}$$

where $\chi = x/L$ and $\xi = 2\eta/b$. U_{ij}, V_{kl}, W_{mn}, Φ_{xij}, and $\Phi_{\eta kl}$ are undetermined coefficients. The Rayleigh–Ritz method imposes a necessary condition that trial functions must satisfy the geometric boundary conditions arbitrarily. The cantilever blade is clamped at $\chi = 0$ ($u = v = w = dw/dx = 0$) along the spanwise direction and is completely free at $\chi = 1$, $\xi = -1$, and $\xi = 1$. The indices i, k, and m in (33), (46), and (47) begin with $i = 1$, $k = 1$, and $m = 2$; it can be ensured that all terms of polynomials satisfied the geometric boundary conditions.

To solve the free frequency of the blade, (32) and (33) are substituted into (26) and (30) in order to obtain the maximum strain energy ($U_{\max}$) and kinetic energy ($T_{\max}$) by setting $t = 0$. The Rayleigh–Ritz method requires the minimization of the function $\prod = T_{\max} - U_{\max}$, which can be achieved by taking the derivatives

$$\frac{\partial \prod}{U_{ij}} = 0, \qquad \frac{\partial \prod}{\Phi_{x_{ij}}} = 0 \quad (i = 1, 2, \dots, I; j = 0, 1, \dots, J), \tag{34a}$$

$$\frac{\partial \prod}{V_{kl}} = 0, \tag{34b}$$

$$\frac{\partial \prod}{\Phi_{\eta_{kl}}} = 0 \quad (k = 1, 2, \dots, K, l = 0, 1, \dots, L), \qquad \frac{\partial \prod}{W_{mn}} = 0 \quad (m = 2, 3, \dots, M, n = 0, 1, \dots, N). \tag{34c}$$

There are a total of $2 \times I \times (J+1) + 2 \times K \times (L+1) + (M-1) \times (N+1)$ equations, which can be described by

$$\{\mathbf{K} - \omega_N^2 \mathbf{M}\}[\Delta] = 0, \tag{35}$$

where $\mathbf{K}$ and $\mathbf{M}$ are the stiffness and mass matrices, respectively; $[\Delta]$ is the vector of undetermined coefficients, given by

$$[\Delta] = \begin{bmatrix} U_{ij} & V_{kl} & W_{mn} & \Phi_{x_{ij}} & \Phi_{\eta_{kl}} \end{bmatrix}^T. \tag{36}$$

The generalized eigenvalues (natural frequency) can be obtained by setting the coefficient matrix of (35) equal to zero, and the corresponding mode shape can be determined by substituting the eigenvector $[\Delta]$ back into (33).

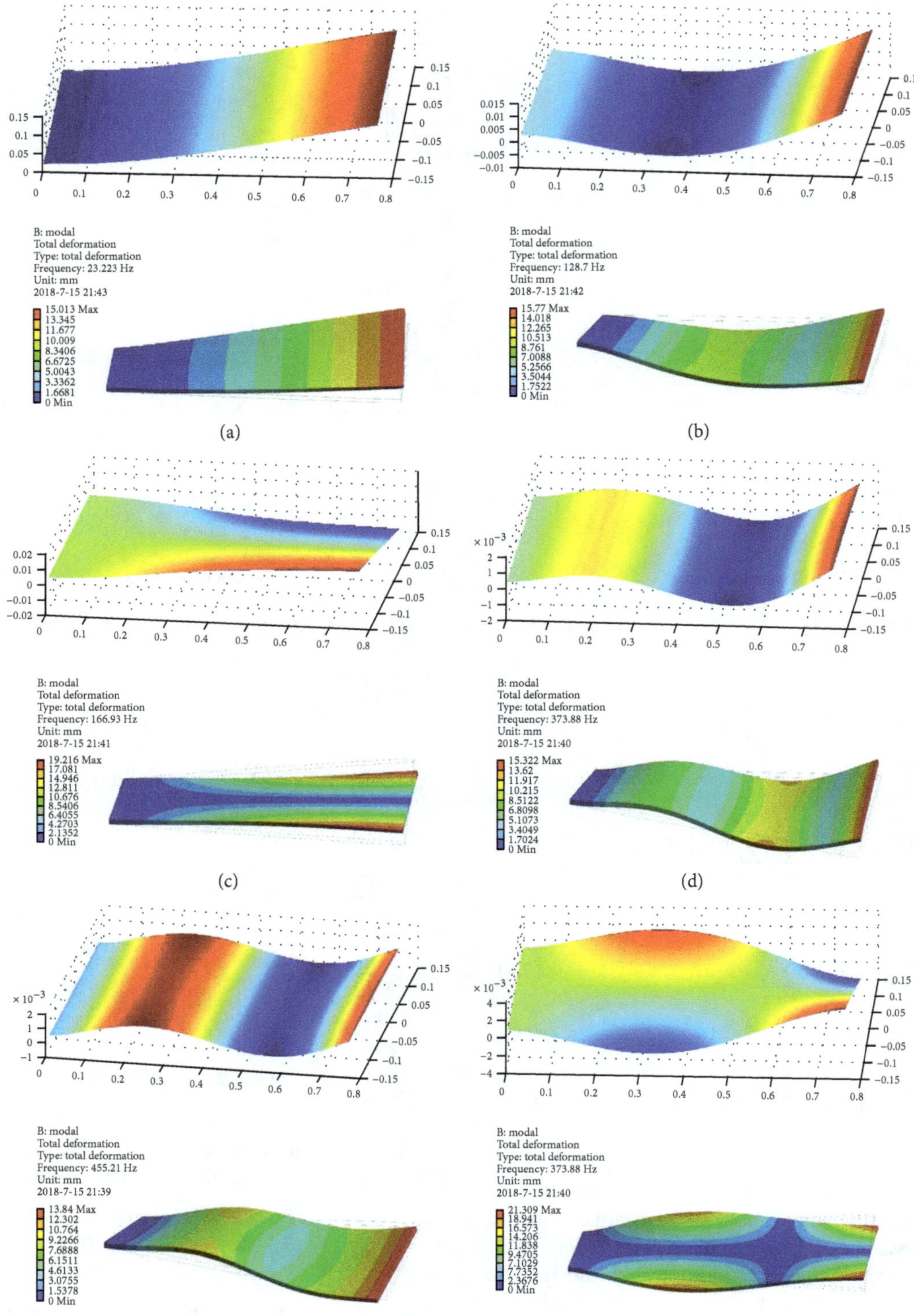

FIGURE 9: Mode shapes of the FG sandwich blade. (A1) 1st mode shape by Matlab; (B1) 2nd mode shape by Matlab; (C1) 3rd mode shape by Matlab; (D1) 4th mode shape by Matlab; (E1) 5th mode shape by Matlab; (F1) 6th mode shape by Matlab; (A2) 1st mode shape by ANSYS; (B2) 2nd mode shape by ANSYS; (C2) 3rd mode shape by ANSYS; (D2) 4th mode shape by ANSYS; (E2) 5th mode shape by ANSYS; (F2) 6th mode shape by ANSYS.

4. Numerical Results and Analyses

4.1. Convergence Analysis. A typical example of pretwisted blade with chord length $b = 0.254$ m, aspect ratio $\bar{l} = L/b$ from 1 to 5, thickness ratio $\bar{h} = b/h$ from 10 to 100, twist angle θ from 0° to 60°, layer thickness ratio $\bar{h}_l = (h - h_m)/2h_m$ from 0.25 to 4, scalar parameter of volume fraction of FGM n from 0.1 to 10, stagger angle φ from 0° to 90°, and disk radius ratio $\bar{r} = R/b$ is used.

TABLE 6: Nondimensional frequency parameters for the first six modes of FG sandwich blade with varying twist angle ($\bar{l}=3$, $\bar{h}=20$, $n=2$, and $\bar{h}_l=1/2$).

Mode	0°	5°	10°	15°	20°	25°	30°	45°	50°	60°
1st	4.5233	4.5342	4.5342	4.5206	4.5041	4.4935	4.4743	4.4020	4.3605	4.3245
2nd	27.9633	28.0799	27.9178	27.4621	26.8716	26.1715	25.4167	22.9643	22.1860	20.7649
3rd	28.3106	28.2050	28.5154	29.1565	29.9955	31.3439	33.0633	38.0040	39.9970	44.6692
4th	75.9386	75.7447	75.3807	74.8841	74.5120	73.8517	73.0937	70.8379	70.1630	68.9817
5th	79.3035	79.7009	80.7576	82.0284	84.0329	86.7774	89.2792	98.6081	101.9785	107.9219
6th	87.5852	87.9114	88.9680	90.5827	92.8839	95.9725	99.5237	112.9271	118.0741	130.0957

All the numerical results are presented in the form of the nondimensional frequency parameter $\overline{\omega}=\omega_N L^2\sqrt{\rho h/D}=\omega_N L^2/h\sqrt{12\rho(1-\mu^2)/E}$ for the homogeneous material blade, where $D=Eh^3/12(1-\mu^2)$ is the blade flexural rigidity per unit length and $\overline{\omega}=\omega_N L^2/h\sqrt{12\rho_m(1-\mu_m^2)/E_m}$ for the functionally graded sandwich blade in the following section, and the nondimensional rotation velocity is $\overline{\Omega}=\Omega/\omega_1$, where ω_1 is the first static circular frequency ($\Omega=0$) of the functionally graded sandwich blade.

According to [5], the number of terms chosen for each component of the displacements is the same. The results of the homogeneous material blade and functionally graded sandwich blade are given in Tables 1 and 2. The maximum difference between the $5\times 9\times 9=405$ term and the $5\times 8\times 8=320$ term solutions for the first six modes of the homogeneous material blade is no more than 0.28%, and 0.21% for the FG sandwich blade. But the time spent in computing has multiplied. Moreover, the eigenvalues are more likely to be ill-conditioned with the increase in the size of the matrix in numerical calculation. Therefore, it has been decided to use $5\times 8\times 8=320$ terms in the subsequent analysis.

The data used for comparison below "ANSYS" in all tables are computed by the finite element model in ANSYS Workbench (version 17.2). The middle surface model of the blade is established, and the finite element analysis is performed by using the SHELL181 linear quadrilateral element type. The layered section object in ANSYS mechanical is used to define the layered section of composite modeling. In order to get accurate results as much as possible, 20 layers in each skin (total of 41 layers in the thickness direction) are defined to simulate the cross section of the functionally graded sandwich blade.

4.2. Comparative Studies. The homogeneous material blade is used first to demonstrate the numerical accuracy of the analytical model in this paper, and the first six nondimensional frequency parameters are compared with ANSYS and reference data [15] in Table 3. The results are in good agreement when the twist angle $\theta=0°$ to 45° in the case of the aspect ratio $\bar{l}=3$. The results match well also when the twist angle $\theta=0°$ to 30° in the case of the aspect ratio $\bar{l}=1$, but the maximum error (occurring at the 2nd order) is about 8.86% when θ increases to 45°. The reason is that the Lame parameter A is approximated to 1 in order to improve the calculation speed in the process of numerical calculation. The error is smaller in the case of a larger aspect ratio and smaller twist angle, but the error becomes larger as the aspect ratio decreases and the twist angle increases. Thus, the aspect ratio $\bar{l}=3$ is used to obtain more accurate results.

The functionally graded sandwich blade is composed of metal and ceramic materials; their properties are given in Table 4. In Table 5, the first six nondimensional frequency parameters $\overline{\omega}$ of the FG sandwich blade are compared with results in ANSYS Workbench, both showing good agreements, respectively.

The mode shapes corresponding to the first six orders of frequencies are also listed in Table 5. It should be noted that spanwise bending modes are divided into flapwise bending modes (whose displacements are predominantly in the direction normal to the middle surface) and edgewise bending modes (whose displacements are predominantly in the direction tangent to the surface) [15] in this paper. The modes 1B and 2B represent the first and second flapwise bending modes, respectively, and 1EB represents the first edgewise bending mode. The modes 1T and 2T represent the first and second torsional modes, respectively. The chordwise bending modes (CB, having two or more nodal lines approximately parallel to the pretwisted blade axis) are also found in the successive research. The curves such as 1B, 1T, and 2B and so on in Figures 6–8 are all obtained by using ANSYS Workbench. The curves such as 1st, 2nd, and 3rd are all obtained through numerical calculation.

The first six mode shapes of the FG sandwich blade with $\bar{l}=3$, $\bar{h}=20$, $n=2$, $\bar{h}_l=1/2$, and $\theta=30°$ are given in Figure 9. It can be seen that the results obtained by the two methods are in good agreement. The results in Table 5 show that that the 1st modes are both of 1B mode shape when $\theta=0°$, 15°, and 30°, but the 2nd mode is of 1T mode shape when $\theta=0°$ and 2B mode shape when $\theta=15°$ and 30°. The same happens with the 1EB mode shape. The reason for this phenomenon is that the stiffness of the FG sandwich blade is changed due to the initial twist in (15).

4.3. Free Vibrations of the FG Sandwich Blade. The effect of the twist angle on the vibration characteristics for the pretwist FG sandwich blade with geometric parameters as $\bar{l}=3$, $\bar{h}=20$, $n=2$, and $\bar{h}_l=1/2$ is studied. Table 6 gives the first six nondimensional frequency parameters $\overline{\omega}$ of the FG sandwich blade with varying twist angles (θ from 0° to 60°), and the relationship between the nondimensional frequency parameters and twist angles for the FG sandwich blade is

Table 7: Nondimensional frequency parameters for the first six modes of the FG sandwich blade with varying thickness ratio ($\bar{l} = 3$, $n = 2$, $\bar{h}_l = 1/2$, and $\theta = 30°$).

Mode	10	20	30	40	50	60	70	80	90	100
1st	4.4578	4.4743	4.4745	4.4759	4.4903	4.5064	4.5022	4.5115	4.5146	4.5213
2nd	24.2636	25.4167	25.5518	25.6696	25.7090	25.7337	25.7686	25.8024	25.8433	25.8342
3rd	28.2468	33.0633	38.1941	44.8357	51.5288	59.2127	67.3645	74.2099	77.6186	77.2360
4th	43.9320	73.0937	75.9033	76.1792	76.4980	76.7083	76.7618	76.3814	82.7484	91.1187
5th	76.6391	89.2792	113.8445	130.9914	147.7835	151.9612	156.271	159.8808	159.7313	159.6019
6th	87.7677	99.5237	126.202	147.8068	155.1631	169.1361	191.4968	207.7979	230.7914	250.1723

Table 8: Nondimensional frequency parameters for the first six modes of the FG sandwich blade with varying aspect ratio ($\bar{h} = 20$, $n = 2$, $\bar{h}_l = 1/2$, and $\theta = 30°$).

Mode	1.00	1.25	1.50	1.75	2.00	2.25	2.50	3.00	4.00	5.00
1st	4.4970	4.5187	4.4946	4.4879	4.4857	4.4847	4.4762	4.4743	4.4643	4.4475
2nd	18.7284	20.8402	22.6114	24.2429	25.4037	25.4146	25.4190	25.4167	25.4069	25.3682
3rd	25.0639	25.2854	25.3453	25.3845	25.9349	27.5549	29.2911	33.0633	40.4134	48.5665
4th	35.6083	49.9937	61.8109	68.2541	70.4597	71.6661	72.3847	73.0937	73.8139	73.9446
5th	45.7901	53.9883	63.0533	69.0347	75.8721	82.0019	87.1364	89.2792	91.5483	93.0731
6th	58.3492	65.8222	76.2076	82.7443	84.4782	85.9505	88.1750	99.5237	123.1095	147.6281

Table 9: Nondimensional frequency parameters for the first six modes of the FG sandwich blade with varying layer thickness ratios ($\bar{l} = 3$, $\bar{h} = 20$, $n = 2$, and $\theta = 30°$).

Mode	1/4	1/3	1/2	1/1	2/1	3/1	4/1
1st	4.2021	4.3068	4.4743	4.7574	4.9764	5.0854	5.1378
2nd	23.6280	24.3536	25.4167	27.0889	28.3771	28.9429	29.2352
3rd	30.4576	31.3180	33.0633	35.0677	36.7356	37.6719	38.4396
4th	68.0698	70.1619	73.0937	78.1350	81.7252	83.5669	84.7660
5th	83.1451	85.6584	89.2792	95.5154	100.6206	103.1477	104.3400
6th	92.4498	95.1984	99.5237	106.5238	111.4889	114.2924	116.4530

shown in Figure 6. It can be seen that the resulting curves are very close to the curves obtained by ANSYS in Figure 6(a). The first mode corresponds to the 1B vibration mode for all twist angles. The 2nd and 3rd nondimensional frequency parameter curves represent the 1T and 2B mode shapes when $\theta = 0°$, and they come close to each other and separate gradually with the twist angle increasing. The two curves are up to the nearest when the twist angle θ is about 5°; the two modes exchange with each other, and the variation tendencies of nondimensional frequency parameters are also interchanged. The variations for 4th to 6th can be found in Figure 6(b).

It should be noted that the FG sandwich blade designed by the geometric parameters corresponding to the frequency turning points will be prone to internal resonance.

The effect of the thickness ratio on the vibration characteristics for the pretwist FG sandwich blade is studied also. The first six nondimensional frequency parameters $\overline{\omega}$ of the FG sandwich blade with varying thickness ratios ($\bar{h}$ from 10 to 100) is given in Table 7. Figure 7 shows the relationship between the nondimensional frequency parameters and the thickness ratio. It can be found that the nondimensional frequency parameters of 1B, 2B, 3B, and 4B modes remain essentially unchanged when the thickness ratio varies from 10 to 100. The nondimensional frequency parameters of 1T, 2T, and 1EB modes tend to increase with increasing thickness ratio. On the other side, the 3rd and 4th frequency parameter loci have an obvious veering at $\bar{h} = 85$, and the mode shapes exchanged between 1T and 3B. The same situation can be seen on the 5th and 6th modes, and higher-order frequency parameter locus veering happens more frequently.

The nondimensional frequency parameters $\overline{\omega}$ for the first six modes of the FG sandwich blade with varying aspect ratios ($\bar{l}$ from 1 to 5) are given in Table 8 to study the effect of the aspect ratio on the vibration characteristics for the pretwist FG sandwich blade. Figure 8 shows the relationships between the first six frequency parameters, the corresponding mode shapes, and the aspect ratio. It is obvious that the 1st mode corresponds to the 1B vibration mode for all aspect ratios, and the value of the frequency parameter varies little. The 2nd and 3rd modes exchange with each other, and the

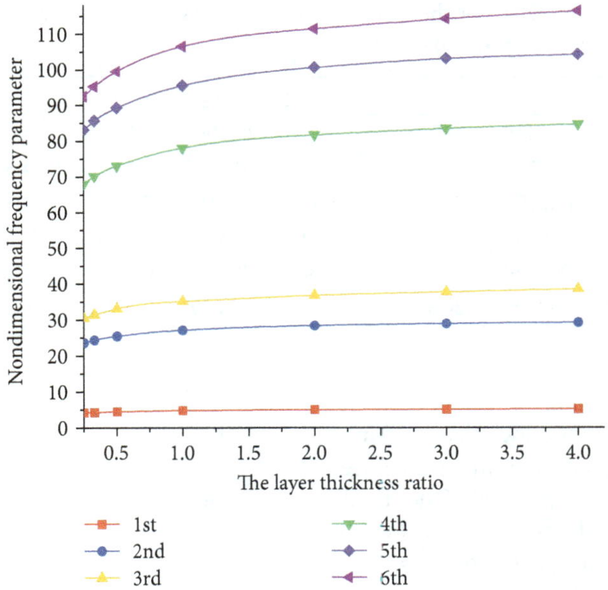

FIGURE 10: Variations of nondimensional frequency parameters with the layer thickness ratio ($\bar{l} = 3$, $\bar{h} = 20$, $n = 2$, and $\theta = 30°$).

TABLE 10: Nondimensional frequency parameters for the first six modes of the FG sandwich blade with varying scalar parameters of volume fraction ($\bar{l} = 3$, $\bar{h} = 20$, $\bar{h}_l = 1/2$, and $\theta = 30°$).

Mode	0.0	0.1	0.2	0.3	0.4	0.5	1.0	2.0	3.0	4.0	5.0
1st	3.3606	3.5795	3.6493	3.8619	3.8665	3.9753	4.2248	4.4743	4.5659	4.6931	4.7617
2nd	19.1226	20.0849	20.7619	21.4358	21.9111	22.3903	23.8825	25.4167	26.1362	26.5972	26.8850
3rd	24.9882	26.0961	26.9426	27.6920	28.2425	28.7613	30.7357	33.0633	33.7313	34.2049	34.8816
4th	55.8597	58.2781	60.3344	62.0908	63.5612	64.7114	68.9569	73.0937	75.1927	76.5550	77.6652
5th	71.5372	73.7012	75.7393	77.3010	78.8343	80.0439	84.5755	89.2792	91.5699	93.1665	93.7638
6th	75.5421	79.2376	81.6705	84.0076	85.8903	86.6503	93.4294	99.5237	102.7955	104.0268	106.0757

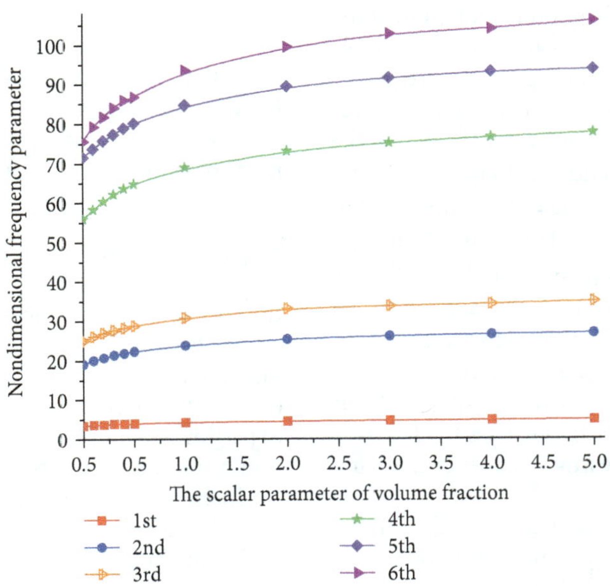

FIGURE 11: Variations of nondimensional frequency parameters with the scalar parameter of volume fraction ($\bar{l} = 3$, $\bar{h} = 20$, $\bar{h}_l = 1/2$, and $\theta = 30°$).

Table 11: Nondimensional frequency parameters for the first six modes of the FG sandwich blade with varying nondimensional rotation velocities ($\bar{l} = 3$, $\bar{h} = 20$, $n = 2$, $\bar{h}_l = 1/2$, $\theta = 30°$, $\bar{r} = 1$, and $\varphi = 45°$).

Mode	0	1	2	3	4	5	6	7	8	9	10	11
1st	4.4743	6.9030	11.2340	15.8419	20.5025	25.1652	29.8098	34.4146	38.9622	43.4424	47.8358	52.1334
2nd	25.4167	28.5235	34.0383	35.1206	36.4308	38.1374	40.2129	42.4223	44.6896	47.3581	50.0889	52.8240
3rd	33.0633	33.0979	35.8589	45.2978	55.2322	64.5633	72.3156	78.0880	82.3864	85.9246	89.1536	92.3135
4th	73.0937	76.2806	80.9215	84.9748	88.5518	92.9797	99.3391	108.0164	118.4814	130.0075	142.1141	154.5548
5th	89.2792	92.7844	96.5525	106.6897	114.5673	121.7465	129.8789	138.6539	147.8984	157.7082	167.8623	178.1525
6th	99.5237	100.9778	103.9269	108.5284	120.9391	137.5384	155.5090	174.2894	193.5633	213.0626	232.5021	250.6452

vibration tendencies of their frequency loci also exchange near $\bar{l} = 2$. The change of the 4th to 6th nondimensional frequency parameters is relatively complicated. The mode shapes include 3B, 2T, 1EB, and 1CB. For example, the 4th to 6th mode shapes are 1CB, 2T, and 1EB when $\bar{l} = 1$; 2T, 1EB, and 1CB when $\bar{l} = 1.5$; and 1EB, 2T, and 3B when $\bar{l} = 1.75$. All in all, the mode shape corresponding to the same order frequency is very sensitive to the variation of small aspect ratios, especially when $\bar{l} < 2$. The tendencies of their frequency parameters become relatively stable until $\bar{l} > 2.5$.

Table 9 and Figure 10 show the variations of the first six nondimensional frequency parameters of the FG sandwich blade with the layer thickness ratio. Table 10 and Figure 11 show the relationships between the first six nondimensional frequency parameters and the scalar parameter of the volume fraction. It can be found that the content of ceramic increases with increasing layer thickness ratio and scalar parameter of the volume fraction, and the frequency parameters of all modes increase, respectively.

Table 11 shows the first six nondimensional frequency parameters of the FG sandwich blade for the nondimensional rotation velocity with the range of $\bar{\Omega}$ 0 to 11 (about 15,000 rpm). A typical Campbell diagram is plotted to study the variation trend of the nondimensional frequency parameters for varying rotation velocities in Figure 12. Frequency locus veering is observed as expected: the 1st and 2nd frequency modes exchange between 1B and 1T around the rotation velocity $\bar{\Omega} = 11$; the 2nd and 3rd frequency modes exchange between 2B and 1T around $\bar{\Omega} = 2$; the 5th and 6th frequency modes exchange between 1EB and 2T around $\bar{\Omega} = 3$; and the 7th and 8th frequency modes exchange between 4B and 3T around $\bar{\Omega} = 1$ and exchange between 3T and 1CB again around the rotation velocity $\bar{\Omega} = 9$. In addition, the 8th frequency locus can be seen veering at $\bar{\Omega} = 4$ and the mode shape is changed from 4B to 1EB. It can be seen clearly that higher-order frequencies tend to have multiple frequency veering with the mode shapes exchanging and coupling.

Figure 13 shows the effects of the stagger angle for the first six nondimensional frequency parameters with varying rotation velocities. It can be found that the stagger angle has a predominant influence in the 1st and 2nd modes, and the effect is very marginal in higher modes. The effect of the stagger angle has little influence on all frequencies when the rotation velocity is low. The reason is that the centrifugal force has little effect on the stiffening and softening when the rotating speed is low, and the influence becomes more and more obvious with the rotating speed increasing.

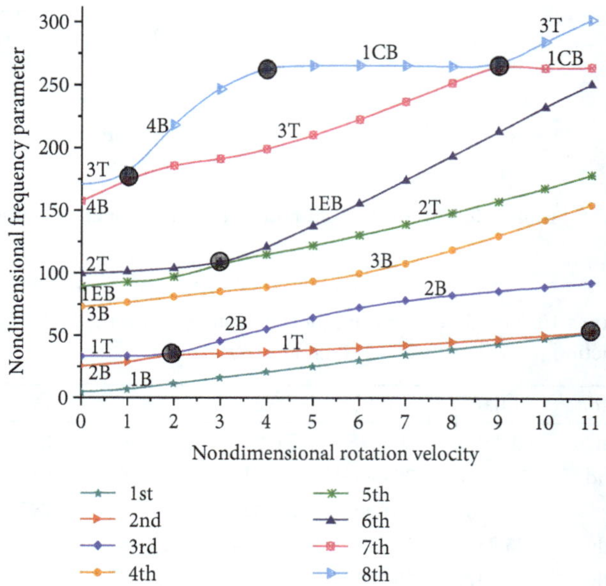

Figure 12: A Campbell diagram of the FG sandwich blade ($\bar{l} = 3$, $\bar{h} = 20$, $n = 2$, $\bar{h}_l = 1/2$, $\theta = 30°$, $\bar{r} = 1$, and $\varphi = 45°$).

5. Conclusion

In this paper, a dynamic model based on the thick shell theory is developed to investigate the free vibration behavior of functionally graded sandwich pretwisted blades. The validation of the homogeneous material and FG sandwich blade is performed by comparison to the literature and ANSYS results, both showing good agreement.

Interesting frequency veering and the associated mode shift phenomena are found and discussed for different twist angles, thickness ratios, and aspect ratios in static natural frequency analysis. The results show that the fundamental mode is flapwise bending in every case. The mode shapes of the 2nd to 6th orders are changing when different geometric parameters are used; especially, the thickness ratio is 15–50

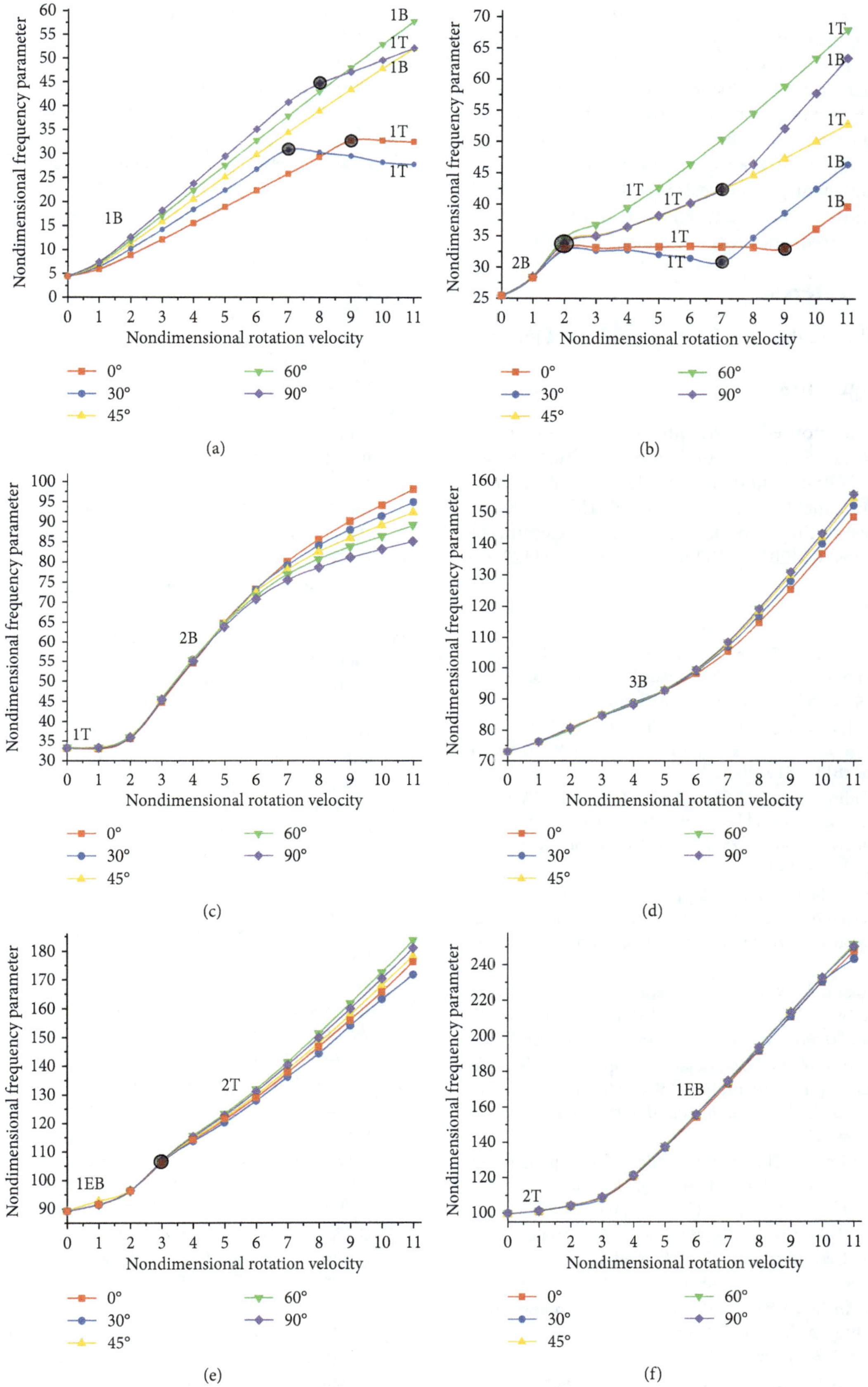

FIGURE 13: Variations of the first six frequency parameters with the rotation velocity and stagger angles ($\bar{l} = 3$, $\bar{h} = 20$, $n = 2$, $\bar{h}_l = 1/2$, $\theta = 30°$, and $\bar{r} = 1$). (a) 1st mode, (b) 2nd mode, (c) 3rd mode, (d) 4th mode, (e) 5th mode, and (f) 6th mode.

and the aspect ratio is 1.0–2.5. Higher-order frequency loci (≥4) may veer more than once between the bending and torsional vibrations of the FG sandwich blade.

The stagger angle is considered to study the dynamic frequency with varying rotation velocities. The results show that the frequency modes have more complicated shapes and stronger coupling in the rotating FG sandwich blade system. The Campbell diagram reveals that the rotating velocity, especially high rotating velocity, has a significant impact on the variations of frequency loci and mode shapes.

Conflicts of Interest

The authors declare that they have no conflicts of interest.

Acknowledgments

The authors acknowledge the financial support of the National Natural Science Foundation of China through Grant no. 11472056, Fundamental Research Program of Shenzhen Municipality no. JCYJ20160608153749600, and Scientific Research Project of Beijing Municipal Commission of Education nos. KM201811232002 and KM201711232002.

References

[1] A. Leissa and K. I. Jacob, "Three-dimensional vibrations of twisted cantilevered parallelepipeds," *Journal of Applied Mechanics*, vol. 53, no. 3, pp. 614–618, 1986.

[2] H. H. Yoo, J. H. Park, and J. Park, "Vibration analysis of rotating pre-twisted blades," *Computers and Structures*, vol. 79, no. 19, pp. 1811–1819, 2001.

[3] N. K. Chandiramani, C. D. Shete, and L. I. Librescu, "Vibration of higher-order-shearable pretwisted rotating composite blades," *International Journal of Mechanical Sciences*, vol. 45, no. 12, pp. 2017–2041, 2003.

[4] E. Carrera, M. Filippi, and E. Zappino, "Free vibration analysis of rotating composite blades via Carrera Unified Formulation," *Composite Structures*, vol. 106, no. 106, pp. 317–325, 2013.

[5] M. S. Qatu and A. W. Leissa, "Vibration studies for laminated composite twisted cantilever plates," *International Journal of Mechanical Sciences*, vol. 33, no. 11, pp. 927–940, 1991.

[6] S. Mohamed Nabi and N. Ganesan, "Comparison of beam and plate theories for free vibrations of metal matrix composite pre-twisted blades," *Journal of Sound and Vibration*, vol. 189, no. 2, pp. 149–160, 1996.

[7] H. H. Yoo and J. Chung, "Dynamics of rectangular plates undergoing prescribed overall motion," *Journal of Sound and Vibration*, vol. 239, no. 1, pp. 123–137, 2001.

[8] X. X. Hu, T. Sakiyama, and K. Itakura, "A new numerical procedure for free vibrations of pretwisted plates," *Archive of Applied Mechanics*, vol. 72, no. 4-5, pp. 330–341, 2002.

[9] S. H. Hashemi, S. Farhadi, and S. Carra, "Free vibration analysis of rotating thick plates," *Journal of Sound and Vibration*, vol. 323, no. 1-2, pp. 366–384, 2009.

[10] S. K. Sinha and K. E. Turner, "Natural frequencies of a pre-twisted blade in a centrifugal force field," *Journal of Sound and Vibration*, vol. 330, no. 11, pp. 2655–2681, 2011.

[11] J. Sun, L. Kari, and I. Lopez Arteaga, "A dynamic rotating blade model at an arbitrary stagger angle based on classical plate theory and the Hamilton's principle," *Journal of Sound and Vibration*, vol. 332, no. 5, pp. 1355–1371, 2013.

[12] J. S. Rao and K. Gupta, "Free vibrations of rotating small aspect ratio pretwisted blades," *Mechanism and Machine Theory*, vol. 22, no. 2, pp. 159–167, 1987.

[13] J. Sun, I. Lopez Arteaga, and L. Kari, "General shell model for a rotating pretwisted blade," *Journal of Sound and Vibration*, vol. 332, no. 22, pp. 5804–5820, 2013.

[14] S. K. Sinha and R. P. Zylka, "Vibration analysis of composite airfoil blade using orthotropic thin shell bending theory," *International Journal of Mechanical Sciences*, vol. 121, no. 2, pp. 90–105, 2017.

[15] R. E. Kielb, A. W. Leissa, J. C. Macbain, and K. S. Carney, "Joint research effort on vibrations of twisted plates, phase 1: final results," Tech. Rep. 1150, NASA Lewis Research Center, Cleveland, OH, USA, 1985.

[16] R. W. Claassen and C. J. Thorne, "Vibrations of a rectangular cantilever plate," *Journal of the Aerospace Sciences*, vol. 29, no. 11, pp. 1300–1305, 1962.

[17] D. Afolabi and O. Mehmed, "On curve veering and flutter of rotating blades," *Journal of Engineering for Gas Turbines and Power*, vol. 116, no. 3, pp. 702–708, 1994.

[18] H. H. Yoo and C. Pierre, "Modal characteristic of a rotating rectangular cantilever plate," *Journal of Sound and Vibration*, vol. 259, no. 1, pp. 81–96, 2003.

[19] H. H. Yoo, S. K. Kim, and D. J. Inman, "Modal analysis of rotating composite cantilever plates," *Journal of Sound and Vibration*, vol. 258, no. 2, pp. 233–246, 2002.

[20] S. Y. Oh, L. Librescu, and O. Song, "Vibration of turbomachinery rotating blades made-up of functionally graded materials and operating in a high temperature field," *Acta Mechanica*, vol. 166, no. 1–4, pp. 69–87, 2003.

[21] M. N. V. Ramesh and N. Mohan Rao, "Free vibration analysis of pre-twisted rotating FGM beams," *International Journal of Mechanics and Materials in Design*, vol. 9, no. 4, pp. 367–383, 2013.

[22] Y. Oh and H. H. Yoo, "Vibration analysis of rotating pretwisted tapered blades made of functionally graded materials," *International Journal of Mechanical Sciences*, vol. 119, pp. 68–79, 2016.

[23] J. L. Mantari and E. V. Granados, "Dynamic analysis of functionally graded plates using a novel FSDT," *Composites Part B: Engineering*, vol. 75, pp. 148–155, 2015.

[24] L. Li and D. G. Zhang, "Free vibration analysis of rotating functionally graded rectangular plates," *Composite Structures*, vol. 136, pp. 493–504, 2016.

[25] J. Sun, I. Lopez Arteaga, and L. Kari, "Dynamic modeling of a multilayer rotating blade via quadratic layerwise theory," *Composite Structures*, vol. 99, pp. 276–287, 2013.

[26] D. Cao, B. Liu, M. Yao, and W. Zhang, "Free vibration analysis of a pre-twisted sandwich blade with thermal barrier coatings layers," *Science China Technological Sciences*, vol. 60, no. 11, pp. 1747–1761, 2017.

[27] A. M. A. Neves, A. J. M. Ferreira, E. Carrera et al., "Static, free vibration and buckling analysis of isotropic and sandwich functionally graded plates using a quasi-3d higher-order shear deformation theory and a meshless technique," *Composites Part B: Engineering*, vol. 44, no. 1, pp. 657–674, 2013.

[28] T. Tsuji and T. Sueoka, "Free vibrations of pre-twisted plates: fundamental theory: vibration, control engineering,

engineering for industry," *JSME International Journal*, vol. 30, no. 264, pp. 958–962, 1987.

[29] M. S. Qatu, R. W. Sullivan, and W. Wang, "Recent research advances on the dynamic analysis of composite shells: 2000–2009," *Composite Structures*, vol. 93, no. 1, pp. 14–31, 2010.

[30] M. Amabili, *Nonlinear Vibrations and Stability of Shells and Plates*, Cambridge University Press, 2008.

[31] M. S. Qatu, *Vibration of Laminated Shells and Plates*, Elsevier, San Diego, CA, USA, 2004.

[32] J. N. Reddy, *Mechanics of Laminated Composite Plates and Shells: Theory and Analysis*, CRC Press, 2004.

10

Aero Engine Component Fault Diagnosis using Multi-Hidden-Layer Extreme Learning Machine with Optimized Structure

Shan Pang,[1] Xinyi Yang,[2] and Xiaofeng Zhang[1]

[1] *College of Information and Electrical Engineering, Ludong University, Yantai 264025, China*
[2] *Department of Aircraft Engineering, Naval Aeronautical and Astronautical University, Yantai 264001, China*

Correspondence should be addressed to Shan Pang; pangshanpp@163.com

Academic Editor: Kenneth M. Sobel

A new aero gas turbine engine gas path component fault diagnosis method based on multi-hidden-layer extreme learning machine with optimized structure (OM-ELM) was proposed. OM-ELM employs quantum-behaved particle swarm optimization to automatically obtain the optimal network structure according to both the root mean square error on training data set and the norm of output weights. The proposed method is applied to handwritten recognition data set and a gas turbine engine diagnostic application and is compared with basic ELM, multi-hidden-layer ELM, and two state-of-the-art deep learning algorithms: deep belief network and the stacked denoising autoencoder. Results show that, with optimized network structure, OM-ELM obtains better test accuracy in both applications and is more robust to sensor noise. Meanwhile it controls the model complexity and needs far less hidden nodes than multi-hidden-layer ELM, thus saving computer memory and making it more efficient to implement. All these advantages make our method an effective and reliable tool for engine component fault diagnosis tool.

1. Introduction

The aero gas turbine engine is susceptible to many problems, including erosion, corrosion, fouling, and foreign object damage during its operation [1]. These problems may cause engine component deterioration thus affecting the engine performance. Therefore, it is very important to develop engine component diagnostics methods using engine performance data to detect and isolate the component faults for the safety of aircrafts and reduce the maintenance cost.

Traditional model-based diagnostics methods, which are often used in practice, require an accurate engine mathematical model and their reliability often decreases as the system nonlinear complexities and modeling uncertainties increase. In essence, the engine component fault diagnosis is a challenging classification problem and could be resolved using neural network-based techniques. Applications of neural networks in engine fault diagnosis have been widely studied in the literature [2–7]. In recent years, a novel learning algorithm for single-hidden-layer neural networks called extreme learning machine (ELM) [8, 9] has been proposed and was applied in engine fault diagnosis. In ELM, the input weights and hidden biases are randomly generated, and the output weights are calculated by Moore-Penrose (MP) generalized inverse. It learns much faster with higher generalization performance than traditional gradient-based learning algorithms such as back-propagation. It also avoids many problems faced by traditional gradient-based learning algorithms such as stopping criteria, learning rate, and local minima problem. Yigang et al. [10] applied ELM to aircraft engine sensor fault diagnosis and the results show that ELM algorithm has higher classification precision and shorter training time than conventional BP neural network method. Li et al. [11] proposed a fusion diagnosis method of aero gas turbine engine component faults based on ELM and Kalman filters. To overcome the drawbacks of ELM, the input weights and input layer biases are optimized by differential evolution. In [12], an ELM with optimized input weights and hidden biases was applied to a gas turbine fan engine diagnostic problem and achieved better results than SVM and BP neural network methods. However, when the number of hidden nodes increases, the optimization of so many input weights and biases becomes more difficult and time consuming.

The basic ELM and most of its variants employ single hidden layer feed-forward networks, which limits its feature abstraction ability and classification performance in some real world applications. Recently, deep learning methods such as the deep belief network (DBN) [13], the stacked denoising autoencoder (SDAE) [14], and deep Boltzmann machine (DBM) [15] have shown better performance than shallow neural networks in machine learning area [16, 17]. Deep network architectures can be exponentially more efficient than shallow ones [18]. The latter may require a large number of hidden neurons to represent highly varying functions [19–21]. While deep architectures can represent these functions more efficiently, they thus outperform shallow models in many applications.

Inspired by the depth structure of deep learning networks, Kasun et al. [22] developed a multilayer learning architecture using ELM-based autoencoder as its building block. The deep architecture extracts features by a multilayer network, and the higher layers represent more abstract information than those from the lower ones. Test on MNIST shows that M-ELM performs on par with DBM and outperforms SDAE and DBN.

However, in [22] M-ELM employs a fixed network structure and tends to need a large-scale model with huge amount of hidden nodes when dealing with difficult classification tasks. It took a 700-700-15000 network structure on MNIST data set and needed to be implemented in a computer with 32 Gbytes RAM. However, such a network with huge number of hidden nodes cannot be implemented in computers with small and medium size RAM. In addition, designing a suitable network needs a lot of trials and experience. Moreover, a fixed network structure is not robust to noise and may perform even worse when the sensor noise level is high.

In order to address the above mentioned issues of M-ELM, in this paper, we proposed an effective multi-hidden-layer extreme learning machine algorithm which selects the optimal network structure automatically and adaptively. The new method adopts QPSO strategy to optimize the network structure according to both RMSE on training data set and the norm of output weights. Results on both MNIST data set and engine fault diagnosis application show that our method outperforms ELM, M-ELM, and other state-of-the-art deep learning methods in testing accuracy and robustness to sensor noise. And the QPSO helps to reduce the number of hidden nodes significantly, thus saving computation resources and making it more efficient to be implemented.

The rest of the paper is organized as follows. Section 2 gives a brief review of ELM, M-ELM, and QPSO algorithm. Section 3 presents the proposed OM-ELM. In Section 4, our method is applied on MNIST data set and compared with other methods. Section 5 compares OM-ELM with other methods on engine component fault diagnostics applications followed by the conclusions in Section 6.

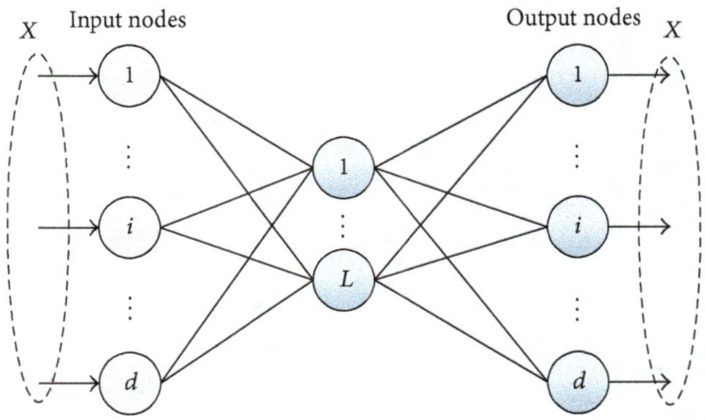

FIGURE 1: ELM-AE structure.

2. Preliminaries

2.1. Multi-Hidden-Layer Extreme Learning Machine. Kasun et al. developed a multi-hidden-layer learning architecture using ELM-based autoencoder (ELM-AE) as its building block for representational learning [22].

M-ELM performs layer-wise unsupervised training for each ELM-AE. However, unlike conventional deep learning algorithms, it does not require fine tuning and the unsupervised training is executed in a batch way. This makes M-ELM run faster than any deep learning algorithm.

2.1.1. ELM-AE. As Figure 1 illustrated, an ELM-AE has input layer, hidden layer, and output layer. Input data is used as output data $t = x$. Random weights and biases of the hidden nodes are chosen to be orthogonal. Orthogonalization of these randomly generated hidden parameters tends to improve ELM-AE's generalization performance [23]. In ELM-AE, the orthogonal random weights and biases of the hidden nodes project the input data to a different or equal dimension space, and they are calculated as

$$\begin{aligned} H &= g(w \cdot x + b), \\ w^T w &= I, \\ b^T b &= I, \end{aligned} \tag{1}$$

where w denotes the orthogonal random weights and b denotes the orthogonal random biases between the input and hidden nodes.

The output weight is calculated as follows:

$$\beta = \left(\frac{I}{C} + H^T H\right)^{-1} H^T X, \tag{2}$$

where H denotes ELM-AE's hidden layer outputs and X is its input and simultaneously output data. I/C is the regularization term, used to improve generalization performance and make the solution more robust.

2.1.2. M-ELM. Figure 2 illustrates the construction of M-ELM. As can be seen from the figure, the output weights with respect to input data are the first layer weights of M-ELM. And the output weights β^{i+1} of ELM-AE, with respect

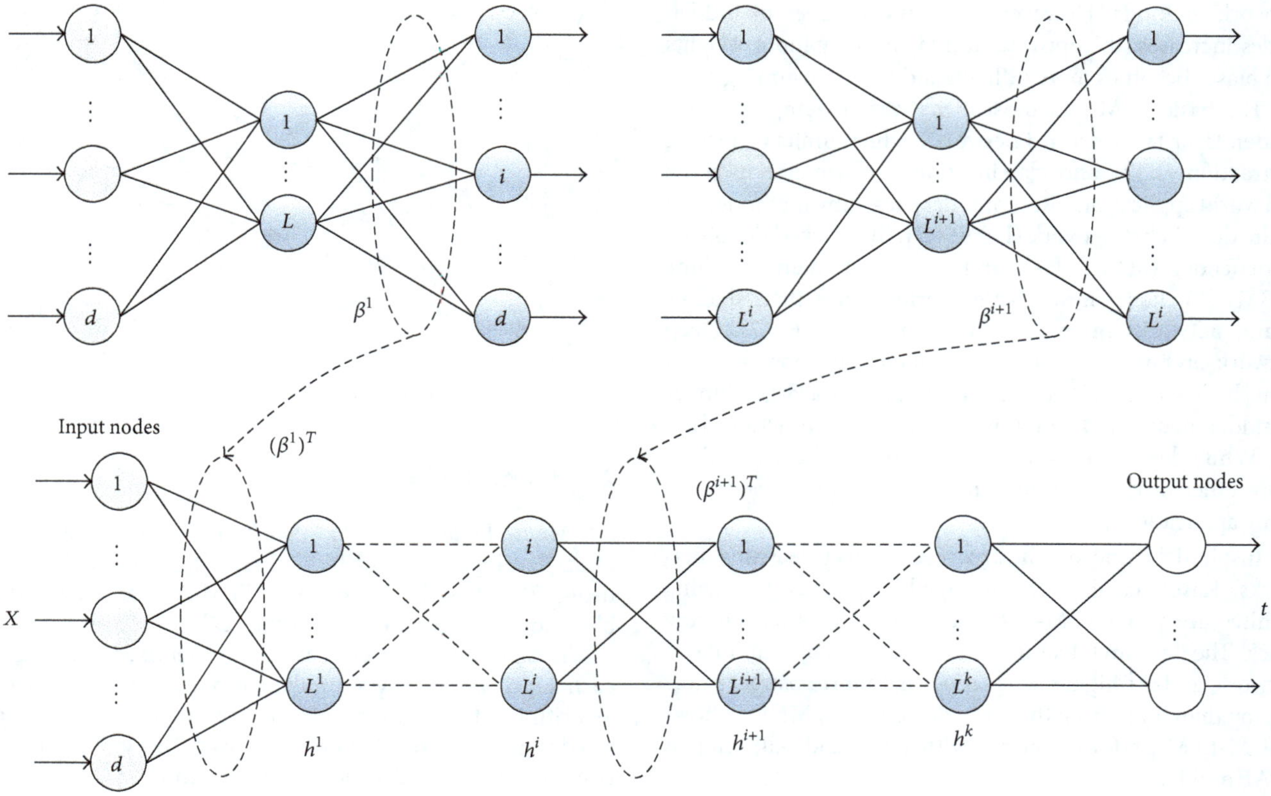

Figure 2: Construction of multi-hidden-layer ELM.

to ith hidden layer output h^i of M-ELM, are the $(i+1)$th layer weights of M-ELM, while the M-ELM output layer weights are calculated using regularized least squares as (2). As the output weights of an ELM-AE are the learned feature, M-ELM realizes a layer-wise feature abstraction like deep learning but without iteration process and fine tuning.

2.2. QPSO. QPSO solves the premature or local convergence problem of PSO and shows better performance than PSO in many applications [23, 24]. In QPSO, the state of a particle y is depicted by Schrodinger wave function $\psi(y,t)$, instead of position and velocity. The position and velocity of the quantum particle cannot be determined simultaneously. The probability of the particle's appearance in a position from probability density function $|\psi(y,t)|^2$, the form of which depends on the potential field the particle lies in, can be only learned. Employing the Monte Carlo method, for the ith particle y_i from the population, the particle moves according to the following iterative equation:

$$y_{i,j}(t+1) = \begin{cases} P_{i,j}(t) - \beta \cdot \left(mBest_j(t) - y_{i,j}(t)\right) \cdot \ln\left(\dfrac{1}{u_{i,j}(t)}\right) & \text{if } k \geq 0.5, \\ P_{i,j}(t) + \beta \cdot \left(mBest_j(t) - y_{i,j}(t)\right) \cdot \ln\left(\dfrac{1}{u_{i,j}(t)}\right) & \text{if } k < 0.5, \end{cases} \quad (3)$$

where $y_{i,j}(t+1)$ is the position of the ith particle with respect to the jth dimension in iteration t. $P_{i,j}$ is the local attractor of ith particle to the jth dimension and is defined as

$$P_{i,j}(t) = \varphi_j(t) \cdot pBest_{i,j}(t) + \left(1 - \varphi_j(t)\right) gBest_j(t), \quad (4)$$

$$mBest_j(t) = \frac{1}{N_p}\sum_{i=1}^{N_p} pBest_{i,j}(t), \quad (5)$$

where N_p is the number of particles and $pBest_i$ represents the best previous position of the ith particle. $gBest$ is the global best position of the particle swarm. $mBest$ is the mean best position defined as the mean of all the best positions of the population and k, u, and φ are random numbers distributed uniformly in $[0,1]$, respectively. Contraction-expansion coefficient β is used to control the convergence speed of the algorithm.

3. OM-ELM

Since M-ELM is not able to design a suitable network structure automatically and adaptively in different applications, it tends to need a large number of hidden nodes to attain a good classification performance, thus costing much more computation time and computer memory. Furthermore, there

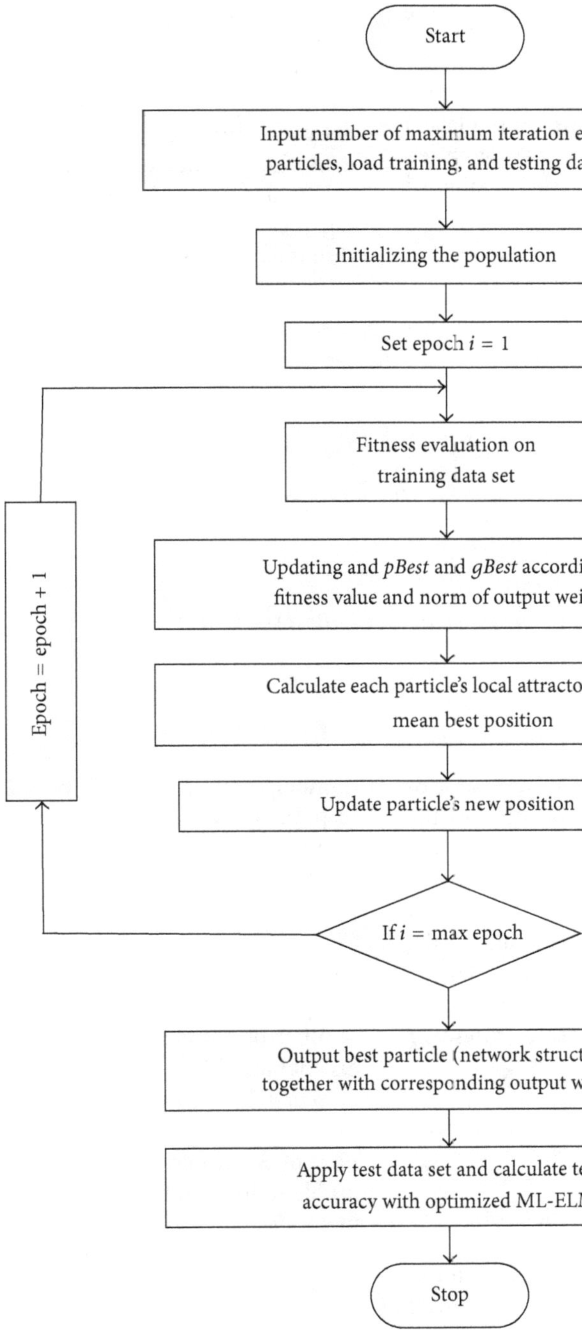

FIGURE 3: Flowchart of OM-ELM.

may exist many unnecessary redundant nodes, which leads to an ill-conditioned hidden output matrix and decreased generalization performance.

To address these problems and achieve an optimal network structure automatically, in this section, we proposed a method named OM-ELM, as illustrated in Figure 3. The method uses QPSO to optimize network structure according to both training accuracy and the norm of output weights to achieve a good generalization performance.

The main steps of the proposed OM-ELM are as follows.

Step 1 (initializing). Firstly, we generate the population of particles $p_i = [n_1, \ldots, n_H]$ randomly, where n_i, $i = 1, \ldots, H$, denotes the number of nodes in the ith hidden layer. Note that it must be integer and thus is rounded to integer by the following equation when it is not during iteration:

$$n_i = \text{round}\left(n_i\right). \tag{6}$$

In order to control complexity of network and save computing resources, n_i is restricted to a certain range with predefined upper and lower bonds according to the applications.

Step 2 (fitness evaluation). The corresponding output weights (the weights between the last hidden layer and output layer) of each particle (a potential network structure) are computed according to (2). Then the fitness of each particle is evaluated by the root mean square error between the desired output and estimated output:

$$f(\cdot) = \sqrt{\frac{\sum_{j=1}^{N} \sum_{i=1}^{K} \beta_i g\left(w_i \cdot x_j + b_i\right) - t_j}{N}}. \quad (7)$$

Step 3 (updating $pBest_i$ and $gBest$). With the fitness values of all particles in population, the best previous position for ith particle $pBest_i$ and the global best position $gBest$ of the current population is updated. As suggested in [25], neural network tends to have better generalization performance with the weights of smaller norm. Therefore, the fitness values along with the norm of output weights are considered together for updating $pBest_i$ and $gBest$. The updating criterion is as follows:

$$pBest_i = \begin{cases} p_i, & \left(f(pBest_i) - f(p_i) > \eta f(pBest_i)\right) \text{ or } \left(\left|f(pBest_i) - f(p_i)\right| < \eta f(pBest_i), \left\|wo_{p_i}\right\| < \left\|wo_{pBest_i}\right\|\right), \\ pBest_i, & \text{else,} \end{cases}$$

$$gBest = \begin{cases} p_i, & \left(f(gBest) - f(p_i) > \eta f(gBest)\right) \text{ or } \left(\left|f(gBest) - f(p_i)\right| < \eta f(gBest), \left\|wo_{p_i}\right\| < \left\|wo_{gBest}\right\|\right), \\ gBest, & \text{else,} \end{cases} \quad (8)$$

where $f(p_i)$, $f(pBest_i)$, and $f(gBest)$ are the fitness value of the ith particle's position, the best previous position of the ith particle, and the global best position of the population. wo_{p_i}, wo_{pBest_i}, and wo_{gBest} are the corresponding output weights (the weights between the last hidden layer and output layer) of the ith particle, the best previous position of the ith particle, and the global best position obtained so far. By this updating criterion, particles with smaller fitness values or smaller norms are more likely to be selected as $pBest_i$ or $gBest$.

Step 4. Calculate each particle's local attractor P_i and mean best position $mBest$ according to (4) and (5).

Step 5. Update particle's new position according to (3).

Steps 2 to 5 are repeated until the maximum number of epochs is reached. Finally, we obtained the optimized network structure and apply it to the testing data set.

4. Comparisons on MNIST Data Set

Before we apply the method to aero gas turbine engine fault diagnosis, in this section, we first applied it on MNIST handwriting data set [26]. The MNIST consists of 60 000 training images and 10 000 testing images of handwriting digits 0–9. As different digits have their unique shapes and different people write the number in their own ways, the MNIST is an ideal data set and commonly used to test the deep learning algorithms' performance.

In this section, we compare the proposed method with other four state-of-the-art classification methods: basic ELM, M-ELM, SDAE, and DBN. In our method, the maximum number of epochs of QPSO is 20, the number of particles in population is 20, and the upper bound of hidden nodes number in each hidden layer is 1500. As we have done some validation tests to pick the ridge parameter C in (2), it is set as 10^8 for all hidden layers.

We first test our method on the MNIST data set and obtained an optimized structure of 75-108-1473. Therefore, to compare fairly, all the other algorithms have roughly the same number of hidden nodes and the same three hidden layers except basic ELM.

For DBN, SDAE, and M-ELM, their hidden layer structure is 400-400-800. And ELM has one hidden layer with 1000 nodes (more nodes may cause a "run out of memory" problem in computer with medium size RAM). And all methods adopt sigmoid activation function. For the two deep learning methods, their learning rate is set as 0.1. The unsupervised pretraining epoch is set as 200 and supervised fine-tuning epoch is set as 400. The training data set is divided into mini-batches each containing 100 samples.

All simulations have been made in MATLAB R2008a environment running on a PC with 3.4 GHz CPU with 2 cores and 4 GB RAM. The results are listed in Table 1.

It can be seen from the table that our method achieved the highest testing accuracy among the state-of-the-art learning methods with similar size of hidden nodes. This testing accuracy is slightly less than the result in [22], but it takes only 1656 nodes, roughly one-tenth of the number used in [22]. Thus it saves much computation memory and is efficient to implement in common computers without large RAM. The computing time is larger than M-ELM and basic ELM as our method needs to evaluate the whole population for iterations. But compared with deep learning methods, OM-ELM saves much time.

TABLE 1: Performance of all methods on MNIST.

	Training accuracy	Testing accuracy	Training time (second)
DBN	0.994	0.984	38345.4
SDAE	0.989	0.983	26986.6
M-ELM	0.973	0.964	24.6
ELM	0.948	0.934	173.1
OM-ELM	0.979	**0.987**	8657.3

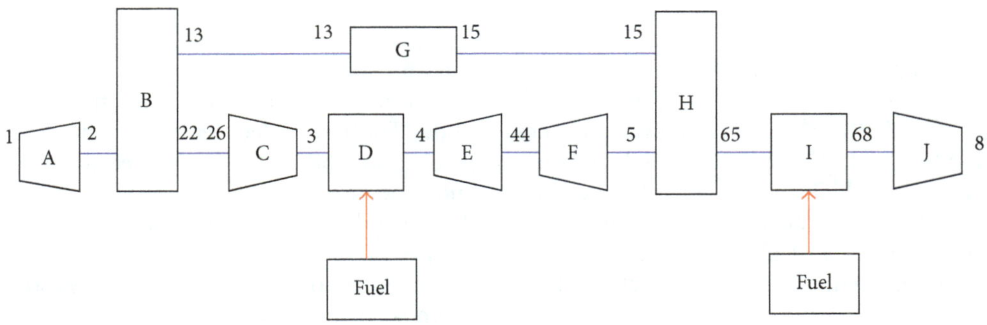

FIGURE 4: Schematic layout of studied turbine fan engine. A, inlet; B, fan; C, high pressure compressor; D, main combustor; E, high pressure turbine; F, low pressure turbine; G, external duct; H, mixer; I, afterburner; J, nozzle.

The good performance on MNIST suggests that our method is a good tool for engine fault diagnosis.

5. Fault Diagnosis Using OM-ELM

5.1. Engine Selection and Modeling. We evaluate the methods on a two-shaft turbine fan engine with a mixer and an afterburner (for confidentiality reasons the engine type is omitted), as is illustrated in Figure 4. This engine has a low by-pass ratio of 0.62.

The gas turbine engine is susceptible to many physical problems and these problems may result in the component fault and reduce the component isentropic efficiency. Thus they result in the deviations of some performance parameters such as rotational speed, pressures, and temperatures across different engine components. It is a practical way to detect and isolate the default component using measured engine performance data. However, the performance data of real faulty engine is very difficult to obtain and often belongs to manufacturer's or users' proprietary information and cannot be accessed easily. Therefore the component fault is usually simulated by engine mathematical model as suggested in [12]. In this study, we simulate the behavior of the engine with component faults using the engine mathematical model developed in MATLAB environment. By implanting certain magnitude of isentropic efficiency deterioration of some certain component to the engine performance model, we can obtain simulated engine performance parameter data with component faults.

5.2. Generating Component Fault Samples. In this study, we mainly focus on four rotating components, and different engine component fault scenarios including single and multifault cases were tested and are listed in Table 2.

TABLE 2: Different fault classes setting.

	C1	C2	C3	C4	C5	C6	C7	C8
LPC	F	—	—	—	F	F	F	F
HPC	—	F	—	—	F	—	F	—
LPT	—	—	F	—	—	F	F	F
HPT	—	—	—	F	—	—	—	F

TABLE 3: Description of the studied operating point.

Fuel flow	Environment setting parameters	
kg/s	Velocity (Mach number)	Altitude (km)
1.142	0.8	5.8

The first four columns represent four single fault cases. They are low pressure compressor (LPC) fault class, high pressure compressor (HPC) fault class, low pressure turbine (LPT) fault case, and high pressure turbine (HPT) fault class. Each class is labeled with an "F". C5 and C6 represent double faults cases, which are "LPC + HPC" and "LPC + LPT" fault class. And the last two columns represent triple faults cases, which are "LPC + HPC + LPT" and "LPC + LPT + HPT" fault class.

According to [12], the engine operating point has no obvious effect on classification accuracies of all fault detection methods; therefore, we chose only one operating point condition. The fuel flow and environment setting parameters of the operating point are listed in Table 3.

The input parameters of the training and test data set are the relative deviations of simulated engine performance parameters with component fault to the "healthy" engine parameters. And these parameters are selected by sensitivity analysis. They are low pressure rotor rotational speed n_1,

TABLE 4: Performance of all methods (noise level $N_l = 0.05$).

	Training accuracy		Testing accuracy		Training time (second)
	Mean	std	Mean	std	
DBN	0.960	0.0241	0.955	0.0345	131.05
SDAE	0.979	0.0048	0.976	0.0263	97.43
M-ELM	0.947	0.0134	0.973	0.0094	0.092
ELM	**0.991**	0.0025	0.914	0.0429	0.043
OM-ELM	0.972	0.0069	**0.985**	0.0218	30.47

high pressure rotor rotational speed n_2, total pressure and total temperature after LPC p_{22}^*, T_{22}^*, total pressure and total temperature after HPC p_3^*, T_3^*, total pressure and total temperature after HPT p_{44}^*, T_{44}^*, and total pressure and total temperature after LPT p_5^*, T_5^*. In this study, all the input parameters have been normalized into the range $[0, 1]$. And the output is different fault classes. For example, $[0\ 1\ 0\ 0\ 0\ 0\ 0\ 0]$ represents the second fault class (HPC fault) in Table 2.

Figure 5 shows the deviation response of engine performance parameters (i.e., the input parameters) against different fault patterns (1% loss in isentropic efficiency). It can be seen that the engine performance deviation responses of HPC and HPT are very similar. Thus it is very difficult to distinguish these two faults for a diagnosis method.

For each single fault class, 50 samples were generated by randomly selecting corresponding component isentropic efficiency deterioration magnitude within the range 1%–5%. For double faults classes, 100 instances were generated for each class by randomly setting the isentropic efficiency deterioration of two faulty components within the range 1%–5% simultaneously. The same method was applied for triple faults classes and each class has 300 samples. Altogether we have 1000 samples.

In real engine applications, there always exist sensor noises. To simulate real engine sensory signals, all input data are contaminated with measurement noise as the following equation:

$$X_n = X + N_l \cdot \sigma \cdot \text{rand}(1, n), \quad (9)$$

where X is clean input parameter, N_l denotes the imposed noise level, and σ is the standard deviation of data set. Meanwhile, for an imposed noise level, we expand the data samples from 1000 to 4000 proportionally. And we chose 3000 samples as training data set (the number of samples in each class is in proportion to their origin data set) and the left 1000 samples are testing data set.

5.3. Engine Component Fault Diagnosis by Five Methods

5.3.1. Parameter Settings. In our method, the maximum number of epochs is 30, the number of particles in population is 20, and the upper bound of hidden nodes number in each hidden layer is 200. The ridge parameter C is set as 10^7 for all hidden layers.

We first test our method on the data set with noise $N_l = 0.05$ and obtain an optimized structure 34-51-129. To compare fairly, the hidden layer structure for DBN, SDAE, and M-ELM is 60-60-90. Thus they have about the same total number of hidden nodes. And ELM has one single hidden layer of 210 nodes. For the two deep learning methods, their learning rate is set as 0.1. The unsupervised pretraining epoch is set as 200 and supervised fine-tuning epoch is set as 500. The training data set is divided into small mini-batches each containing 30 samples.

In order to account for the stochastic nature of these diagnostics methods, all the five methods are run 10 times separately. All simulations have been made in the same environment as in Section 4.

5.3.2. Comparisons of the Five Methods. Performance comparisons of the five methods were first conducted with small noise level $N_l = 0.05$. Table 4 lists the mean performance of training and testing accuracy on all fault classes and training time in 10 runs. Figure 6 is the mean testing accuracies on different fault classes.

It can be seen from Table 4 that basic ELM obtained the least testing accuracy but the highest training accuracy, which means the generalization performance of ELM is not as good as other methods with more hidden layers. This suggests that multi-hidden-layer structure is able to ameliorate the overfitting problem faced by single hidden layer neural network.

Among the four multi-hidden-layer methods, our method achieved the highest mean testing accuracy on both single fault classes and multifault classes. The mean testing accuracy on all fault classes (0.985) is also better than any other method, which is consistent with the results on MNIST. Furthermore, the testing performance is very stable as it achieves the least mean standard deviation. The testing performance of M-ELM is on par with SDAE and better than DBN.

Due to the iteration nature of OM-ELM, it costs much training time compared with basic ELM and M-ELM. But the training of our method is much faster than that of deep learning algorithms, such as DBN and SDAE.

Table 5 presents the confusion matrix of our method in a random run. It can be seen that our method achieved a satisfactory result. It recognized four fault classes with 100 percent accuracy. And the number of misclassified samples is less than other methods.

We also compared the performance of these methods with different noise level. Table 6 lists the mean training accuracy, testing accuracy, and training time in 10 runs with noise level $N_l = 0.1$. Our method still achieved the best testing accuracy among all methods.

TABLE 5: Confusion matrix of our method.

Actual class	Predicted class 1	2	3	4	5	6	7	8
1	**50**	0	0	0	0	0	0	0
2	0	**47**	**1**	**2**	0	0	0	0
3	0	0	**50**	0	0	0	0	0
4	0	**1**	0	**49**	0	0	0	0
5	0	0	0	0	**100**	0	0	0
6	0	0	0	0	0	**100**	0	0
7	0	0	**1**	0	0	**1**	**296**	**2**
8	0	0	0	**1**	**2**	0	**3**	**294**

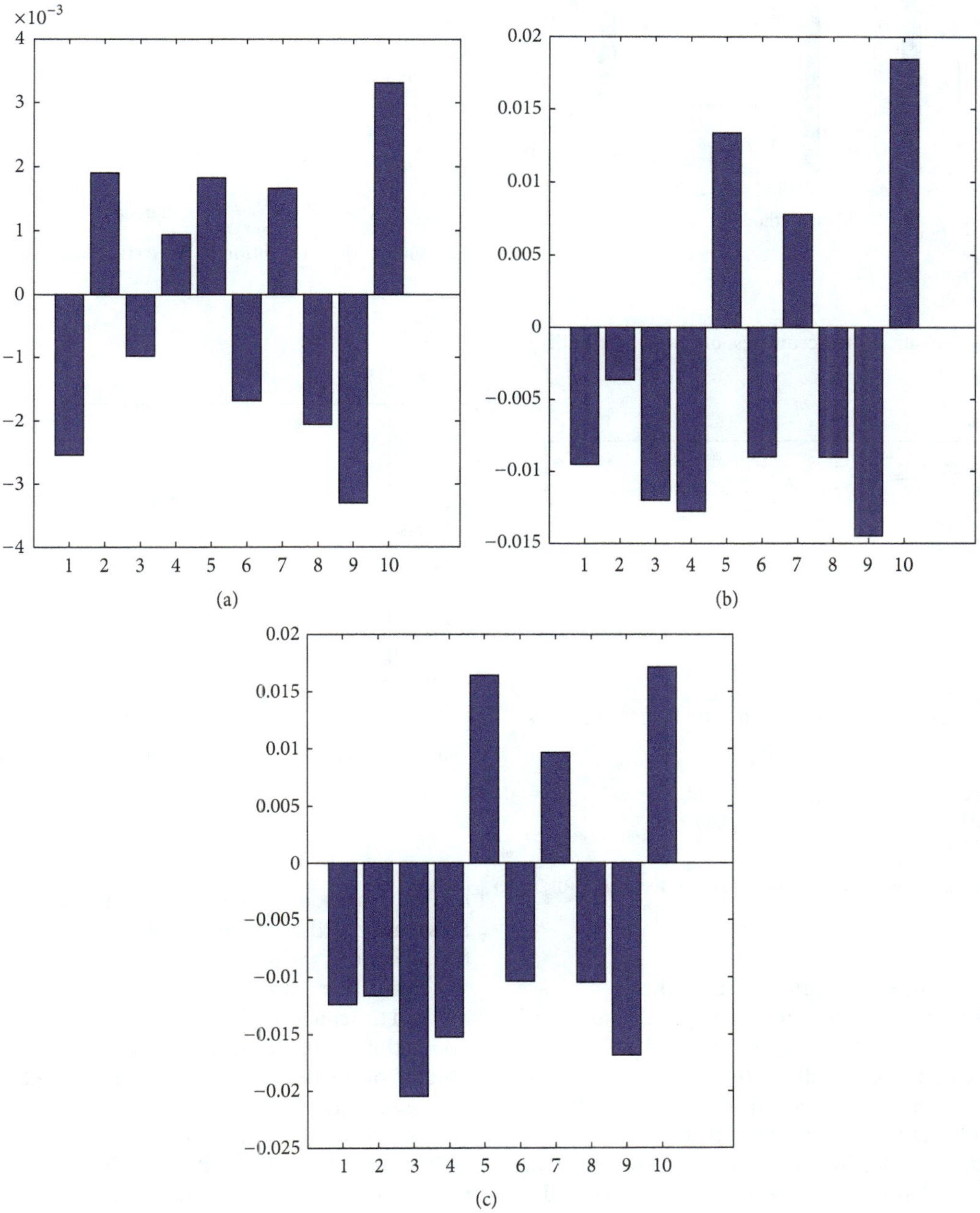

FIGURE 5: Sensitivity comparison of different component fault: (a) LPC, (b) HPC, and (c) HPT.

TABLE 6: Performance of all methods on data set with noise level $N_l = 0.1$.

	Training accuracy		Testing accuracy		Training time (second)
	Mean	std	Mean	std	
DBN	0.953	0.0173	0.948	0.0467	129.87
SDAE	0.969	0.0024	0.964	0.0135	99.24
M-ELM	0.923	0.0039	0.951	0.0195	0.097
ELM	**0.984**	0.0012	0.765	0.0466	0.048
OM-ELM	0.975	0.0097	**0.983**	0.0105	32.85

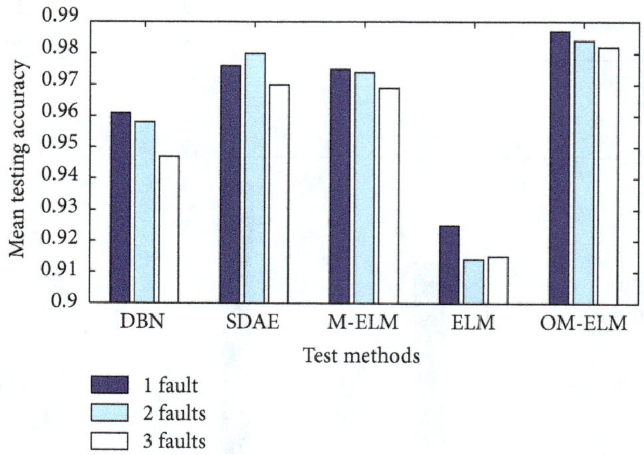

FIGURE 6: Mean classification accuracies of the 6 methods of different conditions.

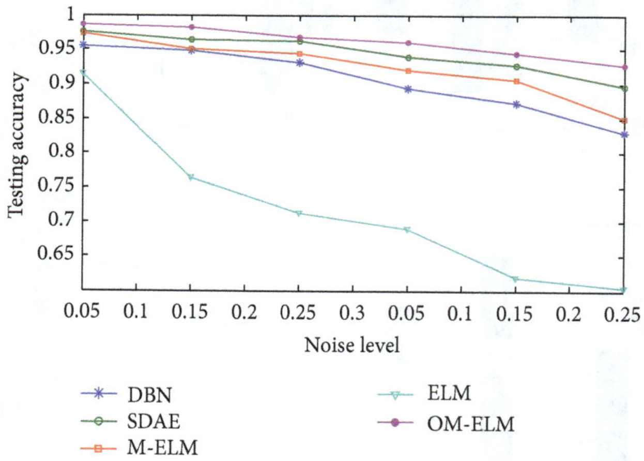

FIGURE 7: Testing accuracy versus noise level of five methods.

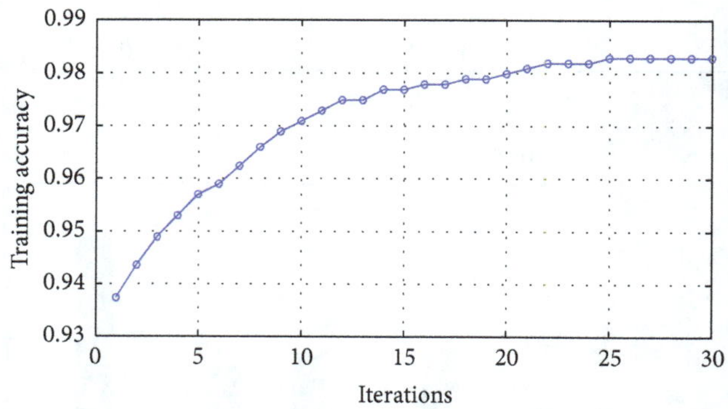

FIGURE 8: Mean evolution of the testing accuracy of our method.

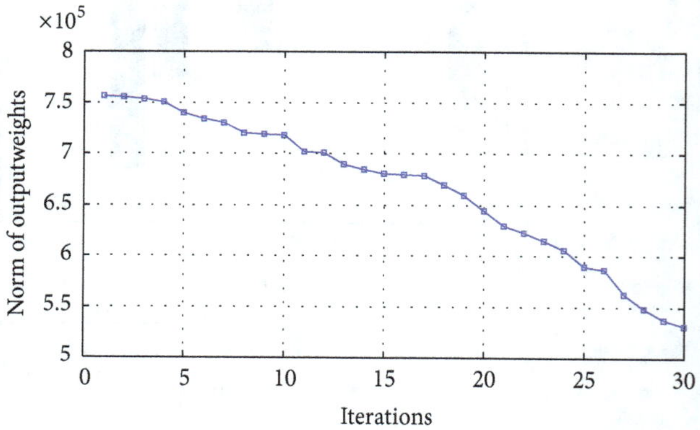

FIGURE 9: Mean evolution of output weight norm of our method.

To study how noise level affects the methods, we have tested the performance of these methods on six noise level conditions: $N_l = 0.05, 0.1, 0.15, 0.2, 0.25, 0.3$. The mean testing accuracy of the methods versus noise level is illustrated in Figure 7. It can be seen from the figure that all methods' testing accuracy decreases with the increasing of noise level. Due to its shallow architecture, basic ELM did not perform as good as other methods with multiple hidden layers. With the optimized network structure, our method was the least affected by sensor noise and achieved the best testing accuracy in all noise level conditions. This suggests that our method is more reliable and robust to the sensor noise and could be more suitable for aero gas turbine engine fault diagnosis tasks.

To further show how the QPSO strategy helps our method to achieve such a good performance, the evolution of populations' mean testing accuracy and the norm of output weights on noise level $N_l = 0.1$ in a single run is listed in Figures 8 and 9, respectively.

As testing accuracy values along with the norm of output weights are all considered in the QPSO updating criterion, the testing accuracy keeps increasing with iterations and the norm decreases with iterations. In the end, the optimized network structure is attained and our method is able to obtain a good classification and generalization capability.

6. Conclusions

In this paper, we have proposed a multi-hidden-layer extreme learning machine with network structure optimized by QPSO. We have evaluated the effectiveness of the method on MNIST data set and a gas turbine fan engine component fault diagnosis. Results on both applications show that our method not only outperforms deep learning algorithms in classification performance and training time, but also is better than M-ELM in terms of testing accuracy and stability with a more robust and reliable network structure.

The good performance is attributed to the QPSO strategy, which is able to automatically optimize the network structure according to RMSE in training set and norm of output weights, thus attaining a good classification performance and generalization capability in different applications. Moreover, it controls the scale of the network, thus significantly reducing hidden nodes required in original M-ELM and making the method more efficient. Also our method is more independent of engine mathematical model and easy to implement. In practice, if there were enough and adequate engine performance data, it would be suitable and effective to use our method for engine fault diagnostics application. But the applicability of method is mainly limited by engine performance data. Sometimes, engine performance data is not enough, and some engines may have fewer measuring parameters. In that case, the application of the developed method could be a future research topic.

Competing Interests

The authors declare no conflict of interests regarding the publication of this paper.

References

[1] X. Yang, W. Shen, S. Pang, B. Li, K. Jiang, and Y. Wang, "A novel gas turbine engine health status estimation method using quantum-behaved particle swarm optimization," *Mathematical Problems in Engineering*, vol. 2014, Article ID 302514, 11 pages, 2014.

[2] S. O. T. Ogaji, Y. G. Li, S. Sampath, and R. Singh, "Gas path fault diagnosis of a turbofan engine from transient data using artificial neural networks," in *Proceedings of the 2003 ASME Turbine and Aeroengine Congress*, ASME Paper No. GT2003-38423, Atlanta, Ga, USA, June 2003.

[3] L. C. Jaw, "Recent advancements in aircraft Engine Health Management (EHM) technologies and recommendations for the next step," in *Proceedings of the 50th ASME International Gas Turbine & Aeroengine Technical Congress*, pp. 683–695, June 2005.

[4] S. Osowski, K. Siwek, and T. Markiewicz, "MLP and SVM networks—a comparative study," in *Proceedings of the 6th Nordic Signal Processing Symposium (NORSIG '04)*, pp. 37–40, June 2004.

[5] M. Zedda and R. Singh, "Fault diagnosis of a turbofan engine using neural networks—a quantitative approach," in *Proceedings of the 34th AIAA/ASME/SAE/ASEE Joint Propulsion Conference and Exhibit*, AIAA 98-3602, Cleveland, Ohio, USA, 1998.

[6] C. Romessis, A. Stamatis, and K. Mathioudakis, "A parametric investigation of the diagnostic ability of probabilistic neural networks on turbo fan engines," ASME 2001-GT-11, 2001.

[7] A. J. Volponi, H. DePold, R. Ganguli, and C. Daguang, "The use of kalman filter and neural network methodologies in gas turbine performance diagnostics: a comparative study," *Journal of Engineering for Gas Turbines and Power*, vol. 125, no. 4, pp. 917–924, 2003.

[8] G.-B. Huang, Q.-Y. Zhu, and C.-K. Siew, "Extreme learning machine: a new learning scheme of feedforward neural networks," in *Proceedings of the IEEE International Joint Conference on Neural Networks*, pp. 985–990, Budapest, Hungary, July 2004.

[9] G.-B. Huang, Q.-Y. Zhu, and C.-K. Siew, "Extreme learning machine: theory and applications," *Neurocomputing*, vol. 70, no. 1-3, pp. 489–501, 2006.

[10] S. Yigang, L. Jingya, and Z. Zhen, "Aircraft engine sensor diagnosis based on extreme learning machine," *Transducer and Microsystem Technologies*, no. 33, pp. 23–26, 2014.

[11] Y. Li, Q. Li, X. Huang, and Y. Zhao, "Research on gas fault fusion diagnosis of aero-engine component," *Acta Aeronautica et Astronautica Sinica*, vol. 35, no. 6, pp. 1612–1622, 2014.

[12] X. Yang, S. Pang, W. Shen, X. Lin, K. Jiang, and Y. Wang, "Aero engine fault diagnosis using an optimized extreme learning machine," *International Journal of Aerospace Engineering*, vol. 2016, Article ID 7892875, 10 pages, 2016.

[13] G. E. Hinton and R. R. Salakhutdinov, "Reducing the dimensionality of data with neural networks," *Science*, vol. 313, no. 5786, pp. 504–507, 2006.

[14] P. Vincent, H. Larochelle, I. Lajoie, Y. Bengio, and P.-A. Manzagol, "Stacked denoising autoencoders: learning useful representations in a deep network with a local denoising criterion," *Journal of Machine Learning Research*, vol. 11, pp. 3371–3408, 2010.

[15] R. Salakhutdinov and H. Larochelle, "Efficient learning of deep boltzmann machines," *Journal of Machine Learning Research*, vol. 9, pp. 693–700, 2010.

[16] D. Yu and L. Deng, "Deep learning and its applications to signal and information processing [exploratory DSP]," *IEEE Signal Processing Magazine*, vol. 28, no. 1, pp. 145–154, 2011.

[17] N. Lopes, B. Ribeiro, and J. Gonçalves, "Restricted Boltzmann machines and deep belief networks on multi-core processors," in *Proceedings of the International Joint Conference on Neural Networks*, pp. 1–7, Brisbane, Australia, June 2012.

[18] N. Le Roux and Y. Bengio, "Deep belief networks are compact universal approximators," *Neural Computation*, vol. 22, no. 8, pp. 2192–2207, 2010.

[19] H. Larochelle, D. Erhan, A. Courville, J. Bergstra, and Y. Bengio, "An empirical evaluation of deep architectures on problems with many factors of variation," in *Proceedings of the 24th International Conference on Machine Learning (ICML '07)*, pp. 473–480, Corvallis, Ore, UA, June 2007.

[20] N. Le Roux and Y. Bengio, "Representational power of restricted Boltzmann machines and deep belief networks," *Neural Computation*, vol. 20, no. 6, pp. 1631–1649, 2008.

[21] Y. Bengio, "Learning deep architectures for AI," *Foundations and Trends in Machine Learning*, vol. 2, no. 1, pp. 1–27, 2009.

[22] L. L. C. Kasun, H. Zhou, G.-B. Huang, and C. M. Vong, "Representational learning with extreme learning machine for big data," *IEEE Intelligent Systems*, vol. 28, no. 6, pp. 31–34, 2013.

[23] J. Sun, C.-H. Lai, W.-B. Xu, Y. Ding, and Z. Chai, "A modified quantum-behaved particle swarm Optimization," in *Proceedings of the 7th International Conference on Computational Science*, pp. 294–301, Beijing, China, May 2007.

[24] M. Xi, J. Sun, and W. Xu, "An improved quantum-behaved particle swarm optimization algorithm with weighted mean best position," *Applied Mathematics and Computation*, vol. 205, no. 2, pp. 751–759, 2008.

[25] Q.-Y. Zhu, A. K. Qin, P. N. Suganthan, and G.-B. Huang, "Evolutionary extreme learning machine," *Pattern Recognition*, vol. 38, no. 10, pp. 1759–1763, 2005.

[26] Y. LeCun, L. Bottou, Y. Bengio, and P. Haffner, "Gradient-based learning applied to document recognition," *Proceedings of the IEEE*, vol. 86, no. 11, pp. 2278–2323, 1998.

11

Effects of Active and Passive Control Techniques on Mach 1.5 Cavity Flow Dynamics

Selin Aradag, Kubra Asena Gelisli, and Elcin Ceren Yaldir

Department of Mechanical Engineering, TOBB University of Economics and Technology, Sogutozu Cad. No. 43, 06560 Ankara, Turkey

Correspondence should be addressed to Selin Aradag; selinaradag@gmail.com

Academic Editor: Mahmut Reyhanoglu

Supersonic flow over cavities has been of interest since 1960s because cavities represent the bomb bays of aircraft. The flow is transient, turbulent, and complicated. Pressure fluctuations inside the cavity can impede successful weapon release. The objective of this study is to use active and passive control methods on supersonic cavity flow numerically to decrease or eliminate pressure oscillations. Jet blowing at several locations on the front and aft walls of the cavity configuration is used as an active control method. Several techniques are used for passive control including using a cover plate to separate the flow dynamics inside and outside of the cavity, trailing edge wall modifications, such as inclination of the trailing edge, and providing curvature to the trailing edge wall. The results of active and passive control techniques are compared with the baseline case in terms of pressure fluctuations, sound pressure levels at the leading edge, trailing edge walls, and cavity floor and in terms of formation of the flow structures and the results are presented. It is observed from the results that modification of the trailing edge wall is the most effective of the control methods tested leading to up to 40 dB reductions in cavity tones.

1. Introduction

Supersonic flow over cavities has been a fluid dynamics phenomenon of interest since the 1960s. Cavities represent the models of bomb bays of aircraft; therefore, they have been crucial for aerodynamic applications. Supersonic aircraft carry the stores inside and the velocity must be decreased to subsonic zone during store separation. This leads to more time over the target leading to increased possibility of being detected by a radar. Carrying the stores outside of the aircraft is a possible solution. However, this increases the radar signature, as well as the aerodynamic loads and heating. Therefore, it is important for supersonic aircraft to carry the stores inside and to be able to release them at supersonic speeds.

Supersonic cavity flow has gained importance due to these necessities. Although performing experiments at these speeds requires specifically designed wind tunnels, the advent of the computational techniques and computer capabilities with the help of computer clusters increased interest towards this flow. The flow is three-dimensional, turbulent, and very complicated in nature due to its high speed, different modes of vortical phenomenon, and high speed waves.

There are several studies in literature that explain the supersonic flow structure inside the cavities based on rigorous experiments or time consuming computations. It can be said that the cavity dynamics is now known and well documented and can be explained by combining several fluid dynamics phenomenon, although it is complicated and random in nature [1]. The pressure fluctuations inside the cavity are very high, the peaks can be three times the mean pressure and this increases the sound pressure levels in the cavity which may cause the unsuccessful release of the stores. In some extreme cases, the stores can return and hit the aircraft.

The next step in cavity flow research is the application of new or established control techniques on the flow structure and to observe the effects of these techniques on the flow, in terms of pressure fluctuations and sound pressure levels. If the pressure fluctuations can be eliminated or at least suppressed, this can be a meaningful step on the road to supersonic era. There are several studies in literature on different active and passive control techniques to suppress these pressure oscillations. Lawson and Barakos present a review of numerical simulations for high speed, turbulent cavity flows [2]. Several experimental and numerical control techniques are to decrease the pressure oscillations inside the cavity and are summarized in their review. Cattafesta III et al. also provide a review on active control techniques for pressure oscillations in the cavities focusing on practical applications [3]. Rizzetta et al. [4] performed Large Eddy Simulations (LES) of supersonic cavity flow, including the simulations with applied control techniques. They also provided comparisons with experimental data where available. Comte et al. carried out simulations of high-subsonic cavity flow controlled with a spanwise rod [5]. Levasseur et al. [6] performed LES of flow on unstructured meshes by implementing passive control techniques on subsonic cavities, as well as experiments. After validation of the baseline case with experiments, they concentrated on two passive acoustic flow control methods: rod in cross-flow and flat-top spoiler. Leading edge blowing was proved to be useful for the alleviation of cavity loads by Arunajatesan et al. [7]. Aradag et al. [8], Yilmaz and Aradag [9], and Lazar et al. [10] concentrated on localized laser energy deposition for active cavity flow control. Another interesting passive control technique proposed by Lee et al. [11] is a subsonic cavity installed at the leading edge of the supersonic cavity. They also worked on the determination of the effects of a triangular bump on the oscillations.

Experimental studies of flow control for supersonic cavities have also been a topic of interest for several years. For example, Ukeiley et al. [12] worked on control of cavity flow using active fluid injection methods and passive modification methods for cavity walls. Alvi and Cattafesta III [13] investigated the effects of zero-net mass-flux actuators on several types of flow including the flow over the cavity. The power of using smoke, shadowgraph, Schlieren images, and Particle Image Velocimetry is investigated. Zhuang et al. used supersonic microjets for Mach 2 cavity flow control [14]. Flow control using plasma actuators was also investigated for supersonic cavities by Narayanaswamy et al. [15]. Another research group who worked on control of supersonic cavity oscillations by microjet injection is Thangamani and Kurian [16]. They observed in their experiments that microjets reduce the coherence of the shear layer but increase the turbulence.

In this research, the objective is to suppress pressure oscillations caused by the supersonic flow over cavities with the help of active and passive control techniques. Jet blowing at several locations on the front and aft walls of the cavity configuration is used as an active control method. Several techniques are used for passive control including, using a cover plate to separate the flow dynamics inside and outside of the cavity, trailing edge wall modifications, such as inclination of the trailing edge, and providing curvature to the trailing edge wall. The results of active and passive control techniques are compared with the baseline case in terms of pressure fluctuations both on the trailing wall and on the cavity floor, sound pressure levels at the leading edge, trailing edge walls, and cavity floor and in terms of formation of the flow structures as presented by density, pressure, temperature, streamwise velocity contours, and streamlines.

Table 1: Numerical parameters.

Total pressure	66.4 kPa
Total temperature	218 K
Mach number	1.5
Reynolds number	1.09×10^6
Cavity length	0.012065 m
Cavity depth	0.0238 m
Cavity width	0.0635 m
Boundary layer thickness	0.0051 m

2. Methodology

Baseline case used in the computations is taken from the experiments of Kaufman et al. [17]. Length to depth ratio of the cavity is 5.07 and the Mach number is 1.5. Table 1 shows the parameters used in the CFD analysis.

A representative sketch of the geometry of the cavity under consideration is shown in Figure 1(a). Figure 1(b) shows the geometry for the computations with the dimensions in millimeters.

The details of the baseline computations are given in a previous study of our group [18]. The computations are performed using commercial Ansys Fluent software. For time resolution, second-order implicit method is used with 20 inner iterations for each time step. Second-order upwind method is used for space discretization. Density based algorithm is utilized since the flow is compressible. k-ω two-equation turbulence model is used in Reynolds averaged Navier-Stokes (RANS) equations to obtain the transient, turbulent flow characteristics.

Outlet and upper wall boundary conditions are selected as pressure far field, and these boundaries are placed far enough to avoid reflection. No slip wall boundary condition is used for the cavity walls. The solution of the two-dimensional steady-state Navier-Stokes equations over a flat plate is used as inflow boundary condition to make sure the flow is fully developed at the entrance of the cavity. Boundary layer thickness of the incoming flow matches the experimental values. Per Kaufman et al.'s experiments [17], the boundary layer thickness is 0.0051 m. The boundary layer thickness used as the inlet boundary condition matches the experiments as shown in Figure 2.

Flow characteristics for the baseline case (Mach 1.5 supersonic cavity flow with $L/D = 5.07$) are shown in Figure 3. Cavity flow is very complex in nature, comprising vortex-wall, shear layer-wall, and shock wave interactions causing pressure fluctuations. The flow is transient, however periodic in nature. The period of the fluctuations is called "Rossiter period" named after the pioneer of the cavity flow, Rossiter [19]. The trailing edge wall of the cavity witnesses more pressure fluctuations and higher sound pressure levels (SPL) because of the wall-shear layer interactions.

The following flow phenomena are observed in one Rossiter period of the periodic cavity flow:

(1) The free stream flow separates inside the cavity and reattaches at the trailing edge wall.

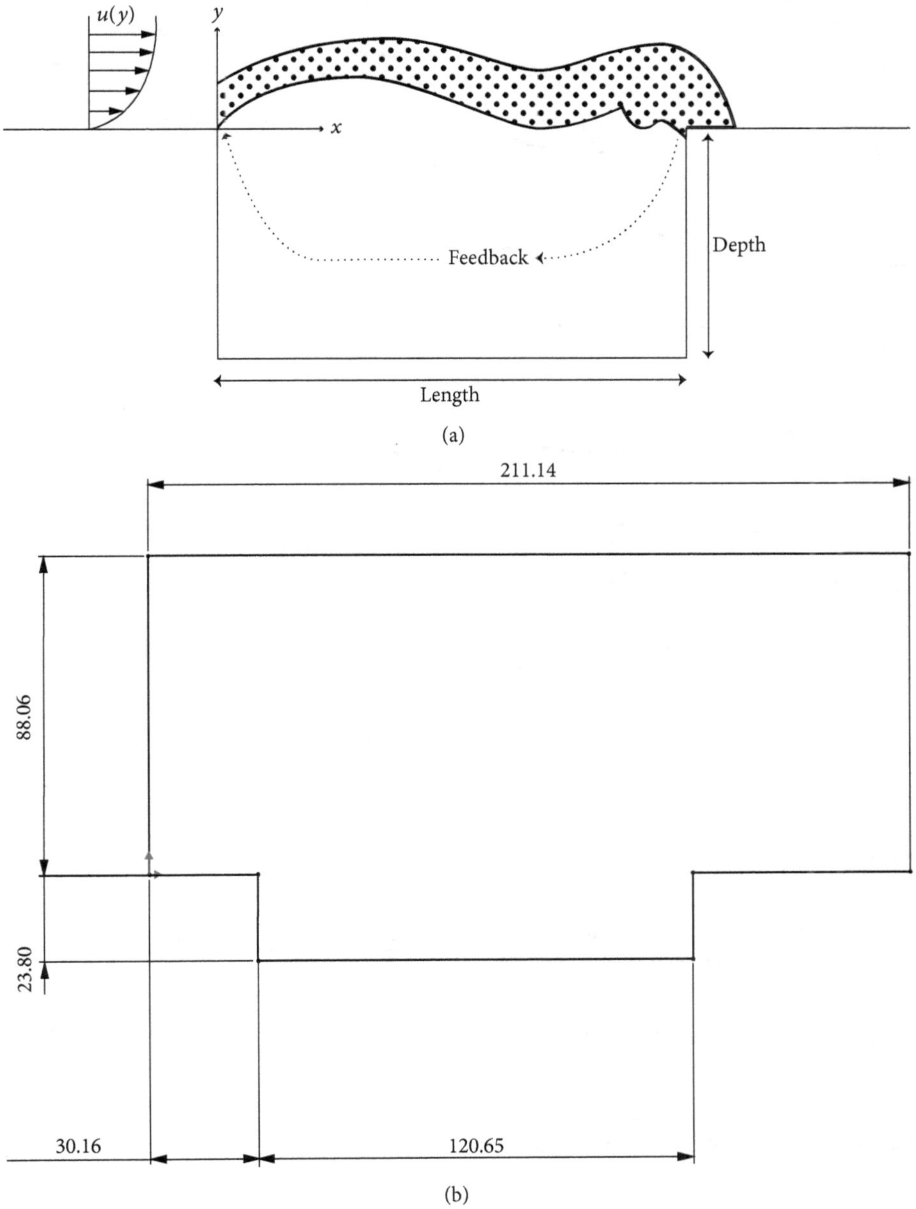

FIGURE 1: (a) Schematic view of the cavity. (b) Dimensions of the numerical domain.

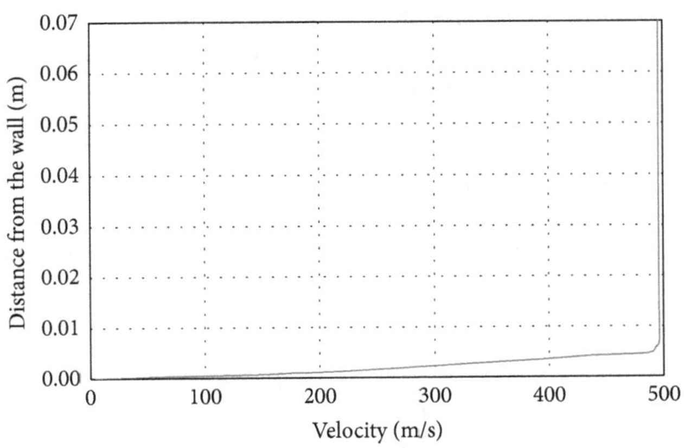

FIGURE 2: Boundary layer velocity profile used for the inlet boundary condition.

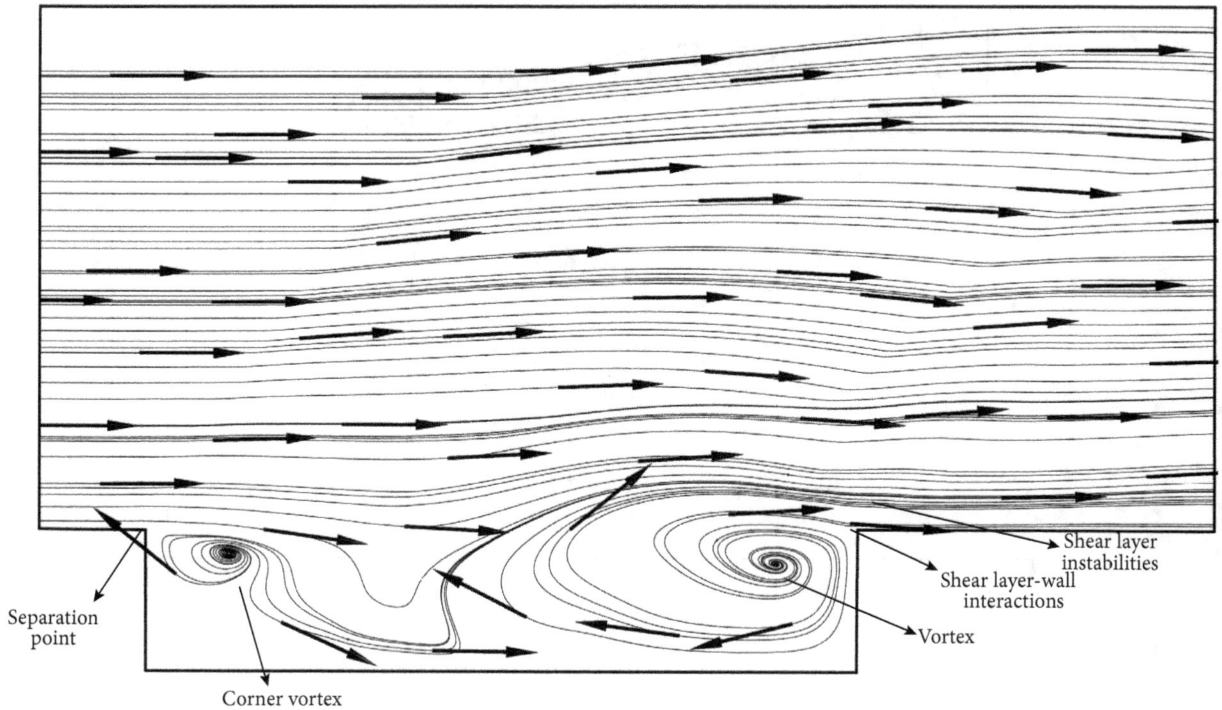

FIGURE 3: Supersonic cavity flow dynamics.

(2) The shear layer reattaches with the flow from inside the cavity and this causes vortices inside the cavity.

(3) The pressures inside the cavity decrease because of the vortices and the shear layer is bended and mass leaves the inside of the cavity.

(4) The mass transfer causes pressure fluctuations at the trailing wall of the cavity.

(5) The flow reaches back to the leading wall of the cavity and the pressures inside the cavity rise and mass is transferred to the outside zone.

Time step in the computations is 1 microsecond and the computations are performed for 14 periods corresponding to 12,000 time steps. A time step of 10 microseconds is also used and the results are compared with the results for a time step of 1 microsecond to select the proper time step. It is seen from the results that the periodic nature of the flow cannot be observed with a time step of 10 milliseconds. The pressures do not fluctuate, which is not physical.

Grid refinement study is also an essential part of this work as for all computational research. Six different grids are generated and the results are compared to each other and the experimental data of Kaufman et al. [17] to select the proper grid to be used in further flow control simulations. Table 2 shows the properties of the computational meshes used in the study.

If the pressure fluctuations can be eliminated or at least suppressed, this can be a meaningful step for flow control research. Passive and active flow control techniques are applied to the flow to decrease the pressure fluctuations and consequently the sound pressure levels (SPL) inside and on the walls of the cavity to enable successful store release. The advantage of passive control techniques is the fact that they are simple and inexpensive. However, once they are applied to the flow, they cannot be changed when it is necessary, since they include geometry modifications. Active control techniques are applied in real time only when it is necessary; that is, the flow structure needs to be modified. At other times, they can be turned off, which makes them expensive but powerful. Jet blowing through four different locations and the combinations of these four locations are the active flow control method utilized. In terms of passive control, using a cover plate to separate the flow dynamics inside and outside of the cavity, trailing edge wall modifications, such as inclination of the trailing edge (with different slopes), and providing curvature to the trailing edge wall are used. The test cases for flow control are summarized in Table 3. Figure 4 shows these modifications on a sketch for clarity.

TABLE 2: The properties of the computational meshes used in the study.

Grid	Number of grid cells	Average $y+$	Minimum $y+$
1	4055	41.8	3.31
2	16104	11.9	0.31
3	36321	3.4	0.07
4	54721	3.0	0.05
5	64410	3.1	0.03
6	337356	1.21	<0.001

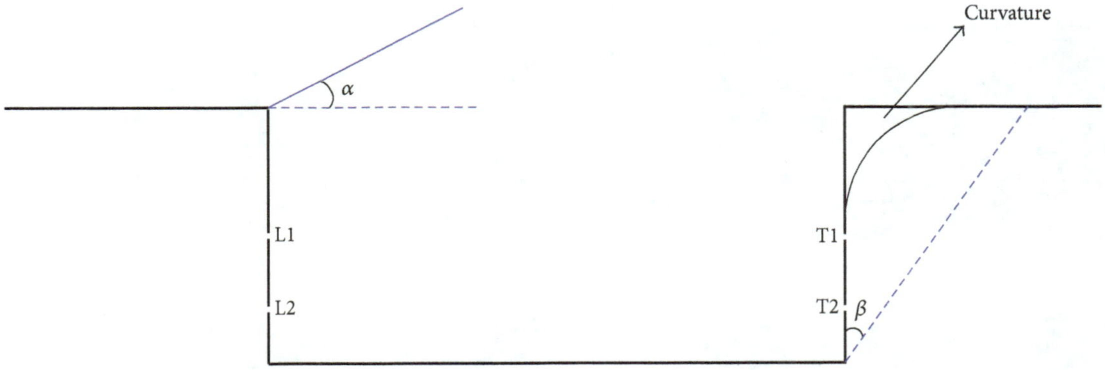

FIGURE 4: Modifications for passive and active flow control.

TABLE 3: Active and passive control cases.

Control method	Case 1	Case 2	Case 3	Case 4	Case 5
Trailing edge wall modification (TWM)	$\beta = 30^\circ$	$\beta = 45^\circ$	$\beta = 60^\circ$	—	—
Trailing edge curvature (TC)	$r = 5$ mm	$r = 10$ mm	$r = 15$ mm	$r = 20$ mm	—
Cover plate (CP)	$\alpha = 0^\circ$	$\alpha = 15^\circ$	$\alpha = 30^\circ$	$\alpha = 45^\circ$	—
Jet blowing (JB)	T1 only	T1 and T2	L1 only	L1 and L2	All active

3. Results

The results for the baseline case will be presented in the first section of this part and then the results for the control cases will be shown in detail.

3.1. Results of the Baseline Computations. Table 4 shows the results for the frequency of the first mode of oscillations. The prediction from the Rossiter formulation for this frequency is 1109 Hz. As seen in Table 3, grids 2, 3, 4, and 5 have similar results in terms of frequencies. The first and the last grids deviate.

The fourth grid was selected for further computations, both based on the flow structures and frequencies. The prediction for the second mode frequency is 2218 Hz. The second mode frequency is predicted as 2167 Hz with computations with an error of 2.3% calculated based on Rossiter's theoretical formula. The experimental data of Kaufman et al. [17] shows the sound pressure levels at six locations on cavity floor. Table 5 shows the experimental results, their location, and computational predictions for the same location with the selected grid.

Pressure fluctuations for the baseline case together with the streamlines are shown in Figure 5. The formation of vortices due to shear layer reattachment, pressure fluctuations inside the cavity during the whole feedback cycle, bending of the shear layer causing mass to leave the cavity, and the formation of high pressure zones especially at the trailing edge wall of the cavity are shown in the figure for one Rossiter period. The pressures are much higher at the trailing edge wall than the leading edge wall. A circulation region fills the cavity. Shocks and vortices interact which causes pressure fluctuations. Figure 6 shows the density contours for the baseline case which clearly demonstrate the supersonic nature of the flow. The feedback mechanism inside the cavity is also observed in density contours as in the pressure contours. Kelvin Helmholtz instability of the shear layer can also be detected in the density contours. The temperature contours for several time steps are presented in Figure 7. The x-velocity contours and streamlines are shown in Figure 8. They also demonstrate the vortex region that is formed inside the cavity clearly.

3.2. Results for Controlled Computations. Firstly, the results of different techniques will be presented and then, the results comparing the most effective case of each control method will be shown. The comparisons are made in terms of pressure values at the control point on the trailing edge wall, which is the location where the pressure fluctuations are expected to be maximum and Fast Fourier Transform (FFT) applied to this pressure distribution, and sound pressure levels for the cavity walls (front, back, and bottom walls). The location of the control point on the trailing edge wall is 0.6 cavity depth away from the bottom and 0.4 cavity depth away from the trailing edge of the cavity. Power spectra at several locations on the cavity floor are also presented both for the baseline case and for controlled simulations for comparison purposes. Pressure, density, streamwise velocity contours, and streamlines for various time steps are used for comparison, as well.

Figure 9 shows the pressure fluctuations and power spectrum for the control point for the baseline case with no flow control for comparison purposes. The pressure is nondimensionalized by free stream pressure and the time is nondimensionalized by the Rossiter period which is 857 time steps corresponding to 857 microseconds. As seen from the figure, the pressure oscillations are periodic. The dominant frequency (2167 Hz) corresponds to two Rossiter frequencies

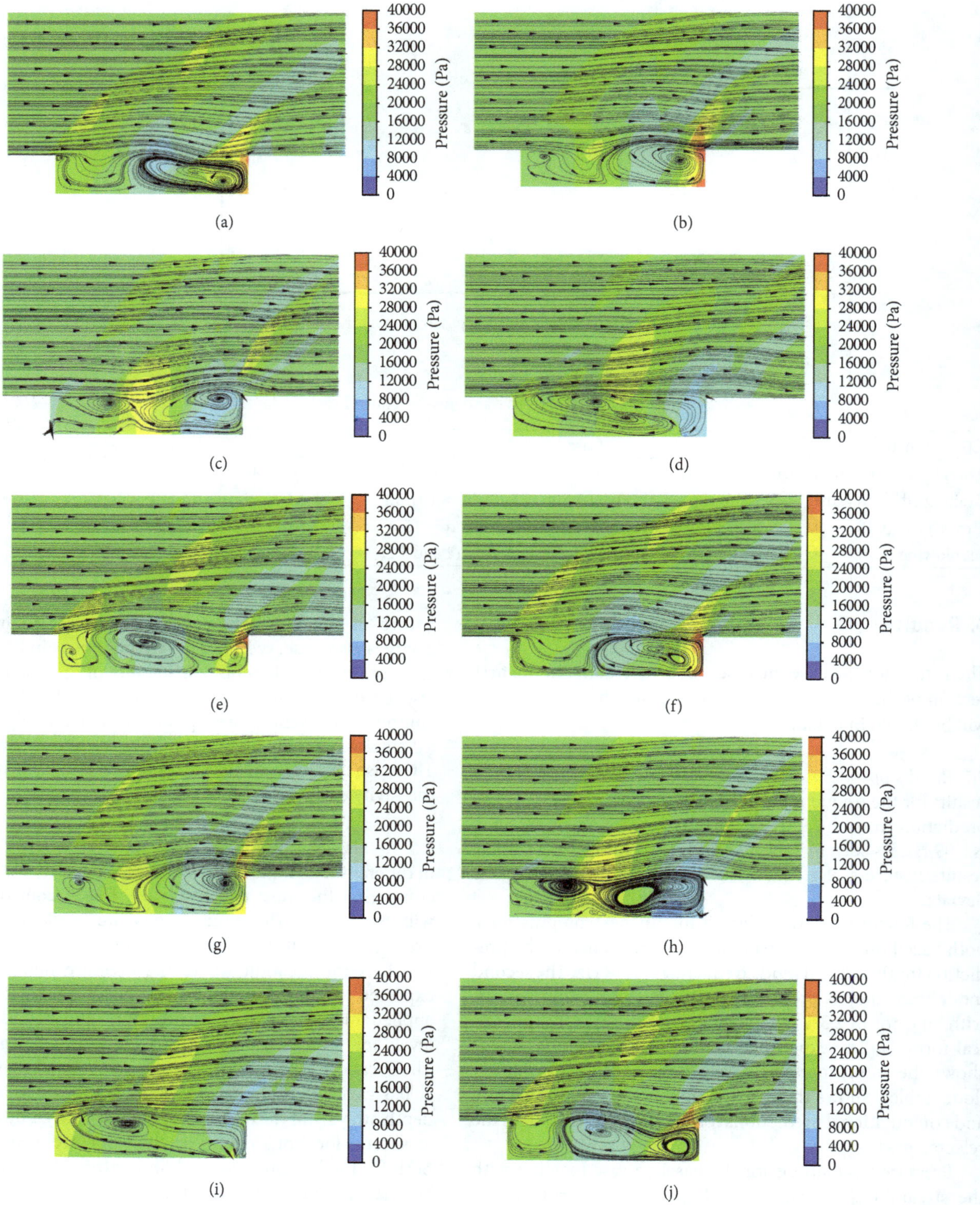

FIGURE 5: Pressure fluctuations of the whole domain for the baseline case (a) $t/t_R = 0$, (b) $t/t_R = 0.11$, (c) $t/t_R = 0.22$, (d) $t/t_R = 0.33$, (e) $t/t_R = 0.44$, (f) $t/t_R = 0.55$, (g) $t/t_R = 0.66$, (h) $t/t_R = 0.77$, (i) $t/t_R = 0.88$, and (j) $t/t_R = 1$.

which is predicted as 2218 Hz by Rossiter's formula. The difference between the computational results and Rossiter's formula is 2.3%.

Figure 10 shows the pressure distribution and power spectra at three different locations on the cavity floor, where nondimensional streamwise distance from the leading edge (x/L) is 0, 0.4, and 0.8. As seen in this figure, the pressure distribution and power spectrum at the cavity floor show a similar trend with the control point.

3.3. Cover Plate. A cover plate shown in Figure 11 is applied to the cavity as a passive control method. The slope of the plate is

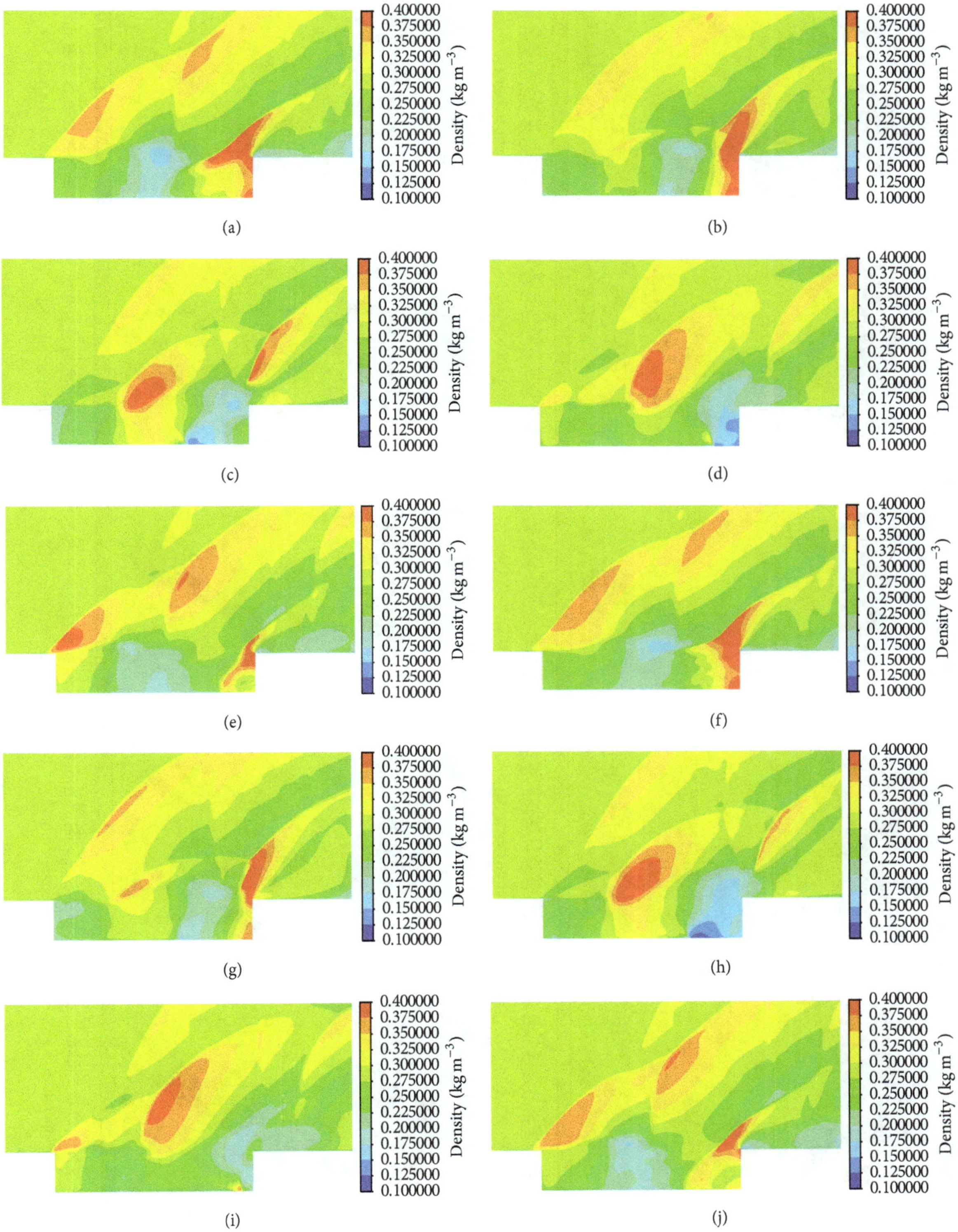

FIGURE 6: Density contours for the baseline case (a) $t/t_R = 0$, (b) $t/t_R = 0.11$, (c) $t/t_R = 0.22$, (d) $t/t_R = 0.33$, (e) $t/t_R = 0.44$, (f) $t/t_R = 0.55$, (g) $t/t_R = 0.66$, (h) $t/t_R = 0.77$, (i) $t/t_R = 0.88$, and (j) $t/t_R = 1$.

0 (CP1), 15 (CP2), 30 (CP3), and 45 (CP4) degrees as shown in detail in Table 3. The plate with no slope is shown in Figure 11 for demonstration purposes. The length of the plate is 25% of the cavity length and the thickness is 2 mm.

Figure 12(a) shows the pressure distribution at the control point for all cover plate control cases. Figure 12(b) shows the FFT analysis of this pressure distribution to observe the dominant frequencies. The FFT analysis of all cover plate control

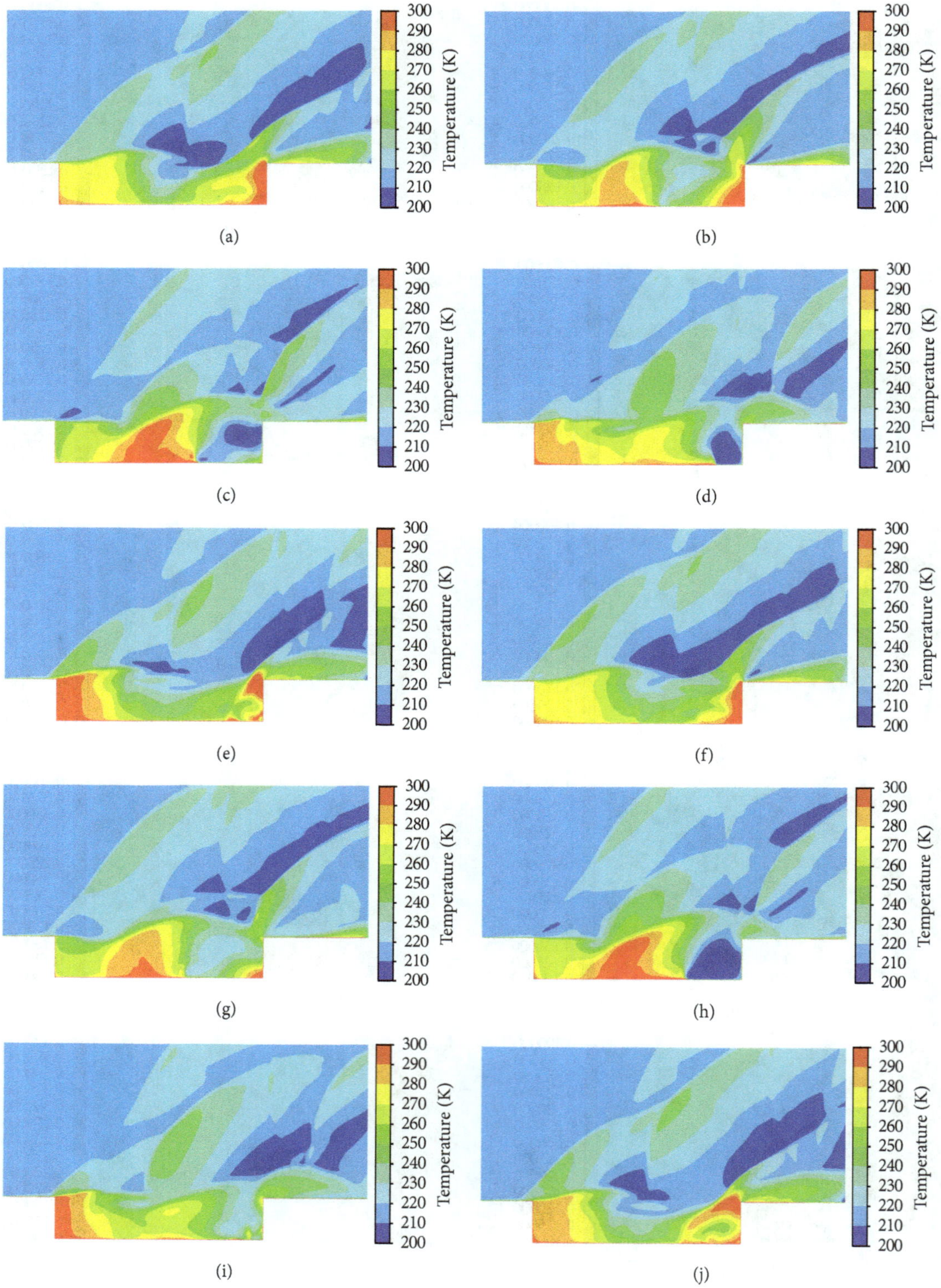

FIGURE 7: Temperature contours for the baseline case (a) $t/t_R = 0$, (b) $t/t_R = 0.11$, (c) $t/t_R = 0.22$, (d) $t/t_R = 0.33$, (e) $t/t_R = 0.44$, (f) $t/t_R = 0.55$, (g) $t/t_R = 0.66$, (h) $t/t_R = 0.77$, (i) $t/t_R = 0.88$, and (j) $t/t_R = 1$.

cases provide similar results; therefore only the FFT results of the CP4 case (45 degrees) are shown for demonstration purposes. Figure 9 clearly shows a dominant frequency at 2167 Hz (corresponding to two Rossiter periods); however, the dominant frequency in Figure 12 is more than 3000 Hz which corresponds to approximately three Rossiter periods. The pressure distributions and dominant frequencies are altered with the use of cover plate. When the slope of the cover plate increases, the magnitude of pressure oscillations decreases, but the periodicity of the flow remains. However,

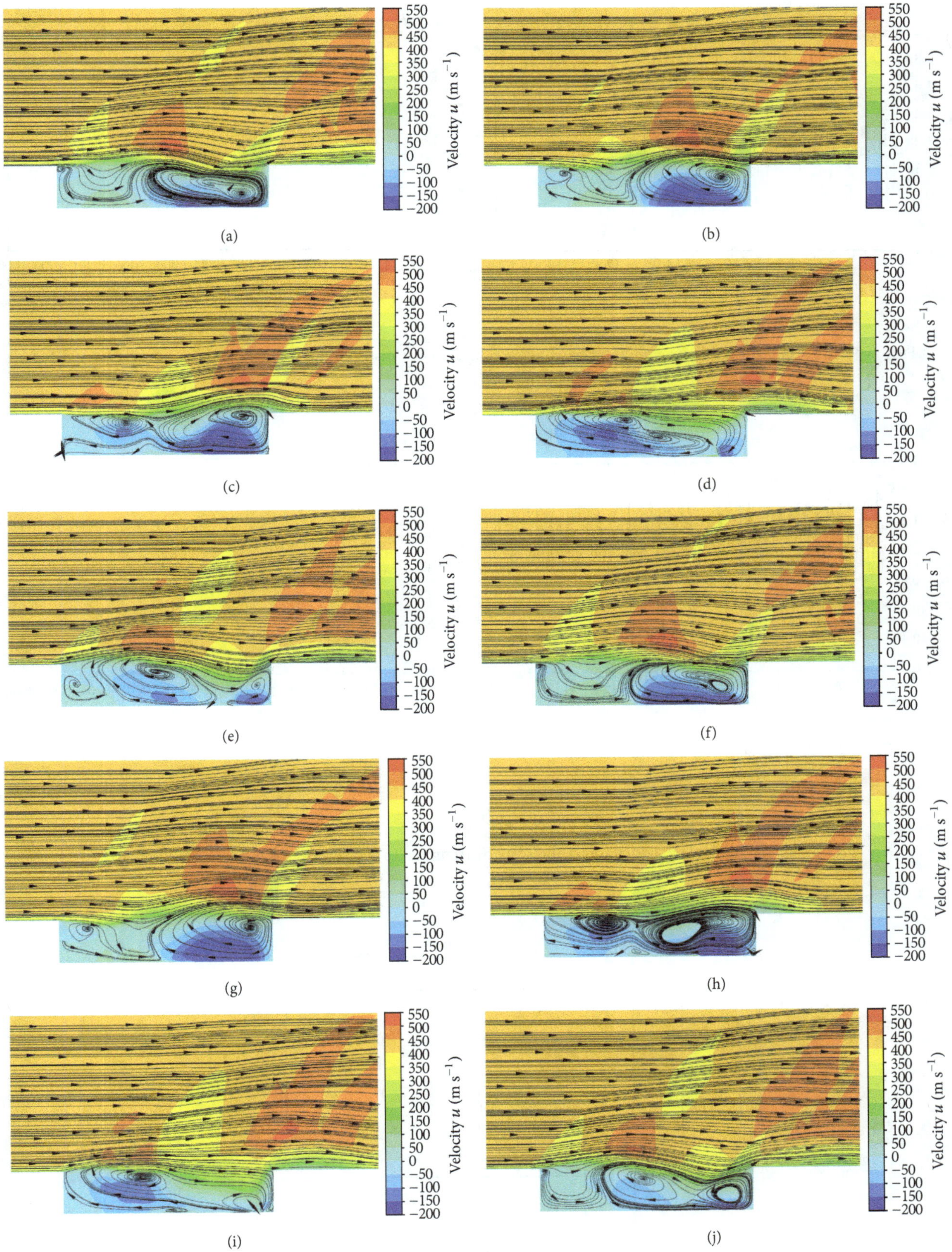

FIGURE 8: Contours of velocity in streamwise direction and streamlines for the baseline case (a) $t/t_R = 0$, (b) $t/t_R = 0.11$, (c) $t/t_R = 0.22$, (d) $t/t_R = 0.33$, (e) $t/t_R = 0.44$, (f) $t/t_R = 0.55$, (g) $t/t_R = 0.66$, (h) $t/t_R = 0.77$, (i) $t/t_R = 0.88$, and (j) $t/t_R = 1$.

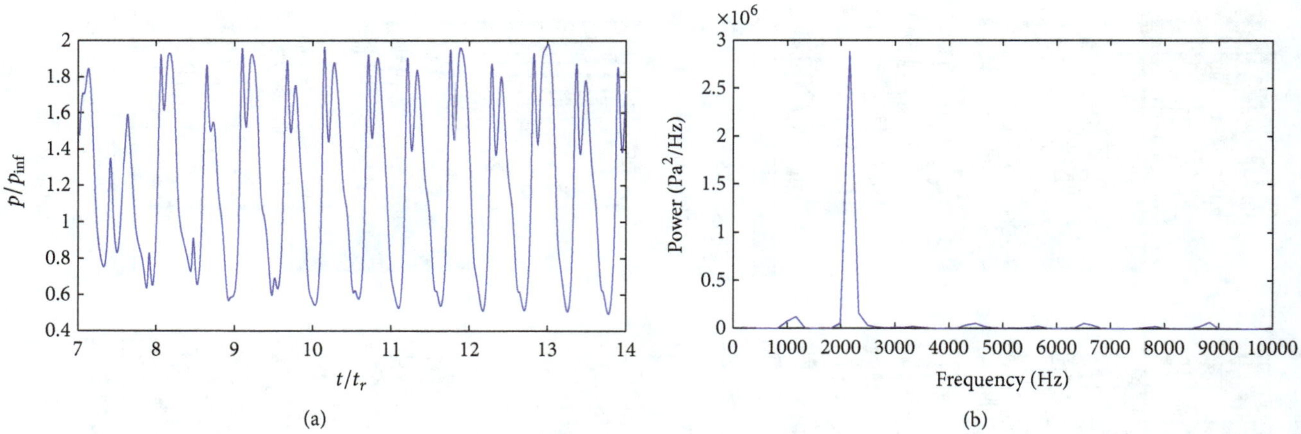

FIGURE 9: The results for the control point for baseline case. (a) Pressure distribution. (b) FFT analysis of the pressure data.

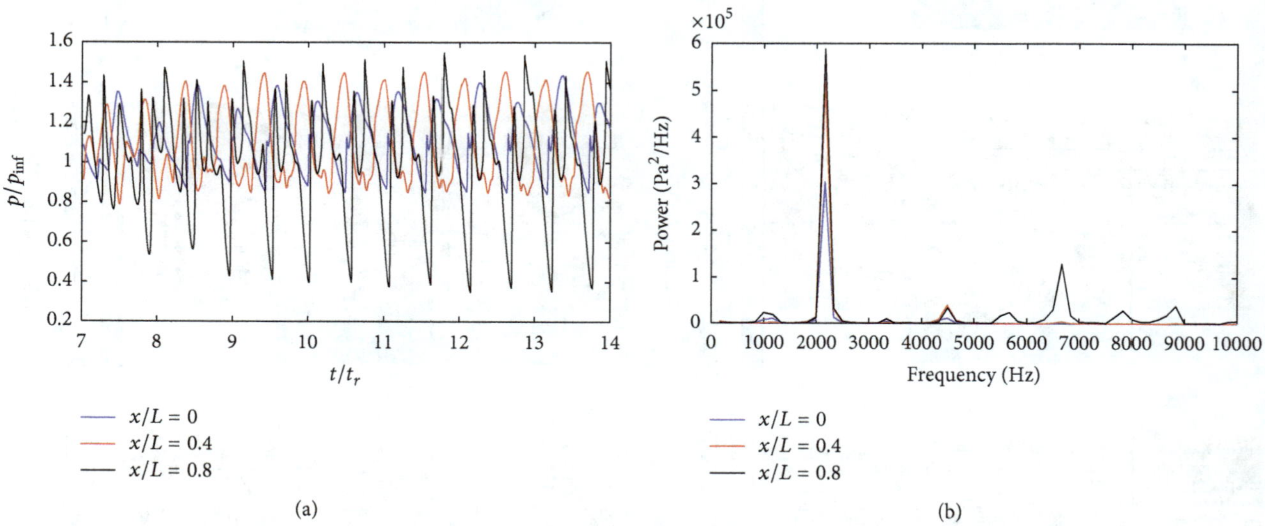

FIGURE 10: The results for cavity floor. (a) Pressure distribution. (b) Power spectrum for the baseline case.

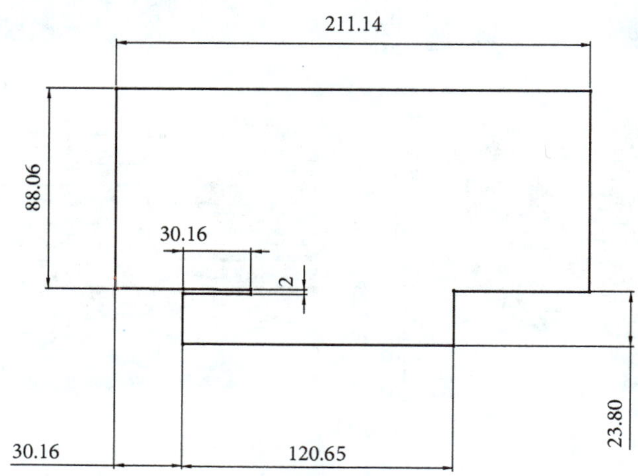

FIGURE 11: Cover plate geometry.

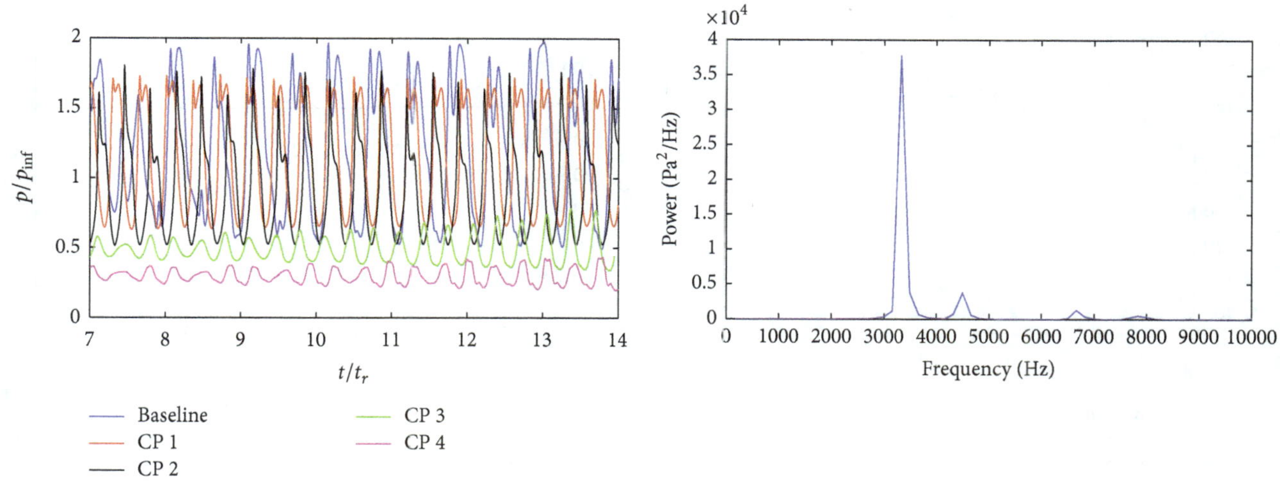

(a) (b)

FIGURE 12: The results for the control point. (a) Pressure distribution for the baseline case and the cases with cover plate. (b) FFT analysis for the pressure data obtained from the control point (CP4).

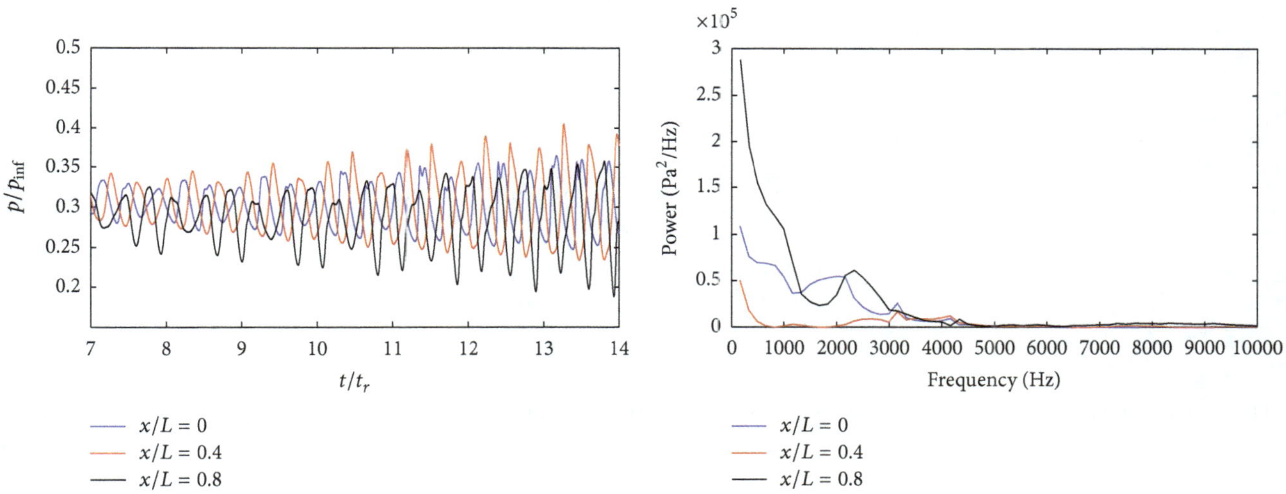

(a) (b)

FIGURE 13: The results for cavity floor (x/L = 0, 0.4, and 0.8). (a) Pressure distribution. (b) Power spectrum (CP4).

there is a clear shift in periods, as also shown in the FFT graph. Similar trend is shown for several locations at the cavity floor (x/L = 0, 0.4, and 0.8) in terms of pressure history and power spectra in Figure 13, for CP4 case.

Figure 14 shows the sound pressure levels in terms of decibels for the bottom wall, leading edge wall, and trailing edge wall when cover plates are used. The trend in pressure distribution at the control point is also observed in the sound pressure level (SPL) distributions. Increasing the cover plate slope decreases the magnitude of the pressure oscillations; therefore, it also decreases the SPL values for the cavity walls.

Figures 15 and 16 show the density contours and x-velocity contours with streamlines, respectively, for CP4 configuration which is the most effective configuration of control cases with cover plate. The cover plate acts as a barrier for the incoming flow and breaks the periodic feedback structure inside the cavity. The shear layer is distorted and the flows before and after the cover plate act like two different zones. The distortion in the shear layer gives more space to the vortex inside the cavity. This lets the flow have a lower streamwise velocity inside the cavity and the velocity increase takes place further away from the inside, as observed in Figure 16 that shows the streamlines and x-velocity contours. The densities are very high before the cover plate. However, they are much lower at the region behind the plate. The reason is most probably the change in the feedback mechanism due to the plate. When Figures 5, 6, and 8 for the baseline case are examined, it is observed that the periodic nature of the flow is due to the formation and impingement of the vortices inside the cavity to the trailing edge wall. However, with the cover plate, this is not the case. This loop is broken and the vortices do not collapse on the trailing edge wall; instead more flow

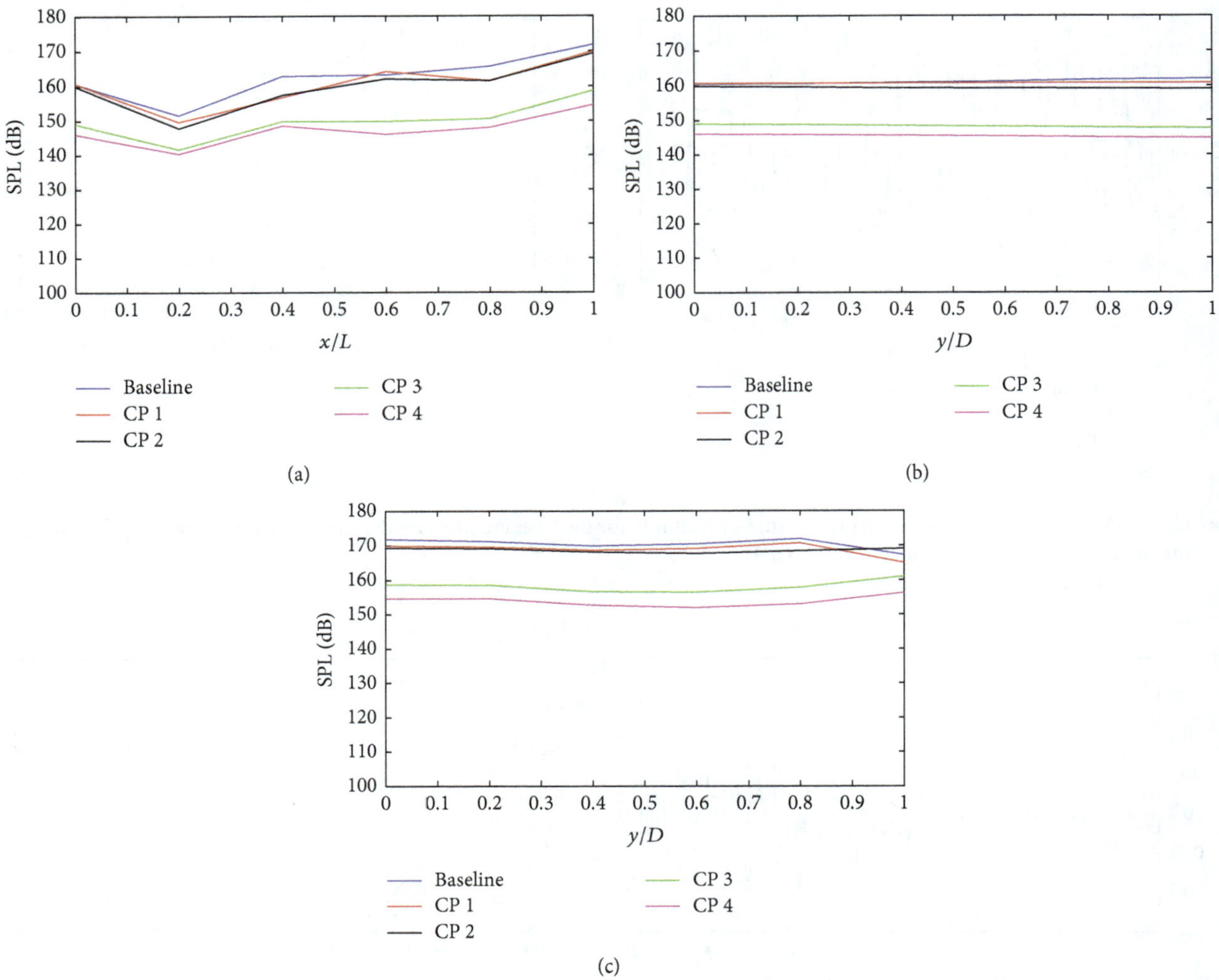

FIGURE 14: Sound pressure levels for cavity walls for cover plate cases. (a) Bottom wall. (b) Leading edge wall. (c) Trailing edge wall.

TABLE 4: Frequency predictions with different grid resolutions.

Grid	Rossiter frequency (1st mode)	% difference with Rossiter frequency formulation
1	1230	11
2	1195	7.7
3	1037	6.5
4	1166	5.22
5	1050	5.3
6	1010	9

TABLE 5: Comparison of computations and experiments for cavity floor sound pressure levels.

Cavity floor location (x/L)	SPL (experiment [17]) (dB)	SPL (computation)	% difference
0	152	160	5.3
0.2	147	150	2.0
0.4	153	161	5.2
0.6	154	160	3.9
0.8	155	160	3.2
1.0	162	168	3.7

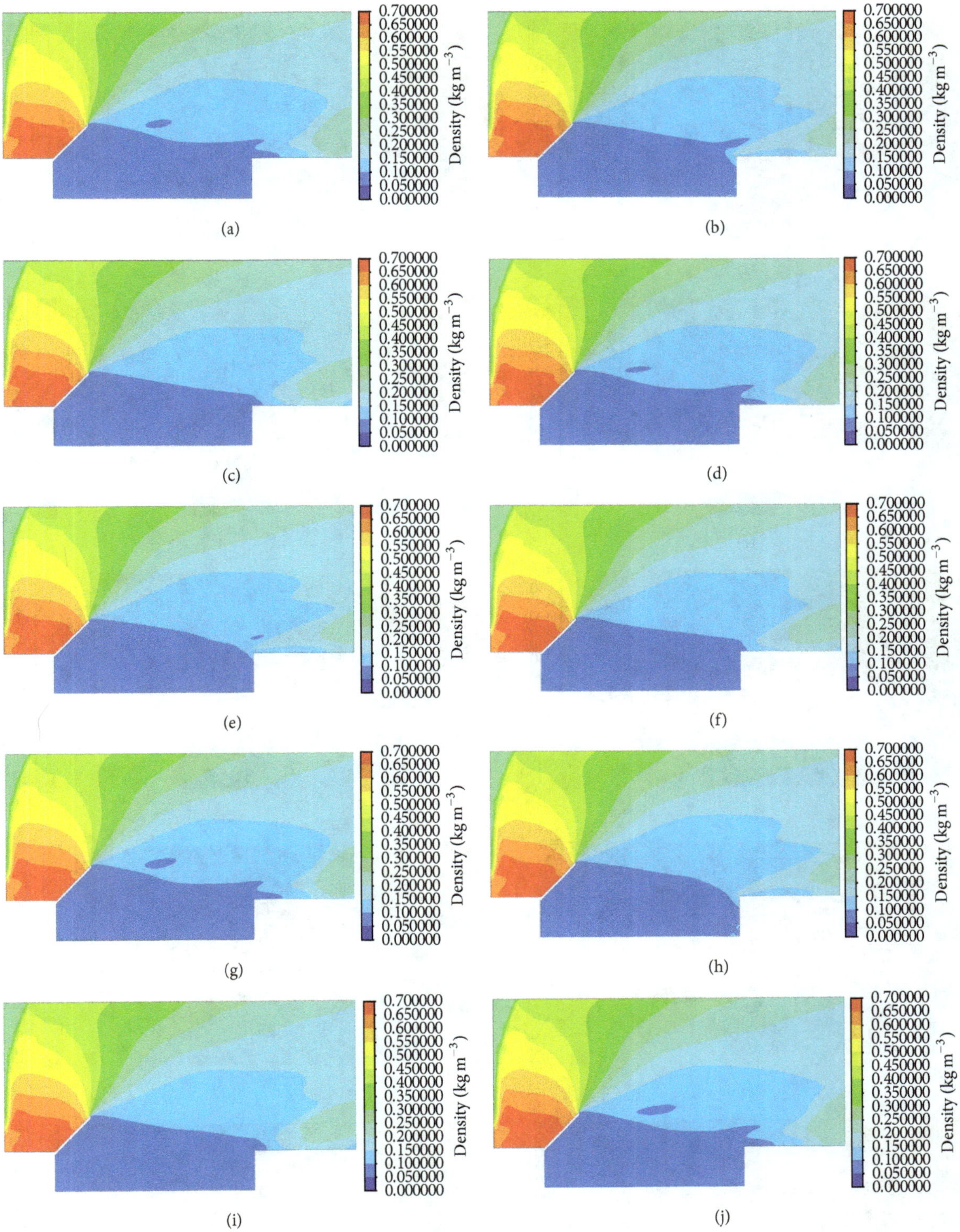

FIGURE 15: Density contours for CP4 case (a) $t/t_R = 0$, (b) $t/t_R = 0.11$, (c) $t/t_R = 0.22$, (d) $t/t_R = 0.33$, (e) $t/t_R = 0.44$, (f) $t/t_R = 0.55$, (g) $t/t_R = 0.66$, (h) $t/t_R = 0.77$, (i) $t/t_R = 0.88$, and (j) $t/t_R = 1$.

goes smoothly over the cavity. As observed from all the graphs shown, CP4 which has a slope of 45° is the most effective of all. As the angle increases, the blockage of the entrance to cavity decreases. Also, increased angles provide more space to the vortex inside the cavity and more space for shear layer deformation, which makes it more convenient to use higher slopes. However, if the slope is increased more than necessary, it can block the incoming flow directly and may act as a shock; therefore, it is reasonable to have the best results in terms of pressure fluctuations with CP4.

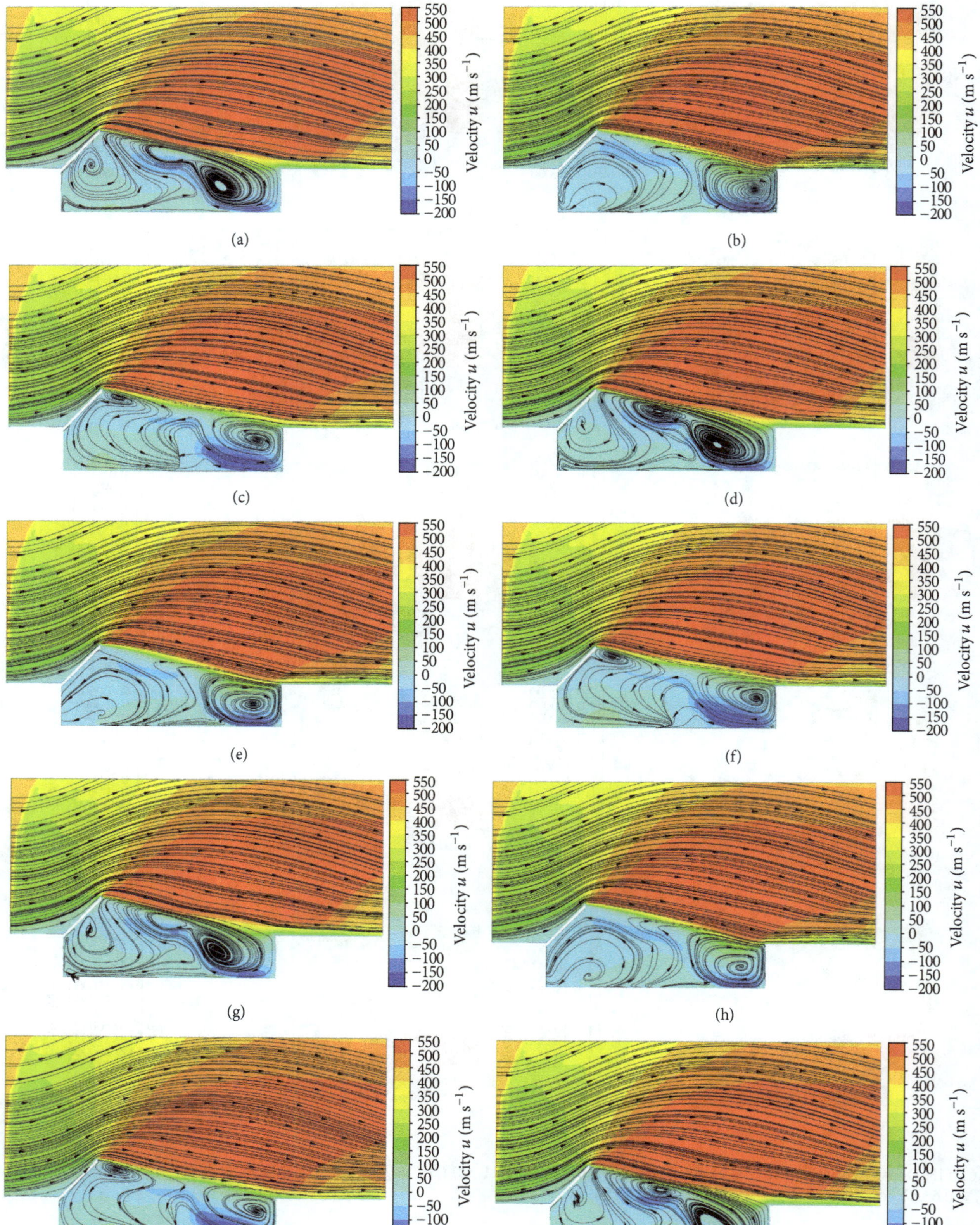

FIGURE 16: X-velocity contours and streamlines for CP4 case (a) $t/t_R = 0$, (b) $t/t_R = 0.11$, (c) $t/t_R = 0.22$, (d) $t/t_R = 0.33$, (e) $t/t_R = 0.44$, (f) $t/t_R = 0.55$, (g) $t/t_R = 0.66$, (h) $t/t_R = 0.77$, (i) $t/t_R = 0.88$, and (j) $t/t_R = 1$.

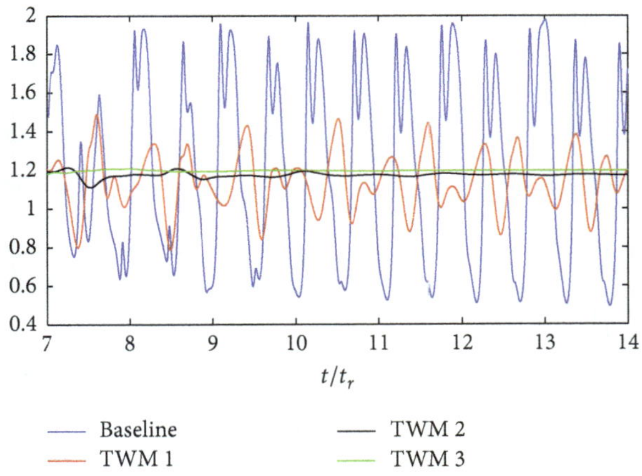

FIGURE 17: Pressure distribution at the control point with trailing edge wall modifications.

3.4. Trailing Edge Wall Modifications. The trailing edge wall of the cavity is modified by providing a slope of 30 (TWM1), 45 (TWM2), and 60 (TWM3) degrees. The details of the cases are shown in Table 3. Figure 17 shows the pressure distribution for the control point, whereas Figure 18 shows the SPL distribution for the bottom, leading edge, and trailing edge walls of the cavity. The pressure fluctuations (both magnitude and frequency) decrease drastically when the trailing edge wall of the cavity is modified.

Figure 19 shows the pressure fluctuations and power spectra at three different locations on cavity floor for the most effective trailing edge modification case: TWM3. Similar trend is observed as shown in previous figures: the pressure oscillations disappear on the cavity floor, as well.

Time dependent pressure fluctuations for the whole domain after trailing edge wall modification 3 (TWM3) which is the most effective of TWM cases are shown in Figure 20. The contours are provided with the same pressure levels for the baseline case (Figure 5) for comparison purposes. As clearly observed from the pressure contours and streamlines, the flow for the controlled case is not periodic; however, the baseline case shows the pressure oscillations and time dependent flow structure. Figures 21 and 22 show the density contours and x-velocity contours with streamlines for trailing edge wall modification 3 configuration. As seen in density, pressure, and x-velocity contours, as well as the streamlines, the flow structure is not periodic any more when the trailing edge wall is modified with a slope of 60 degrees. In the baseline case, a vortex develops inside the cavity and the movement of this vortex and interactions of it with the shear layer increase the pressure fluctuations inside. However, when the trailing edge wall is modified, the flow does not hit a straight trailing edge wall and go back to the leading edge, strengthening the vortex inside the cavity. On the contrary, more fluid goes over the sloped wall smoothly without causing pressures to change, leading the pressure oscillations to be alleviated. When the sound pressure levels are examined, it is observed that leading edge wall SPL values do not change drastically. However, since the trailing wall is smooth and has a slope (TWM 3), the SPL values on this wall decrease and the periodic nature of the flow disappears.

Figure 23 shows the density contours for baseline and all trailing edge wall modification cases (TWM 1, TWM 2, and TWM 3) for only one time value ($t/t_R = 0.55$). The baseline case clearly has the highest densities on its straight trailing edge wall. Introducing a slope at the trailing edge wall smoothens out the flow. Increasing the slope of the wall as much as possible lets more fluid go over the smooth wall and exit the cavity instead of helping to strengthen the vortex inside the cavity, which explains why TWM 3 with the highest slope is the most effective in alleviating pressure fluctuations.

3.5. Trailing Edge Curvature. As a third passive control method, the trailing edge wall radius is changed. The details are presented in Table 3. A radius of 5 (TC1), 10 (TC2), 15 (TC3), and 20 mm (TC4) is used to provide the curvature as shown in Figure 24 for $r = 20$ mm, for demonstration purposes.

Figure 25 shows the pressure distribution for the control point. As seen clearly in the graph, the magnitude of the pressure oscillations decreases drastically when the trailing edge wall is provided with a curvature to let the air flow smoothly over the cavity. Increasing the radius of the curvature helps this smoothening; therefore, the magnitude of the oscillations further decreases with increased curvature.

A representative result of the FFT analysis is shown in Figure 26 for a radius of 15 mm. The graph shows that the peaks are for the same frequencies as the baseline case (Figure 9); however, the magnitudes of the power accumulated at these frequencies differ shifting the dominant mode to higher frequencies.

The SPL values drop to 135 decibels for the bottom and front walls and 150 decibels for the trailing edge wall when curvature is provided to the trailing edge wall, which corresponds to 16% for the bottom and front walls and 11.5% for the trailing edge wall.

As seen both in pressure fluctuations and in power spectra, trailing edge curvature is not as effective as trailing edge wall modification by providing a slope. Periodic nature of the flow does not change; but the magnitude of the fluctuations decreases. Smoothening of the trailing edge wall basically provides a similar effect as introducing a trailing edge slope by letting the flow go over the cavity instead of hitting the edge and strengthening the vortex inside. Figure 27 shows the density contours for baseline and trailing edge curvature cases. As seen from the contours, increasing the radius smoothens the shear layer and incoming flow and lets more air flow over the cavity instead of hitting the wall, which then, in turn, helps to decrease the pressure oscillations caused by vortex-shear layer interactions.

3.6. Active Control: Jet Blowing. Jet blowing is utilized as an active control method. The location of the holes for jet blowing, and the dimensions of the geometry are shown in Figure 28. All the dimensions are given in millimeters. As shown in Table 3, there are five different cases. The jet blowing velocity is taken as 10% of the free stream velocity for all cases.

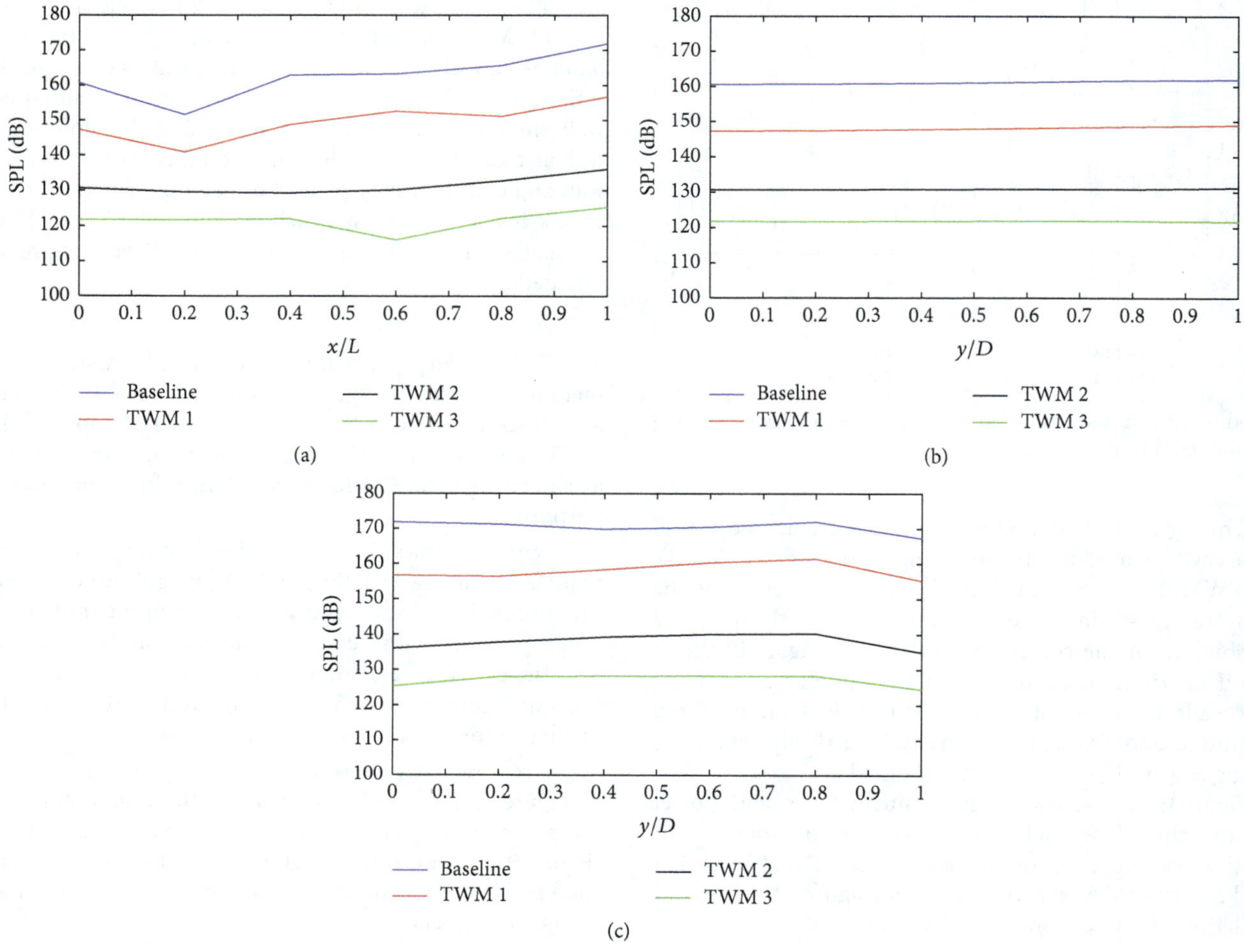

FIGURE 18: Sound pressure levels for cavity walls for TWM cases. (a) Bottom wall. (b) Leading edge wall. (c) Trailing edge wall.

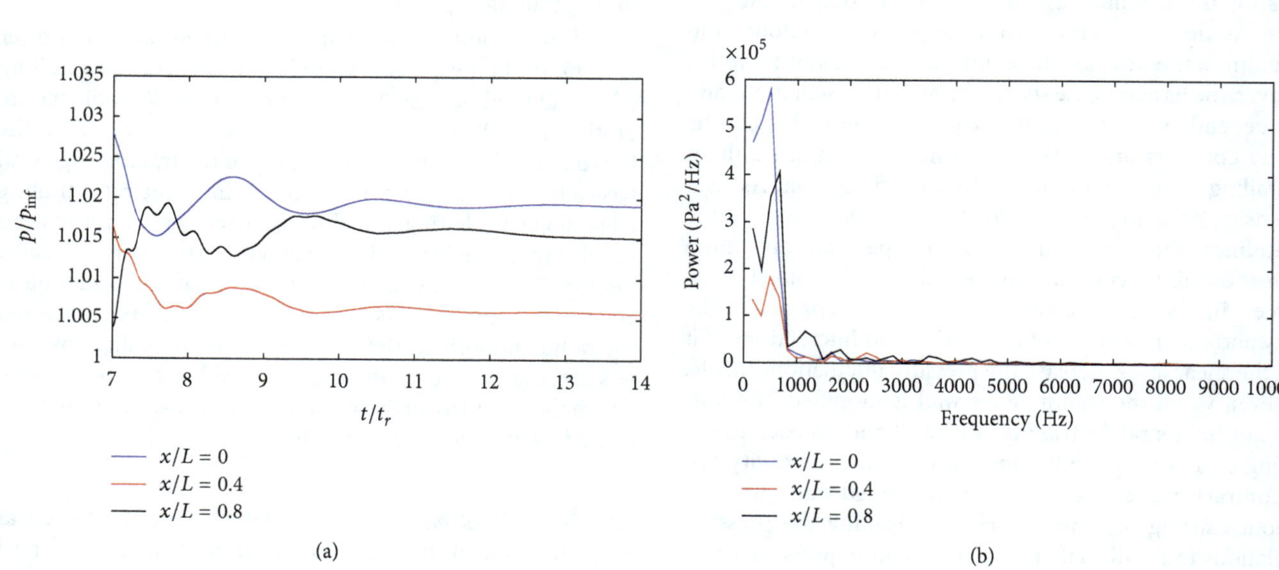

FIGURE 19: The results for cavity floor ($x/L = 0$, 0.4, and 0.8). (a) Pressure distribution. (b) Power spectrum (TWM 3).

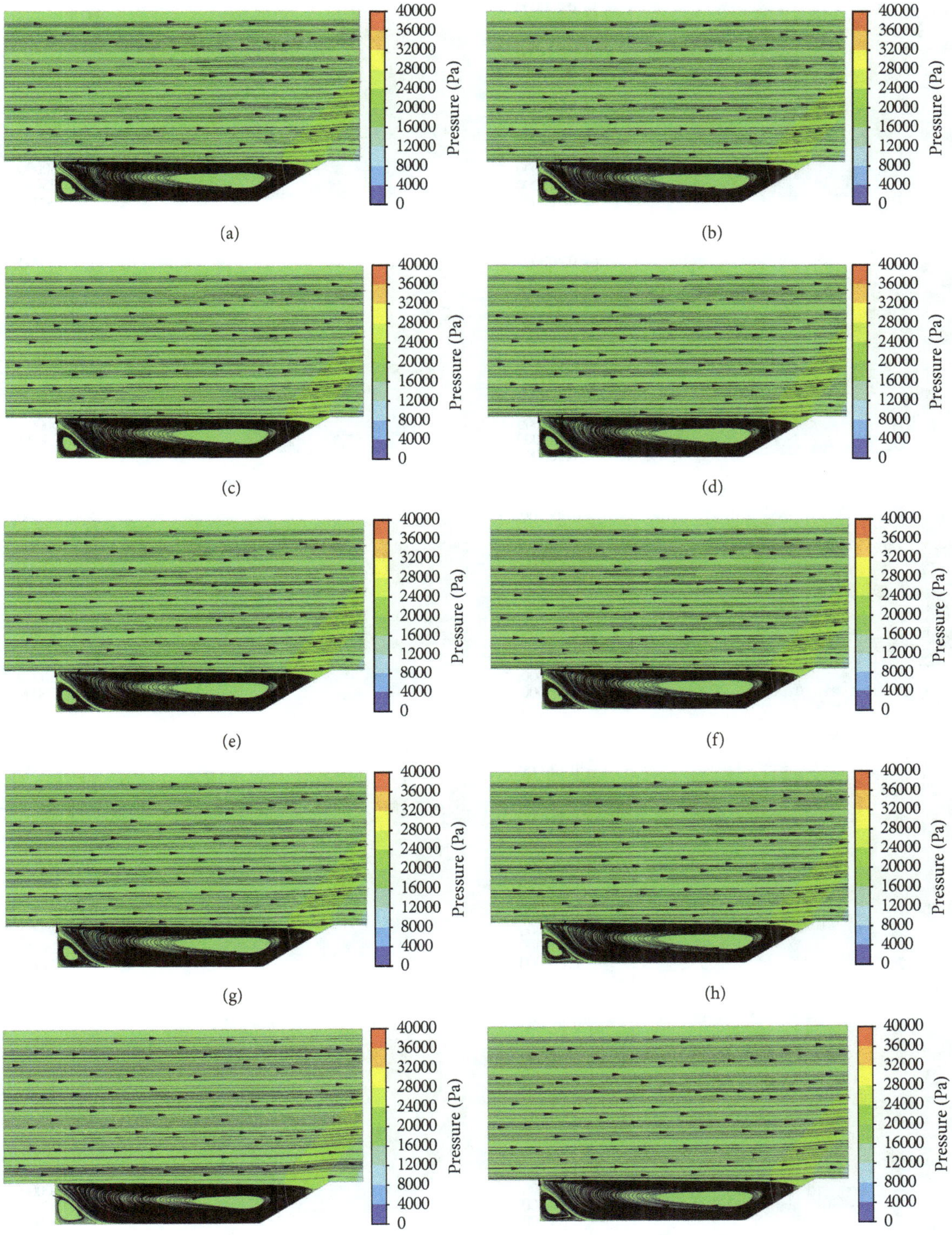

FIGURE 20: Pressure fluctuations of the whole domain (TWM3) (a) $t/t_R = 0$, (b) $t/t_R = 0.11$, (c) $t/t_R = 0.22$, (d) $t/t_R = 0.33$, (e) $t/t_R = 0.44$, (f) $t/t_R = 0.55$, (g) $t/t_R = 0.66$, (h) $t/t_R = 0.77$, (i) $t/t_R = 0.88$, and (j) $t/t_R = 1$.

Air is blown through the holes from outside to inside. Table 6 shows the details of the cases.

The results for the sound pressure levels are shown for the cavity walls in Figure 29. It is seen from the graphs that active control is not as effective as passive control techniques since it only provides around 5 dB of decrease in pressure fluctuations. However, active control methods have an inherent advantage of being able to adapt to changing conditions,

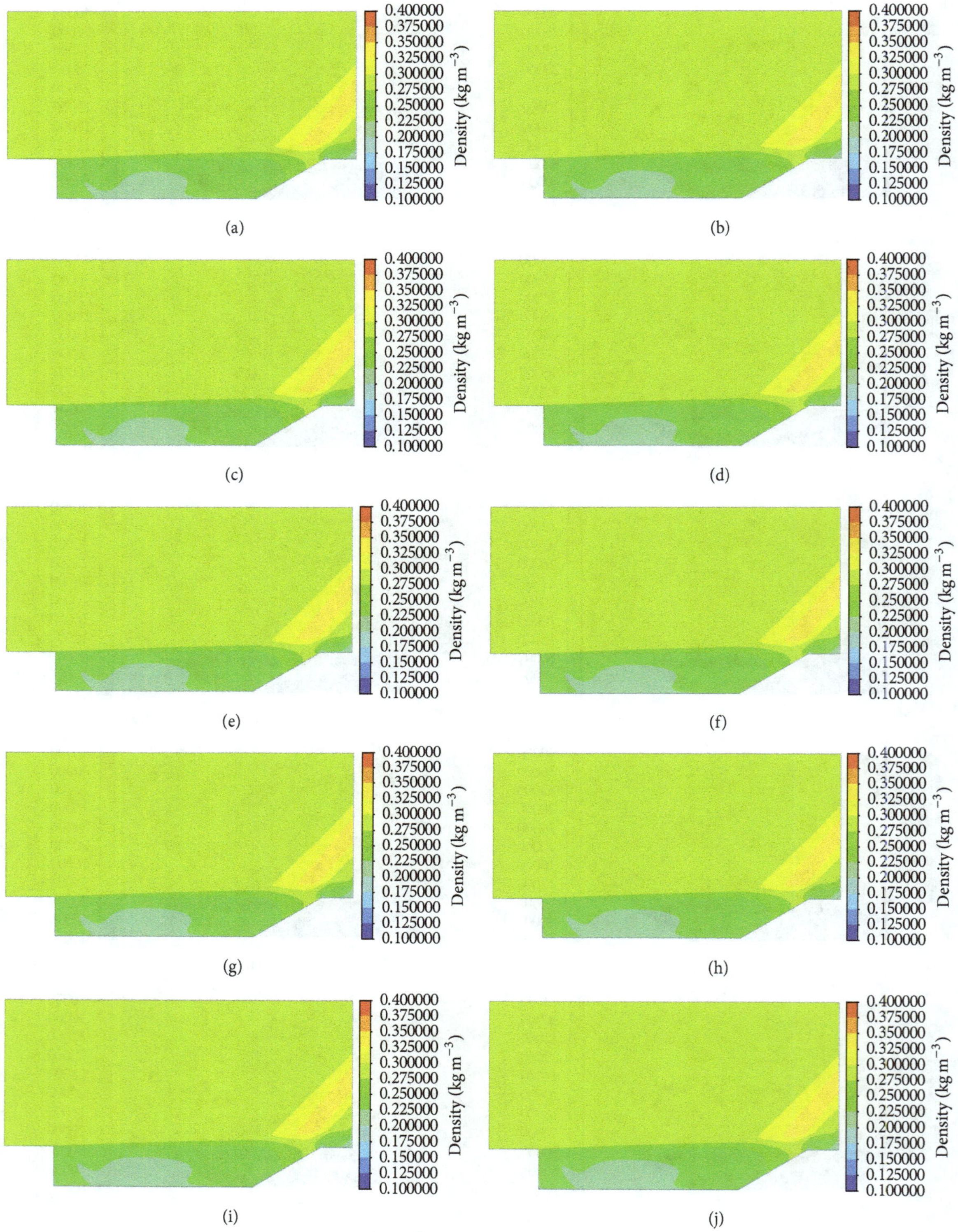

FIGURE 21: Density contours for TWM3 case (a) $t/t_R = 0$, (b) $t/t_R = 0.11$, (c) $t/t_R = 0.22$, (d) $t/t_R = 0.33$, (e) $t/t_R = 0.44$, (f) $t/t_R = 0.55$, (g) $t/t_R = 0.66$, (h) $t/t_R = 0.77$, (i) $t/t_R = 0.88$, and (j) $t/t_R = 1$.

whereas passive control methods are more effective; but they are not adjustable. Once implemented, they cannot be changed during flight.

3.7. Comparison of the Most Effective Cases of Each Control Method. In this part of the article, the results of the best cases of each control method are compared with the results of other control methods. The best cases are chosen as the most effective cases where the magnitude and frequency of the pressure oscillations decrease. Figure 30 shows the pressure oscillations for the selected control cases for the control point. As clearly observed from the graph, the magnitude of the

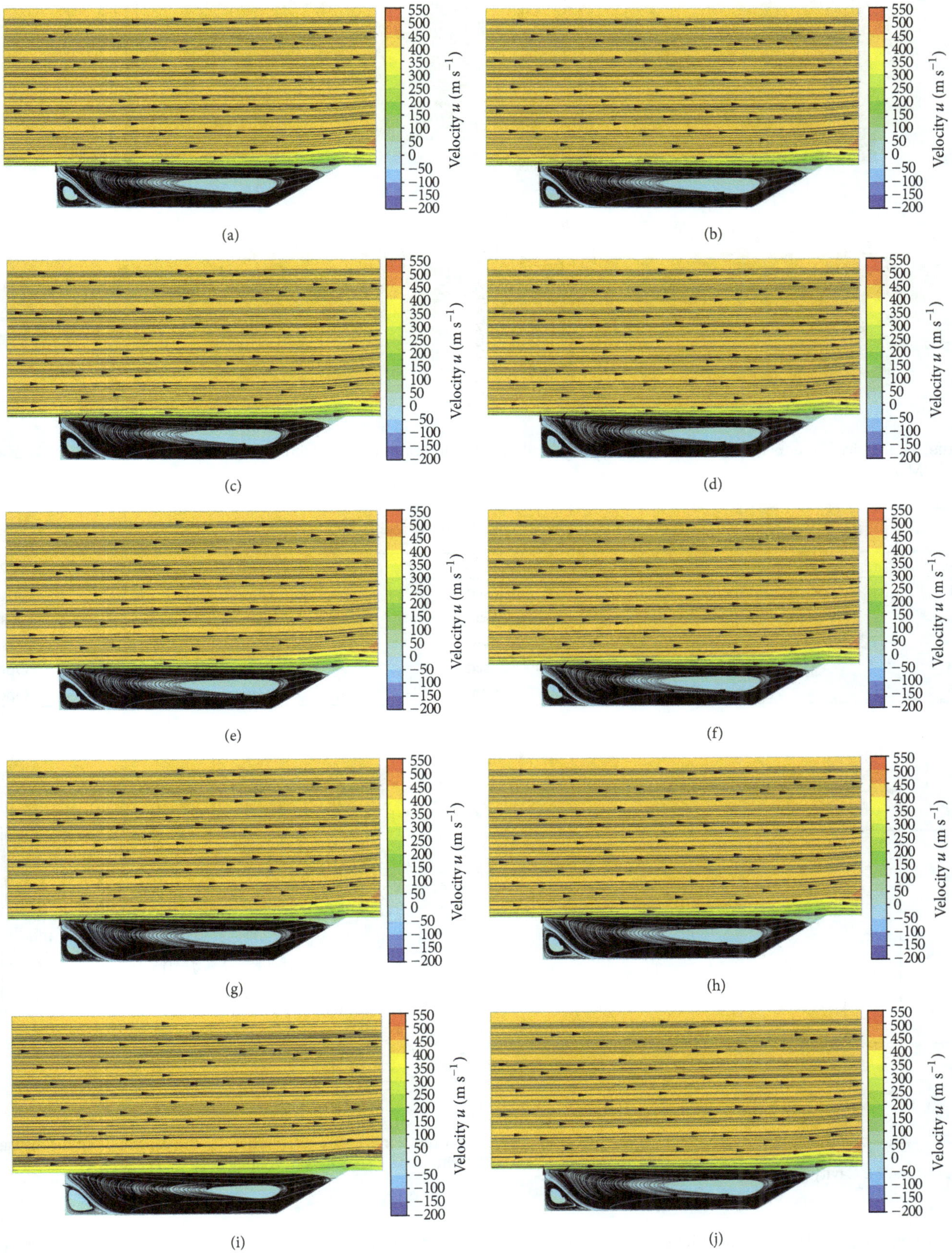

FIGURE 22: X-velocity contours and streamlines for TWM3 case (a) $t/t_R = 0$, (b) $t/t_R = 0.11$, (c) $t/t_R = 0.22$, (d) $t/t_R = 0.33$, (e) $t/t_R = 0.44$, (f) $t/t_R = 0.55$, (g) $t/t_R = 0.66$, (h) $t/t_R = 0.77$, (i) $t/t_R = 0.88$, and (j) $t/t_R = 1$.

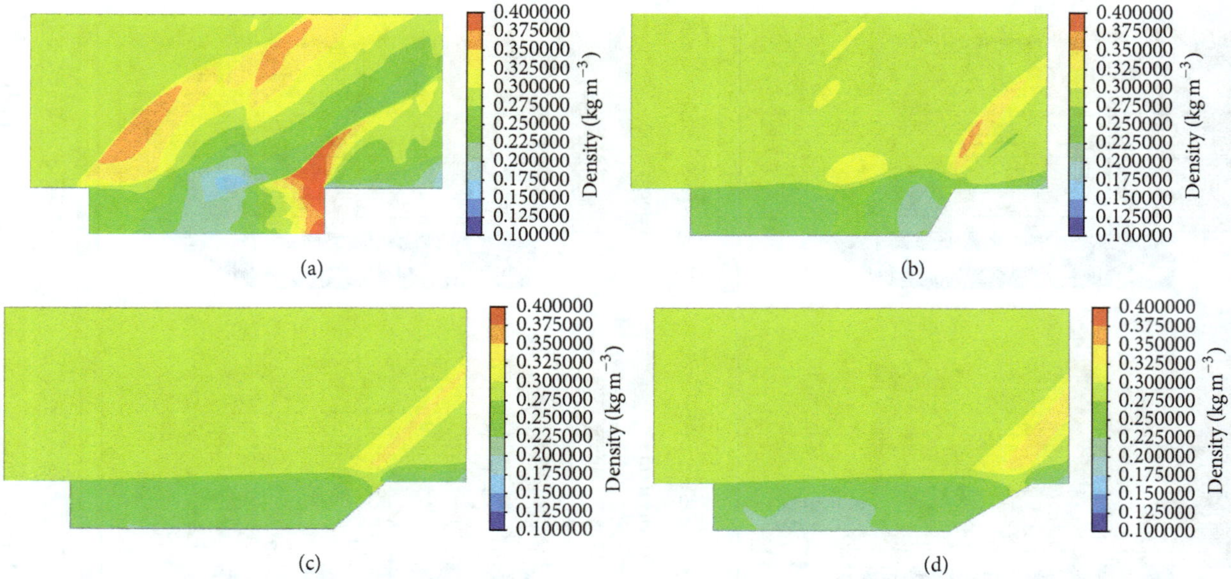

FIGURE 23: Density contours for baseline and trailing edge wall modification cases (t/t_R = 0.55) (a) baseline, (b) TWM 1, (c) TWM 2, and (d) TWM 3.

TABLE 6: Jet blowing cases.

Case	T1	T2	L1	L2
JB 1	Open	Closed	Closed	Closed
JB 2	Open	Open	Closed	Closed
JB 3	Closed	Closed	Open	Closed
JB 4	Closed	Closed	Open	Open
JB 5	Open	Open	Open	Open

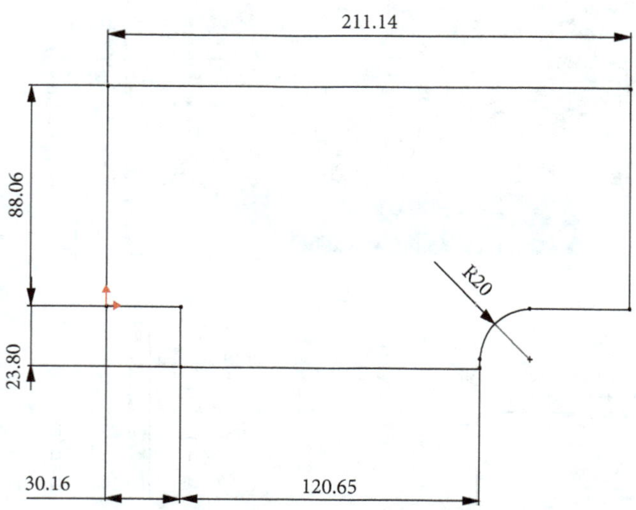

FIGURE 24: Cavity geometry with a curved trailing edge.

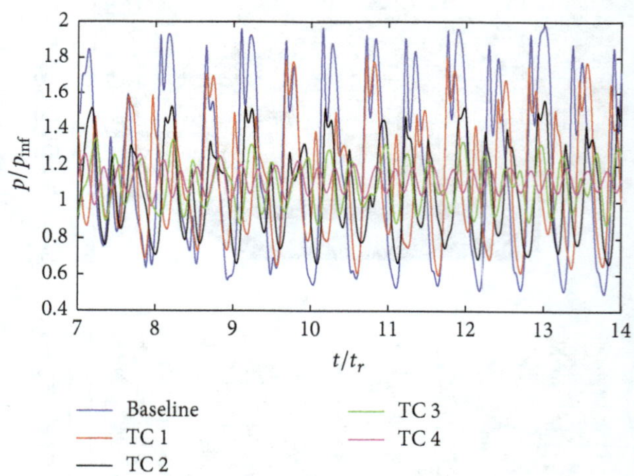

FIGURE 25: Pressure distribution for the control point for the cases with passive control with trailing edge wall curvature.

pressure oscillations decreases with jet blowing; however the oscillations are still periodic. The cover plate decreases the magnitude of the pressure oscillations drastically since it disrupts the feedback loop of the oscillations by providing a physical barrier between the inside and outside of the cavity, which also disturbs the free stream flow. When a curvature is given to the trailing edge wall, again, the magnitudes of the pressure oscillations decrease, still providing periodic flow, however, smoothening the flow. On the contrary, altering the trailing edge wall changes the periodic nature of the

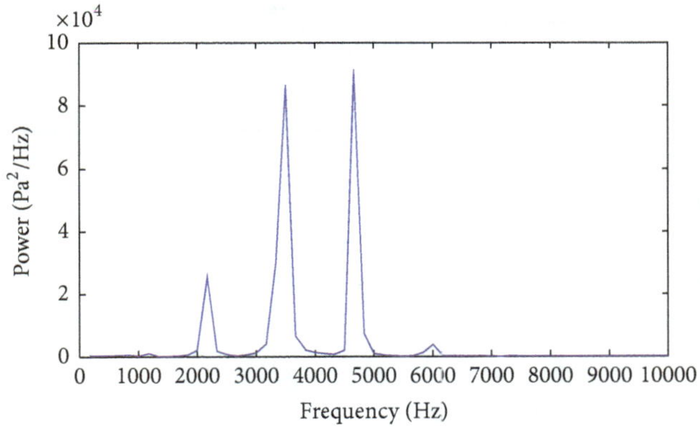

FIGURE 26: FFT of pressure for the control point (TC3).

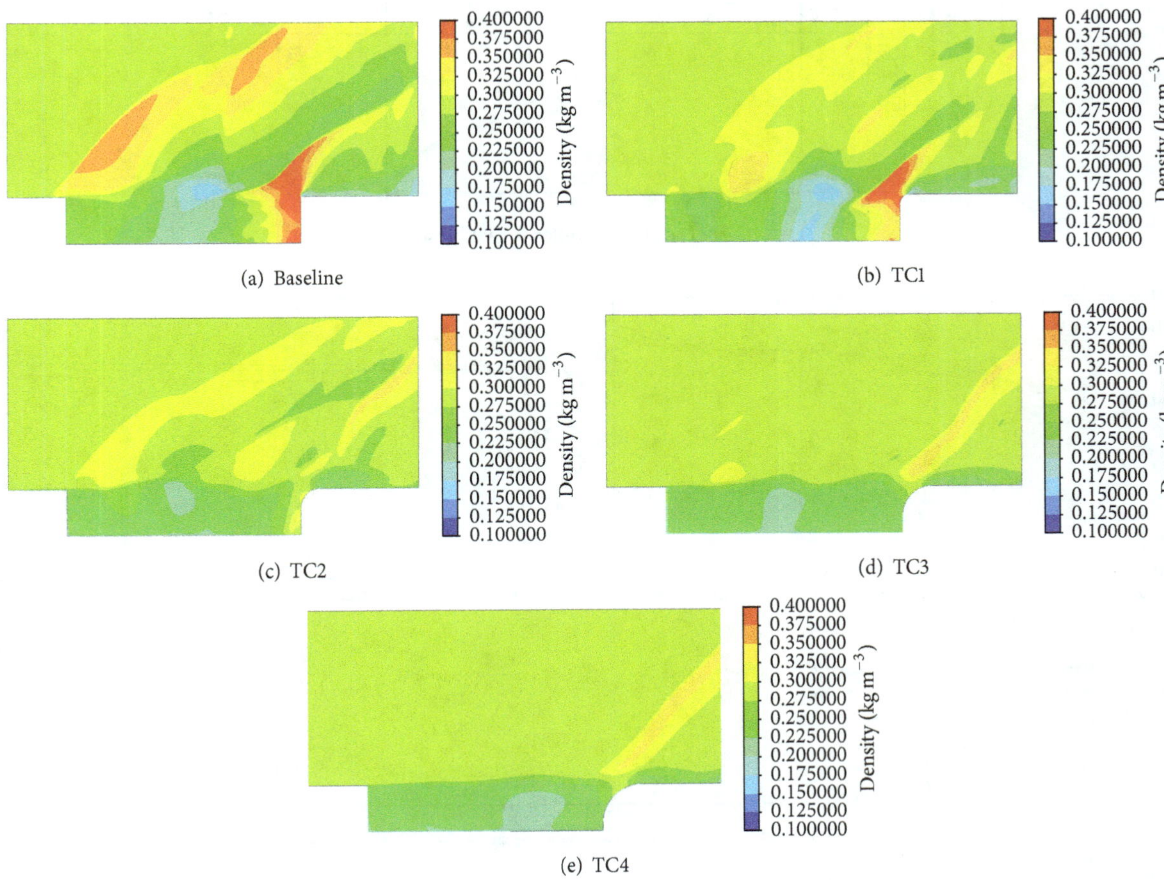

FIGURE 27: Density contours for baseline and trailing edge curvature cases ($t/t_R = 0.55$).

cavity flow. All these control methods prove to be effective in suppressing pressure oscillations.

Table 7 shows the average reductions in sound pressure levels for the cavity walls. As seen in the table, when the trailing edge wall is modified the sound pressure levels for the cavity walls can be increased up to 40 dB. Cover plate and trailing edge wall curvature are also effective methods that can decrease the SPL levels by 20 dB. Although jet blowing does not seem as effective as the passive control methods, a 5 dB decrease is also high in terms of active flow control, since the flow is not irreversibly modified with active methods as in passive control techniques.

4. Discussion and Conclusions

Active and passive control techniques are utilized to suppress the pressure oscillations inside cavities caused by the supersonic flow past them. Jet blowing at several locations on the front and aft walls of the cavity configuration is used as an active control method. Several techniques are used for

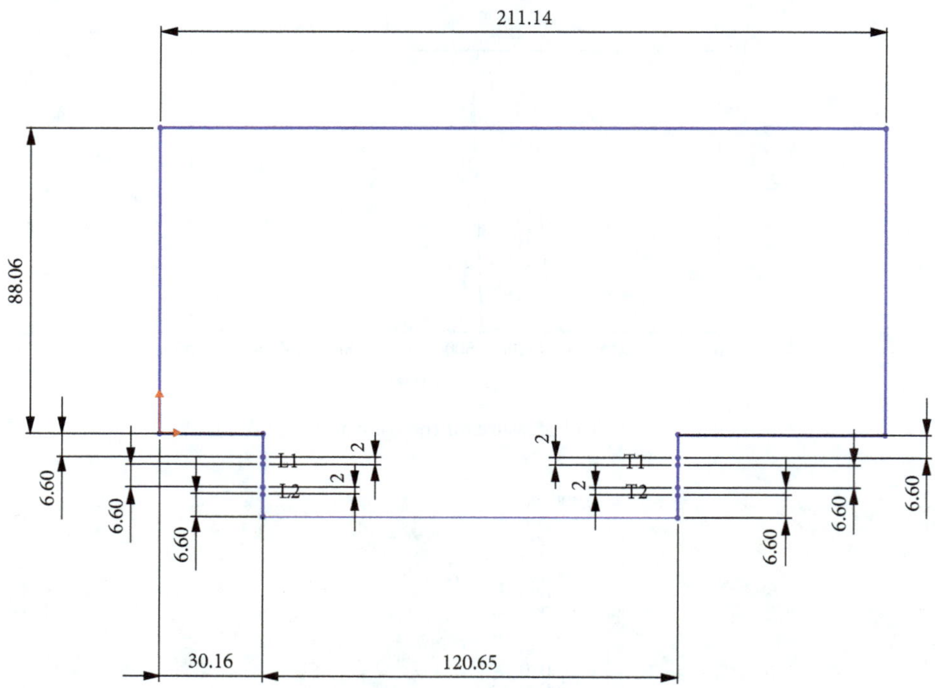

FIGURE 28: Location of holes for jet blowing.

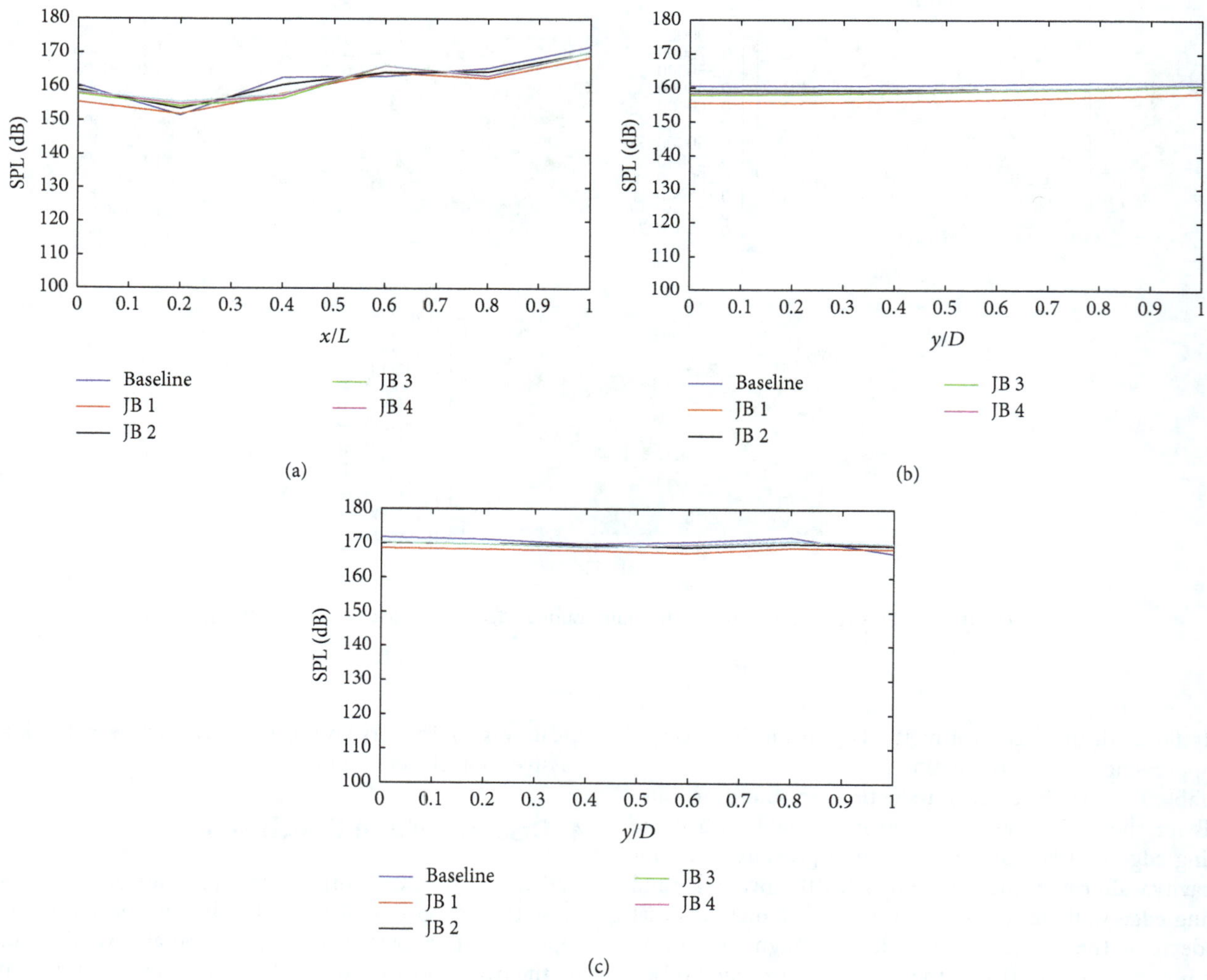

FIGURE 29: SPL values for (a) bottom wall, (b) leading edge wall, and (c) trailing edge wall for active control cases.

TABLE 7: Comparison of sound pressure levels for the most effective cases and the baseline.

Method	Bottom wall average SPL (dB)	% reduction	LE wall average SPL (dB)	% reduction	TE wall average SPL (dB)	% reduction
Baseline	162.6	—	161.2	—	170.4	—
CP 4	147.3	9.4	145.4	9.8	153.8	9.7
TWM 3	131.4	19.2	130.8	18.9	138.0	19.0
TC 4	141.2	13.2	135.4	16.0	150.3	11.8
JB1	160.2	1.5	156.7	2.8	168.3	1.2

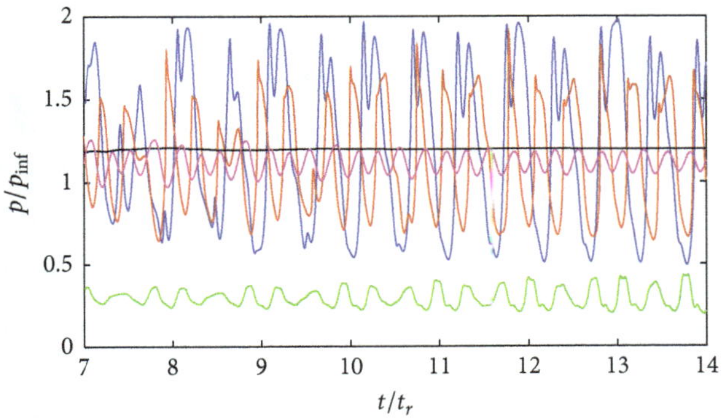

FIGURE 30: Pressure distribution comparison for the most effective control cases.

passive control including, using a cover plate to separate the flow dynamics inside and outside of the cavity, trailing edge wall modifications, such as inclination of the trailing edge, and providing curvature to the trailing edge wall. The results of active and passive control techniques are compared with the baseline case in terms of pressure fluctuations, sound pressure levels at the leading edge, trailing edge walls, and cavity floor and in terms of formation of flow structures as observed from pressure, density contours, and x-velocity contours superimposed with streamlines.

Per the results presented, it is shown that the most effective passive control method is the modification of the trailing edge wall among all the methods tested. Approximately 20% reduction is observed for the average sound pressure levels on the cavity walls including the trailing edge cavity wall. When the pressure fluctuations are examined, it is seen that the periodic nature of the flow is eliminated when the trailing edge wall is modified by introducing a slope. Changing the curvature of the trailing edge wall is also effective, although it is not as effective as introducing a slope. The average SPL values for the cavity walls decrease by 12-13% with this method.

Introducing a cover plate from the leading edge of the cavity is also effective in suppressing the oscillations inside the cavity. 10% reduction in sound pressure levels is observed with a cover plate, by altering the slope of the plate. The plate separates the inside and outside of the cavity, thereby leading to two separate zones for the formation of the vortices inside.

As an active control method, jet blowing is introduced from several locations on the cavity walls. Most effective jet blowing configuration is blowing from the trailing edge wall with 10% of the free stream velocity among the cases that are numerically tested. Active control is not as effective as passive control techniques since it only provides around 5 dB of decrease in pressure fluctuations. However, active control methods have an inherent advantage of being able to adapt to changing conditions, whereas passive control methods are more effective; but they are not adjustable. Once implemented, they cannot be changed during flight.

This study shows that supersonic cavity flow pressure oscillations which is crucial for weapon bay aerodynamics can be suppressed with the help of active and passive control techniques if they are properly applied to the geometry.

Conflicts of Interest

The authors declare that there are no conflicts of interest regarding the publication of this paper.

References

[1] S. Aradag, H. Kim, and D. D. Knight, "Two and three dimensional simulations of supersonic cavity flow," *Engineering Applications of Computational Fluid Mechanics*, vol. 4, no. 4, pp. 612–621, 2010.

[2] S. J. Lawson and G. N. Barakos, "Review of numerical simulations for high-speed, turbulent cavity flows," *Progress in Aerospace Sciences*, vol. 47, no. 3, pp. 186–216, 2011.

[3] L. N. Cattafesta III, Q. Song, D. R. Williams, C. W. Rowley, and F. S. Alvi, "Active control of flow-induced cavity oscillations," *Progress in Aerospace Sciences*, vol. 44, no. 7-8, pp. 479–502, 2008.

[4] D. P. Rizzetta, M. R. Visbal, and P. E. Morgan, "A high-order compact finite-difference scheme for large-eddy simulation of active flow control," *Progress in Aerospace Sciences*, vol. 44, no. 6, pp. 397–426, 2008.

[5] P. Comte, F. Daude, and I. Mary, "Simulation of the reduction of unsteadiness in a passively controlled transonic cavity flow," *Journal of Fluids and Structures*, vol. 24, no. 8, pp. 1252–1261, 2008.

[6] V. Levasseur, P. Sagaut, M. Mallet, and F. Chalot, "Unstructured Large Eddy Simulation of the passive control of the flow in a weapon bay," *Journal of Fluids and Structures*, vol. 24, no. 8, pp. 1204–1215, 2008.

[7] S. Arunajatesan, C. Kannepalli, N. Sinha et al., "Suppression of cavity loads using leading-edge blowing," *AIAA Journal*, vol. 47, no. 5, pp. 1132–1144, 2009.

[8] S. Aradag, H. Yan, and D. Knight, "The effects of laser energy deposition on supersonic cavity flow," *Journal of Thermal Science and Technology*, vol. 29, no. 2, pp. 67–73, 2009.

[9] I. Yilmaz and S. Aradag, "Numerical laser energy deposition on supersonic cavity flow and sensor placement strategies to control the flow," *The Scientific World Journal*, vol. 2013, Article ID 141342, 8 pages, 2013.

[10] E. Lazar, G. Elliotts, and N. Glumac, "Control of the shear layer above a supersonic cavity using energy deposition," *AIAA Journal*, vol. 46, no. 12, pp. 2987–2997, 2008.

[11] Y. Lee, M. Kang, H. Kim, and T. Setoguchi, "Passive control techniques to alleviate supersonic cavity flow oscillation," *Journal of Propulsion and Power*, vol. 24, no. 4, pp. 697–703, 2008.

[12] L. Ukeiley, M. Sheehan, F. Coiffet, F. Alvi, S. Arunajatesan, and B. Jansen, "Control of pressure loads in geometrically complex cavities," *Journal of Aircraft*, vol. 45, no. 3, pp. 1014–1024, 2008.

[13] F. S. Alvi and L. N. Cattafesta III, "The art and science of flow control—case studies using flow visualization methods," *European Physical Journal: Special Topics*, vol. 182, no. 1, pp. 97–112, 2010.

[14] N. Zhuang, F. S. Alvi, M. B. Alkislar, and C. Shih, "Supersonic cavity flows and their control," *AIAA Journal*, vol. 44, no. 9, pp. 2118–2128, 2006.

[15] V. Narayanaswamy, L. L. Raja, and N. T. Clemens, "Characterization of a high-frequency pulsed-plasma jet actuator for supersonic flow control," *AIAA Journal*, vol. 48, no. 2, pp. 297–305, 2010.

[16] V. Thangamani and J. Kurian, "Control of cavity oscillations in a supersonic flow by microjet injection," *Journal of Aircraft*, vol. 50, no. 4, pp. 1305–1308, 2013.

[17] L. G. Kaufman, A. Maciulaitis, and R. L. Clark, "Mach 0.6 to 3.0 flows over rectangular cavities," Tech. Rep. AFWAL-TR-82-3112, Air Force Wright Aeronautical Labs, 1983.

[18] I. Yilmaz, E. Ayli, and S. Aradag, "Investigation of the effects of length to depth ratio on open supersonic cavities using CFD and proper orthogonal decomposition," *The Scientific World Journal*, vol. 2013, Article ID 810175, 12 pages, 2013.

[19] J. E. Rossiter, "Wind tunnel experiments on the flow over rectangular cavities at subsonic and transonic speeds," Tech. Rep. 64037, Royal Aircraft Establishment, Surrey, UK, 1964.

12

Three-Dimensional Numerical Analysis of LOX/ Kerosene Engine Exhaust Plume Flow Field Characteristics

Hong-hua Cai, Wan-sheng Nie, Xin-lei Yang, Rui Wu, and Ling-yu Su

Department of Space Equipment, Equipment Academy, Beijing 101416, China

Correspondence should be addressed to Wan-sheng Nie; nws1969@126.com

Academic Editor: Wen Bao

Aiming at calculating and studying the flow field characteristics of engine exhaust plume and comparative analyzing the effects of different chemical reaction mechanisms on the engine exhaust plume flow field characteristics, a method considering fully the combustion state influence is put forward, which is applied to exhaust plume flow field calculation of multinozzle engine. On this basis, a three-dimensional numerical analysis of the effects of different chemical reaction mechanisms on LOX/ kerosene engine exhaust plume flow field characteristics was carried out. It is found that multistep chemical reaction can accurately describe the combustion process in the LOX/kerosene engine, the average chamber pressure from the calculation is 4.63% greater than that of the test, and the average chamber temperature from the calculation is 3.34% greater than that from the thermodynamic calculation. The exhaust plumes of single nozzle and double nozzle calculated using the global chemical reaction are longer than those using the multistep chemical reaction; the highest temperature and the highest velocity on the plume axis calculated using the former are greater than that using the latter. The important influence of chemical reaction mechanism must be considered in the study of the fixing structure of double nozzle engine on the rocket body.

1. Introduction

Engine exhaust plume brings a variety of problems for the design and launch of rocket, such as [1] (1) the bottom convective heat transfer problems; (2) the bottom radiation heat transfer problems; (3) the radar signal attenuation problem; and (4) the ground thermal shock problem during the take-off of the rocket. The study of rocket engine exhaust plume flow field characteristics has the following important application value [2]: (1) the obtaining mature theory and method which can be used to help study tracking large missile [3, 4] and (2) studying the attenuation effect of exhaust plume on radio frequency communication signal [5].

Using CFD (Computational Fluid Dynamics) software for exhaust plume flow field calculation has the significant advantages, such as three-dimensional numerical simulation, choosing the turbulence model, obtaining detailed flow field parameters, and so on; it has been widely developed and applied. William and Calhoon's simulation [6] analyzed the rocket exhaust plume flow field using the structured and compressible N-S flow solver software GASP (the General Aerodynamic Simulation Program); the simulation study of Saturn V heavy-lift launch vehicle plume was carried out [7] using the CFD calculation procedures—OVERFLOW. Staged calculation method [2] was used to calculate exhaust plume flow fields: first, get the stagnation temperature and other parameters which can be the inlet boundary condition of second stage using the CEA (Chemical Equilibrium with Applications) from NASA for the zero-dimensional numerical analysis and then calculate exhaust plume flow fields with CFD software FLUENT. Three-dimensional numerical simulation of pure gas-phase single-nozzle [8] and four-nozzle [9] exhaust flame flow fields was calculated using the method of large eddy simulation (LES) under the premise of ignoring the chemical reaction influence. FLUENT was used to validate the simplified flow field model of Tomahawk

TABLE 1: Single-step global chemical reaction.

Reaction	A	E_a	n_{KERO}	n_{O2}
$C_{12}H_{23} + 17.75O_2 \rightarrow 12CO_2 + 12H_2O$	2.587*E*9	1.256*E*8	0.25	1.5

cruise missiles BGM-109 [10]. The method of exhaust plume flow field calculation above is conducted without considering the influence of internal combustion state on exhaust plume flow field characteristics. Cai et al. and Feng et al. [11, 12] developed an integrated calculation method of the inner flow field and the plume field of the engine; the effects of combustion chamber combustion model and the nozzle inside surface type on the plume flow field characteristics were studied. The integrated calculation method considered the effect of engine internal combustion state, but it needs a greater number of repeated calculations when calculating the multinozzle engine exhaust plume flow field.

In this paper, a method fully considering the combustion state influence based on the CFD software FLUENT is put forward, which is applied to exhaust plume flow field calculation of multinozzle engine. First of all, the simulation of engine internal flow field is carried out, and the nozzle throat section parameters are as the inlet boundary. Then the calculation of single-nozzle and multinozzle engine exhaust plume flow fields is carried out. Based on the method, numerical analysis of effects of different chemical reaction mechanisms on single-nozzle and double-nozzle LOX/kerosene engine exhaust plume flow field characteristics is carried out.

2. Mathematical and Physical Model

2.1. Governing Equations. Multicomponent chemical reaction and conservative three-dimensional N-S equations are used as the flow, the exchange of matter and energy, and the control equation of combustion of the model. The following is the general form of it.

$$\frac{\partial \vec{U}}{\partial t} + \frac{\partial(\vec{F} - \vec{F}_v)}{\partial x} + \frac{\partial(\vec{G} - \vec{G}_v)}{\partial y} + \frac{\partial(\vec{H} - \vec{H}_v)}{\partial z} = \vec{J}, \tag{1}$$

where $\vec{U}$ is the conservative variable vector; t is the time variable; $\vec{F}$, $\vec{G}$, and $\vec{H}$ are the convection term vector; $\vec{F}_v$, $\vec{G}_v$, and $\vec{H}_v$ are the viscous term vector; and $\vec{J}$ is the source term vector. The equation above consists quality equation, momentum equation, energy equation, and equation of components in the direction of x, y, and z, respectively.

2.2. Chemical Reaction Mechanism. There are hundreds of components in kerosene, so it is difficult to describe the actual chemical reaction process accurately. In this paper, single-step global chemical reaction mechanism and multistep chemical reaction mechanism are studied.

The first part is the single-step global chemical reaction, and the actual combustion process of kerosene can be simplified by the alternative fuel $C_{12}H_{23}$, by one-step oxidation process; the reaction produces H_2O and CO_2, and the chemical reaction rate K_f is calculated by

TABLE 2: Multistep chemical reaction.

Reaction	A	B	E/R
Paraffin global step			
$C_{12}H_{24} + 6O_2 \rightarrow 12CO + 12H_2$	3.888*E*4	1	1.220*E*4
Naphthene global step			
$C_{12}H_{24} + 6O_2 \rightarrow 12CO + 12H_2$	2.132*E*7	1	1.965*E*4
$H_2 + O_2 = OH + OH$	1.700*E*13	0	2.407*E*4
$OH + H_2 = H_2O + H$	2.190*E*13	0	2.590*E*3
$OH + OH = O + H_2O$	6.023*E*12	0	5.500*E*2
$O + H_2 = H + OH$	1.800*E*10	1.0	4.480*E*3
$H + O_2 = O + OH$	1.220*E*17	−0.9	8.369*E*3
$M + O + H = OH + M$	1.000*E*16	1	0
$M + O + O = O_2 + M$	2.550*E*18	0	5939*E*4
$M + H + H = H_2 + M$	5.000*E*15	−1.0	0
$M + H + OH = H_2O + M$	8.400*E*21	0	0
$CO + OH = H + CO_2$	4.000*E*12	−2.0	4.030*E*3
$CO + O_2 = CO_2 + O$	3.000*E*12	0	2.500*E*4
$CO + O + M = CO_2 + M$	6.000*E*13	0	0

$$K_f = Ae^{-E_a/T}[\text{KERO}]^{n_{\text{KERO}}}[O_2]^{n_{O_2}}, \tag{2}$$

where T is the temperature, the unit is in K; [KERO] is the molar concentration of kerosene, the unit is in mol/cm^3; and $[O_2]$ is the molar concentration of oxygen, the unit is in mol/cm^3. The specific values of other parameters are shown in Table 1.

The second part is the multistep chemical reaction mechanism (9 components and 14-step chemical reaction), which is given by [13, 14]. The chemical reaction rate K_f is calculated by

$$K_f = AT^{B}e^{-E/RT}. \tag{3}$$

The specific values of the parameters in the formula are shown in Table 2, and as the third carrier, M does not participate in the chemical reaction.

2.3. Turbulent Combustion Model. Compressible Reynolds-averaged Navier-Stokes (RANS) equations are adopted in the calculation to describe the turbulent flow of flow field. Realizable k-ε double equation model is adopted, as it can better simulate large curvature and vortex flow and easier to converge which referred to [15].

The eddy dissipation/finite rate model is adopted in the calculation of single-step global chemical reaction. The eddy dissipation concept model is adopted in the calculation of multistep chemical reaction. In order to effectively avoid the rounding error of big index former factor and activation energy in the rigid mechanism, double precision solver is

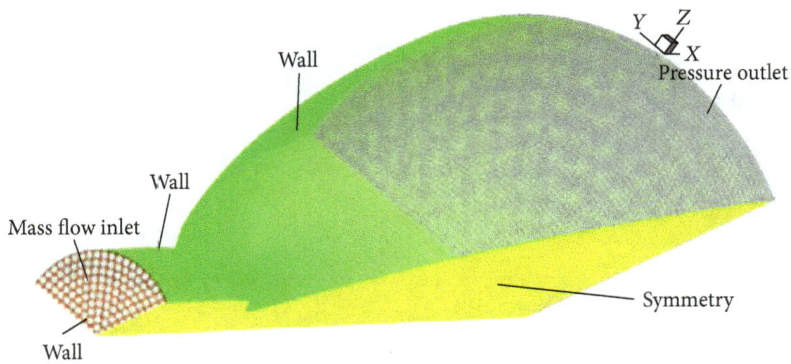

FIGURE 1: Mesh and boundaries of engine internal flow field.

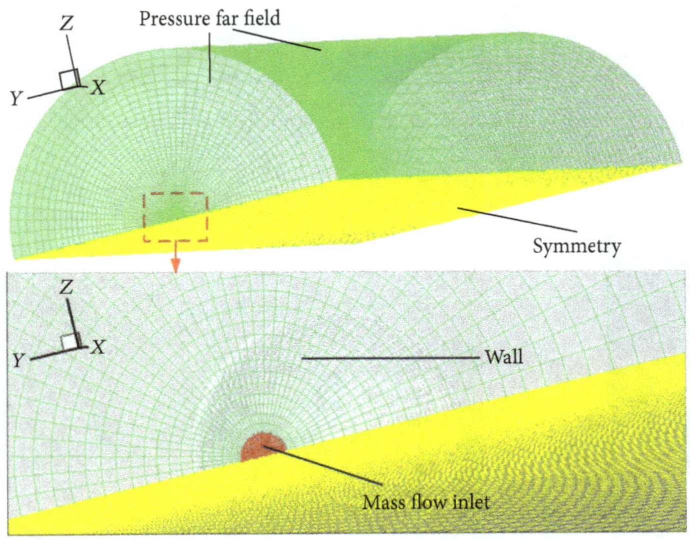

FIGURE 2: Mesh and boundaries of plume flow field of single-nozzle engine.

adopted in the calculation, and the complete multicomponent diffusion and thermal diffusion are considered.

2.4. Disperse and Solution of the Equation. The equations are dispersed in terms of finite volume method, and two order upwind schemes are adopted. Solver based on the pressure is adopted, and the algorithm is PISO (the pressure implicit with splitting of operators) algorithm.

2.5. Grid and Boundary Conditions. Grid generation software is used to deal with the three-dimensional calculation area of the engine internal flow field, the flow field of the single-nozzle engine and the double-nozzle engine; grids and boundary conditions are shown in Figures 1, 2, and 3. In order to save computing resources, according to the symmetry of the engine structure and the calculation area, the internal flow field of the engine is calculated with 1/3 symmetry, and the single-nozzle and double-nozzle exhaust plume flow fields are calculated with 1/2 symmetry. Engine structure is shown in [16], the distance between the two nozzle axes is 1.5 m, and the installation angle is 6°. Calculation areas of single-nozzle and double-nozzle exhaust plumes are the length 130 m and the radius 15 m and the length 150 m and the radius 15 m, respectively.

Boundary conditions of the engine internal flow field are defined as follows: the mass flow inlet boundary is 1.178 kg/s and the mixing ratio of oxygen/kerosene is 2.4; the engine wall boundary is no-slip wall (refer to the LOX/kerosene engine cooling system design requirements provided by [17]), and the engine cooling process is simplified into a wall temperature of 800 K; and the pressure outlet boundary is of 300 K and 101,325 Pa. Boundary conditions of the exhaust plume are defined as follows: the mass flow inlet boundary is determined by the engine internal flow field calculation results; the pressure far field boundary is of 300 K and 101,325 Pa; and the wall boundary is no-slip wall.

3. Calculation Results and Analysis of the Internal Flow Field of the Engine

First, grid independence verification is carried out to eliminate the influence of grid quality on the calculation results. 800 thousand, 1.6 million, and 2.4 million grids were used to calculate the international liquid oxygen/kerosene engine combustion flow; the results from the 800 thousand grids are significantly different from that from the 1.6 million grids and the 2.4 million grids and ultimately determine the 1.6 million grids because of the large computing resources

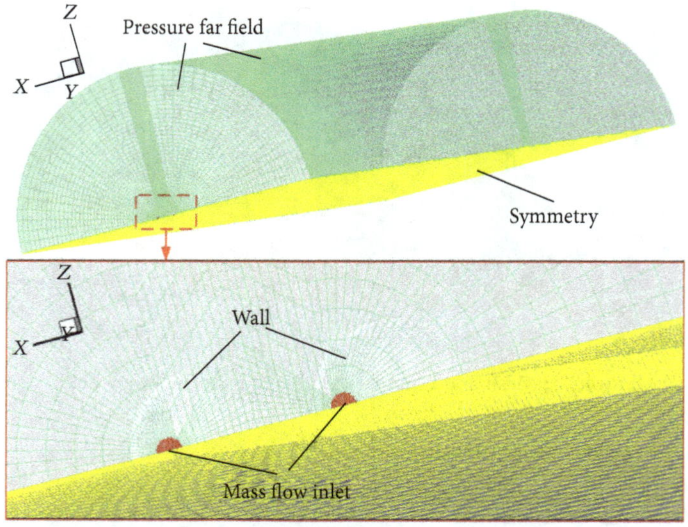

FIGURE 3: Mesh and boundaries of plume flow field of double-nozzle engine.

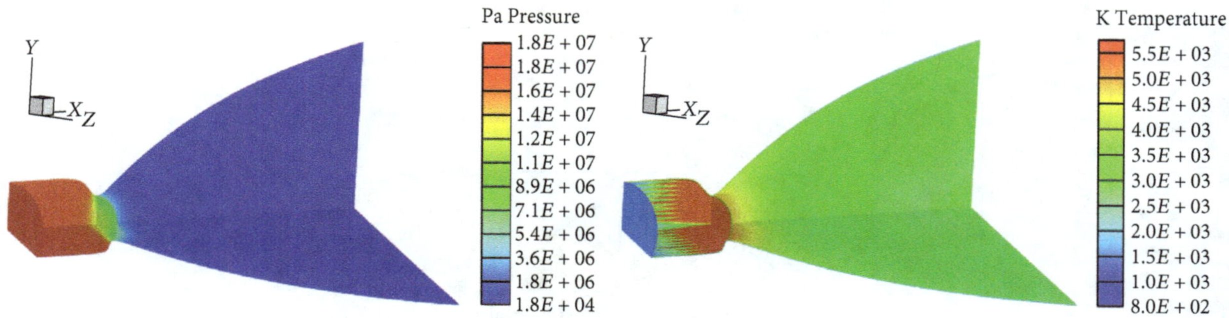

(a) Pressure

(b) Temperature

FIGURE 4: Flow fields in the engine using the single-step global chemical reaction.

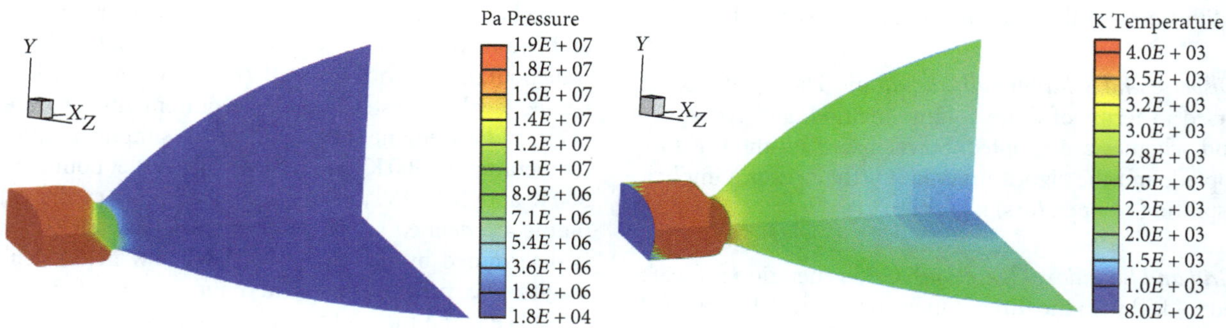

(a) Pressure

(b) Temperature

FIGURE 5: Flow fields in the engine using the multistep chemical reaction.

of the 2.4 million grids. Internal flow field pressure and temperature distributions obtained using the 1.6 million grids by the single-step global chemical reaction are shown in Figures 4(a) and 4(b); the average pressure of combustion chamber is 17.82 MPa, which is 0.68% greater than 17.7 MPa obtained by model engine test, and the result is relatively accurate. The average temperature of combustion chamber is 5782 K, which is 51.71% greater than 3811.17 K calculated by the thermodynamic calculation of the model engine combustion chamber; this is the result due to ignoring the middle of multistep endothermic reaction process when the single-step global chemical reaction is adopted.

Internal flow field pressure and temperature distributions obtained by the multistep are shown in Figures 5(a) and 5(b). The average pressure of combustion chamber is 18.52 MPa, which is 4.63% greater than 17.7 MPa obtained

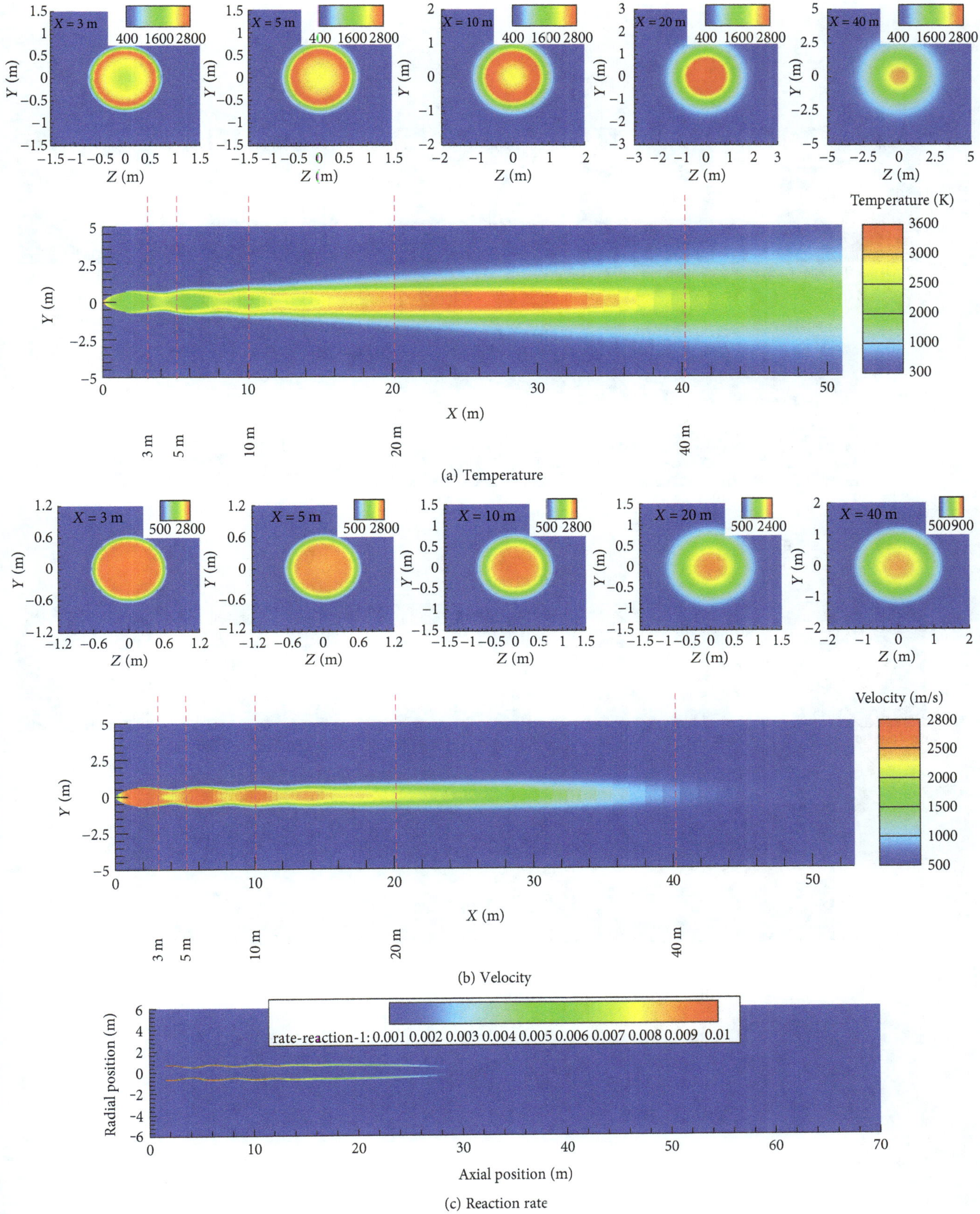

Figure 6: Plume flow fields of the single-nozzle engine (single step).

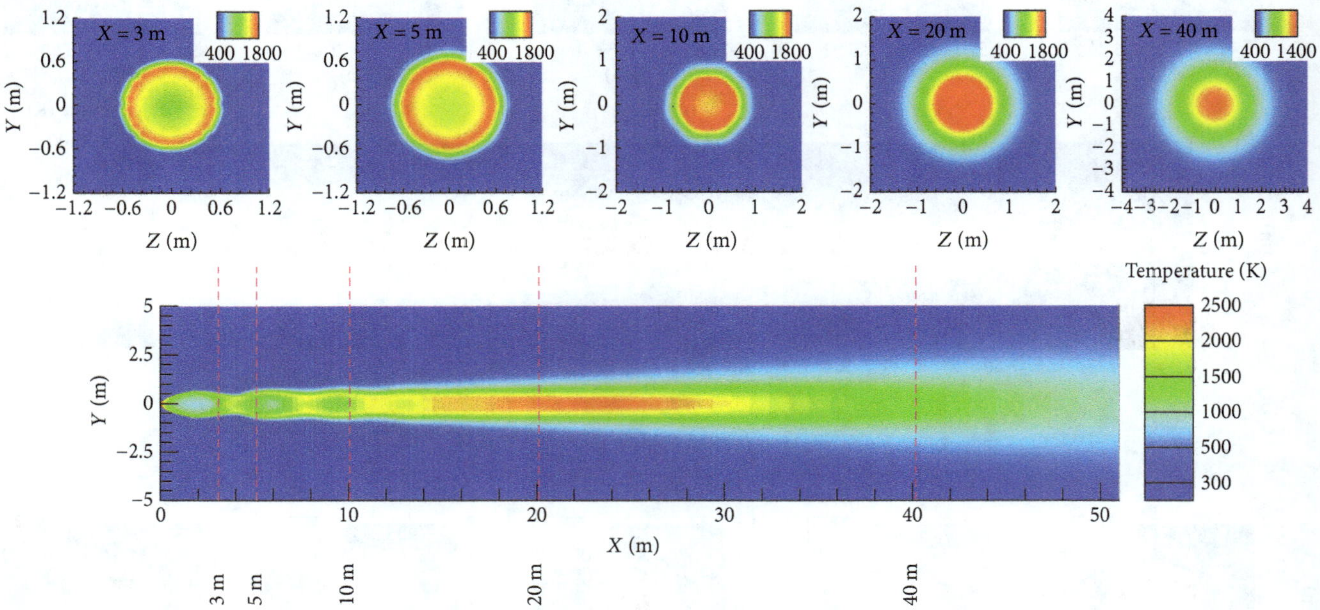

(a) Temperature

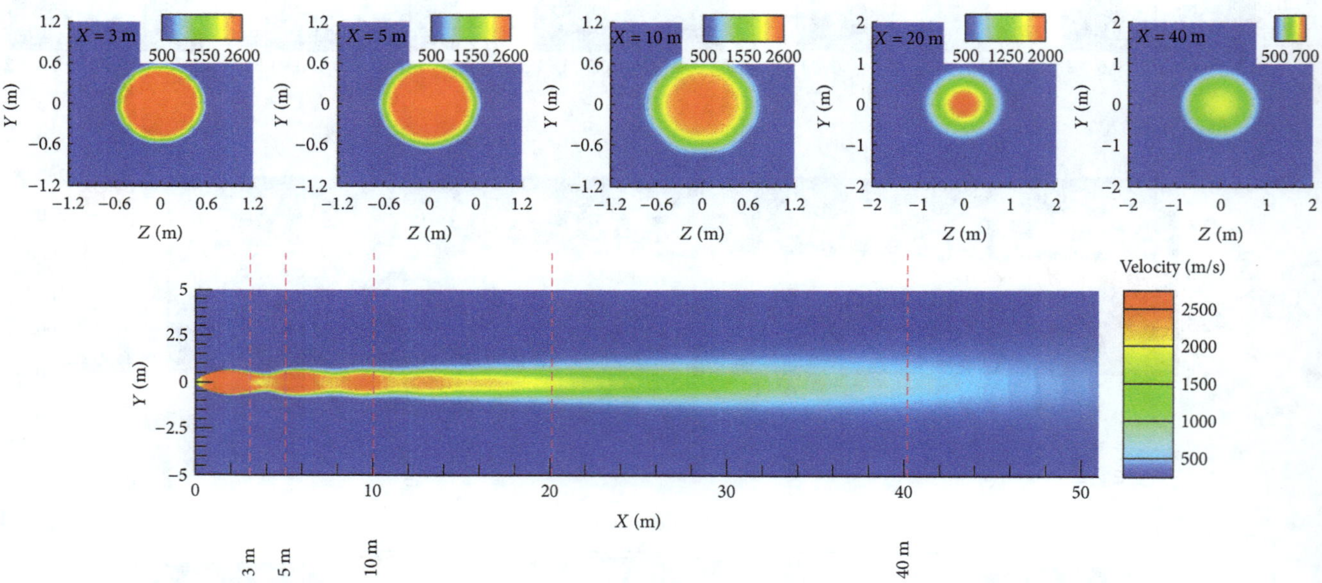

(b) Velocity

Figure 7: Continued.

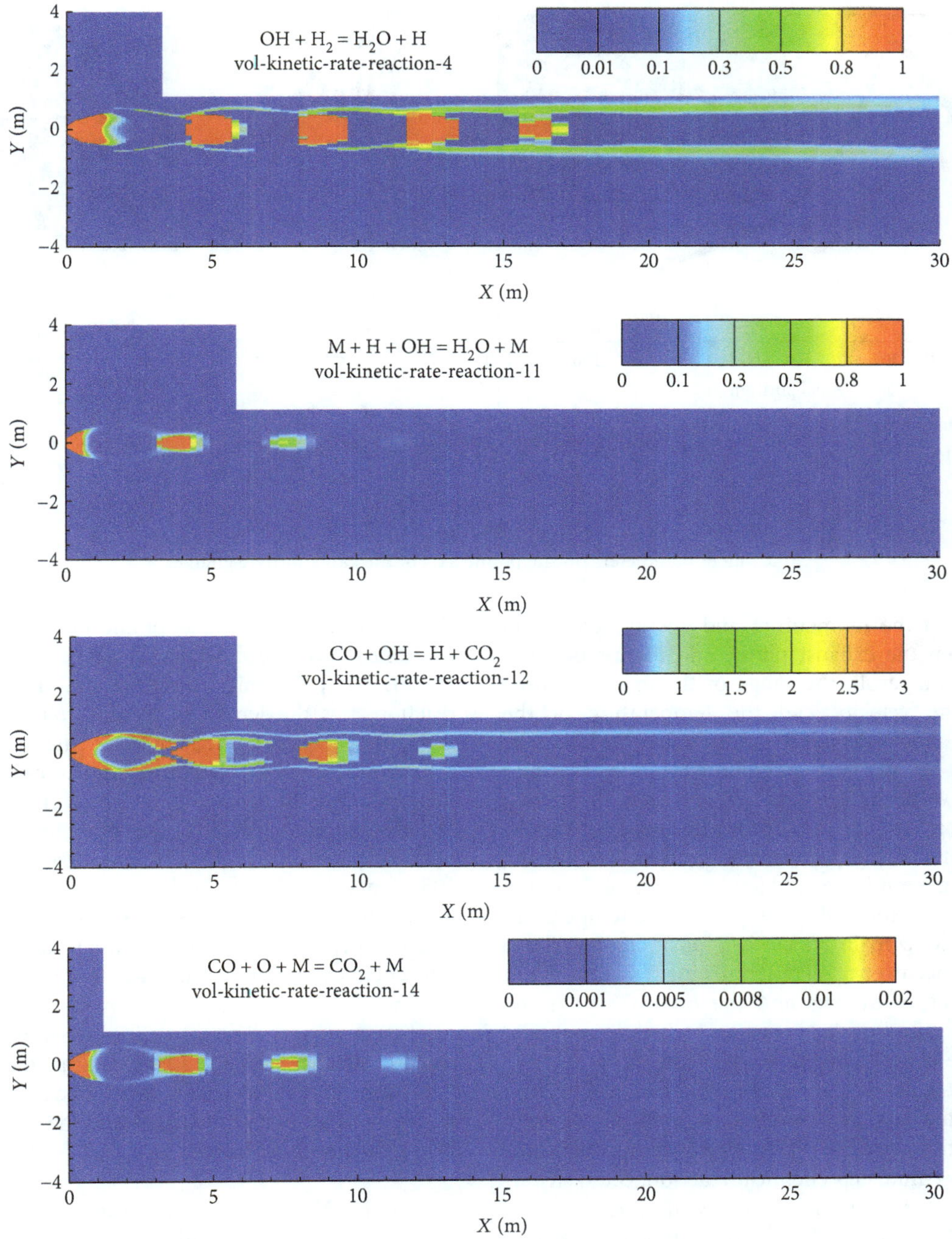

(c) Reaction rate

Figure 7: Plume flow fields of the single-nozzle engine (multistep).

by model engine test, and the average temperature of combustion chamber is 3938.53 K, which is 3.34% greater than 3811.17 K calculated by the thermodynamic calculation of the model engine combustion chamber, so the result of internal flow field pressure and temperature are relatively accurate.

4. Calculation Results and Analysis of the Engine Exhaust Plume Flow Field

4.1. Single-Nozzle Engine Exhaust Plume Flow Field. First, grid independence verification is carried out using 300 thousand, 650 thousand, and 960 thousand grids, and 650 thousand grids is used in considering the accuracy and the cost. The parameter distribution of internal flow field calculated by the single-step chemical reaction is used as the inlet boundary condition, and the plume afterburning reaction is also described using the single-step global chemical reaction. The distributions of temperature, velocity, and reaction rate in the exhaust plume flow field on the different distance cross sections of the axial line and the symmetrical plane are shown in Figures 6(a), 6(b), and 6(c). Exhaust plume gas is ejected from the nozzle into the air of the environment in the infinite region; the air medium is sucked by the plume because of the

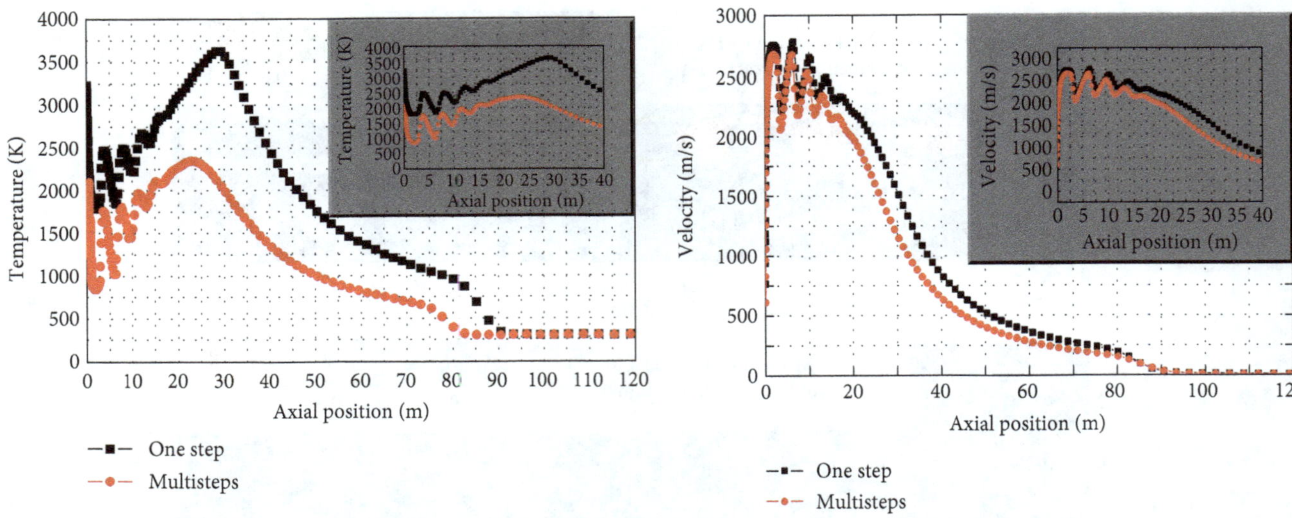

(a) Temperature

(b) Velocity

FIGURE 8: Parameters on the plume axis of the single-nozzle engine.

viscous effect, and the momentum and energy conversion exchange between the exhaust plume and the air medium is conducted. As a result, the range of the exhaust plume expands and the velocity and the temperature of the exhaust plume both decrease continuously. Along the radial direction, the plume temperature first increases and then decreases with the ambient temperature; the plume velocity decreases to the environment speed gradually. The first reason is the momentum conversion of the exhaust plume gas and the ambient air mixing, and the exhaust plume gas velocity decreases while the temperature increases; the second main reason is a lot of energy release because of the occurrence of afterburning reaction when the plume and air is mixing, and the mixing area is the main region of afterburning reaction, which causes the temperature rise sharply.

The parameter distribution of internal flow field calculated by multistep chemical reaction is used as the inlet boundary condition, and the plume afterburning reaction is also described by multistep chemical reaction; the distributions of temperature, velocity, and reaction rate in the exhaust plume flow field on the different distance cross sections of the axial line and the symmetrical plane are shown in Figures 7(a) and 7(b). The plume flow field structure is the same with that calculated with the one-step chemical reaction model, but the overall velocity and temperature of the plume reduce significantly; this is because the single-step chemical reaction model ignores a lot of endothermic reaction process, which leads to the temperature rise sharply. The chemical reaction rate in the exhaust plume flow field is shown in Figure 7(c), all intermediate chemical reactions occur in the core area of the whole plume, and the distribution is like the continuous shock wave; this is the important difference between multistep chemical reaction model and single step chemical reaction model to describe afterburning chemical reaction.

The parameter distribution in the single-nozzle engine obtained by different chemical reaction mechanisms is shown in Figure 8; similar plume flow field structure and the same number of shock waves can be obtained by methods of single-step chemical reaction and multistep chemical reaction, but the length of exhaust plume of the former is 8 m longer than that of the latter (the judgment standard is the temperature on the plume axis decreases to the ambient temperature or the velocity on the plume axis decreases to the environment velocity), the peak temperature and the peak velocity were 1269.47 K (54.02% of the latter), and 105.94 m/s (3.96% of the latter) greater than that of the latter. The reasons are as follows: first, a lot of endothermic reaction processes are ignored when adopting the single-step chemical reaction; thus, the temperature and velocity obtained are higher. Second, exhaust plume afterburning reaction occurs in different regions when different chemical reaction mechanisms are adopted; thus, the exhaust plume flow field temperature and velocity calculated by the former are higher than that of the latter, and the exhaust plume flow field structure also change.

4.2. Double-Nozzle Engine Exhaust Plume Flow Field. First, grid independence verification is carried out using 470 thousand, 750 thousand, and 1.4 million thousand grids, and 750 thousand grids is used in considering the accuracy and the cost. The parameter distribution of internal flow field calculated by single-step chemical reaction is used as the inlet boundary condition, and the plume afterburning reaction is also described by single-step global chemical reaction; the distributions of temperature, velocity, and reaction rate in the exhaust plume flow field on the different distance cross sections of the axial line and the symmetrical plane are shown in Figures 9(a), 9(b), and 9(c). At first, each of the two engine exhaust plumes maintains an independent shock structure, and then, with the increase of the axial distance, the continuous mixing of the exhaust plume and the ambient air leads to the continuous expansion of the exhaust plume; at last, the two exhaust plumes melt into a plume as a whole. The afterburning reaction is the most obvious both in the interface

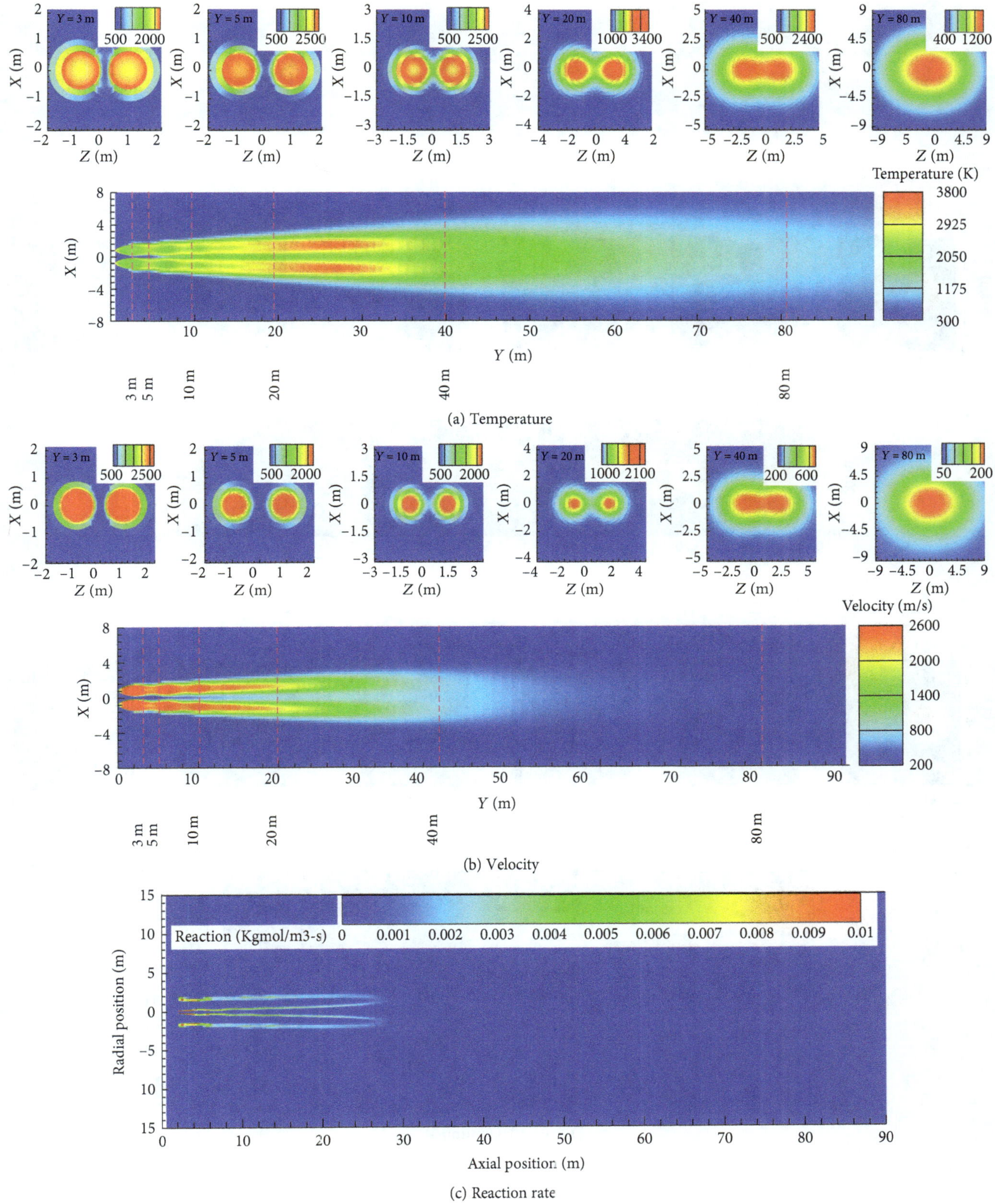

FIGURE 9: Plume flow fields of the double-nozzle engine (single step).

between exhaust plume gas and ambient air and also the intersection of two nozzle plume interactions.

The parameter distribution of internal flow field calculated by multistep chemical reaction is used as the inlet boundary condition, and the plume afterburning reaction is also described by multistep chemical reaction. The distributions of temperature, velocity, and reaction rate in the exhaust plume flow field on the different distance cross

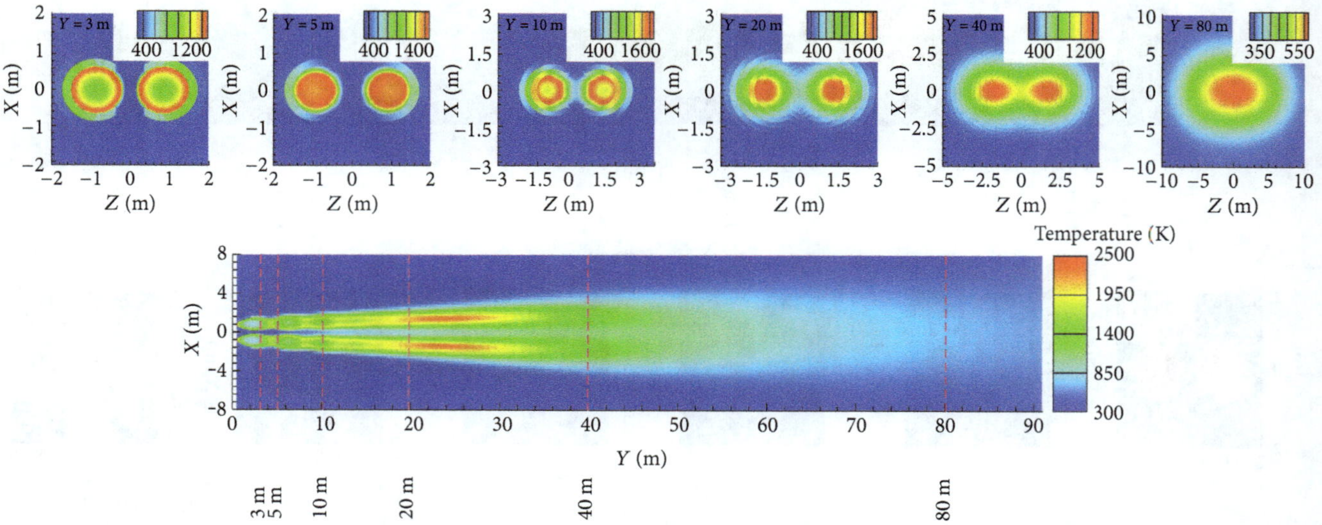

(a) Temperature

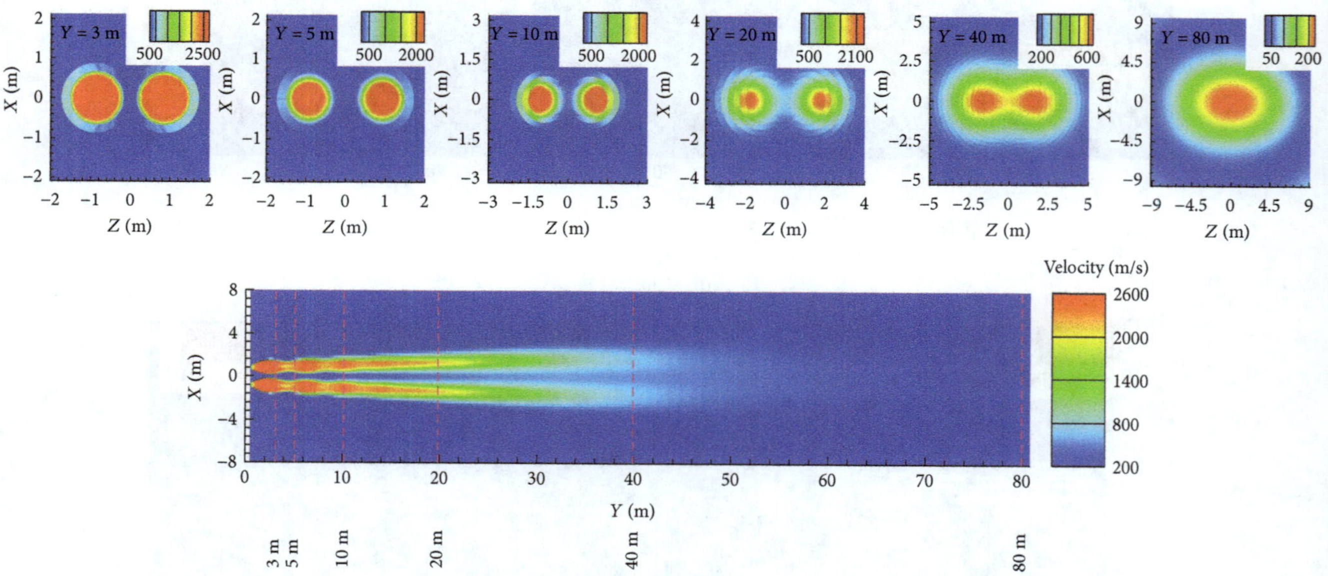

(b) Velocity

FIGURE 10: Continued.

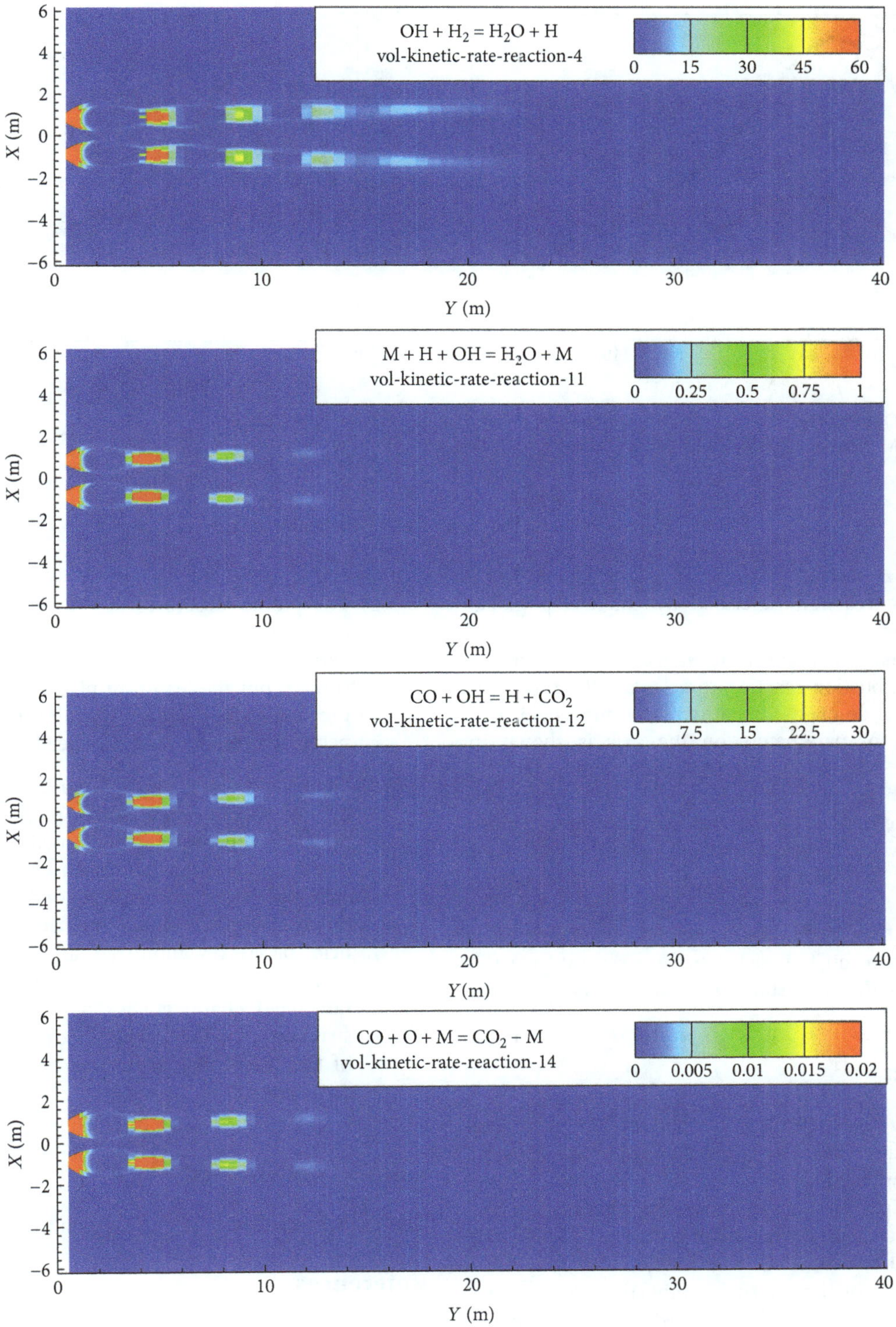

(c) Reaction rate

FIGURE 10: Plume flow fields of the double-nozzle engine (multistep).

sections of the axial line and the symmetrical plane are shown in Figures 10(a) and 10(b). The exhaust plume flow field structure obtained is similar to that obtained by single-step global chemical reaction; however, as the temperature and the velocity of the exhaust plume gas calculated by the multi-step chemical reaction are relatively low, the range of the exhaust plume increases slowly with the increase of axial distance, the interaction between the two exhaust plumes is relatively weak, and the calculation results fully reflect the effect of two engine mounting angles on avoiding the interaction between double engine exhaust plume. The chemical reaction rate in the exhaust plume flow field is shown in Figure 10(c), all intermediate chemical reactions occur in the core area of the whole plume, and it is different from

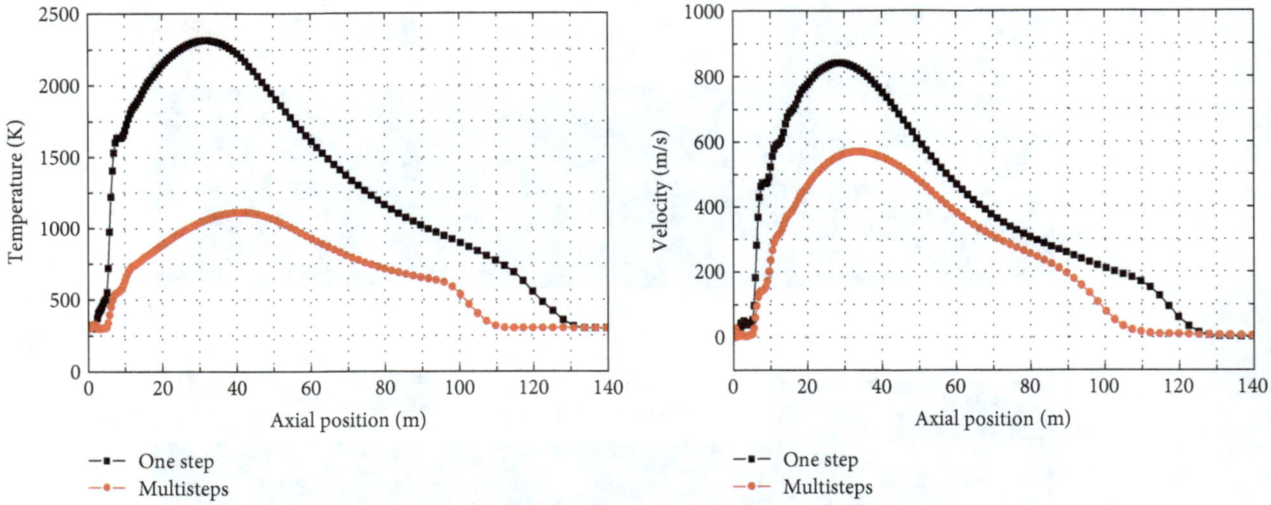

(a) Temperature

(b) Velocity

FIGURE 11: Parameters on the plume axis of the double-nozzle engine.

the afterburning reaction described by single-step global chemical reaction.

Different chemical reaction mechanisms are adopted to describe the combustion reaction to calculate the flow field of the double-nozzle engine exhaust plume flow field, and the distribution of parameters on the axis is shown in Figure 11; the length of exhaust plume obtained by single-step chemical reaction is longer than that obtained by multistep chemical reaction, and the peak temperature and the peak velocity are 1205.55 K (about 108.81% of the latter) and 271.76 m/s (about 47.71% of the latter), respectively. In conclusion, when analyzing the double-nozzle plume flow field characteristics or the engine mounting structure on the rocket body, we must give full consideration to the important influence of the chemical reaction model.

5. Conclusion

In this paper, a computational method of LOX/kerosene engine plume flow field is proposed. First, the internal combustion flow field of the engine is calculated, and the parameters of the nozzle throat are obtained as the inlet boundary of plume flow field simulation. Then, the plume flow field is simulated. The method can fully consider the influence of the internal combustion state on the plume flow field and avoid the repeated calculation of the internal flow field of the engine when calculating the exhaust plume of a multinozzle engine. The effects of different chemical reaction mechanisms on the single-nozzle and the double-nozzle LOX/kerosene engine exhaust plumes are analyzed based on the method. The conclusions are as follows:

(1) Multistep chemical reaction can be more accurate compared with one-step chemical reaction on the description of LOX/kerosene engine internal combustion process.

(2) The plume structures of the single-nozzle and the double-nozzle LOX/kerosene engine exhaust plumes with different chemical reaction mechanisms are similar, but the length of plume flow field based on the single-step chemical reaction is longer, the temperature is higher, and the speed is faster.

(3) Double-nozzle engine plume flow field is formed by two sets of single-nozzle plume flow field. Due to the temperature and velocity, distributions of single-nozzle engine plume flow field based on multistep and one-step chemical reaction are completely different and the double-nozzle engine plume flow field based on multistep and one-step chemical reaction is different. Analysis of double-nozzle plume flow field characteristics or the engine mounting structure on the rocket body and the influence of the chemical reaction mechanism must be fully considered.

Conflicts of Interest

The authors declare that they have no conflicts of interest.

References

[1] G. Terry, S. David, and P. Robert, "Analysis of liquid rocket propellant engine exhaust plumes," in *AIAA 5th Thermophysics conference*, Los Angeles, CA, USA, June 29–July 1, 1970.

[2] D. R. Gonzalez, P. Walman, M. Sanford, J. Monat, and J. Carney, "Characterization of rocket-plume fluid-dynamic environment using numerical and experimental approaches," *Journal of Spacecraft and Rockets*, vol. 50, no. 3, pp. 527–539, 2013.

[3] G. P. Sutton and O. Biblarz, *Rocket propulsion elements*, John Wiley & Sons, New York, NY, USA, 7th edition, 2001.

[4] F. S. Simmons, *Rocket Exhaust Plume Phenomenology*, The Aerospace Press, AIAA, Reston, VA, USA, 1st edition, 2000.

[5] A. Mathur, "Rocket plume attenuation model," in *24th AIAA International Communications Satellite Conference*, San Diego, CA, USA, June 2006.

[6] H. William and J. Calhoon, "Computational assessment of afterburning cessation mechanisms in fuel-rich rocket exhaust plumes," *Journal of Propulsion and Power*, vol. 17, no. 1, pp. 111–119, 2001.

[7] G. Marshall, H. Jeffrey, and K. Cetin, "Best practices for CFD simulations of launch vehicle ascent with plumes - OVERFLOW perspective," in *49th AIAA Aerospace Science Meeting including the New Horizons Forum and Aerospace Exposition*, Orlando, FL, USA, January 2011.

[8] S.-c. Hu, F.-t. Bao, Z. Wang, and Q. Cai, "Numerical simulation of jet flow field and noise in nozzle," *Journal of Solid Rocket Technology*, vol. 34, no. 4, pp. 436–439, 2011.

[9] S.-c. Hu, F.-t. Bao, and Y. Zhao, "Numerical simulation for multi-jet aeroacoustics," *Journal of Propulsion Technology*, vol. 33, no. 3, pp. 430–435, 2012.

[10] X.-f. Xie, Z.-b. Ren, and X.-y. Cao, "Model building and analysis on infrared radiation characteristics of exhaust plume of cruise missils," *Opto-Electronic Engineering*, vol. 36, no. 4, pp. 70–74, 2009.

[11] H.-h. Cai, N. Wan-sheng, and Z. Gang, "Impact of the inner nozzle surface on plume characteristics of LOX/kerosene engine," *Missile and Space Vehicles*, vol. 5, pp. 35–38, 2015.

[12] S.-j. Feng, N. Wan-sheng, and X. Qing-fen, "Influence of combustion model in chamber on exhaust plume flow field and radiance," *Journal of Rocket Propulsion*, vol. 32, no. 2, pp. 6–10, 2006.

[13] T. S. Wang, R. C. Farmer, and R. B. Edelman, "Turbulent combustion kinetics for complex hydrocarbon fuels," in *AIAA 26th Aerospace Sciences Meeting*, Reno, USA, January 1988.

[14] R. B. Edelman and P. T. Harsha, "Laminar and turbulent gas dynamics in combustors current status," in *34th Fluid Mechanics & Heat Transfer*, Cleveland, OH, USA, March 1977.

[15] B. T. Vu, N. Bachchan, O. Peroomian, and V. Akdag, "Multiphase modeling of water injection on flame deflector," in *21st AIAA Computational Fluid Dynamics Conference*, San Diego, CA, USA, June 2013.

[16] C. Hong-hua, N. Wan-sheng, and H. Zhi-yong, "Impact effect of liquid rocket motor exhaust plume on launching platform," *Journal of Aerospace Power*, vol. 31, no. 8, pp. 1995–2001, 2016.

[17] P. Pascal, F. Thomas, and V. Hilda, "Lox/methane and Lox/kerosene high thrust engine trade-off," in *37th AIAA/ASME/SAE/ASEE Joint Propulsion*, Salt Lake City, UT, USA, July 2001.

13

Three-Dimensional CST Parameterization Method Applied in Aircraft Aeroelastic Analysis

Hua Su, Chunlin Gong, and Liangxian Gu

Shaanxi Aerospace Flight Vehicle Design Key Laboratory, Northwestern Polytechnical University, Xi'an, Shaanxi 710072, China

Correspondence should be addressed to Hua Su; su@nwpu.edu.cn

Academic Editor: Kenneth M. Sobel

Class/shape transformation (CST) method has advantages of adjustable design variables and powerful parametric geometric shape design ability and has been widely used in aerodynamic design and optimization processes. Three-dimensional CST is an extension for complex aircraft and can generate diverse three-dimensional aircraft and the corresponding mesh automatically and quickly. This paper proposes a parametric structural modeling method based on gridding feature extraction from the aerodynamic mesh generated by the three-dimensional CST method. This novel method can create parametric structural model for fuselage and wing and keep the coordination between the aerodynamic mesh and the structural mesh. Based on the generated aerodynamic model and structural model, an automatic process for aeroelastic modeling and solving is presented with the panel method for aerodynamic solver and NASTRAN for structural solver. A reusable launch vehicle (RLV) is used to illustrate the process for aeroelastic modeling and solving. The result shows that this method can generate aeroelastic model for diverse complex three-dimensional aircraft automatically and reduce the difficulty of aeroelastic analysis dramatically. It provides an effective approach to make use of the aeroelastic analysis at the conceptual design phase for modern aircraft.

1. Introduction

In the wake of requirements for high lift-drag ratio aerodynamic shape and light-weight structure, the aeroelastic phenomena caused by interaction between fluid and structure have a growing influence on the integrated performance of modern aircraft [1, 2]. Aeroelastic analysis becomes an important process for modern aircraft design [3, 4]. Especially in the conceptual design phase, main performance of an aircraft is determined during this phase and, therefore, how to carry out the aeroelastic analysis quickly and stably to improve the integrated performance and the design rationality of the aircraft scheme will help a lot in the following aircraft design.

Aeroelastic analysis is given attention by lots of researchers [5]. Many tools such as ZAERO [6], ENSAERO [7], and NeoCASS [8] are developed to perform aeroelastic analysis based on the frequency domain analysis method and time domain analysis method and have been widely applied on high aspect ratio wing, unmanned aerial vehicle, and hypersonic aircraft. Although there exist some mature aeroelastic analysis and solving method, the aerodynamic modeling and structural modeling are still a complex and time-consuming process. First, the parametric geometric shape and modeling tools should work together closely and automatically; second, the aerodynamic solver and structural solver should coordinate with each other to ensure that the aeroelastic analysis procedure is executed consistently and accurately. In the conceptual design phase, the design scheme of aircraft usually needs constant modification to improve performance. The size parameters and structural topology of the scheme need to change frequently. It is difficult for the traditional CAD-based geometry modeling method to satisfy the need of rapid geometry iteration and large range modification. Aeroelastic analysis applied in the conceptual design phase faces the following problems:

(1) The aerodynamic model and structural model established in many aeroelastic literatures were complicated; the sizing of the aerodynamic shape and structural layout is difficult and time-consuming, which makes it hard to meet the demand for rapid modification at the conceptual design phase.

(2) Because of the differences between aerodynamic modeling and structural modeling in mechanism, the aerodynamic mesh and structural mesh are not compatible. The data conversion is used additionally to interchange aerodynamic force and structural deformation between the aerodynamic model and the structural model. Extra computation overhead is needed, and some errors are introduced inevitably.

(3) Quantity and position of the structural inner elements, such as beam, bulkhead, spar, and rib, are hard to be modified using the traditional CAD-based geometry modeling method. The structural layout is vested and cannot obtain the optimal scheme by topology optimization.

These problems hinder the application of aeroelastic analysis in the conceptual design phase. More effective aerodynamic modeling and structural modeling method should be studied to simplify and improve the aeroelastic modeling and analysis process. The CST method is an analytical method developed by Kulfan [9, 10]. It combines a class function representing a specific class of shapes and a shape function defining the deviation from the class function. These provide an efficient shape parameterization ability on complex geometry using fewer design variables. There are some comparisons with other shape parameterization methods [11, 12], showing that the CST method has advantages in smoothness, mathematical efficiency, fitting performances, and Intuitiveness. CST has been widely used to parameterize and optimize two-dimensional airfoil [13, 14] and simple 3D aircraft [15, 16]. Liu [17, 18] presents a multiblock CST method to model the hypersonic aircraft with complex shape, which can join adjacent surfaces smoothly and retain the good properties of the CST method. Leal [19] proposes an aerostructural optimization method for determining in a preliminary manner morphing wing configurations that provide benefits during various disparate flight conditions with CST parameterization.

The authors in [20, 21] proposed a universal three-dimensional CST method for geometry modeling of complex aircraft. It generates complex three-dimensional geometric shape to support various aircraft aerodynamic shape modeling, which gives a simple and effective way to aerodynamic optimization. This novel three-dimensional CST method is extended in aeroelastic analysis of complex aircraft in this article. A novel parametric structural modeling method is proposed based on gridding feature extraction from the aerodynamic mesh generated by the three-dimensional CST method. An automatic and effective way for aeroelastic modeling and analysis is also established. The structure of this article is as follows. First, the basic principle of the three-dimensional CST method is introduced briefly; then the aircraft characteristic components library is established, including fuselage, wing, and empennage; on this basis, the structural modeling method is presented in detail, and the aeroelastic modeling and analysis process is constructed; finally, a static aeroelastic analysis example is used to verify the proposed aeroelastic modeling and analysis process.

2. Three-Dimensional CST Method

CST method has an efficient and brief shape description for the two-dimensional airfoils and simple three-dimensional geometries. This section gives an improvement to expand the original two-dimensional CST method to three dimensions. A universal three-dimensional CST method is proposed to provide adjustable parameterization ability to complex three-dimensional geometry.

2.1. Basic Cross Section Definition. Most of the aircraft geometry can be described by innumerable cross sections along the axial direction. Combining an analytical function (the class function) with a parametric curve (the shape function), the basic cross section can be defined by the B-spline CST method [22] as follows:

$$\zeta(\psi)\big|_{\text{upp,low}} = C_{N2}^{N1}(\psi)\cdot S(\psi) + \Delta\zeta_N(\psi), \tag{1}$$

where $\psi = z/L_w$. z is the lateral coordinate. ψ is the normalized lateral coordinate. L_w is the total length of the basic cross section in the lateral direction. $\zeta(\psi)$ is the normal height in the lateral position. $\Delta\zeta_N(\psi)$ is the eccentric distance in the lateral position. $C(\psi)$ and $S(\psi)$ are the class function and the shape function defined in CST method. $N1$ and $N2$ are control parameters of the class function. For the symmetrical section, $N1$ is equal to $N2$. The class function $C(\psi)$ is defined as follows:

$$C_{N2}^{N1}(\psi) = \psi^{N1}[1-\psi]^{N2}. \tag{2}$$

The basic cross section is assembled by the upper $\zeta(\psi)$ and the inverted lower $\zeta(\psi)$. If we set $\Delta\zeta_N(\psi)$ to 0, ignore the shape function, and change the control parameters of the class function simultaneously, different sections can be generated. Various cross section shapes can be generated by changing control parameters of the class function.

2.2. B-Spline Shape Function. The B-spline function can adjust the range of influence by selecting different orders of sub-Bernstein polynomials. It has better local control ability and computation efficiency than Bernstein polynomials. So it has been chosen as the shape function of the CST method in the basic cross section definition. The B-spline function is combined by the massive low-order Bezier curves to approximate the high-order curve. It is defined as follows:

$$S(\psi) = \sum_{i=0}^{n} b_i N_k^i(\psi), \tag{3}$$

where b_i, $i = 0, 1, \ldots, n$ is the weight factor. $k+1$ order B-spine basic function is defined as the piecewise polynomials at node vector $T = \{t_0, t_1, \ldots, t_{n+k+1}\}$, $t_i \le t_{i+1}$. $N_k^i(\psi)$ is the B-spline basic function defined at $\{t_i\}_{i=0}^{n+k+1}$, which can be obtained by recurrence formula De Boor-Cox.

$$N_1^i(\psi) = \begin{cases} 1, & t_i \le \psi \le t_{i+1} \\ 0, & \text{otherwise} \end{cases}$$

$$N_p^i(\psi) = \frac{\psi - t_i}{t_{i+p+1} - t_i} N_{p-1}^i(\psi) + \frac{t_{i+p+1} - \psi}{t_{i+p+1} - t_{i+1}} N_{p-1}^{i+1}(\psi) \quad (4)$$

$$\frac{0}{0} = 0.$$

The basic cross section defined with B-spline function is as follows:

$$\zeta(\psi)\big|_{\text{upp,low}} = C_{N2}^{N1}(\psi) \cdot \sum_{i=0}^{n} b_i N_k^i(\psi) + \Delta\zeta_N(\psi). \quad (5)$$

2.3. Three-Dimensional CST Method. The three-dimensional geometry can be considered as a series of original cross sections arranged parallel along the axial direction. By defining proper cross sections using the method mentioned above, we can get an analytical surface to express geometric shape. In order to introduce an analytical axial rule, the coefficient b is replaced by the same B-spline CST definition in formula (5). So the two-dimensional CST function can be expanded to a three-dimensional version, which is shown as follows:

$$b_i = C_{M2}^{M1}(\eta) \cdot S(\eta) + \Delta\zeta_M(\eta) \quad (6)$$

$$S(\eta) = \sum_{j=0}^{m} b_j N_k^j(\eta), \quad (7)$$

where $\eta = x/L$. x is the axial coordinate. η is the normalized axial coordinate. L is the total length of the generated surface in the axial direction. Bringing formula (6) to formula (5), we have the definition of the three-dimensional analytical surface.

$$\zeta(\psi,\eta)\big|_{\text{upp,low}} = C_{N2}^{N1}(\psi) C_{M2}^{M1}(\eta) \sum_{i=0}^{n} \sum_{j=0}^{m} b_{i,j} N_k^i(\psi) N_k^j(\eta) + \Delta\zeta_{M,N}(\psi,\eta). \quad (8)$$

Formula (8) is the analytical expression of the extensional three-dimensional surface, where ζ is the third-dimensional coordinate value along the y direction based on the two-dimensional normalized coordinate ψ and η. $C(\psi)$ *and* $C(\eta)$ are the CST's class functions with control parameter $N1$, $N2$, $M1$ and $M2$. $N(\psi)$ *and* $N(\eta)$ are the basic functions of the B-spline, which constitute the CST's shape function. $b_{i,j}$ is the discrete control weight factor in the up and low surfaces. Lower case n and m are the orders of B-spline function, which are expressed as the numbers of lateral and axial control points. The total points of the geometric surface are $(n+1) \times (m+1)$. More control points mean more design parameters and better parametric geometric shape design ability. When $n = 0$ and $m = 0$, the discrete control weight factor b is equal to 1, the shape function $S(\psi,\eta)$ is also equal to 1, and then entire geometric surface is controlled by the control factors $N1$, $N2$, $M1$, and $M2$ of the class function. $\Delta\zeta_{M,N}(\psi,\eta)$ is the eccentric distance in normal position, and the default value is 0.

The geometric surface defined in formula (8) can be described as the third-dimensional coordinate value calculated by the two-dimensional mesh points dispersed as the lateral and axial control points. ψ and η are the normalized coordinates defined in $[0,1] \times [0,1]$. We also need to define the profile Z in the lateral direction with respect to the axial coordinate η. The profile Z is expressed the same as the B-spline CST method in formula (5).

$$Z(\eta) = C_{T2}^{T1}(\eta) \sum_{t=0}^{w} b_t N_k^t(\eta). \quad (9)$$

b_t is the discrete control weight factor in the lateral coordinate. From the above definition, we obtain the three-dimensional analytical surface, which can be transformed into global Cartesian coordinate as formula (10).

$$\begin{aligned} X(\eta) &= \eta, \\ Y(\psi,\eta) &= C_{N2}^{N1}(\psi) C_{M2}^{M1}(\eta) \sum_{i=0}^{n} \sum_{j=0}^{m} b_{i,j} N_k^i(\psi) N_k^j(\eta) \\ &\quad + \Delta\zeta_{M,N}(\psi,\eta)\big|_{\text{upp,low}}, \\ Z(\eta) &= C_{T2}^{T1}(\eta) \sum_{t=0}^{w} b_t N_k^t(\eta), \end{aligned} \quad (10)$$

where $\psi = [0,1]$, $\eta = [0,1]$. The outward boundary box is used to control the size of the geometric surface, and then we get the full design parameters of the entire geometric surface, which includes the following:

Axial and lateral length: L, L_w

Vertical height: L_{H1}, L_{H2}

Sectional section control factor: $N1$, $N2$

Vertical section control factor: $M1$, $M2$

Lateral section control factor: $T1$, $T2$

Surface weight factor: $b_{i,j}$

Lateral weight factor: b_t

L, L_w, L_{H1}, L_{H2} are the size parameters of the geometric surface. $N1$, $N2$, $M1$, $M2$, $T1$, $T2$ are the control parameters of the class function. $b_{i,j}$ and b_t are the discrete control weight factors of the shape function. When $b_{i,j}$ and b_t are equal to 1, the number of total parameters is constant and also minimum. When the $b_{i,j}$ and b_t are expressed as matrix, some control points with the same size of the matrix elements are located in the geometric surface to improve the complexity of the geometric shape. So the size of design parameters could be adjusted dynamically. With these weight factors, the three-dimensional CST method has more flexible parametric geometric shape design ability.

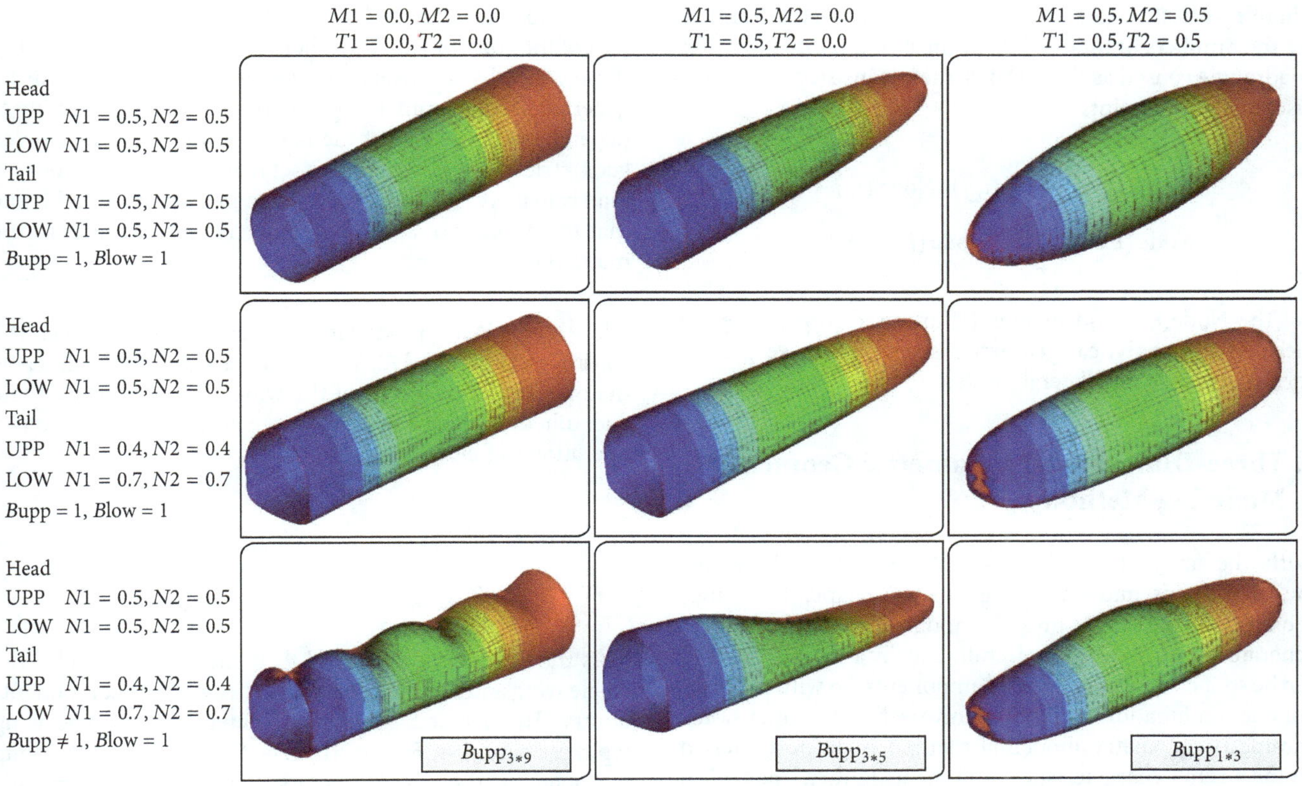

FIGURE 1: Geometric shapes with different control parameters generated by the three-dimensional CST method.

2.4. Basic Geometric Shape. The basic geometric shape is defined by the upper geometric surface and the inverted lower geometric surface. Figure 1 gives some geometric shapes generated by the above three-dimensional CST method with different control parameters. *Bupp*, *Blow*, and *Bt* are the matrix generated by $b_{i,j}$ of the upper surface, $b_{i,j}$ of the lower surface, and b_t of the lateral coordinate. The shapes in the same lines are generated with the same value of $N1$ and $N2$, but with different $M1$, $M2$, $T1$, and $T2$. The shapes in the same columns are generated with the same values $M1$, $M2$, $T1$, and $T2$, but with different $N1$, $N2$, and *Bupp*. The modified shapes in the third line are generated with different sizes of control weight factors of the upper surface compared with the shapes in the second line.

From the various geometric shapes and cross sections shown in Figure 1, the following features can be found for the three-dimensional CST method:

(1) The geometric shape can be expressed as an extruded body made up of a series of basic cross sections with different control factors along the axial direction. The transformation rule of the body is decided by the control factors $N1_{\text{head}}$, $N2_{\text{head}}$ at head and $N1_{\text{tail}}$, $N2_{\text{tail}}$ at tail.

(2) The shape profile at the axial and lateral direction is decided by the control factors $M1$, $M2$, $T1$, and $T2$. It allows continuous control of every single basic cross section. When M and T are equal to 0 simultaneously, the corresponding cross section is opened. When M and T are greater than 0, the corresponding cross section is closed to a point. This feature can be used to generate the closed head and the opening body.

(3) *Bupp*, *Blow*, and *Bt* are matrix with arbitrary size. They produce a global correction or a local adjustment to the geometric shape. The control ability is decided by the size of the matrix.

The shape profile is parameterized by the section control factors, and the geometric shape can be adjusted globally or locally by the weight factor matrix. These features give the three-dimensional CST method a comprehensive parametric geometric shape design ability with adjustable control parameters. Designers could have a more flexible design pattern to parameterize various geometric model.

2.5. Mesh Discretization. A generic mesh discretization method is utilized to generate the corresponding mesh, which will be used as the aerodynamic mesh and structural mesh. The procedure for the three-dimensional CST method is likely to generate structured mesh surface along the ψ and η direction. With these characteristics, the surface is dispersed into some control points uniformly. Then a two-dimensional matrix made up of control points is generated, which can be expressed as follows:

$$\begin{aligned}\text{Node}_{\text{upp}}(i,j) &= (X(\eta), Y(\psi,\eta), Z(\eta)),\\ \text{Node}_{\text{low}}(i,j) &= (X(\eta), Y(\psi,\eta), Z(\eta)),\end{aligned} \tag{11}$$

where $i = 1, 2, \ldots, M - 1$, $j = 1, 2, \ldots, N - 1$. M and N are the numbers of total points in the ψ and η direction. The quadrangle is used as the mesh element consisted with these discrete control points

$$\begin{aligned}\text{Element}(t) = {} & [\text{Node}(i, j), \text{Node}(i, j+1), \\ & \text{Node}(i+1, j+1), \text{Node}(i+1, j)] .\end{aligned} \tag{12}$$

The Node(i, j) and Element(t) are used to generate the mesh of the analytical geometric shape. Then we get the corresponding quadrilateral mesh.

3. Three-Dimensional Parametric Geometry Modeling Method

With the above three-dimensional CST method, we get the continuous and smooth geometric shape. But single geometric shape cannot provide enough degree of freedom to generate an entire complex aircraft. Usually a complex aircraft can be split into some standard components. So with the component combination method, a universal three-dimensional parametric geometry modeling method is proposed based on the aircraft characteristic components library. The typical aircraft characteristic components include fuselage, wing, and empennage, which are introduced as follows.

3.1. Aircraft Characteristic Components Library

3.1.1. Fuselage. The fuselage can be divided into head, head-body, and head-body-tail, three types according to the structural cabins or aerodynamic shape features. By setting the control factors $M1$, $M2$, $T1$, and $T2$ to 1 or 0, it is easy to generate the semiclosed head and tail, and the opened body. The control factors $N1$ and $N2$ must be kept the same value at the joint position between two parts to ensure surface continuation. The detailed control parameters and the three fuselage types are shown in Figure 2(a).

The control factors of the head include

$$\begin{aligned}N_{\text{head}} &= [N1_{\text{head\_upp}}, N2_{\text{head\_upp}}, N1_{\text{head\_low}}, N2_{\text{head\_low}}], \\ M_{\text{head}} &= [M1_{\text{head\_upp}}, M2_{\text{head\_upp}}, M1_{\text{head\_low}}, M2_{\text{head\_low}}], \\ T_{\text{head}} &= [T1_{\text{head\_upp}}, T2_{\text{head\_upp}}, T1_{\text{head\_low}}, T2_{\text{head\_low}}].\end{aligned} \tag{13}$$

The weight factors of the head include

$$\begin{aligned}B &= [B_{\text{head\_upp}}, B_{\text{head\_low}}], \\ B_t &= [B_{t\text{\_head\_upp}}, B_{t\text{\_head\_low}}].\end{aligned} \tag{14}$$

The formulas (13) and (14) are the generic forms of a single part of the component. The others are similar too. By varying these control parameters, various geometric shapes can be generated. The parametric geometric shape design ability is dependent on the size of the matrix B and B_t. For complex geometric shape, designers can set B and B_t to some arbitrary matrix to increase the degree of freedom. But one should note that the number of control parameters also increases as the matrix becomes larger.

3.1.2. Wing. Wing modeling is similar to the fuselage. The control factors $M1$, $M2$, $T1$, and $T2$ are set to 0 to describe the opening tip and root of the wing. The sweepback angle and dihedral angle are also parameterized by mesh offset operation as follows:

$$\begin{aligned}Y_{i,j,\text{correct}} &= Y_{i,j} + X_{i,j} * \tan(\theta_y), \\ Z_{i,j,\text{correct}} &= Z_{i,j} + X_{i,j} * \tan(\theta_z).\end{aligned} \tag{15}$$

θ_y is the sweepback angle and θ_z is the dihedral angle. The single wing and double wing are modeled in the components library. The detailed control parameters and the two wing types are shown in Figure 2(b). The tip and root of the wing are open, so the control factors of the cross section only include N_{in} and N_{out}.

3.1.3. Empennage. Empennage modeling is the same as the wing. The single tail and double tail are modeled in the components library and shown in Figure 2(c).

3.1.4. Others. Some typical components are introduced above, which can be used to create many kinds of common aircraft. There may be other requested components such as engine and nozzle. These can be some simplification of the basic three types components library. For example, the nozzle may be some simplification of the single part fuselage with opening head and tail. For other special components, designers can model it using the three-dimensional CST method and package to the aircraft characteristic components library for reuse.

3.2. Three-Dimensional Geometry Modeling. The aircraft characteristic components library is introduced in Section 3.1. These components are rotated and moved to the proper position to assemble entire complex aircraft. For the discrete control points P of component generated in Section 2.5,

$$P = \begin{bmatrix} x_1 & y_1 & z_1 \\ x_2 & y_2 & z_2 \\ \vdots & \vdots & \vdots \\ x_n & y_n & z_n \end{bmatrix}. \tag{16}$$

The rotation matrix M_r and transfer matrix M_d are as follows:

$$M_r = \begin{bmatrix} \cos(\theta_z)\cos(\theta_y) & \sin(\theta_z) & -\cos(\theta_z)\sin(\theta_y) \\ -\sin(\theta_z)\cos(\theta_y)\cos(\theta_x)+\sin(\theta_y)\sin(\theta_x) & \cos(\theta_z)\cos(\theta_x) & \sin(\theta_z)\sin(\theta_y)\cos(\theta_x)+\cos(\theta_y)\sin(\theta_x) \\ \sin(\theta_z)\cos(\theta_y)\sin(\theta_x)+\sin(\theta_y)\cos(\theta_x) & -\cos(\theta_z)\sin(\theta_x) & -\sin(\theta_z)\sin(\theta_y)\sin(\theta_x)+\cos(\theta_y)\cos(\theta_x) \end{bmatrix}, \quad (17)$$

$$M_d = [dx \ \ dy \ \ dz]. \quad (18)$$

With these matrices, the modified control points P_{new} are evaluated in formula (18). Then the modified component mesh is generated by these points using the mesh discretization method discussed in Section 2.5.

$$P_{new} = PM_r + M_d. \quad (19)$$

4. Universal Structural Modeling Method

Although structural finite element analysis has been widely used in industrial department and research institution, the structure parameterized modeling is still a difficult phase. The most common structural parameterization method is based on parametric CAD model. It is complicated to model complex aircraft and also difficult to support the changing of the structural layout. These limitations cannot satisfy the needs of rapid modification and iterations in the conceptual design phase of the modern aircraft design. This section will provide a detailed illustration of a novel structural modeling method based on the three-dimensional parametric geometry modeling method mentioned above. Structural model is constructed based on the aircraft characteristic components library. Three typical structural models are elucidated below.

4.1. Fuselage. Fuselage is the main body of an aircraft. It is usually designed as a thin-walled structure and constituted by longitudinal stiffeners (like beam and stringer), transverse stiffeners (like bulkhead), and surface skin. A fuselage contains fuel tank, payload, electronic instrument, and other equipment. Also it is as a sole central part connected with wings, empennages, engines, and other components to make up the entire aircraft. The main structural forms of fuselage include girder structure, longeron structure, and monocoque shell structure. Based on the above assumption, all of these structural forms can be simplified as three basic elements: beam/stringer, bulkhead, and skin. The detailed modeling method of these three elements is shown as follows.

4.1.1. Beam/Stringer. Beam and stringer are the longitudinal stiffeners of fuselage structure, used for undertaking axial load coming from fuselage bending, and also used for transferring the load of outer surface skin to bulkhead. Here, the beam and stringer are simplified to one-dimensional beam element. The cross section shape of the beam and the stringer, like I-section, H-section circle-section and so on, can be ignored at the modeling phase and will be considered by the structural solver.

Assuming the surface mesh of fuselage is continuous and compatible among all parts of the component, the discrete nodes of the upper surface or the lower surface can be expressed as

$$\text{Node}_{\text{Fuselage}}(i, j), \quad (20)$$

where $i = 1, 2, \ldots, M$, $j = 1, 2, \ldots, N$. M and N are the numbers of control points along the ψ and η direction. Vector B is used to locate the position of the beam and the stringer along transverse direction.

$$B = [b_1 \ \ b_2 \ \ \cdots \ \ b_n], \quad (21)$$

where $0 \le b_1 < b_2 < \cdots < b_n \le 1$. B is normalized to $[0, 1]$. b_k represents the normalized lateral position of the kth beam/stringer. For the discrete nodes, it can be handled as the nearest point $\text{Node}_{\text{Fuselage}}(k, j)$, which has the minimum distance to the b_k position in the jth line of the discrete surface consisted of $\text{Node}_{\text{Fuselage}}(i, j)$. So the node set of the beam/stringer can be defined as

$$\text{Node}_B^{(k)} = \text{Node}(k, j), \quad j = 1, 2, \ldots, N. \quad (22)$$

And the element set of the beam/stringer can be defined as

$$\text{Element}_B^{(k)} = [\text{Node}(k, j) \ \ \text{Node}(k, j+1)], \quad j = 1, 2, \ldots, N-1. \quad (23)$$

Then the mesh of beam/stringer can be extracted from the surface mesh of the fuselage. Figure 3 shows the beam/stringer structural layout scheme of a hypersonic aircraft.

4.1.2. Bulkhead. Bulkhead can be divided into three types: normal bulkhead, reinforced bulkhead, and connected bulkhead. It supports the skin of the fuselage to maintain geometric shape and also undertakes some concentrated load. The definition of bulkhead is similar to the beam/stringer. One-dimensional beam element is selected to simplify the bulkhead model. The cross section shape of the beam also is ignored at the modeling phase and will be considered by the structural solver. Vector F is used to locate the position of the bulkhead.

$$F = [f_1 \ \ f_2 \ \ \cdots \ \ f_n], \quad (24)$$

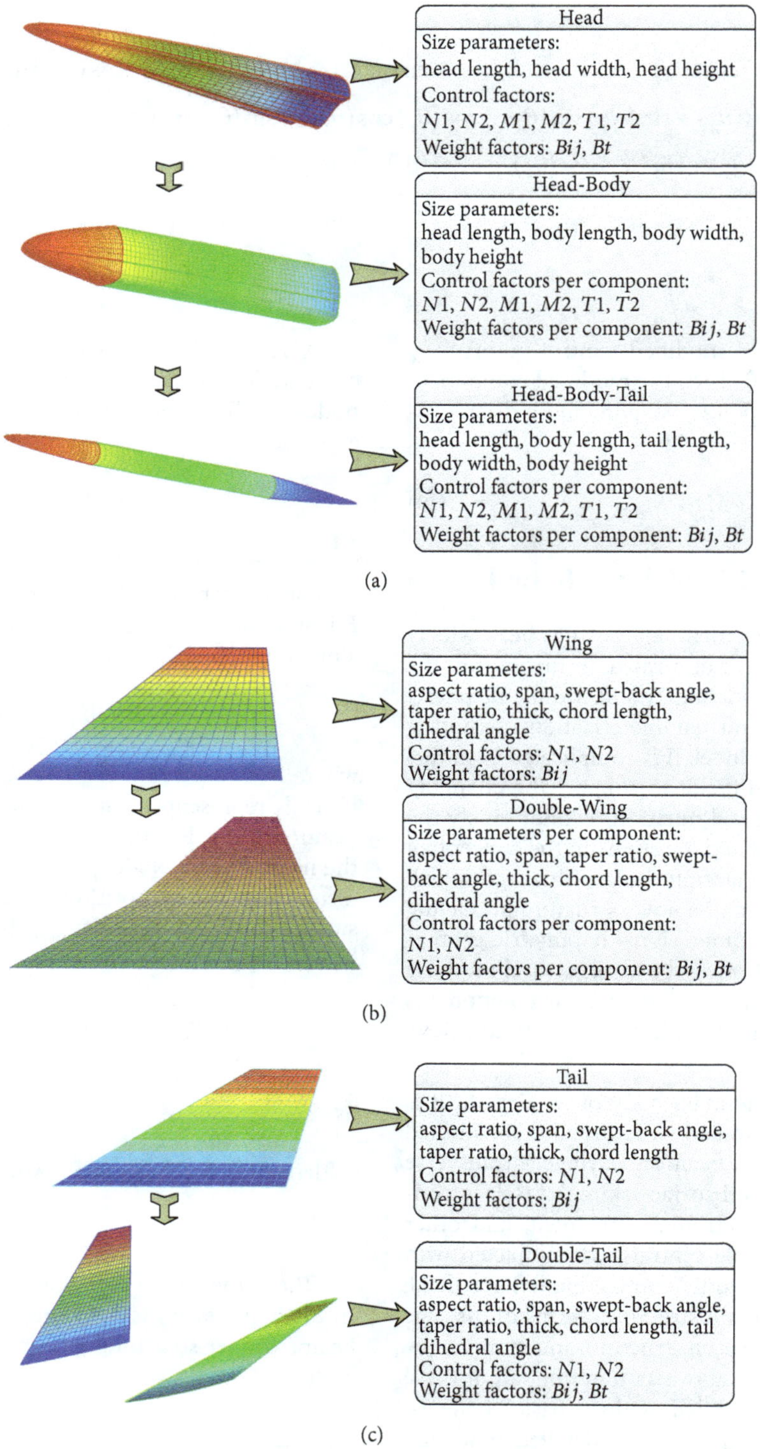

FIGURE 2: Basic components in the aircraft characteristic components library.

where $0 \le f_1 < f_2 < \cdots < f_n \le 1$. F is normalized to $[0, 1]$. f_k represents the normalized axial position of the kth bulkhead. For the discrete nodes, it can be handled as the nearest point $\text{Node}_{\text{Fuselage}}(i, k)$, which has the minimum distance to the f_k position in the ith line of the discrete surface consisting of $\text{Node}_{\text{Fuselage}}(i, j)$. So the node set and the element set of bulkhead can be defined as

$$\begin{aligned} \text{Node}_F^{(k)} &= \text{Node}\,(i, k)\,, \quad i = 1, 2, \ldots, M, \\ \text{Element}_F^{(k)} & \\ &= \left[\text{Node}_{\text{Fuselage}}\,(i, k) \;\; \text{Node}_{\text{Fuselage}}\,(i+1, k)\right], \\ & \qquad i = 1, 2, \ldots, M - 1. \end{aligned} \tag{25}$$

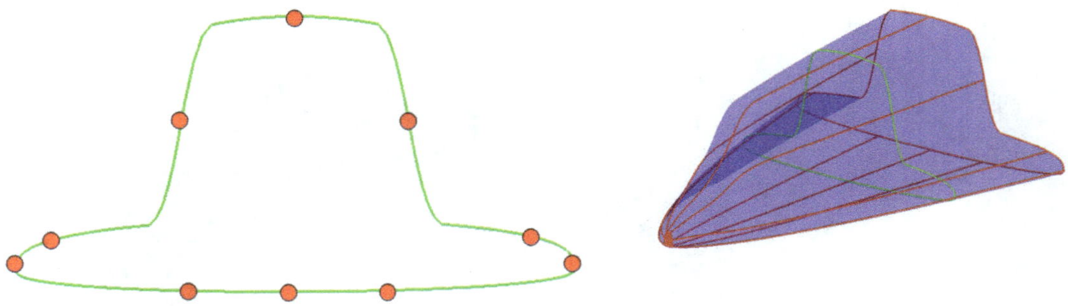

(a) Location points of the beam/stringer (b) Beam/stringer structural layout scheme

FIGURE 3: The beam/stringer structural layout scheme of a hypersonic aircraft.

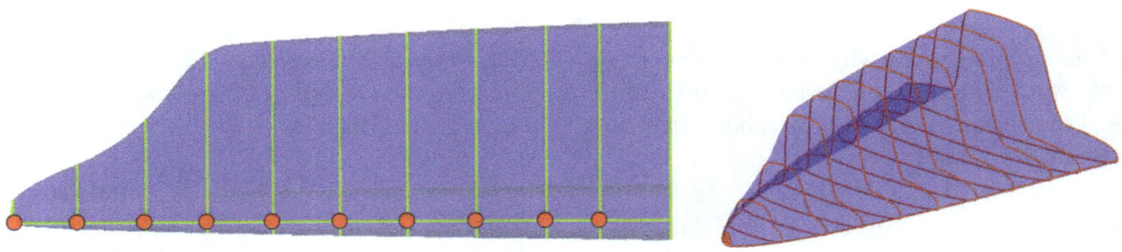

(a) Location points of the bulkhead (b) Bulkhead structural layout scheme

FIGURE 4: The bulkhead structural layout scheme of a hypersonic aircraft.

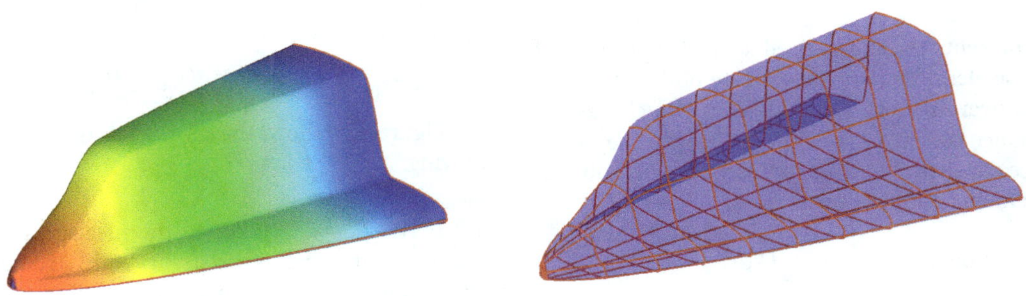

(a) Aerodynamic model of the fuselage (b) Structural model of the fuselage

FIGURE 5: The aerodynamic model and structural model of a hypersonic aircraft.

Then the mesh of the bulkhead can be extracted from the surface mesh of the fuselage. Figure 4 shows the bulkhead structural layout scheme of a hypersonic aircraft.

4.1.3. Fuselage Skin. Fuselage skin is used to maintain the geometric shape of the fuselage. It should be continuous and smooth to support the aerodynamic solver and structural solver. The discrete surface mesh of fuselage is used for both the structural mesh and the aerodynamic mesh. It has the following advantages: (a) the structural model and the aerodynamic model use the same surface mesh coming from mesh discretization in Section 2.5; the data conversion between these models are no longer needed; (b) the mesh of the beam/stringer and the bulkhead are both extracted from the discrete surface mesh; they are coordinated with surface skin mesh naturally. Based on the above definition, the whole structural model of the fuselage can be generated as Figure 5.

4.2. Wing. Wing structure consists of skin, spar, stringer, rib, and connector. The following simplifications are used to simplify the wing structural model: (1) the effects of the connectors are ignored; (2) the effects of the flanges of spar,

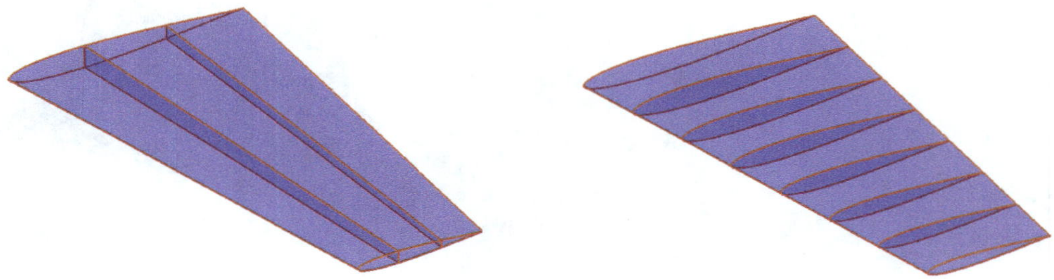

(a) Spanwise structure (b) Chordwise structure

FIGURE 6: The spanwise and chordwise structural layout scheme of wing.

stringer, and rib are imputed to the wing skin. Based on these assumptions, the wing structure can be simplified to some two-dimensional shell elements. Three typical basic elements are used to construct the wing structure.

4.2.1. Spanwise Elements. Spanwise elements include spar and stringer. Assuming these elements are arranged along the chordwise direction rigorously. For the discrete nodes Node(i, j) of the wing, vector Ws is used to locate the position of spanwise elements.

$$Ws = \begin{bmatrix} s_1 & s_2 & \cdots & s_n \end{bmatrix}, \quad (26)$$

where $0 \le s_1 < s_2 < \cdots < s_n \le 1$. Ws is normalized to $[0, 1]$. S_k represents the normalized spanwise position of the kth spanwise elements. For the discrete nodes, it can be handled as the nearest point $\text{Node}_{\text{Wing}}(k, j)$, which has the minimum distance to the S_k position in the jth line of the discrete surface consisting of $\text{Node}_{\text{Wing}}(i, j)$. So the node set of spanwise elements can be defined as

$$\text{Node}_{Ws}^{(k)} = \text{Node}_{\text{Wing}}(k, j), \quad (27)$$

where $j = 1, 2, \ldots, N$. N is the number of the spanwise points. Wt is used to define the node distribution in the thickness direction.

$$Wt = \begin{bmatrix} t_1 & t_2 & \cdots & t_s \end{bmatrix}, \quad (28)$$

where $0 \le t_1 < t_2 < \cdots < t_s \le 1$. Wt is normalized to $[0, 1]$ too. The layouts of the upper surface points and the lower surface points are the same. So by arranging some middle points between the upper surface points and the lower surface points by some node distribution, the total nodes of the wing spanwise section can be generated as

$$\text{Node}_{Ws}^{(k)} = \text{linespace}_{Wt}^{(k)}\left(\text{Node}_{W\text{supp}}^{(k)}, \text{Node}_{W\text{slow}}^{(k)}\right). \quad (29)$$

$\text{linespace}_{Wt}^{(k)}$ means to arrange points in the regularity of distribution Wt. The element set can be generated by connecting these nodes. Figure 6(a) shows the spanwise structural layout scheme of wing.

4.2.2. Chordwise Elements. Rib is the chordwise element. Similar to the spanwise elements, vector Wc is used to define the position of the rib,

$$Wc = \begin{bmatrix} c_1 & c_2 & \cdots & c_n \end{bmatrix}, \quad (30)$$

where $0 \le c_1 < c_2 < \cdots < c_n \le 1$. Wc is normalized to $[0, 1]$. The modeling process is the same as the spanwise elements; the node set can be generated as

$$\text{Node}_{Wc}^{(k)} = \text{Node}_{\text{wing}}(i, k), \quad (31)$$

where $i = 1, 2, \ldots, M$. M is the number of the chordwise points. Wt is also used to define the node distribution in the thickness direction. So the total nodes of the wing chordwise section can be generated as

$$\text{Node}_{Wc}^{(k)} = \text{linespace}_{Wt}^{(k)}\left(\text{Node}_{W\text{cupp}}^{(k)}, \text{Node}_{W\text{clow}}^{(k)}\right). \quad (32)$$

Figure 6(b) shows the chordwise structural layout scheme of wing.

4.2.3. Wing Skin. Similar to the fuselage skin, the discrete mesh of the wing is used for both structural mesh and aerodynamic mesh to maintain consistency. Based on the above definition, the entire structural model of the wing can be generated as Figure 7.

4.3. Empennage. Structural model of the empennage is the same as the wing. Spanwise elements, chordwise elements, and skin are used to construct the empennage structure.

4.4. Others. Most of the common components have the similar structural layouts to the fuselage and wing. They can be structured by the method mentioned above.

5. Aeroelastic Modeling and Analysis Process

The process of aeroelastic modeling and analysis includes three steps: three-dimensional CST modeling, aerodynamic/structural modeling, and aeroelastic analysis. The process of aeroelastic modeling and analysis is shown in Figure 8, and the detailed illustration is as follows.

(a) Aerodynamic model

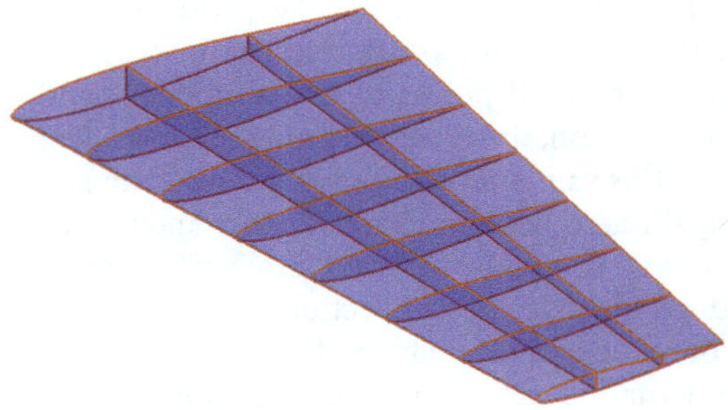

(b) Structural model

FIGURE 7: The aerodynamic model and structural model of wing.

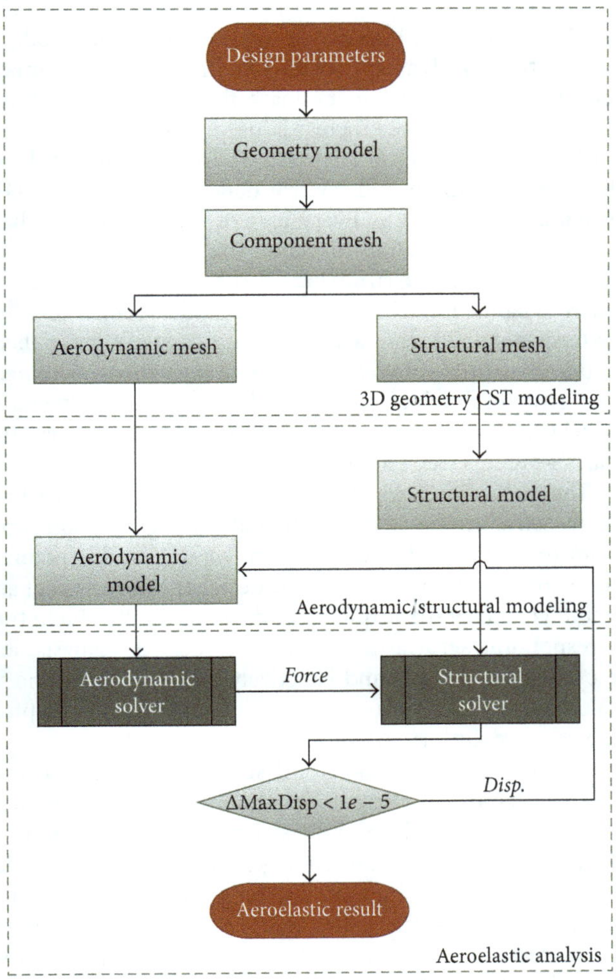

FIGURE 8: The process of aeroelastic modeling and analysis.

5.1. Three-Dimensional CST Modeling. Section 3 lists all design parameters of the three common components. There are four types of design parameters available to use to control the geometric shape: layout parameter, size parameter, shape parameter, and local control parameter. The layout parameter and the size parameter are global design parameters. The former changes the position and the posture of the components, and the latter changes the shape size of the components. They can be used to control the global sizing of aircraft. The shape parameter and local control parameter are local design parameters. The former changes the sketch of the main cross section, and the latter adjusts the details of the component. They can be used to further control the detailed shape of the aircraft. Local control parameter can be any size of matrix. The larger the matrix, the better the parametric geometric shape design ability. According to the requirement of the geometry modeling complexity, the proper design parameters can be selected as the design variables; others may stay constant or vary with respect to the selected parameters. With these design variables, geometry model and corresponding mesh of the component can be generated automatically.

5.2. Aerodynamic/Structural Modeling. A complex aircraft is constituted by several components. Applying the first step, the mesh of these components can be generated, respectively, to make up the entire aircraft. For each of these components, the structural mesh also can be obtained in accordance with the structural modeling method introduced in Section 4. But there are two problems that needed to be solved to generate the integrated aerodynamic/structural model.

The first problem is the redundant mesh caused by the mutual nesting between the connected components. This mesh could influence the precision of aeroelastic analysis. In the worst condition, it may cause some error to the aerodynamic solver and the structural solver. PINPOLYHEDRON, an open-source tool, is used to remove this nested mesh. PINPOLYHEDRON is a C++ code. It provides a novel algorithm to test whether points are inside/outside/on a polyhedron defined by triangular faces and vertices. It can be used for various complicated models such as nonconvex volumes, multimaterial bodies, and so on, and there is no assumption about orientation of the face normal. Above all, the algorithm is very efficient especially for large-scale problems. In this research, the fuselage is as the main polyhedron. If a point of other components is detected inside the main fuselage, this point and the relevant mesh are both removed from the component. Looping all points of the component and removing the nested mesh, then the remaining mesh is taken as the valid mesh. This valid mesh is used as the aerodynamic mesh to evaluate the aerodynamic characteristics and is used as the structural mesh to analyze the structural performance.

The second problem is the connection of the relevant components. For aerodynamic analysis, the panel method is used as the aerodynamic solver, so the connection between components is not necessary. For structure analysis, the connection must be modeled to guarantee the force transmission between the connected components, assuming a structural connection exists only between the bulkhead of the fuselage and the beam/spar of other components. The following method is used to generate these structural connections.

(1) For each bulkhead of fuselage, evaluate the midpoint Pm_i by counting the average coordinate of the points set Sm_i on the bulkhead.

(2) For each beam/spar, find the nearest elements to the fuselage, mark as Es_j, evaluate the midpoint of Es_j, and mark as Ps_j.

(3) For each beam/spar, find the nearest bulkhead by $\min_i |Pm_i - Ps_j|$, and mark as Pm^*.

(4) Sort the points set Sm^* on the bulkhead Pm^* by counting the distance from Ps_j, the sorted points set mark as $Sm_r{}^*$.

(5) The first N_r points of the sorted points $Sm_r{}^*$ are used as the connection points set S_1. The connection points set of the beam mark S_2 and N_r is the number of the connection points S_2.

(6) Generate some quadrilaterals by connecting the relevant points between the connection points set S_1 and connection points set S_2.

(7) Looping all of the beams of the components, generate nodes set and elements set of the structural connection to obtain the connection mesh.

Figure 9 shows the structural connection between the fuselage and wing generated by the above method. The red line is the bulkheads of the fuselage. The blue line is the beam/stringer of the fuselage. The quadrilaterals with green edges are the structural connection mesh. This method provides a simple way to generate the structural connection mesh between the main fuselage and the connected components.

This structural connection mesh guarantees the grid continuity for force transmission. However, the compelling connection may reduce the quality of these mesh and influence the connection stiffness between the fuselage and the wing. These influences can be corrected by modifying the material attribute or adding a spring element. In this paper, these impacts are ignored temporarily.

5.3. Aeroelastic Analysis. The aerodynamic solver and the structural solver interplay in the aeroelastic analysis process.

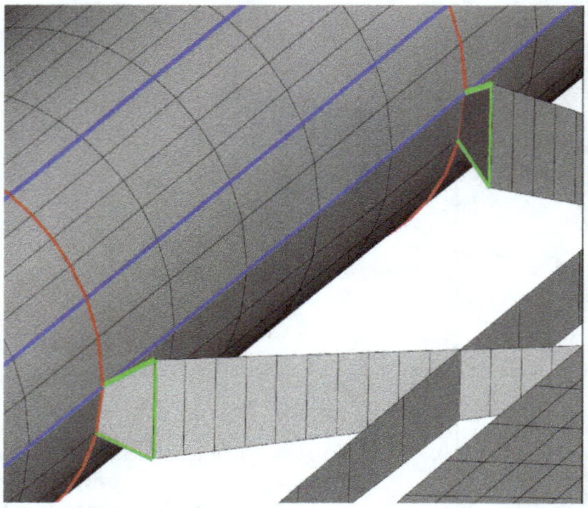

FIGURE 9: The structural connection mesh between the fuselage and the wing.

AeroCalc, a C++ in-house code with the panel method, is used as the aerodynamic solver. The modified Newtonian impact theory is used to evaluate the windward surface, and the Prandtl-Meyer theory is used to evaluate the leeward surface. NASTRAN is used as the structural solver. The structural mesh and control data are written in BDF format as the script file. Every part of the components has its own property section to define thickness, cross section parameters, and material attributes. The outside surface mesh of aerodynamic model and structural model is consistent, so the force obtained from aerodynamic solver can be applied to the structural model and the displacement obtained from structural solver can be applied to the aerodynamic model without any conversion. These simplify the aeroelastic analysis process tremendously.

The three-dimensional CST modeling method is used as a parametric geometry modeling and mesh generation processor to automatically generate aerodynamic and structural model for aeroelastic analysis. The designer can construct an aircraft geometry shape quickly and automatically obtain the corresponding aeroelastic model. An aeroelastic analysis can be carried out quickly and effectively to give a preliminary evaluation of the aeroelastic effect. This could be very helpful for aircraft preliminary design.

The aerodynamic model and structural model use the same surface mesh. The aerodynamic force can be applied to the structural model and the structural displacement can be added to aerodynamic mesh to generate new aerodynamic model directly. This simplification will limit the mesh adaptation for aerodynamic model and structural model, which may increase the computational cost and reduce analysis precision. But it could save the tedious aeroelastic meshing and modeling process, which is time-consuming, especially for complex aircraft. Also these automatic aeroelastic modeling and analysis process can be used for aeroelastic optimization.

6. Static Aeroelastic Analysis Example

A reusable launch vehicle (RLV) is used to illustrate the process of aeroelastic analysis. Three basic components, a

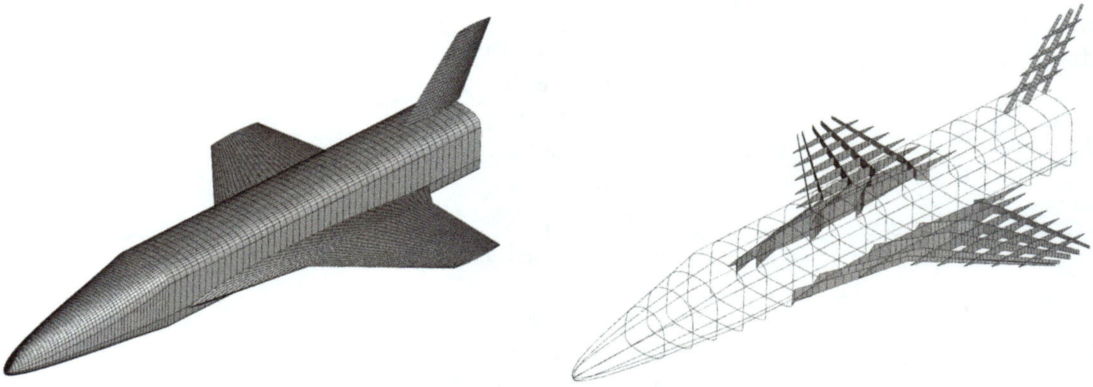

(a) Aerodynamic model (b) Structural model

FIGURE 10: The aerodynamic model and the structural model of the RLV.

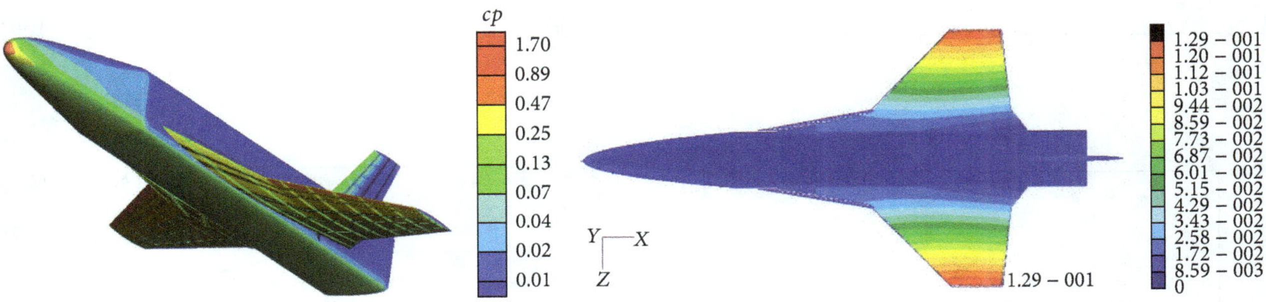

(a) Pressure coefficient distribution (b) Structural displacement distribution

FIGURE 11: Pressure coefficient distribution and structural displacement distribution at the evaluation points.

head-body fuselage, a double wing, and a tail, are utilized to build up the RLV geometry. The main parameters of the RLV are listed in Table 1, and the trimmed aerodynamic model is shown in Figure 10(a). Eighteen bulkheads and eight beams are arranged in the head-body fuselage. Five spars and seven ribs are arranged in the double wing. Three spars and seven ribs are arranged in the vertical tail. Ten structural connections are used to connect the fuselage and the wing. Three structural connections are used to connect the fuselage and the tail. The integrated structural model is shown in Figure 10(b).

The aluminum alloys are used as structural material for the structural frame and skin, with a density of 2,700 $\mathrm{Kg/m^3}$ and an elastic modulus of 72 GPa. The total number of the aerodynamic elements is 15506, which is controlled by the number of points for mesh discretization in the ψ and η direction of each surface. The total number of the structural elements is 18691, including the skin mesh, which is the same as the aerodynamic mesh except the nested mesh, the structural component mesh, and the structural connection mesh.

The maximum dynamic pressure trajectory point during reentry is chosen as the aeroelastic evaluation point with $Ma = 3.6$ and $\alpha = 6.5$ degree. The pressure coefficient distribution evaluated by the panel method at the maximum dynamic pressure point is shown in Figure 11(a). The corresponding aerodynamic force is applied to the structure surface directly without interpolation. The structural mesh and the displacement distribution are shown in Figure 11(b). The relative error of the maximum deformation in the Y direction is used as the convergence criteria during the aeroelastic iterations and it is set to $1e-5$.

The aeroelastic analysis of the RLV converges shortly after five iterations. The iteration history of the lift coefficient Cl, the drag coefficient Cd, and the maximum deformation in the Y direction dY are shown in Figure 12. In the final converged condition, the Cl is 1.3 percent lower and the Cd is 0.37 percent lower, which lead to a 0.94 percent decrease in the lift-to-drag ratio than the initial undeformed condition. The maximum structure deformation in the Y direction occurs at the wingtip position. The maximum deformation at the first structure analysis and the last structure analysis are 0.1287 m and 0.1271 m. The wingtip deformation is shown in Figure 13. The blue line shows the undeformed wing tip, the orange line shows the deformed wing tip after the first structure analysis, and the red line shows the deformed wing tip at the convergence. The change of the wingtip deformation is

Table 1: Main parameters of the RLV.

Components	Design parameters	Value
Head-body	Head length	5,000.0 mm
	Body length	11,431.0 mm
	Body width	1,821.0 mm
	Body height upp	1,284.0 mm
	Body height low	405.0 mm
Double wing	Inner wing span length	650.0 mm
	Outer wing span length	2,600.0 mm
	Inner wing root chord length	8,800.0 mm
	Inner wing taper ratio	1.923
	Outer wing taper ratio	2.7
	Inner wing sweep back angle	80.0 deg
	Outer wing sweep back angle	45.0 deg
Tail	Tail span length	1,925.0 mm
	Tail root chord length	2,075.0 mm
	Tail taper ratio	1.724
	Tail sweep back angle	45.0 deg

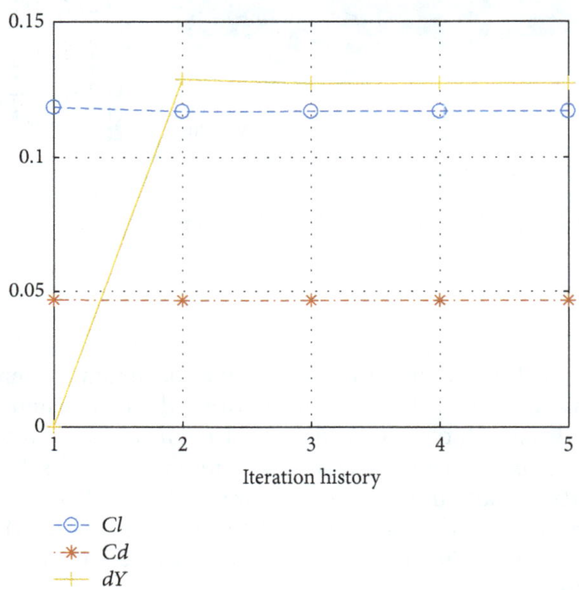

Figure 12: Iteration history of the lift coefficient Cl, the drag coefficient Cd and the maximum deformation in the Y direction dY.

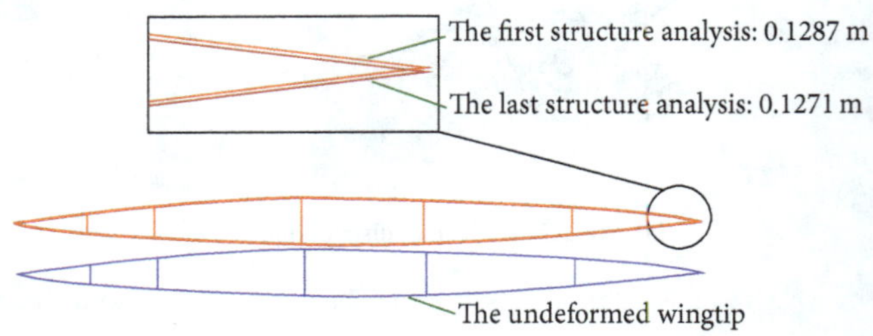

Figure 13: The wingtip deformation.

Table 2: The cost time of the aeroelastic iterations.

	Time per iteration (s)	Total time (s)
Aerodynamic analysis	5.1	25.5
Structure analysis	10.3	50.4
Aeroelastic analysis	16.5	82.5

not obvious during the aeroelastic iterations, which reaches a stable level after the second step.

The three-dimensional CST parameterization method and the aeroelastic analysis process are written in MATLAB code. The geometry modeling costs less than one second, and the generation of aerodynamic mesh and structural mesh cost 5.5 s. The analysis time of aerodynamic analysis and structure analysis per iteration and the total analysis time are shown in Table 2. The whole aeroelastic modeling and analysis process is very fast and efficient. To generate an entire complex RLV and its analysis models only needs several seconds. This will simplify and shorten the aeroelastic analysis and make it easy to use in the aircraft conceptual design phase.

7. Conclusion and Future Work

In this paper, a novel aerodynamic and structural modeling method based on the three-dimensional CST is developed to provide a fast and simple way to carry out an entire aeroelastic analysis process. The aerodynamic model is generated by trimming the embedded aircraft characteristic components. The structural model is generated by arranging proper inner parts from the basic mesh topology. The consistency between the aerodynamic model and the structural model is satisfied naturally. Finally, an efficient aeroelastic analysis process is created and tested by a RLV. In conclusion,

(1) the three-dimensional parametric geometry modeling method gives a universal way to generate geometry model of common three-dimensional complex aircraft. It inherits the advantages of the original CST method and has fast and stable parametric geometric shape design ability. An aircraft can be modeled with a few control parameters in a few minutes and discretized to surface mesh. More common aircraft characteristic components library will be introduced in future work to give a broader support to various kinds of aircraft;

(2) a universal aeroelastic modeling and analysis process is introduced. By using the three-dimensional parametric geometry modeling method, the aerodynamic model and the structural model can be generated together to keep the mesh consistency. Fluid structure interaction (FSI) can be ignored to simplify the aeroelastic analysis process. This simplification may limit the mesh adaptation for aerodynamic model and structural model, but it gives a fast and simple way to carry out an aeroelastic analysis process for complex aircraft, which is very useful in the conceptual design phase;

(3) the structural modeling method in this article gives an automatic and fast structural layout parameterization and generation approach. Topology of the structure can be parameterized from a wide range stably. This could be used as parameterized structural model for topology optimization;

(4) the generated mesh by the three-dimensional parametric geometry modeling method will be of poor quality at the position where the slope of the surface along axial direction and lateral direction is too large. These are caused by the uniform mesh discretization along the axial direction and lateral direction. These narrow grids may reduce the precision of the aeroelastic analysis. Nonuniform mesh discretization and mesh repairing method will be studied in future work to improve mesh quality.

Conflicts of Interest

The authors declare that there are no conflicts of interest regarding the publication of this paper.

Acknowledgments

This research was supported by a fund from the National Natural Science Foundation of China (no. 51505385), the Shanghai Aerospace Science and Technology Innovation Foundation (no. SAST2015010), and the Defense Basic Research Program (no. JCKY2016204B102 and no. JCKY2016208C001). The authors are also thankful to Shaanxi Aerospace Flight Vehicle Design Key Laboratory of NPU.

References

[1] D. P. Raymer, *Aircraft Design: A Conceptual Approach*, AIAA Education Series, 4th edition, 2006.

[2] S. A. Brandt, R. J. Stiles, J. Bertin et al., *Intorduction to Aeronautics: A Design Perspective*, vol. 2nd of *AIAA Education Series*, 2004.

[3] B. Thuruthimattam, P. Friedmann, K. Powell, and J. McNamara, "Aeroelasticity of a generic hypersonic vehicle," in *Proceedings of the 43rd AIAA/ASME/ASCE/AHS/ASC Structures, Structural Dynamics, and Materials Conference*, Denver, Colorado, April 2002.

[4] K. K. Gupta, L. S. Voelker, C. Bach, T. Doyle, and E. Hahn, "CFD-based aeroelastic analysis of the X-43 hypersonic flight vehicle," in *Proceedings of the 39th Aerospace Sciences Meeting and Exhibit*, Reno, Nev, USA, January 2001.

[5] J. Heeg, P. Chwalowski, J. P. Florance, C. D. Wieseman, D. M. Schuster, and B. Perry Jr., "Overview of the aeroelastic prediction workshop," in *Proceedings of the 51st AIAA Aerospace Sciences Meeting including the New Horizons Forum and Aerospace Exposition*, Grapevine, Tex, USA, January 2013.

[6] D. H. Lee and P. C. Chen, "Nonlinear aeroelastic studies on a folding wing configuration with free-play hinge nonlinearity," in *Proceedings of the 47th AIAA/ASME/ASCE/AHS/ASC Structures, Structural Dynamics, and Materials Conference*, American Institute of Aeronautics and Astronautics, Newport, Rhode Island, May 2006.

[7] D. Yeh, "Preliminary findings in certification of ENSAERO code for rigid and flexible configuration," in *Proceedings of the Fluid Dynamics Conference*, American Institute of Aeronautics and Astronautics, Colorado Springs, Colo, USA, June 1994.

[8] C. Luca, R. Sergio, and T. Lorenzo, Neocass: An Integrated Tool for Structural Sizing, Aeroelastic Analysis and MDO at Conceptual Design Level, American Institute of Aeronautics and Astronautics, 2010.

[9] B. M. Kulfan, "Universal parametric geometry representation method," *Journal of Aircraft*, vol. 45, no. 1, pp. 142–158, 2008.

[10] B. Kulfan, "A universal parametric geometry representation method—'CST'," in *Proceedings of the 45th AIAA Aerospace Sciences Meeting and Exhibit*, American Institute of Aeronautics and Astronautics Inc., Reno, Nev, USA, January 2007.

[11] V. Sripawadkul, M. Padulo, and M. Guenov, "A comparison of airfoil shape parameterization techniques for early design optimization," in *Proceedings of the 13th AIAA/ISSMO Multidisciplinary Analysis and Optimization Conference (MAO '10)*, American Institute of Aeronautics and Astronautics, Fort Worth, Tex, USA, September 2010.

[12] S. Nadarajah, P. Castonguay, and A. Mousavi, "Survey of shape parameterization techniques and its effect on three-dimensional aerodynamic shape optimization," in *Proceedings of the 18th AIAA Computational Fluid Dynamics Conference*, American Institute of Aeronautics and Astronautics Inc., Miami, Fla, USA, June 2007.

[13] G. L. Mura and N. Qin, "Local class shape transformation parameterization (L-CST) for airfoils," in *Proceedings of the 55th AIAA Aerospace Sciences Meeting*, Institute of Aeronautics and Astronautics, Grapevine, Tex, USA, January 2017.

[14] F. Zhu and N. Qin, "Intuitive class/shape function parameterization for airfoils," *AIAA Journal*, vol. 52, no. 1, pp. 17–25, 2014.

[15] E. D. Olson, "Three-dimensional piecewise-continuous class-shape transformation of wings," in *Proceedings of the 16th AIAA/ISSMO Multidisciplinary Analysis and Optimization Conference*,

Institute of Aeronautics and Astronautics, Dallas, Tex, USA, June 2015.

[16] M. H. Straathof and M. J. L. Van Tooren, "Adjoint optimization of a wing using the class-shape-refinement-transformation method," *Journal of Aircraft*, vol. 49, no. 4, pp. 1091–1100, 2012.

[17] C. Liu, Y. Duan, J. Cai, and J. Wang, "Application of the 3D multi-block CST method to hypersonic aircraft optimization," *Aerospace Science and Technology*, vol. 50, pp. 295–303, 2016.

[18] C. Liu, Y. Duan, J. Cai, and G. Yang, "Applications of multi-block CST method for quasi-waverider design," in *Proceedings of the 52nd Aerospace Sciences Meeting*, Institute of Aeronautics and Astronautics, Maryland, Md, USA, January 2014.

[19] P. B. Leal, D. J. Hartl, and C. L. Bertagne, "Aero-structural optimization of shape memory alloy-based wing morphing via a class/shape transformation approach," in *Proceedings of the 23nd AIAA/AHS Adaptive Structures Conference*, Kissimmee, Fla, USA, January 2015.

[20] S. Hua, G. Liangxian, and G. Chunlin, "The research on geometry modeling method based on three-dimensional CST parameterization technology," in *Proceedings of the 16th AIAA/ISSMO Multidisciplinary Analysis and Optimization Conference*, American Institute of Aeronautics and Astronautics, Dallas, Tex, USA, June 2015.

[21] H. Su, C.-L. Gong, and L.-X. Gu, "Two-level aerodynamic shape optimization strategy based on three-dimensional CST modeling method," *Journal of Solid Rocket Technology*, vol. 37, no. 1, pp. 1–6, 2014.

[22] M. H. Straathof and M. J. L. van Tooren, "Extension to the class-shape-transformation method based on B-splines," *AIAA Journal*, vol. 49, no. 4, pp. 780–790, 2011.

14

Effect of Reynolds Number on Aerodynamics of Airfoil with Gurney Flap

Shubham Jain,[1] Nekkanti Sitaram,[2] and Sriram Krishnaswamy[3]

[1] *Department of Aerospace Engineering, PEC University of Technology, Sector 12, Chandigarh 160012, India*
[2] *Department of Mechanical Engineering, IIT Madras, Chennai, Tamil Nadu 600036, India*
[3] *Department of Mechanical Engineering, BITS Pilani, Hyderabad Campus, Jawahar Nagar, Shameerpet Mandal, Hyderabad, Telangana 500078, India*

Correspondence should be addressed to Shubham Jain; shubhamjain.jld@gmail.com

Academic Editor: Jechin Han

Steady state, two-dimensional computational investigations performed on NACA 0012 airfoil to analyze the effect of variation in Reynolds number on the aerodynamics of the airfoil without and with a Gurney flap of height of 3% chord are presented in this paper. RANS based one-equation Spalart-Allmaras model is used for the computations. Both lift and drag coefficients increase with Gurney flap compared to those without Gurney flap at all Reynolds numbers at all angles of attack. The zero lift angle of attack seems to become more negative as Reynolds number increases due to effective increase of the airfoil camber. However the stall angle of attack decreased by 2° for the airfoil with Gurney flap. Lift coefficient decreases rapidly and drag coefficient increases rapidly when Reynolds number is decreased below critical range. This occurs due to change in flow pattern near Gurney flap at low Reynolds numbers.

1. Introduction

A Gurney flap (GF) is a microtab fitted perpendicular to the airfoil near the trailing edge on its pressure surface which increases the lift by altering the Kutta condition and increasing effective camber. They are extensively used on helicopter stabilizers [1]. Considerable efforts are carried out to study the effects of Gurney flap on the airfoil aerodynamics [2–4]. Many investigations are undertaken to determine the effect of various parameters of Gurney flap such as height, position, mounting angle, and configurations. But most of the investigations are performed at high Reynolds numbers above critical range. Mueller and Batill [5], Selig et al. [6], Brown and Filippone [7], and Traub and Agarwal [8] have studied the performance of airfoils at low Reynolds numbers but a systematic investigation is still not available. For smaller UAVs and MAVs, airfoil chord and velocity of aircraft are lower; hence, they operate at low Reynolds numbers. Reynolds numbers between $\mathrm{Re} = 3 \times 10^4$ and 7×10^4 are of great interest to MAV designers [9]. Most UAVs operate at $\mathrm{Re} = 5 \times 10^4$ and 3×10^5 [10]. Gurney flaps are also used in turbomachinery to improve their performance at low Reynolds number due to their inexpensive method of lift enhancement [11].

A semiempirical formula linking flap height to free stream velocity and the airfoil chord is proposed by Brown and Filippone [7]:

$$h_{\mathrm{opt}} = 37.155 \left(\frac{\mathrm{Ch}^{0.8}}{U^{0.2}} \right), \tag{1}$$

where h_{opt} is optimum GF height (mm), Ch is chord length (m), and U is free stream velocity (m/s).

Keeping this in mind, computations on the effect of six Reynolds numbers from high Reynolds number to low Reynolds number well below the critical range are carried out to analyze the performance of the airfoil.

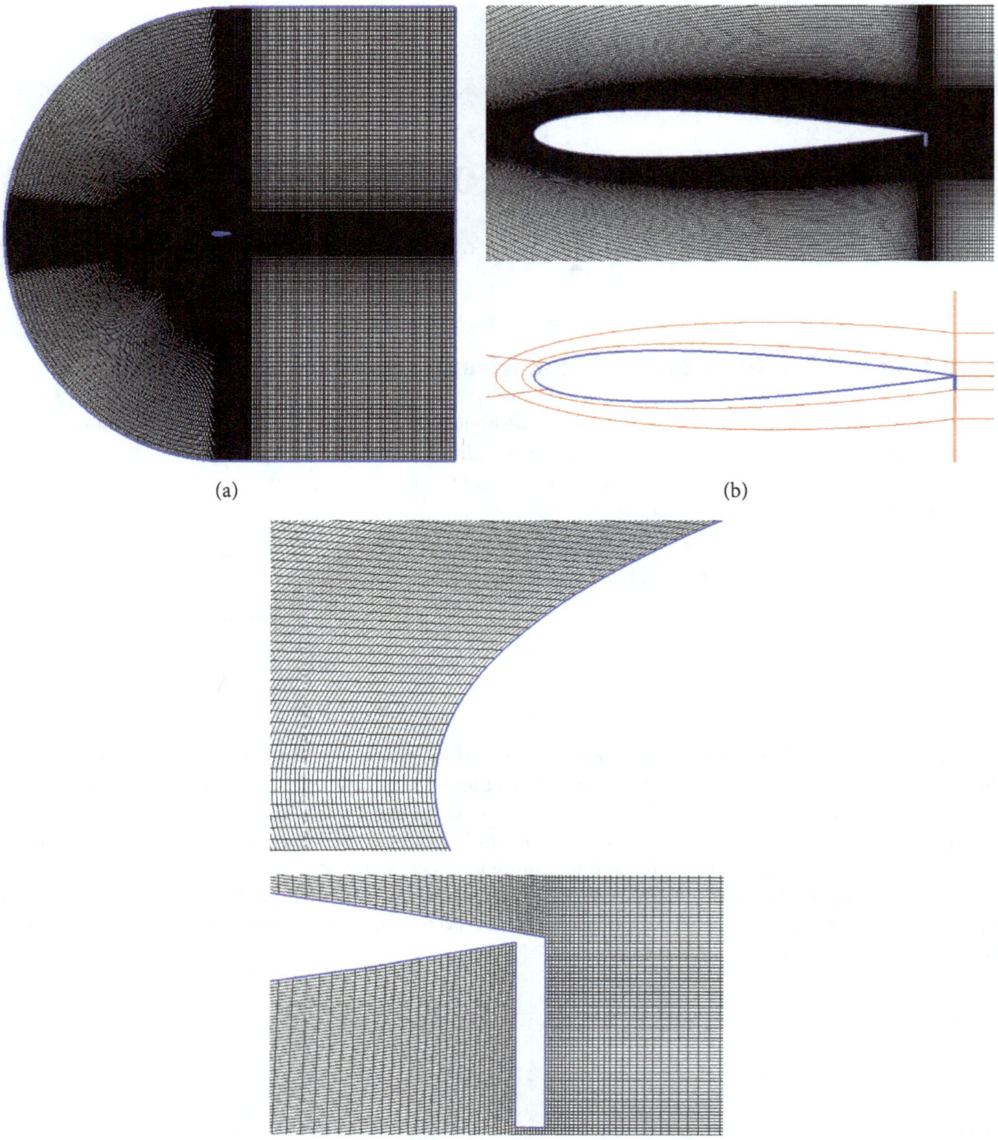

FIGURE 1: (a) Entire domain of the grid used. (b) Near surface grid and blocking near the airfoil surfaces. (c) Boundary layer elements near the leading and trailing edges of the airfoil.

2. Computational Methodology

2.1. Geometry and Grid Generation. The airfoil considered in this study is NACA 0012 airfoil with chord length of 5 cm. C-type domain and grid are created in ICEM-CFD with far-field boundaries 12.5 chords away from trailing edge in all directions (Figure 1(a)). Grid is generated in ICEM-CFD with at least 300,000 nodes. Flow around the airfoil with GF is highly complex with high intensity vortices. Hence, first layer with cell width of 0.025 mm is generated up to 3% chord length around the airfoil. Second layer with cell width of 0.035 mm up to next 7% of the airfoil chord is generated around the airfoil as illustrated in the blocking strategy shown in Figure 1(b). Thereafter cell growth ratio of 1.05 is set up to far-field. Wall y^+ values along the airfoil surface are kept below 1 to account for low Reynolds number flow regime. Boundary layer elements near the leading and trailing edges of the airfoil are shown in Figure 1(c).

2.2. Turbulence Modeling. Previous studies show that out of two equation RANS based models of FLUENT, κ-ε RNG model is best suited for the present airfoil study [12, 13]. However these computations are carried out at a higher Reynolds number of 2.1×10^6. RNG theory accounts for low Reynolds number effects. However effective use of this feature depends on appropriate treatment [14]. Hence it was decided to explore one-equation Spalart-Allmaras (SA) turbulence model for the present computations at low Reynolds numbers and compare these results with those of κ-ε RNG model for

TABLE 1: Comparison of C_L predicted by κ-ε RNG and SA models at Re = 2.1 × 10^6.

α	RNG	SA	Expt.	% difference between RNG and Expt.	SA and Expt.	RNG and SA
0°	0.542	0.538	0.406	25.1	24.5	0.7
4°	0.990	0.997	0.868	12.3	12.9	−0.7
8°	1.405	1.435	1.285	8.5	10.5	−2.1
12°	1.756	1.827	1.646	6.3	9.9	−4.0

improved predictions. Validation is done on an airfoil with 1 m chord length.

2.3. Boundary Conditions and Solver Settings. The airfoil boundary is assigned as solid-wall with no-slip condition while inlet is assigned as velocity inlet and outlet is assigned as pressure-outlet conditions. Density based implicit solving scheme is used with the flow medium being air and Mach number less than 0.3. Hence the fluid is assumed to be incompressible with constant density of 1.225 kg/m^3 and dynamic viscosity of 1.7894 × 10^{-5} kg/m-s. Under relaxation factors for all the transport variables are set to 0.8. Solution initialization is computed from velocity inlet followed by FMG initialization with solution steering. Equations are solved until a convergence criterion of 10^{-5} for all the residuals is satisfied.

3. Results and Discussions

3.1. Validation. Lift and drag coefficients, C_L and C_D, for the airfoil of 1 m chord are computed with κ-ε RNG and SA turbulence models and compared at high Reynolds numbers (2.1 and 3.0 × 10^6) and at a lower Reynolds number (5.0 × 10^5). The results are presented as a function of angle of attack in Figure 2. Comparison is also made with available experimental results at Re = 2.1 × 10^6 [3].

For high Reynolds number, C_L values obtained by both models are very close (Table 1). Masoud et al. [15] also found good agreement with experimental values of lift and drag coefficients for NACA 23018 airfoil using Spalart-Allmaras (SA) turbulence model. At the lower Reynolds number, the C_L values predicted by κ-ε RNG model decrease rapidly more than SA model. This difference also increases at higher angles of attack as shown in Figure 2. Similar trends are noticed for drag coefficient. Both computational models predict higher drag than experimental results. However the C_D values predicted by the SA model are close to the experimental values (Table 2) and almost 40%–50% lower than κ-ε RNG model at Re = 5.0 × 10^5. Based on the validation studies, it was decided to use SA model for all computations at low Reynolds numbers. It is quite possible that improved predictions particularly drag coefficient may be obtained using advanced turbulence models. This is demonstrated by Arko and McQuilling [16], who used advanced turbulence models of Abe et al. [17, 18] and Kato and Launder [19]. However FLUENT does not have these turbulence models available.

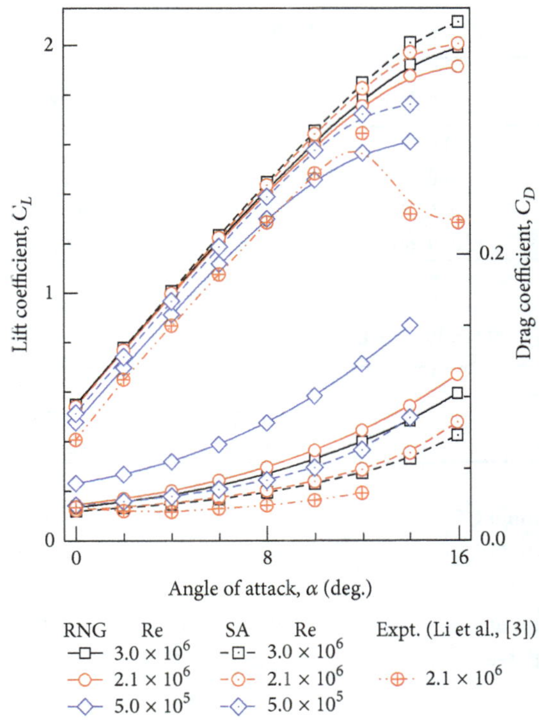

FIGURE 2: Variation of C_L and C_D with AoA for κ-ε RNG and SA turbulence models.

3.2. Effect of Reynolds Number on Aerodynamics of the Airfoil without and with Gurney Flap. Computations are carried out for the airfoil without and with a GF of height of 3% chord at six different Reynolds numbers (Re = 3.0 × 10^5, 1.5 × 10^5, 1.0 × 10^5, 8 × 10^4, 5 × 10^4 and 3 × 10^4). Based on the equation for optimum GF given by Traub and Agarwal [8], the optimum GF height varies from 2.7 to 4.2% of airfoil chord as Reynolds number decreases from 3.0 × 10^5 to 3.0 × 10^4. Hence a GF of height of 3% of airfoil chord is chosen for computations at all Reynolds numbers. At low Reynolds numbers, SA turbulence model works better than κ-ε RNG turbulence model. Hence the former model is used for all the following investigations.

3.2.1. Lift Coefficient. The lift coefficient C_L versus AoA characteristics for the airfoil without and with Gurney flap is presented in Figure 3. At all Reynolds numbers, the lift coefficient for the airfoil with GF is always higher than the lift coefficient for the airfoil without GF at all angles of attack below the stall angle. The lift coefficient obtained at AoA = 8° for the airfoil with GF is higher than maximum lift coefficient obtained for the airfoil without GF. As Reynolds number decreases, the lift coefficient also decreases for both cases: without and with GF. The extent to which Reynolds number affects the performance varies for both the cases and is presented in Table 3. The zero lift angle seems to become more negative as Reynolds number increases (it is estimated to be −2 to −3° as Reynolds number increases from 3.0 × 10^4 to 3.0 × 10^5). This suggests that the effect of GF is to increase the effective camber of the airfoil. However the effective camber decreases as Reynolds number decreases.

Table 2: Comparison of C_D predicted by κ-ε RNG and SA models at Re = 2.1×10^6.

α	RNG	SA	Expt.	% difference between RNG and Expt.	SA and Expt.	RNG and SA
0°	0.0250	0.0212	0.0233	6.8	−9.9	15.2
4°	0.0347	0.0264	0.0199	42.7	24.6	23.9
8°	0.0510	0.0348	0.0245	52.0	29.6	31.8
12°	0.0766	0.0496	0.0331	56.8	33.3	35.2

Table 3: Percentage Decrement in C_L per every 10,000 decrease in Reynolds number for different ranges of Reynolds number.

AoA, α	Reynolds number range: 3.0×10^5 to 1.5×10^5	1.5×10^5 to 1.0×10^5	1.0×10^5 to 8.0×10^4	8.0×10^4 to 5.0×10^4	5.0×10^4 to 3.0×10^4
4°					
Without GF	0.12	0.25	0.37	0.57	1.12
With GF	0.18	0.49	0.86	1.57	3.41
8°					
Without GF	0.15	0.31	0.45	0.72	1.30
With GF	0.20	0.50	0.82	1.40	2.74
12°					
Without GF	0.25	0.50	0.66	0.98	1.62
With GF	0.42	0.85	1.17	1.65	2.51

For the airfoil without GF, variation in C_L values with Reynolds number is negligible at low angles of attack and increases slightly with increasing angle of attack. When Reynolds number is changed from 1.5×10^5 to 1.0×10^5, C_L decrement for AoA = 4°, 8° and 12° is 1.2%, 1.6%, and 2.5%, respectively. However, for the airfoil with Gurney flap, decrement in C_L is almost the same at lower and intermediate angles of attack. The lift coefficient decreases rapidly near stall. Changing the Reynolds number from 1.5×10^5 to 1.0×10^5, decrement in C_L values the airfoil with GF at AoA = 4° and AoA = 8° is 2.5%, whereas it at AoA = 12° is 4.2%.

Table 3 represents the percentage decrement in C_L values for every 10,000 decreases in Reynolds number for different ranges of Reynolds number for the airfoil without and with GF. For AoA = 12°, percentage decrease in C_L for the airfoil without GF for every 10,000 decreases in Reynolds number near critical Reynolds number range ($Re_{cr} \approx 1.5 \times 10^5$) is 0.25. Below critical range the percentage decrement increases from 0.6 for the Reynolds number range of 1.0×10^5 to 8.0×10^4 to 1.62 at very low Reynolds number range of 5.0×10^4 to 3.0×10^4. For AoA = 12°, percentage decrease in C_L for the airfoil with GF for every 10,000 decreases in Reynolds number near critical range is 0.42. Below critical range the percentage decrement increases from 1.17 for the Reynolds number range of 1.0×10^5 to 8.0×10^4 to 2.51 at very low Reynolds number range of 5.0×10^4 to 3.0×10^4. Hence it may be concluded that Gurney flap has adverse effects on lift coefficient as Reynolds number is decreased. However the lift coefficient for the airfoil with GF is always higher than that for the airfoil without GF at all Reynolds number and at all angles of attack.

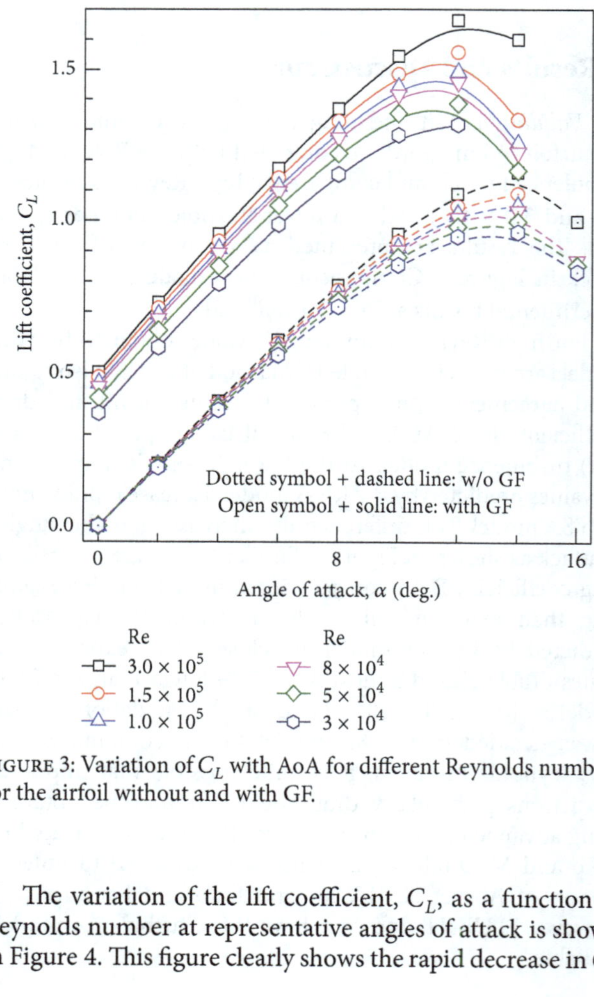

Figure 3: Variation of C_L with AoA for different Reynolds numbers for the airfoil without and with GF.

The variation of the lift coefficient, C_L, as a function of Reynolds number at representative angles of attack is shown in Figure 4. This figure clearly shows the rapid decrease in C_L

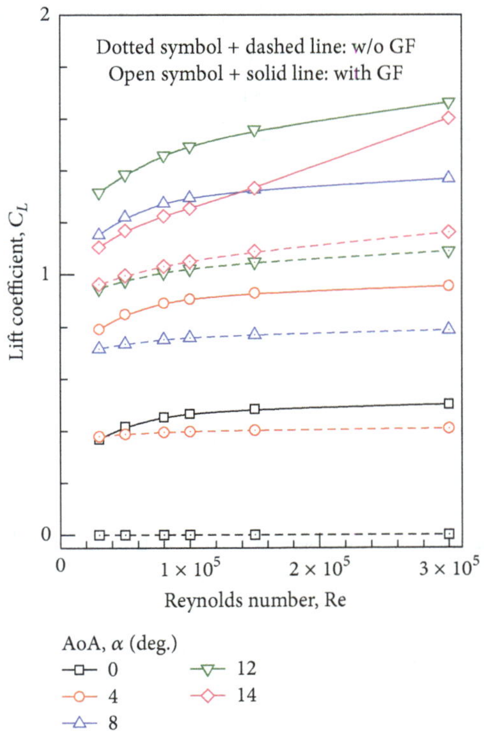

FIGURE 4: Variation of C_L with Reynolds number for the airfoil without and with GF.

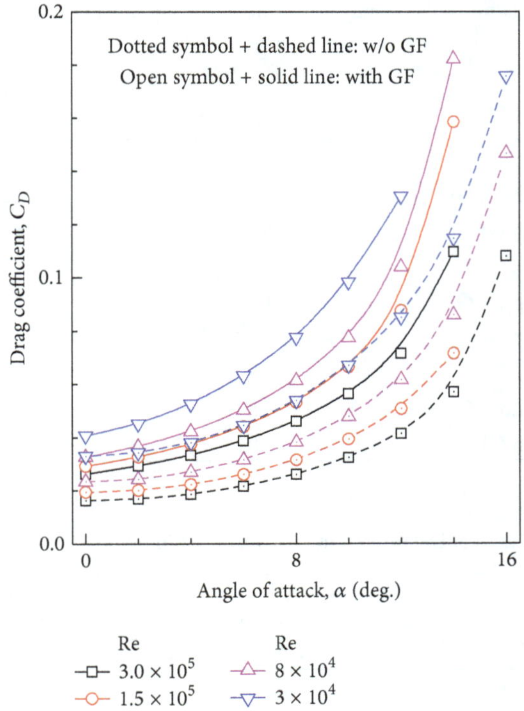

FIGURE 5: Variation of C_D with AoA for different Reynolds numbers for the airfoil without and with GF.

below the critical Reynolds number. This figure also shows the lift enhancement capability of Gurney flap at all Reynolds numbers. It is also noted that the performance of the airfoil without GF is less affected by variation in Reynolds number.

3.2.2. Drag Coefficients. Apart from decreasing C_L with decreasing Reynolds number, C_D is also found to increase as shown in Figure 5. For the sake of clarity, drag coefficient is presented at four Reynolds numbers only, namely, 3.0×10^5, 1.5×10^5, 8.0×10^4, and 3.0×10^4. Magnitude by which C_D increases with decreasing Reynolds number is nearly independent of angle of attack. Increase in C_D is very steep below critical Reynolds number as clearly shown in Figure 6. For higher Reynolds number range, the slope of C_D increment is more for high AoA. Apart from lift enhancement capability of Gurney flap, increased drag penalty is also clearly visible. The C_D obtained at AoA = 12° for the airfoil without GF is obtained at only AoA = 8° with GF. However, the increase in C_D is less for the airfoil with GF when compared to the airfoil without GF.

The percentage increment in C_D values for every 10,000 decreases in Reynolds number for different ranges of Reynolds number for the airfoil without and with GF is presented in Table 4. For AoA = 12°, percentage increase in C_D for the airfoil for every 10,000 decreases in Reynolds number near critical range is 1.39, whereas below critical range the value increases from 3.19 for the Reynolds number range of 1.0×10^5 to 8.0×10^4 to 10.07 at very low Reynolds number range of 5×10^4 to 3×10^4. The respective values for the airfoil with GF are 1.3, 2.62, and 6.60. Hence it may be concluded that Gurney flap has certain beneficial effects on drag coefficient as Reynolds number is decreased.

3.2.3. Lift-to-Drag Ratio. The lift-to-drag ratio is presented as C_L/C_D at different angles of attack in Figure 7. From the figure it is evident that the lift-to-drag ratio for the airfoil with GF is higher than that for the airfoil without GF up to angle of attack of 8°. At this angle of attack, the lift-to-drag ratios for the airfoil without and with GF are almost equal at all Reynolds numbers. Beyond this angle of attack, the lift-to-drag ratio for the airfoil with GF is lower than that for the airfoil without GF, indicating additional drag penalty due to GF. The difference between the lift-to-drag ratios for the airfoil without and with GF seems to be constant at all Reynolds numbers. However the difference between the lift-to-drag ratios for the airfoil without and with GF at angle of attack of 14° is substantially higher than that 12°. From Figure 3, it is evident that the airfoil with GF stalls at angle of attack of 12° while the airfoil without GF stalls at angle of attack of 14°. This is the reason for the large difference at angle of attack of 14°. As discussed earlier, the effective camber of the airfoil with GF increases causing earlier stall.

The decrease in C_L/C_D ratio per every 10,000 decreases in Reynolds number for different ranges of Reynolds number is presented in Table 5. This decrement seems to be nearly constant for the airfoil without and with GF at all Reynolds number ranges. However as the Reynolds number range decreases the lift/drag ratio is decreasing at a faster rate. When Reynolds number decreases from 3.0×10^5 to 1.5×10^5, the decrease in C_L/C_D ratio per every 10,000 decreases

TABLE 4: Percentage Increment in C_D per every 10,000 decrease in Reynolds number for different ranges of Reynolds number.

AoA, α	Reynolds number range 3.0×10^5 to 1.5×10^5	1.5×10^5 to 1.0×10^5	1.0×10^5 to 8.0×10^4	8.0×10^4 to 5.0×10^4	5.0×10^4 to 3.0×10^4
4°					
Without GF	1.27	2.52	3.59	5.71	10.30
With GF	0.73	1.51	2.18	3.45	6.40
8°					
Without GF	1.30	2.58	3.66	5.78	10.26
With GF	0.83	1.62	2.31	3.56	6.25
12°					
Without GF	1.39	2.65	3.69	5.81	10.07
With GF	1.03	1.91	2.62	3.94	6.60

TABLE 5: Decrement in lift-to-drag ratio per every 10,000 decrease in Reynolds number for different ranges of Reynolds number.

AoA, α	Reynolds number range 3.0×10^5 to 1.5×10^5	1.5×10^5 to 1.0×10^5	1.0×10^5 to 8.0×10^4	8.0×10^4 to 5.0×10^4	5.0×10^4 to 3.0×10^4
4°					
Without GF	0.262	0.454	0.595	0.793	1.161
With GF	0.255	0.483	0.663	0.990	1.551
8°					
Without GF	0.386	0.641	0.824	1.095	1.552
With GF	0.317	0.551	0.734	0.987	1.465
12°					
Without GF	0.375	0.572	0.698	0.910	1.256
With GF	0.363	0.512	0.593	0.710	0.910

in Reynolds number varies from 0.262 to 0.375 for the airfoil without GF as the angle of attack increases from 4° to 12°. The corresponding values for the airfoil with GF are 0.255 and 0.363. When Reynolds number decreases from 3.0×10^5 to 1.5×10^5, the corresponding values for the airfoil without and with GF are 1.161 and 1.256 and 1.551 and 0.910, respectively.

3.2.4. Lift-to-Drag Ratio versus Lift Coefficient. The lift-to-drag ratio as a function of lift coefficient for the airfoil without and with GF is presented in Figure 8. For the sake of clarity, the lift-to-drag ration is presented at four Reynolds numbers only (Re = 3.0×10^4, 8×10^4, 1.5×10^5 and 3.0×10^5). The lift-to-drag ratio increases as Reynolds number increases for both the airfoils without and with GF. At lower lift coefficients, there is a drag penalty associated with GF. This drag penalty increases with Reynolds number. At higher lift coefficients, the lift-to-drag ratio increased. Hence the lift coefficient increases for a given lift-to-drag ratio. However at the highest lift coefficient, the lift-to-drag ratio is substantially reduced indicating large drag penalty as the airfoil approaches stall angle. This is true for the airfoil without GF and with GF at all Reynolds numbers.

3.2.5. Static Pressure Distribution. Static pressure distribution in terms of nondimensional coefficient C_P is presented in Figure 9. For the sake of clarity, the static pressure distribution is presented for three Reynolds numbers only (Re = 3.0×10^4, 8×10^4 and 1.5×10^5) for the airfoil with and without Gurney flap at AoA = 12°, which is close to the stall angle. For the airfoil without Gurney flap, static pressure distributions are almost overlapping with each other for different Reynolds numbers. The effect of Reynolds number is clearly visible for the airfoil with Gurney flap. As the Reynolds number decreases, suction and pressure are continuously decreasing all along both the respective surfaces.

Maximum pressure build-up is almost the same for the airfoils with and without GF. Value of maximum C_P for variation in Reynolds number from Re = 3.0×10^4 to 3.0×10^5 varies from 1.11 to 1.02 for the airfoil without GF and from 1.09 to 1.02 for the airfoil with GF. The maximum suction decreases as Reynolds number decreases. However, the change is very small for the airfoil without GF. Maximum suction varies from −3.4 to −4.6 for Re = 3.0×10^4 to 3.0×10^5 for the airfoil without GF, whereas, for the same variation in Reynolds number, C_P values vary from −4.5 to −6.8 for the airfoil with GF. On the pressure surface, the static pressure

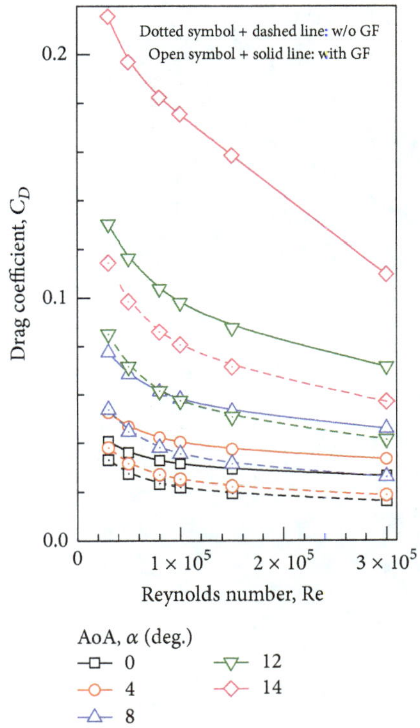

FIGURE 6: Variation of C_D with Reynolds number for the airfoil without and with GF.

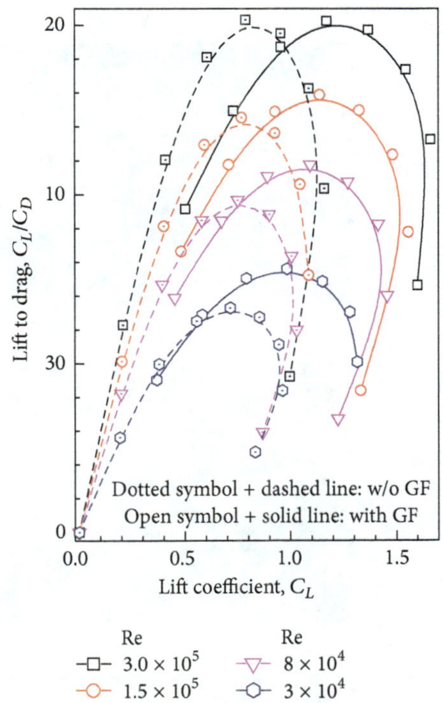

FIGURE 8: Variation of lift-to-drag ratio with lift coefficient at different Reynolds numbers for the airfoil without and with GF.

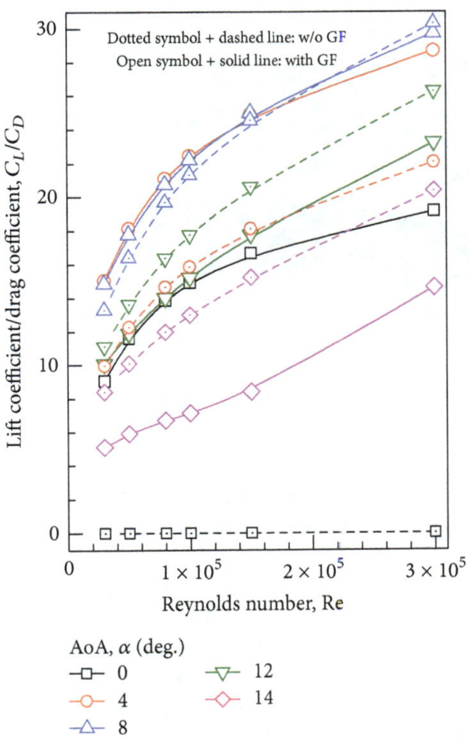

FIGURE 7: Variation of lift-to-drag ratio with Reynolds number for the airfoil without and with GF.

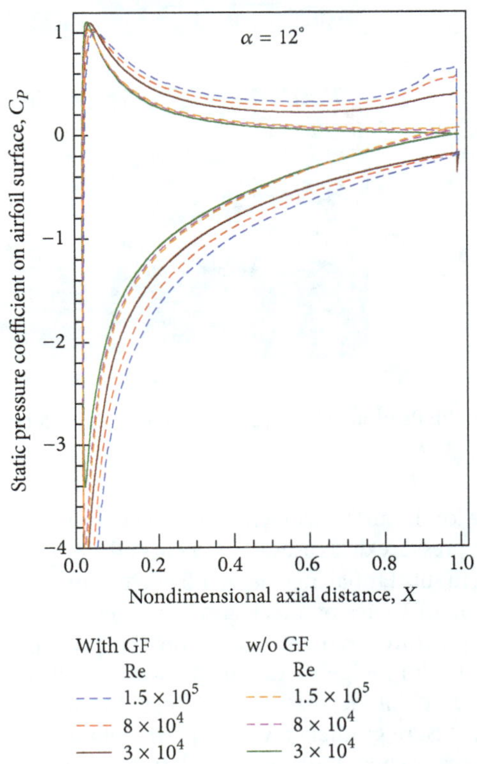

FIGURE 9: Static pressure distributions for different Reynolds numbers at AoA = 12° for the airfoil without and with GF.

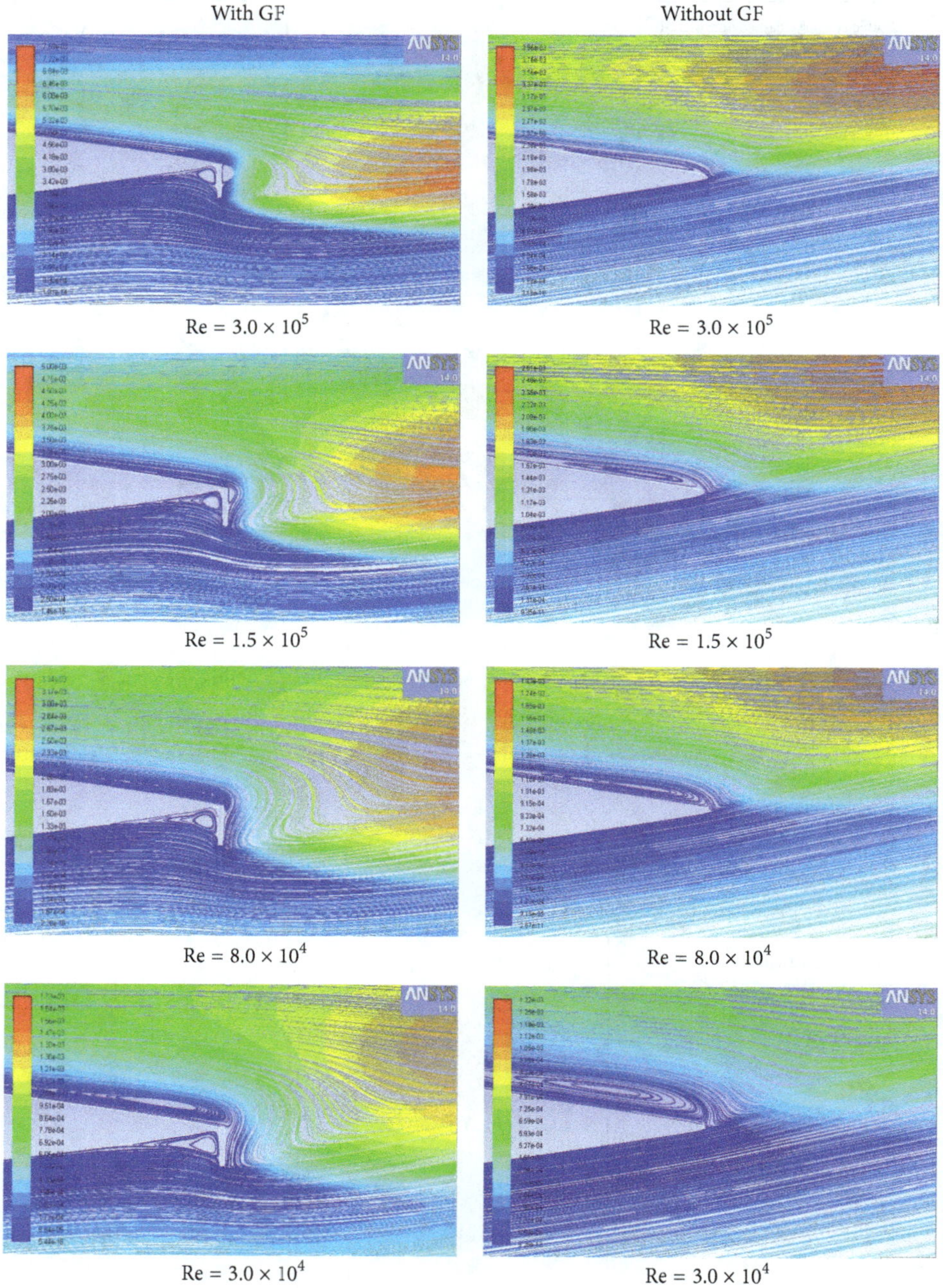

FIGURE 10: Pathlines and turbulent viscosity for different Reynolds numbers at AoA = 12° of the airfoil with and without Gurney flap.

increases for the airfoil with GF, while, on the suction surface, the static pressure decreases. This occurs all along the surface resulting in substantial increase in lift coefficient for the airfoil with GF at all angles of attack and at all Reynolds numbers. Adverse pressure gradient occurs on the pressure surface near the trailing edge of the airfoil with GF. This adverse pressure gradient decreases with Reynolds number. Such adverse pressure gradients were experimentally observed by previous researchers [20]. A recirculating vortex that occurs just upstream of Gurney flap may be the reason for this adverse pressure gradient [2]. Static pressure coefficients are reduced near the Gurney flap as Reynolds number decreases. Unlike the sudden rise in pressure just before Gurney flap on pressure surface at high Reynolds numbers, rate of pressure build-up is not rapid at low Reynolds numbers.

3.2.6. Pathlines. Pathlines are highly affected by the variation in the Reynolds number and the formation of vortices experience drastic change. Pathlines superimposed with the contours of turbulent viscosity at AoA = 12° for the airfoil without and with GF at four different Reynolds numbers presented in Figure 10. Unlike at very high Reynolds number (of the order of Re = 10^6), only one vortex is completely formed in the wake of Gurney flap at Re = 3.0×10^5 instead of two vortices. As the Reynolds number further decreases, the only vortex in wake also start to disappear.

A laminar separation bubble starts its formation. As the Reynolds number is decreased, the flow loses its ability to make transition into turbulent flow in the attached boundary layer, hence forming a laminar separation bubble. Below Re = 1.0×10^5, no vortex is present behind the Gurney flap. The increased pressure on pressure surface is only due to vortex ahead of the Gurney flap which might explain absence of sudden pressure increase near the Gurney flap as shown in static pressure distribution. This laminar separation bubble increases the effective thickness of the airfoil, thereby increasing the pressure drag over the region, which explains the increase in drag at low Reynolds number and degraded performance of the airfoil at low Reynolds numbers. For each Reynolds number, flow is turned towards the Gurney flap whereas, due to absence of any suction, the flow leaves at the airfoil at higher angle without Gurney flap. However the turning of flow towards GF is reduced at lower Reynolds numbers.

4. Conclusions

A computational investigation on the effects of Reynolds number on the aerodynamics of NACA0012 airfoil without and with Gurney flap of height of 3% airfoil chord has been carried out. ANSYS FLUENT commercial CFD code with one-equation Spalart-Allmaras turbulence model is used for the six Reynolds numbers varying from 3.0×10^5 to 3.0×10^4. From this investigation, the following major conclusions are drawn.

(1) Reynolds number plays a very major role in the airfoil aerodynamics for the NACA0012 airfoil without and with Gurney flap. Lift decreases and drag increases when Reynolds number is decreased.

(2) For the airfoil with GF, Reynolds number has adverse effects on lift coefficient, while drag coefficient of the airfoil with GF has some beneficial effects compared to the airfoil without GF.

(3) For high Reynolds number above critical range, decrease in C_L and increase in C_D are negligible.

(4) As the Reynolds number is decreased below the critical Reynolds number range, C_L decreases and C_D increases at very steep rate and the performance degrades rapidly.

(5) For lower Reynolds numbers, the two vortices behind the Gurney flap vanish. The Gurney flap seems to increase the effective camber of the airfoil, causing negative zero lift angle and reduced stall angle.

Nomenclature

AoA, α: Angle of attack (deg.)
C_D: Drag coefficient = $2L/\rho U^2$
Ch: Airfoil chord (m)
C_L: Lift coefficient = $2D/\rho U^2$
C_P: Static pressure coefficient = $2(p_s - p_{s2})/\rho U^2$
D: Drag force (N)
h: Optimum Gurney flap (m)
H: Gurney flap height as percentage of chord
L: Lift force (N)
p_s: Static pressure (Pa)
Re: Reynolds number = UCh/ν
U: Freestream velocity (m/s)
ε: Rate of dissipation of turbulent kinetic energy (m^2/s^3)
κ: Turbulent kinetic energy (m^2/s^2)
ρ: Air density (kgm/m^3)
ν: Kinematic viscosity (m^2/s).

Subscripts

2: Exit.

Abbreviations

CFD: Computational fluid dynamics
GF: Gurney flap
κ-ε RNG: κ-ε RNG turbulence model
RANS: Reynolds averaged Navier Stokes equations
SA: Spalart-Allmaras turbulence model.

Conflict of Interests

The authors declare that there is no conflict of interests regarding the publication of this paper.

References

[1] R. W. Prouty, *Helicopter Aerodynamics*, vol. 2, Philips Publishing, 1985.

[2] R. H. Liebeck, "Design of subsonic aerofoils for high lift," *Journal of Aircraft*, vol. 1, no. 9, pp. 547–561, 1978.

[3] Y. C. Li, J. J. Wang, and P. F. Zhang, "Effects of Gurney flaps on a NACA0012 airfoil," *Flow, Turbulence and Combustion*, vol. 68, no. 1, pp. 27–39, 2002.

[4] J. J. Wang, Y. C. Li, and K.-S. Choi, "Gurney flap-Lift enhancement, mechanisms and applications," *Progress in Aerospace Sciences*, vol. 44, no. 1, pp. 22–47, 2008.

[5] T. J. Mueller and S. M. Batill, "Experimental studies of separation on a two-dimensional the airfoil at low Reynolds numbers," *AIAA journal*, vol. 20, no. 4, pp. 457–463, 1982.

[6] M. S. Selig, J. F. Donovan, and D. B. Fraser, *Airfoils at Low Speeds*, Soartech 8, Soartech Publications, Virginia Beach, Va, USA, 1989.

[7] L. Brown and A. Filippone, "Aerofoil at low speeds with Gurney flaps," *The Aeronautical Journal*, vol. 107, no. 1075, pp. 539–546, 2003.

[8] L. W. Traub and G. Agarwal, "Aerodynamic characteristics of a gurney/jet flap at low reynolds numbers," *Journal of Aircraft*, vol. 45, no. 2, pp. 424–429, 2008.

[9] T. J. Mueller, "Aerodynamic measurements at low Reynolds numbers for fixed wing micro-air vehicles," in *Proceedings of the Special Course on Development and Operations of UAVs for Military and Civil Applications*, Rhode-Saint-Genèse, Belgium, September 1999.

[10] G. K. Ananda, P. Sukumar, and M. S. Selig, "Low-to-moderate aspect ratio wings tested at low Reynolds numbers," AIAA Paper 3026, 2012.

[11] M. Suresh and N. Sitaram, "Gurney flap applications for aerodynamic flow control," in *Proceedings of the 9th International Conference on Mechanical Engineering*, Paper no. ICME 11-FL-40, Dhaka, Bangladesh, December 2011.

[12] K. Sriram, S. Jain, and N. Sitaram, "Grid and turbulence model based exhaustive analysis of NACA 0012 airfoil," *Journal of Advanced Research in Applied Mechanics and Computational Fluid Dynamics*, vol. 1, no. 1, pp. 13–18, 2014.

[13] S. Jain, N. Sitaram, and S. Krishnaswamy, "Computational investigations on the effects of gurney flap on airfoil aerodynamics," *International Scholarly Research Notices*, vol. 2015, Article ID 402358, 11 pages, 2015.

[14] Ansys, *ANSYS Fluent Theory Guide*, Ansys, Canonsburg, Pa, USA, 2011.

[15] J. N. Masoud, S. Habibollah, and A. Mekanik, "Numerical investigation of aerodynamic characteristics of NACA 23018 airfoil with a gurney flap," *International Journal of Mechanical Engineering and Robotics Research*, vol. 1, no. 3, pp. 341–349, 2012.

[16] B. M. Arko and M. McQuilling, "Computational study of high-lift low-pressure turbine cascade aerodynamics at low reynolds number," *Journal of Propulsion and Power*, vol. 29, no. 2, pp. 446–459, 2013.

[17] K. Abe, T. Kondoh, and Y. Nagano, "A new turbulence model for predicting fluid flow and heat transfer in separating and reattaching flows—I. Flow field calculations," *International Journal of Heat and Mass Transfer*, vol. 37, no. 1, pp. 139–151, 1994.

[18] K. Abe, T. Kondoh, and Y. Nagano, "A new turbulence model for predicting fluid flow and heat transfer in separating and reattaching flows-II. Thermal field calculations," *International Journal of Heat and Mass Transfer*, vol. 38, no. 8, pp. 1467–1481, 1995.

[19] M. Kato and B. E. Launder, "The modeling of turbulent flow around stationary and vibrating square cylinders," in *Proceedings of the 9th Symposium on Turbulent Shear Flows*, pp. 10.4.1–10.4.6, Kyoto University, Kyoto, Japan, August 1993.

[20] R. Myose, I. Heron, and M. Papadakis, "The post-stall effect of gurney flaps on a NACA-0011 airfoil," in *Proceedings of the Aerospace Atlantic Conference*, SAE Paper 961316, Dayton, Ohio, USA, May 1996.

15

External Aerodynamics Simulations in a Rotating Frame of Reference

Filomena Cariglino,[1] Nicola Ceresola,[2] and Renzo Arina[1]

[1] *Politecnico di Torino, Corso Duca degli Abruzzi 24, 10129 Torino, Italy*
[2] *Alenia Aermacchi S.p.A., Corso Francia 426, 10100 Torino, Italy*

Correspondence should be addressed to Filomena Cariglino; filomena.cariglino@polito.it

Academic Editor: James J. McGuirk

This paper presents the development of a tool integrated in the UNS3D code, proprietary of Alenia Aermacchi, for the simulation of external aerodynamic flow in a rotating reference frame, with the main objective of predicting propeller-aircraft integration effects. The equations in a rotating frame of reference have been formulated in terms of the absolute velocity components; in this way, the artificial dissipation needed for convergence is lessened, as the Coriolis source term is only introduced in the momentum equation. An Explicit Algebraic Reynolds Stress turbulence model is used. The first assessment of effectiveness of this method is made computing stability derivatives of a NACA 0012 airfoil. Finally, steady Navier-Stokes and Euler simulations of a four-blade single-rotating propeller are presented, demonstrating the efficiency of the chosen approach in terms of computational cost.

1. Introduction

The main objective of the present work is to calculate the flow field around a propeller. Unsteady Reynolds averaged Navier-Stokes equations (RANS) represent the state of the art for the numerical prediction of the viscous flow around propellers [1, 2]. However, unsteady Navier-Stokes simulations for such complex flows are computationally expensive. Therefore, these simulations are generally carried out only using reduced order methodologies [3]. In our case, to simulate the unsteady viscous flow around a propeller, we use the Navier-Stokes, or Euler, equations in a noninertial (rotating) reference frame in which the propeller is at rest. In this way, the simulations can be performed in steady-state mode. The advantage of this method is the relatively lower computational cost needed to obtain a CFD solution in comparison to a fully unsteady simulation. Moreover, the present method permits the insertion of the propeller as a building block into the full computational model of an aircraft, to investigate propeller-aircraft integration issues from an aerodynamics point of view.

When the Navier-Stokes equations for a steady viscous compressible flow are written in a rotating frame of reference, there are two choices regarding the velocity vector components. Either they can be the components with respect to the absolute (inertial) frame, hereafter called the absolute velocity components [4], or they can be the components with respect to the rotating (noninertial) frame, hereafter called the relative velocity components [5]. In this paper, the first formulation has been chosen because it has the advantage of enabling a steady-state formulation as the flow field can be viewed as a steady state in the rotating frame. In order to fully use the UNS3D code which is written in the absolute velocity components, source terms are added to take the coordinate-system rotation into account in the governing equations.

To assess this method, two applications are presented: the first application is the determination of the stability derivatives for a NACA 0012 airfoil and the second is the simulation of the flow field around a rotating propeller. The first application allows us to assess the method by comparing the results to a reference test case, for which several authors have obtained a solution [6–9]. In this case, the flow field is computed using Euler equations. The second application is a more complex test case, which has the purpose to evaluate the accuracy, efficiency, and robustness of the current method to predict the complex flow field of a rotating propeller. For this test case, a comparison with experimental results for cruise conditions in terms of thrust coefficient is also made.

2. Governing Equations

The governing equations which are numerically resolved using the UNS3D code are summarized below. The Navier-Stokes equations in a noninertial frame of reference are expressed as

$$\frac{\partial \vec{X}}{\partial t} + \nabla \vec{F_I} + \nabla \vec{F_V} = \vec{Q_X}, \tag{1}$$

where $\vec{X}$ is defined as

$$\vec{X} = \begin{bmatrix} \rho \\ \rho \vec{v} \\ \rho E \end{bmatrix}. \tag{2}$$

$\vec{F_I}$ and $\vec{F_V}$ are the respective flux vectors:

$$F_{Ii} = \begin{bmatrix} \rho v_i \\ \rho v_1 v_i + p\delta_{1i} \\ \rho v_2 v_i + p\delta_{2i} \\ \rho v_3 v_i + p\delta_{3i} \\ (\rho E + p) v_i \end{bmatrix}, \tag{3}$$

$$-F_{Vi} = \begin{bmatrix} 0 \\ \tau_{i1} \\ \tau_{i2} \\ \tau_{i3} \\ q_i + v_j \tau_{ij} \end{bmatrix}, \tag{4}$$

and the source terms are contained within $\vec{Q_X}$:

$$\vec{Q_X} = \begin{bmatrix} 0 \\ \rho \vec{f}_e \\ W_f \end{bmatrix}. \tag{5}$$

W_f is equal to the following relation:

$$W_f = \rho \vec{f_e} \cdot \vec{v}. \tag{6}$$

2.1. Formulation in Rotating Frame for the Absolute Velocity. Normally the governing equations for rotating systems are solved for the relative velocity components in the relative frame of reference. However, in an external aerodynamic context such as the propeller-wing configuration, two main advantages can be derived through taking instead the absolute velocity components in the relative reference frame. The first, and most important one, is that the grid block containing the propeller can be interfaced more easily within the full aircraft model. The second is that a smaller numerical effect of source terms is present, as will be shown below. To express (1) in terms of a relative reference frame, the following relations for substantial and local derivatives are used:

$$\begin{aligned} \frac{Db}{Dt} &= \frac{D'b}{Dt'}, \\ \frac{DB}{Dt} &= \frac{D'B}{Dt'} + \vec{\omega} \times \vec{B}, \\ \frac{\partial b}{\partial t} &= \frac{\partial' b}{\partial t'} - (\vec{\omega} \times \vec{r}) \cdot \nabla b, \\ \frac{\partial B}{\partial t} &= \frac{\partial' B}{\partial t'} - (\vec{\omega} \times \vec{r}) \cdot \nabla \vec{B} + \vec{\omega} \times \vec{B}, \end{aligned} \tag{7}$$

where the prime $'$ denotes the operation with respect to the relative reference frame. By using relations (7), the right-hand side of (1) becomes

$$\vec{Q_X} = \begin{bmatrix} 0 \\ -\rho (\vec{\omega} \times \vec{v}) \\ 0 \end{bmatrix}, \tag{8}$$

where $\vec{\omega}$ is defined as

$$\vec{\omega} = \begin{pmatrix} \omega_x \\ \omega_y \\ \omega_z \end{pmatrix}. \tag{9}$$

With this formulation, the source term vector (see (8)) contains only the contribution of the Coriolis force and the contribution of the centrifugal force is omitted. In this way the magnitude of the source term is greatly reduced and a smaller amount of artificial dissipation is required to ensure convergence.

To take into account the rotation of the coordinate system, (3) is modified in the following way:

$$F_{Ii} = \begin{bmatrix} \rho v_i \\ \rho v_1 (v_i - v_{b_i}) + p\delta_{1i} \\ \rho v_2 (v_i - v_{b_i}) + p\delta_{2i} \\ \rho v_3 (v_i - v_{b_i}) + p\delta_{3i} \\ (\rho E + p) v_i \end{bmatrix}, \tag{10}$$

where $\vec{v_b}$ is defined as

$$\vec{v_b} = \vec{\omega} \times \vec{r}. \tag{11}$$

It is noteworthy to mention that the above expression for the Navier-Stokes equations is valid only for a rotating reference frame. Linear acceleration components should be added in case one wants to express the equations of motion in a general noninertial frame.

In addition, it should be pointed out that it is not possible in the present context to impose any radial equilibrium condition to the pressure, as it is normally done in turbomachinery.

An important thing to be noted here is that, except for the source term $\vec{Q_X}$, the functional form of the noninertial

Navier-Stokes equations is the same as the functional form of the standard conservative equations defined for inertial reference frames and including the Algebraic Lagrangian Eulerian (ALE) approach for generalized motion of the grid. Therefore, it is possible to implement a conservative formulation in terms of the conservative variables $\vec{X}$ defined in (2) and the introduction of the ALE approach permits a local application of the noninertial frame of reference as a building block in a more complex configuration framework, without any interface between the noninertial and inertial part of the same mesh, because this formulation guarantees the flux conservation. In applications, it is sufficient to specify as input the bounds of the rotational region and in this way a rotational speed is imposed at the nodes inside this region.

2.2. Enforcement of Compatibility Conditions. The numerical source error due to the noninertial reference frame can be examined analytically by imposing the conservation of the freestream. In this case, all the flow derivatives are zero and the velocity vector is

$$\vec{V} = \vec{V_\infty} + \left(\vec{\omega} \times \vec{r}\right), \tag{12}$$

where $\vec{V_\infty}$ is equal to

$$\vec{V_\infty} = \begin{pmatrix} u_\infty \\ v_\infty \\ w_\infty \end{pmatrix}. \tag{13}$$

The continuity, momentum, and energy equations (see (1)) can then be reduced to the following expressions:

$$\rho \left[\vec{\omega} \cdot \left(\nabla \times \vec{r}\right)\right] = 0, \tag{14}$$

$$\rho\vec{V}\left[\vec{\omega} \cdot \left(\nabla \times \vec{r}\right)\right] + \rho\vec{V_\infty} \cdot \nabla\vec{V} = -\rho\left(\vec{\omega} \times \vec{V}\right), \tag{15}$$

$$\rho E \left[\vec{\omega} \cdot \left(\nabla \times \vec{r}\right)\right] = 0. \tag{16}$$

For the first application case, where we have a steady rotation parallel to the y-direction, (14) can be reduced to the following expression:

$$0 = \omega_y \left(\frac{\partial z}{\partial x} - \frac{\partial x}{\partial z}\right) \tag{17}$$

that is identically zero for any nonzero angular velocity ω_y, whereas, for the second case, with a rotation parallel to x-direction, (14) becomes

$$0 = \omega_x \left(-\frac{\partial z}{\partial y} + \frac{\partial y}{\partial z}\right). \tag{18}$$

In both cases, for the numerical formulation, the right-hand side is not exactly zero, however, producing a freestream error.

Using the results of (17) and (18), and denoting the right-hand side as S_Ω, a simple and straightforward source term correction can be applied in (1). In particular, an additional source term $\vec{K}$ can be included to exactly cancel the free-stream error:

$$\frac{\partial \vec{X}}{\partial t} + \nabla\vec{F}_I + \nabla\vec{F}_V = \vec{Q_X} + \vec{K}, \tag{19}$$

where $\vec{K}$ for the NACA 0012 is

$$\vec{K} = S_\Omega \begin{bmatrix} \rho \\ 0 \\ \rho v_2 \\ 0 \\ \rho E \end{bmatrix}, \tag{20}$$

whereas for the propeller (20) becomes

$$\vec{K} = S_\Omega \begin{bmatrix} \rho \\ \rho v_1 \\ 0 \\ 0 \\ \rho E \end{bmatrix}. \tag{21}$$

3. Numerical Method

The computations have been performed using the code UNS3D. The solution algorithm is based on a finite volume, node centred approach operating on a hybrid unstructured grid. The artificial dissipation model is derived from the non-linear scheme of Jameson [10], with no eigenvalue blending. Scalar or matrix dissipation can be chosen.

The Navier-Stokes equations are integrated in time with a second order backward difference scheme and dual time stepping. A five-stage Runge-Kutta scheme is used to drive toward zero the residual at each time step. With the use of residual averaging, a local CFL number of 4.9 could be employed in the multistage subiteration process. The Algebraic Lagrangian Eulerian approach for generalized motion of the grid is included [11].

The Weiss and Smith version of low Mach number preconditioning is implemented in the code [12]. A sensor depending on cell Reynolds number was also introduced to avoid applying the preconditioning inside boundary layers. For the computation of the present test case, its application was found to be beneficial in order to reduce numerical dissipation and enhance convergence.

Matrix dissipation was also found to be beneficial, allowing a strong reduction of the dissipation associated with convective eigenvalues, hence enabling a better resolution of vortices.

3.1. RANS Turbulence Model. The k-ω turbulence model proposed by Hellsten [13] has been employed. The model constants have been calibrated requiring consistent behavior near boundaries between turbulent and laminar flow, inside shear flows and for zero pressure gradient wall flows. In particular, the calibration has been considered taking into account a variable c_μ, as it is the case if an algebraic stress model (EARSM) is included.

The Wallin-Johansson Explicit Algebraic Stress Model (WJ-EARSM) [14] is implemented using Hellsten's k-ω as

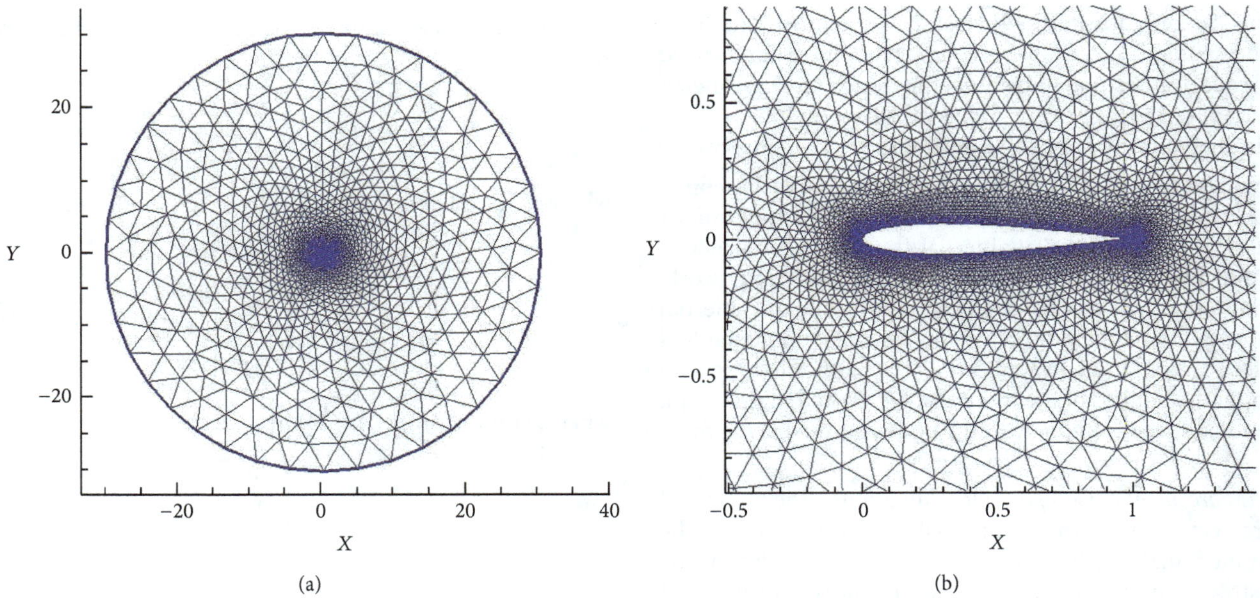

FIGURE 1: Mesh: (a) wide-view and (b) close-view.

the basic RANS model. The model is an exact solution of the corresponding ARSM in two-dimensional mean flow. In three dimensions there is still a complete, while approximate, solution.

The full anisotropic version of the model is used; that is, the anisotropic part of the Reynolds stress tensor is directly introduced in the momentum equations, while the isotropic part is taken into account in the form of an effective variable c_μ.

4. Boundary Conditions

The boundary conditions along solid walls for Navier-Stokes (viscous) flows are different from those for Euler flows. In the case of viscous flows, the velocity of the flow must vanish at the walls, while, in the case of Euler flows, it is only required that the flow does not go through the wall.

As a consequence of the foregoing statement, at the airfoil, the condition of nonpenetration has been imposed, whereas, on the blade surface, no-slip and no-penetration conditions are used by setting the absolute velocity equal to the absolute local blade velocity and the adiabatic wall condition and zero-normal pressure gradient condition at the wall are imposed at the blade surface.

In general, the boundary conditions applied at the far-field boundary are the same for Navier-Stokes and Euler flows; therefore, the far-field boundaries are treated by using characteristic boundary conditions.

5. Numerical Results

5.1. Model Validation: Steady Rotary NACA 0012 Airfoil. To validate the numerical model, the stability derivatives for a NACA 0012 were computed using finite differences and compared with the results obtained by Limache and Cliff [6]. In the experiment, an airfoil is submitted to a steady rotation performed at constant incidence α for a given pitch rate q, generating a steady flow field in a reference frame attached to the airfoil. The radius of the loop is inversely proportional to q. Thus, as q reduces to zero, the radius approaches infinity and steady level flight is recovered.

The results presented below are all computed for an angle of attack equal to zero, so we use the wind-axis reference frame for the computation of the derivatives.

5.2. Numerical Results. In Figure 1 is shown the 2D unstructured grid. The outer boundary is at a distance 30 times the length of the airfoil's chord with respect to the grid center, coincident with the leading edge of the airfoil. The grid is made by 12334 nodes and 12096 elements.

To verify the implementation of 3D Navier-Stokes equations in terms of absolute velocities, we compare results for the NACA 0012 airfoil rotating at a finite q to those produced by Limache [7] simulating inviscid flow around a NACA 0012 airfoil at Mach equal to 0.2 for nondimensional pitch rate $\hat{q}$ equal to 0, 0.01, 0.03, and 0.05. In fact, at the present test conditions (low Mach number and low incidence) we do not expect that the integral quantities computed using viscous and inviscid methods, respectively, differ significantly.

The C_p distributions around the airfoil are shown in Figure 2 where the computed C_p distributions and streamlines of relative velocity are compared with those presented in [7] for $\hat{q} = 0.01$. In Figure 3, the pressure coefficient contours and streamlines of relative velocity in the whole computational domain are shown.

Finally, in Table 1, we compare the values of C_l and C_m from our implementation to the references results. The two implementations match quite well over a range of $\hat{q}$ values.

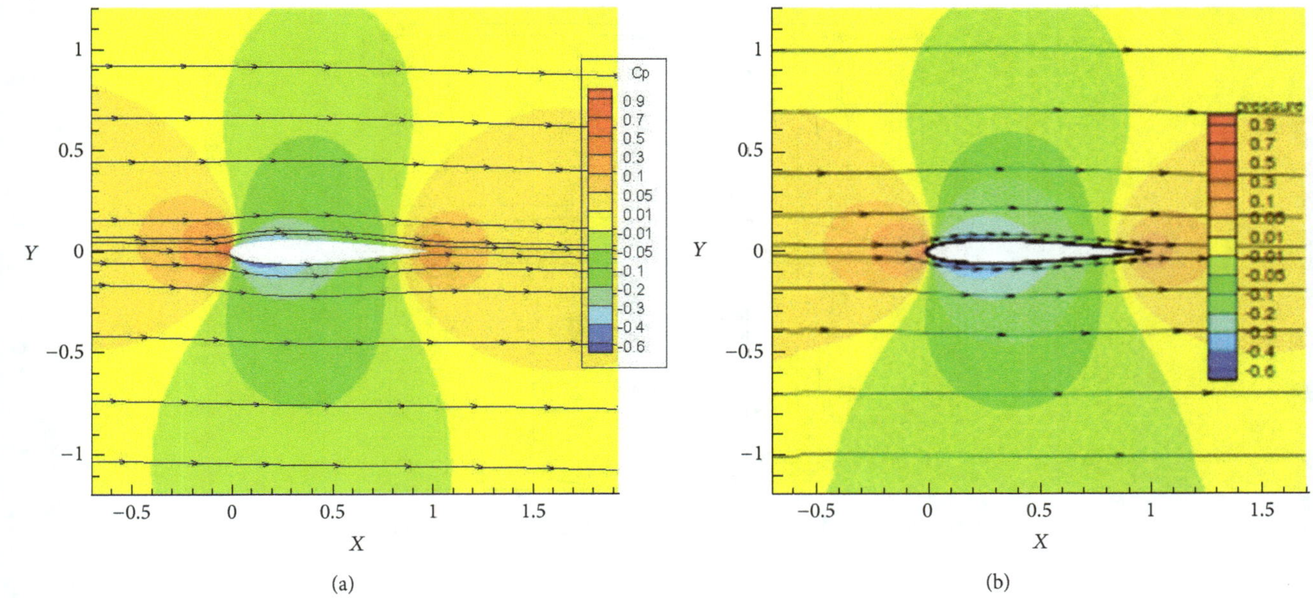

FIGURE 2: C_p contours and streamlines comparison for rotating NACA 0012 airfoil at Mach = 0.2 and $\alpha = 0^\circ$: (a) present work, $\hat{q} = 0.01$, (b) Limache [7], $\hat{q} = 0.01$.

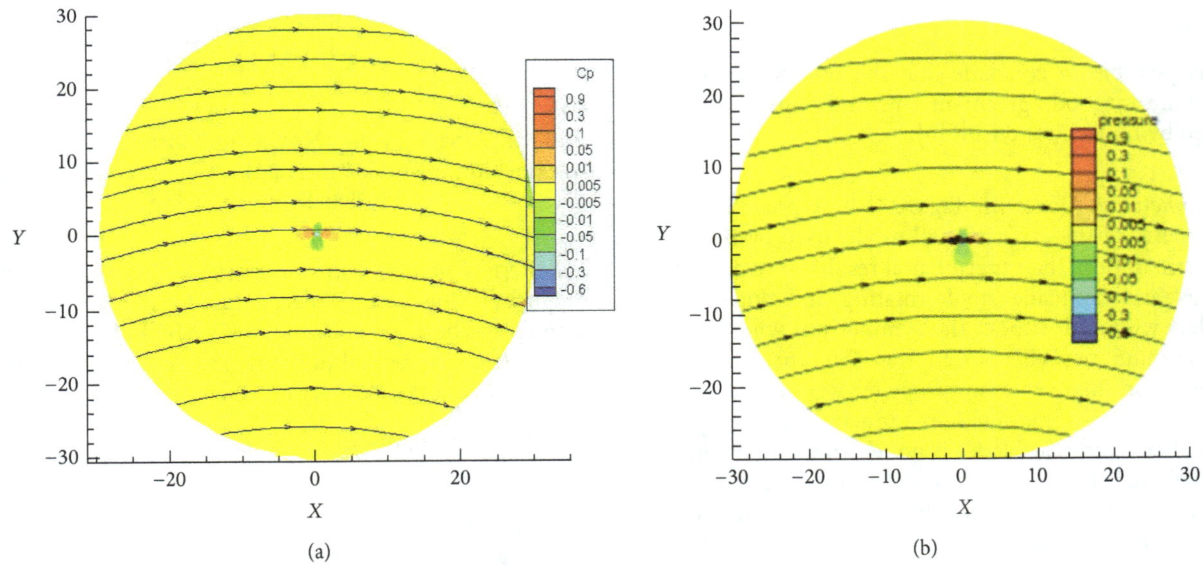

FIGURE 3: C_p contours and streamlines comparison for rotating NACA 0012 airfoil at Mach = 0.2 and $\alpha = 0^\circ$: (a) present work, $\hat{q} = 0.01$, (b) Limache [7], $\hat{q} = 0.01$.

TABLE 1: Comparison of lift and moment coefficients for the NACA 0012 at $\alpha = 0^\circ$ and Mach = 0.2 at various values of $\hat{q}$ (results from Limache [7] are in parentheses).

$\hat{q}$	C_l	C_m
0.01	−0.051 (−0.053)	−0.02 (−0.018)
0.03	−0.153 (−0.157)	−0.06 (−0.053)
0.05	−0.26 (−0.262)	−0.1 (−0.088)

The stability derivatives are calculated using finite differences:

$$C_{(\blacksquare)_q} \approx \frac{C_{(\blacksquare)}(M, \alpha, q + \Delta q) - C_{(\blacksquare)}(M, \alpha, q,)}{\Delta q}. \quad (22)$$

In Table 2, the stability derivatives are compared with the results obtained by Limache.

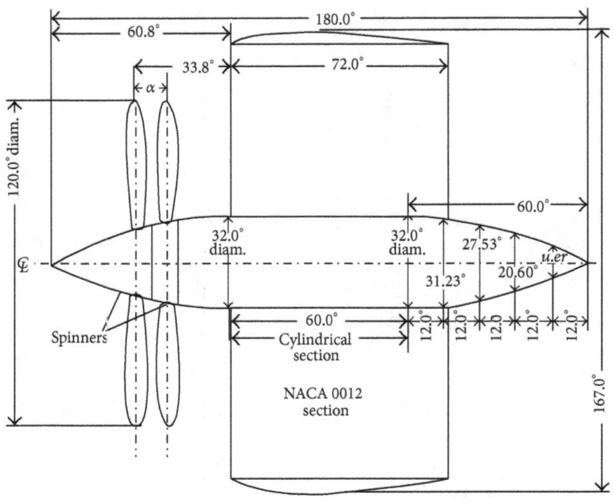

FIGURE 4: Geometrical and experimental model used by Biermann and Haetman [15].

TABLE 2: Comparison of stability derivatives for the NACA 0012 at $\alpha = 0^\circ$ and Mach = 0.2.

Derivatives	UNS3D	Limache and Cliff [6]
C_{lq}	−5.225	−5.250
C_{mq}	−1.932	−1.766

It is possible to conclude that the results obtained by UNS3D are in good agreement with the numerical results obtained by Limache and Cliff [6].

5.3. Geometrical Model. In Figure 4, it is possible to see the geometry of the experimental model used by Biermann and Haetman [15]. The experimental results were performed for four- and six-blade single-rotating and dual-rotating propellers with and without the symmetrical wing in place. The maximum propeller speed was 550 rpm. The results for four-blade single-rotating propeller were made up with two two-way hubs mounted in tandem and the spacing between front and rear blades is not equal and therefore the front blade led the rear by 85.4 deg. In this paper, only the four-blade single-rotating propellers with and without wing are considered for the comparison with the experimental results in terms of thrust coefficient. The propeller, namely, a Hamilton Standard 3155-6, consists of four blades installed on a single way hub in front of a streamline body, or nacelle, housing the engine needed to spin it. The four blades are streamlined using Clark Y profiles and the angle between two blades is 90 deg.

In the report of Biermann and Haetman [15], several blade pitch angles, defined as the angle between the rotation plane and the airfoil chord at 75% of the radius of the propeller, in the range of 20 deg to 65 deg were studied. In our case, it was decided to investigate a propeller with a blade pitch angle of 45 deg. The propeller diameter is 3.08 m. Starting from the geometrical details reported in the mentioned report, a mathematical model describing the propeller has been created with CATIA V5 (Figure 5).

TABLE 3: Operating conditions investigated.

Advance ratio	Velocity, m/s	Rotational speed, rps
1.43	49.1744	11.14
1.5	49.1744	10.6
1.8	49.1744	8.9
2.0	49.1744	8.0
2.4	49.1744	6.6

The wing, shaped using NACA 0012 airfoils, is located in a midposition of the spinner and set at an angle of attack of 0 deg. Wing chord is 1828.8 mm and wing span is 4241.8 mm.

5.4. Results for the Four-Blade Single-Rotating Propeller + Spinner (Viscous). A 3D unstructured grid of the propeller + spinner has been generated with ICEM-CFD (Figure 6). This grid is strongly refined in the region around the blades and on the blade surfaces. It is made by

(i) 5672824 nodes,

(ii) 16265544 elements,

(iii) 21 prismatic layers on solid surfaces to correctly match boundary layer behavior.

In the code UNS3D it is possible to specify an arbitrary velocity for a specific group of nodes within the mesh (ALE formulation). The resulting fluxes are automatically interfaced in order to ensure conservation at the boundary between rotating and nonrotating zones. The solution is then computed specifying a rotational velocity, as described in (11), only for the nodes inside the rotating block (Figure 7) and taking into account the source terms (see (8)) and the correction terms as in (23). It is worth noting that the formulation described above is steady. The solution is indeed an instantaneous snapshot of the flow field, which can be seen as frozen in correspondence with a fixed phase angle.

Five different operating conditions, shown in Table 3, were investigated. The axial undisturbed velocity has been set

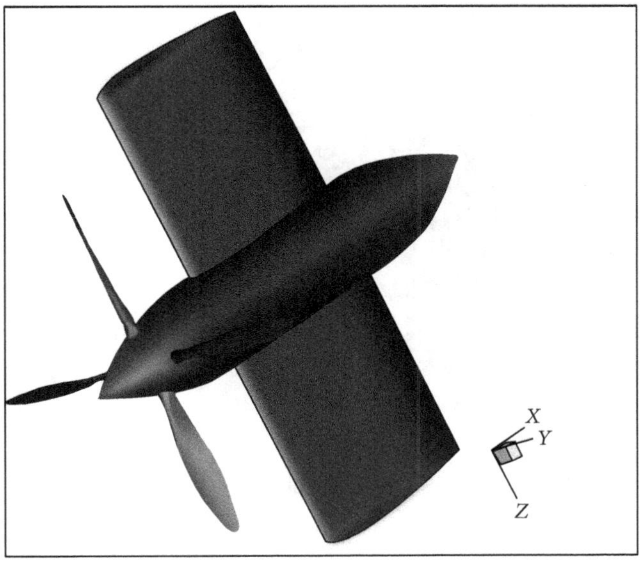

(a)

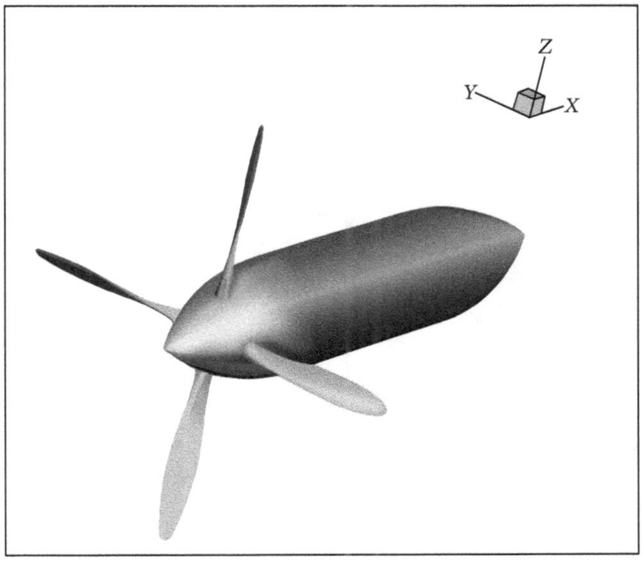

(b)

FIGURE 5: CATIA model: (a) propeller with wing and (b) propeller without wing.

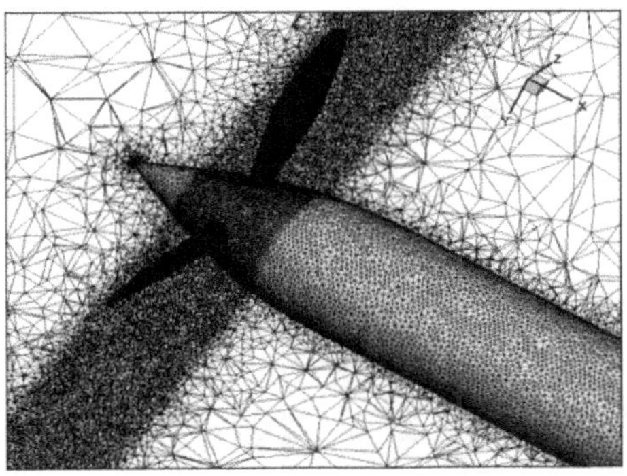

FIGURE 6: Grid generated with CFD-ICEM.

equal to 49.1744 m/s, corresponding to the maximum wind tunnel test speed of 110 mph [15].

In Figure 8, it is possible to note a general increasing level of the relative Mach number from the nacelle surface toward the tip, which is the result of the increasing rotational speed with increasing radius.

As the propeller rotates, it induces swirls in the slipstream and the blade tip vortices pass by periodically. This phenomenon is more evident when the advance ratio decreases. In fact, for lower advance ratios, we see a strong vortex shedding, which starts from each blade and travels downstream with the perturbation velocity creating strong spiral type regions in the rear wake for each blade (Figure 9). Furthermore, strong hub and tip vortices (Figure 10) are continuously shed from the respective blade regions and "absorb" the weaker vorticity regions at inner blade radii producing also spiral type patterns.

In Figures 11 and 12, the azimuthal velocity profiles downstream of the blades are shown at different position of x

TABLE 4: Experimental and computational results for the thrust coefficient.

Advance ratio J	Experimental C_T	Computational C_T
1.43	0.2192	0.217
1.5	0.2175	0.215
1.8	0.1782	0.174
2.0	0.1374	0.1376
2.4	0.0487	0.05

and it is possible to note how the swirl induced by the rotation of the propeller vanishes with increasing distance from the blades. In particular, the fast decay of the slipstream swirl, in the region far from the spinner, is due to the coarse mesh and consequently to the high levels of numerical dissipation in that region.

With the aim of comparing the results with those obtained by Biermann and Haetman [15], the thrust coefficient, defined as

$$C_T = \frac{T}{\rho n^2 D^4}, \quad (23)$$

has been calculated. Following the experimental procedure adopted by Biermann and Haetman [15], the thrust force T has been obtained by integrating the forces along x-direction on all the blade surfaces and subtracting the drag force due to the blades alone in case of zero thrust coefficient.

In Figure 13, the obtained thrust coefficients for five different advance ratios are plotted and compared with those obtained by Biermann and Haetman [15].

The computed thrust coefficients are in good agreement with the experimental values as listed in Table 4.

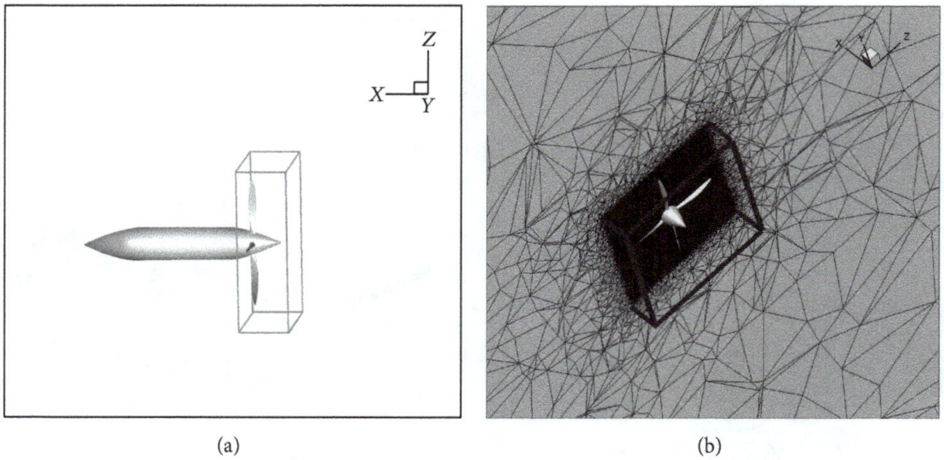

FIGURE 7: Rotating block within the mesh: (a) propeller and (b) mesh inside the block.

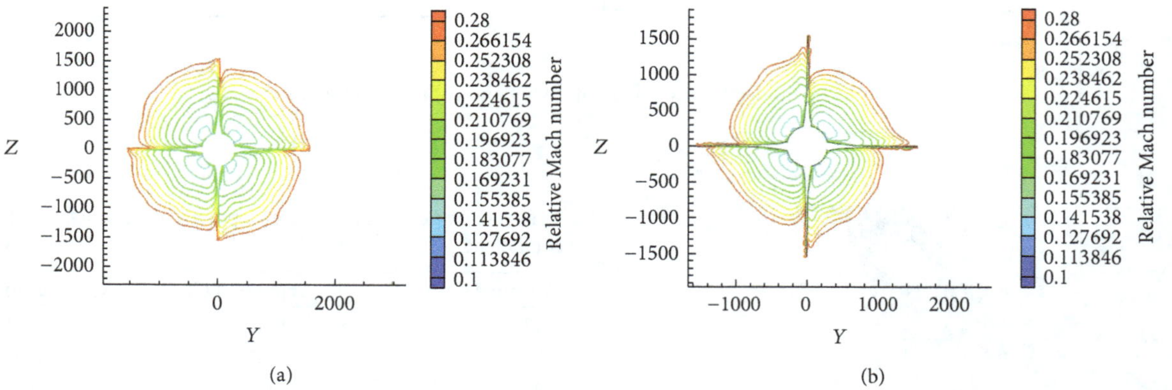

FIGURE 8: Relative Mach number contours in the plane (y, z): (a) $J = 1.8$ and (b) $J = 1.43$.

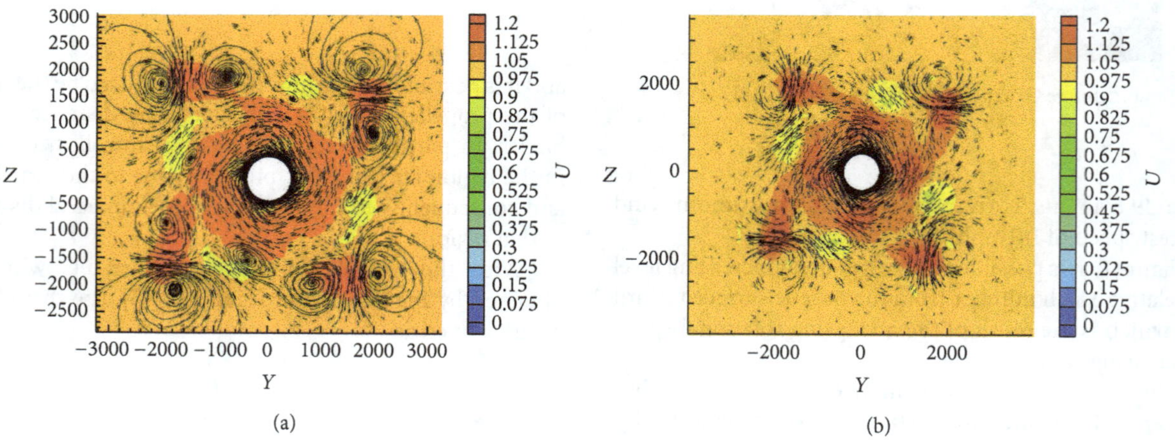

FIGURE 9: X-velocity in the plane $x = 750$ mm: (a) $J = 1.8$ and (b) $J = 1.5$.

As the experimental errors are unknown, it is not possible to determine whether the computed results are within the range of the experimental uncertainty or to give a precise assessment of the quality of the results.

5.5. Results for the Four-Blade Single-Rotating Propeller + Spinner + Wing (Inviscid). Steady Euler results for the propeller + spinner + wing are presented in this section. The grid is generated with ICEM-CFD (Figure 14) and it is made by

(i) 2173935 nodes,

(ii) 10718702 elements.

A rotational velocity has been imposed for the nodes inside the block around the blades, as indicated in Figure 14.

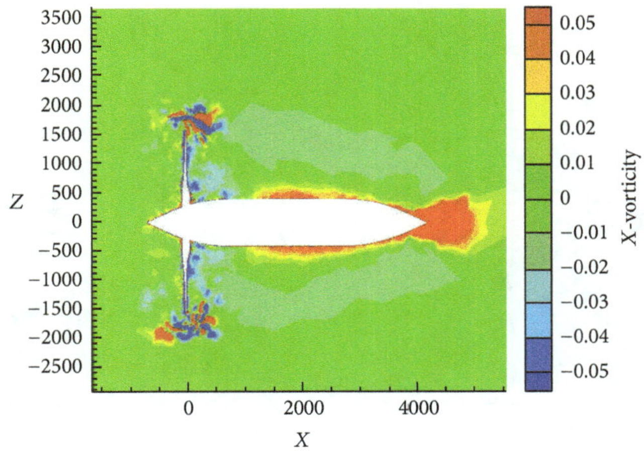

FIGURE 10: X-vorticity in the plane (x, z) for $J = 1.43$.

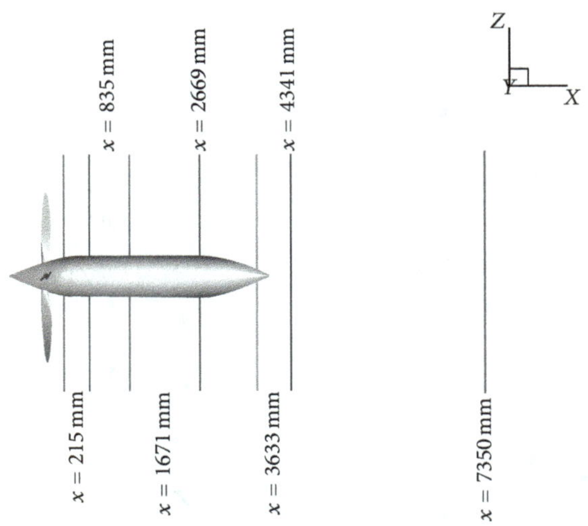

FIGURE 11: Sections for the analysis of the azimuthal velocity profiles.

The investigated operating conditions are identical to the previous case.

The Mach distributions in Figure 15 clearly show the effect of the propeller slipstream that washes the wing. In particular, the impact of the swirl velocity component is very pronounced. It should be noted that the Mach number distributions, much like the pressure distributions, as presented in Figure 16, are affected by both the local propeller induced flow angles and the dynamic pressure increases in the slipstream.

Another phenomenon that is clearly visible, due to the interference between the propeller and the wing, is the rise of vortices around the juncture of spinner and wing (Figure 17). A shedding of these vortices can be evident, which is indicative of the high gradient of spanwise load on the wing.

In fact, the streamwise and spanwise locations of blade vortices are staggered on the upper and lower surfaces of wing and interacted vortices are induced near spinner.

In order to make a further verification of the method the numerical calculations have been compared to experimental data (Figure 18).

TABLE 5: Experimental and computational results for the thrust coefficient.

Advance ratio J	Experimental C_T	Computational C_T
1.43	0.206	0.2
1.5	0.2052	0.196
1.8	0.175	0.18
2.0	0.14	0.135
2.4	0.049	0.054

Again, the computational results are in good agreement with the experiment and the maximum error is around 5% (Table 5). Also in this case the experimental errors are unknown.

6. Conclusions

The Alenia Aermacchi UNS3D code was modified, introducing the capability of flow simulations in a noninertial frame of reference. The modified code was at first applied

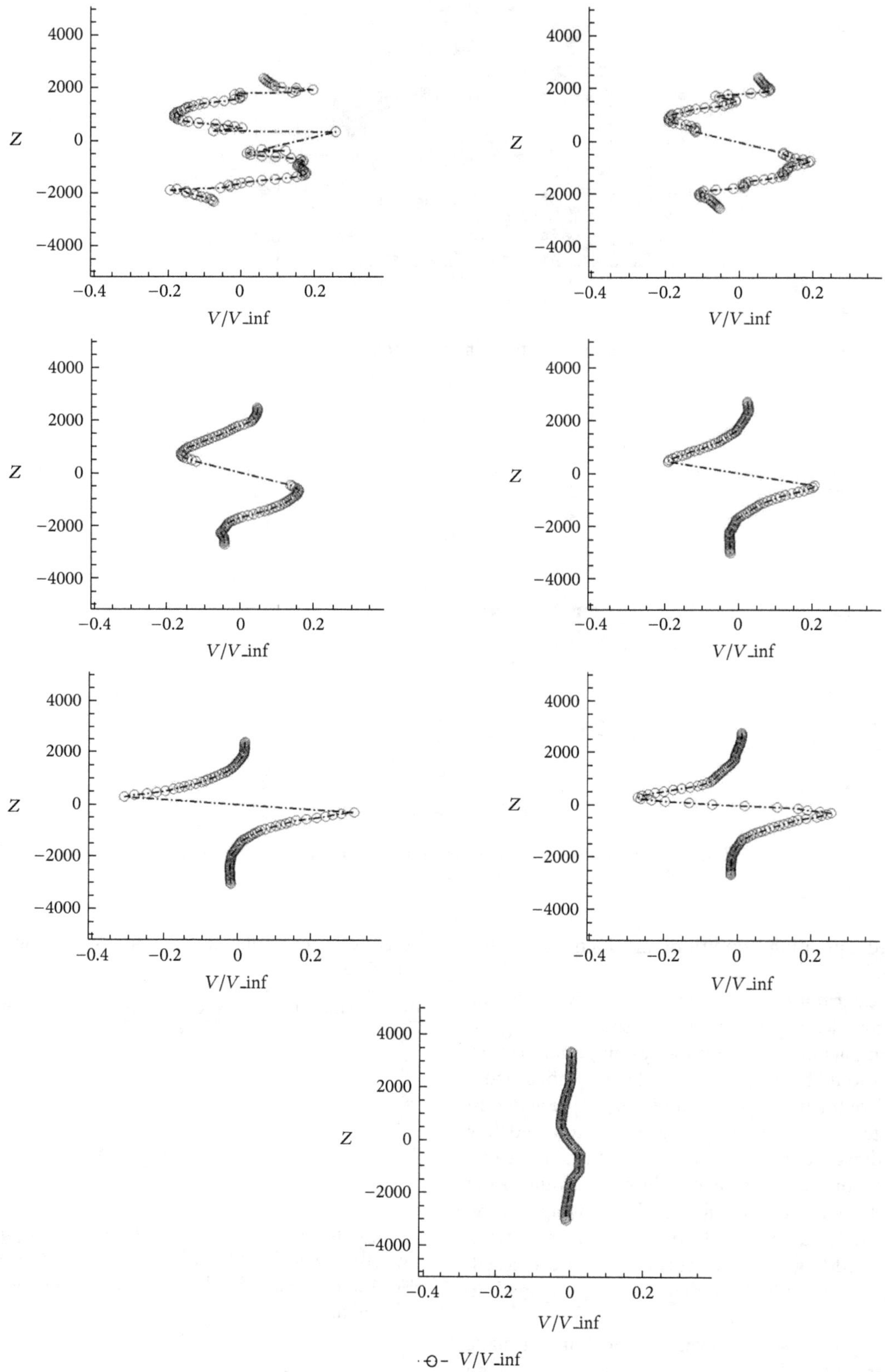

FIGURE 12: Azimuthal velocity profiles downstream the blades at positions of Figure 11 for $J = 1.8$.

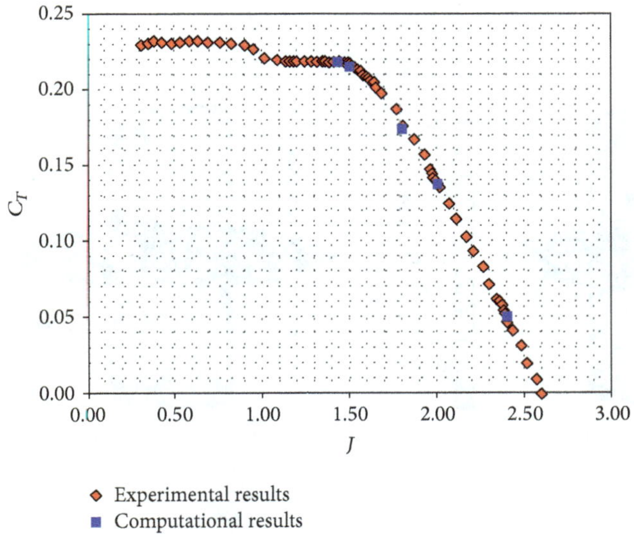

FIGURE 13: Computed versus experimental thrust coefficient.

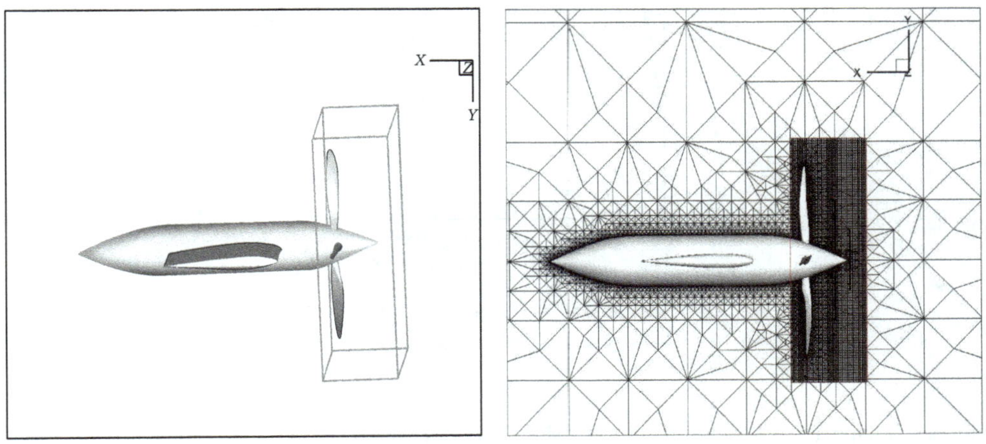

(a) (b)

FIGURE 14: Rotating block within the mesh: (a) propeller and (b) mesh inside the block.

to the computation of damping derivatives of a rotating airfoil and then to the prediction of the performance of a propeller, following the experimental test case described by Biermann and Haetman [15], for different rotational speeds. In the first case, good agreement has been obtained with the numerical results of [7]. In the second, the results are in good agreement with the experimental data within the propeller operating range. The computational results clearly showed the effect of the swirl velocity and the increased total pressure on the spinner and the wing. Therefore, this approach facilitates the identification of typical flow phenomena, like the deformation of the slipstream when passing the wing, being able to model aerodynamic phenomena linked to the propeller-airframe integration.

Nomenclature

b: Generic scalar
$\vec{B}$: Generic vector
c: Airfoil chord
$C_{(\blacksquare)}$: Generic aerodynamic coefficient
C_l: Lift coefficient
C_m: Pitching moment coefficient
C_p: Pressure coefficient
$C_{(\blacksquare)_q}$: Generic stability derivative
C_T: Thrust coefficient
D: Propeller diameter, [m]
E: Total energy per unit of mass, [J/kg]
$\vec{f_e}$: Vector of external forces, [N]
$\vec{F_I}$: Inviscid flux vector
$\vec{F_V}$: Viscous flux vector
J: Advance ratio
M: Mach number
n: Propeller rotational speed, [rps]
p: Pressure, [Pa]
q: Pitch rate

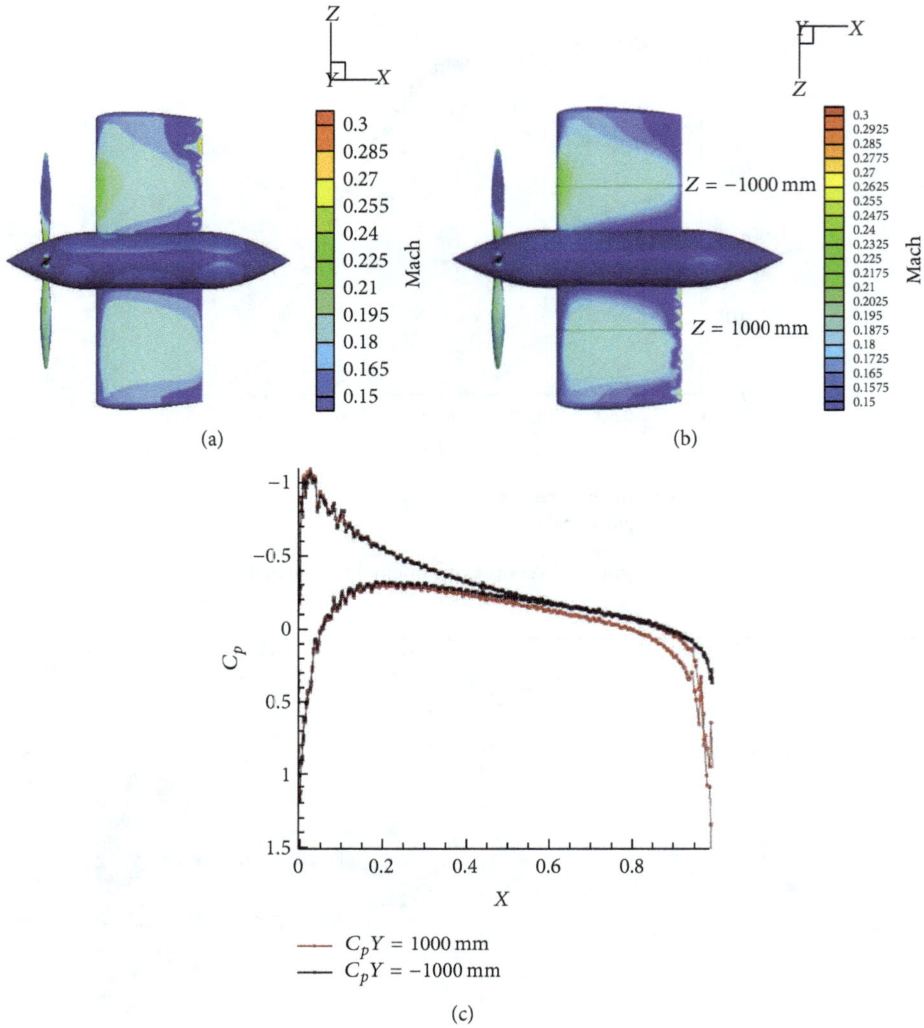

FIGURE 15: Mach number: (a) upper side for $J = 1.8$, (b) lower side for $J = 1.8$, and (c) C_p on surface for $y = 1000$ mm and $y = -1000$ mm.

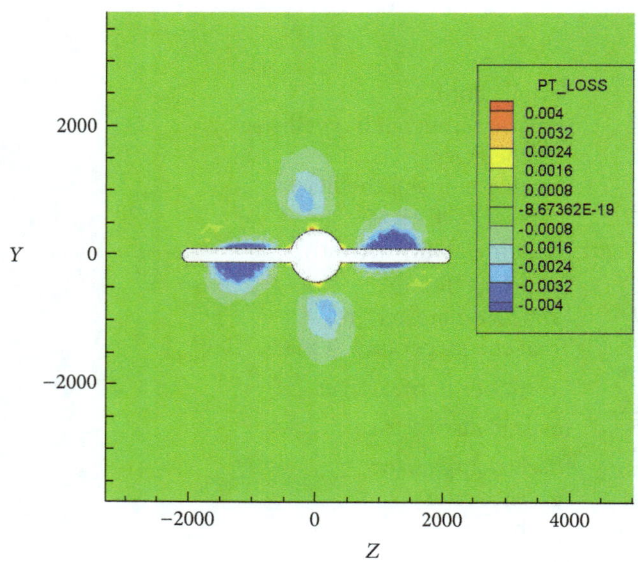

FIGURE 16: Total pressure loss in the plane (y, z) for $J = 1.8$.

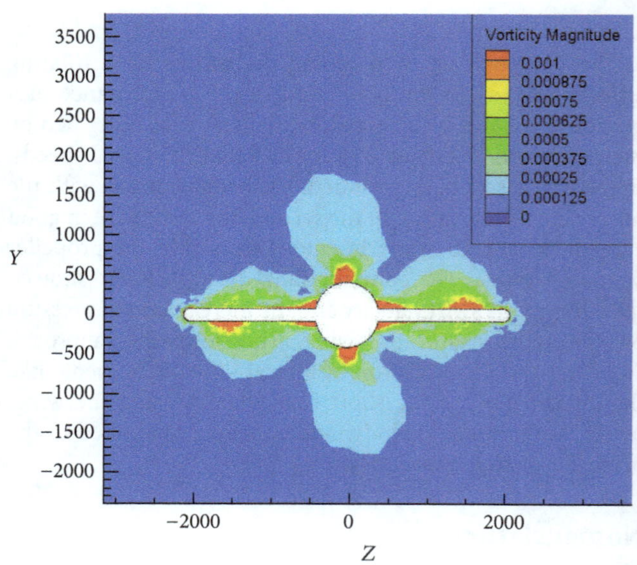

FIGURE 17: Vorticity magnitude in the plane (y, z) for $J = 1.8$.

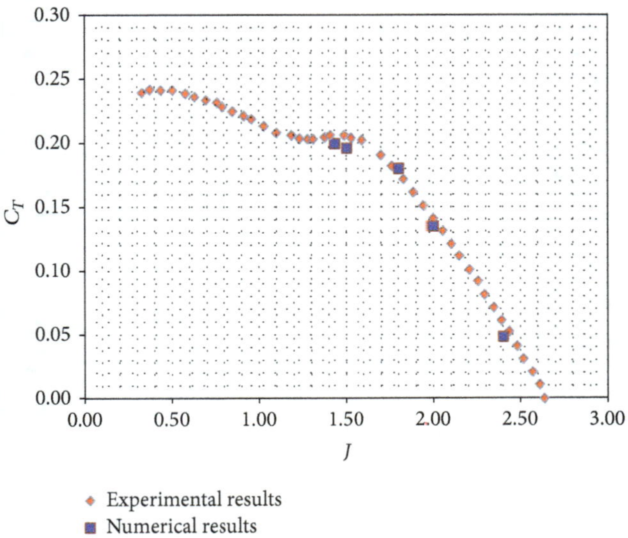

FIGURE 18: Computed versus experimental thrust coefficient.

$\hat{q}$: Normalized pitch rate $(qc/2V)$
q_i: Heat flux, $[\mathrm{W/m^2}]$
$\overrightarrow{Q_X}$: Source vector
$\overrightarrow{r}$: Position vector relative to the rotation center, [m]
t: Time in physical space, [s]
T: Effective thrust, [N]
$u_\infty, v_\infty, w_\infty$: Freestream velocity components, [m/s]
$\overrightarrow{v}, \overrightarrow{V}$: Velocity vector, [m/s]
$\overrightarrow{v_b}$: Rotational speed vector of the coordinate system, [m/s]
v_i: Generic component of the velocity vector, [m/s]
$\overrightarrow{V_\infty}$: Freestream velocity vector, [m/s]
v_1, v_2, v_3: Components of velocity vector, [m/s]
W_f: Work of the external forces, [J]
$\overrightarrow{X}$: Vector of conservative variables
x, y, z: Position vector components relative to the rotation center, [m]
α: Angle of attack
δ_{ij}: Kronecker symbol
ρ: Density, $[\mathrm{kg/m^3}]$
τ_{ij}: Generic component of the shear tensor, [Pa]
$\overrightarrow{\omega}$: Rotational speed vector, [rad/sec]
$\omega_x, \omega_y, \omega_z$: Components of rotational speed vector, [rad/s].

Conflict of Interests

The authors declare that there is no conflict of interests regarding the publication of this paper.

Acknowledgments

The present work has been performed through a close cooperation with Alenia Aermacchi S.p.A. and through the use of Confidential Information and Data, property of Alenia Aermacchi S.p.A. which remains the sole owner of all such relevant IP rights. The results of the present work shall be, therefore, property of Alenia Aermacchi S.p.A.

References

[1] B. Chen and F. Stern, "Computational fluid dynamics of four-quadrant marine-propulsor flow," *Journal of Ship Research*, vol. 43, no. 4, pp. 218–228, 1999.

[2] K. S. Majety, *Solutions to the Navier-Stokes equations in a non-inertial reference frame [M.S. thesis]*, Mississippi State University, 2003.

[3] J. Quilan, J. Eaton, and M. OFlaherty, "Computational investigation of propulsion integration effects on wing propeller installation," in *Proceedings of the 34th Aerospace Sciences Meeting and Exhibit*, AIAA Paper 96-0673, Reno, Nev, USA, January 1996.

[4] R. K. Agarwal and J. E. Deese, "Euler calculations for flowfield of a helicopter rotor in Hover," in *AIM 4th Applied Aerodynamics Conference*, 1987.

[5] M. D. Hathway, R. M. Chriss, J. R. Wood, and A. J. Strazisar, "Experimental and computational investigation of the NASA low-speed centrifugal compressor flow field," in *Proceedings of the 37th ASME International Gas Turbine Conference*, 1993.

[6] A. C. Limache and E. M. Cliff, "Aerodynamic sensitivity theory for rotary stability derivatives," *Journal of Aircraft*, vol. 37, no. 4, pp. 676–683, 2000.

[7] A. C. Limache, *Aerodynamic modeling using computational fluid dynamics and sensitivity equations [Ph.D. thesis]*, Virginia Polytechnic Institute and State University, Blacksburg, Va, USA, 2000.

[8] M. A. Park, L. L. Green, R. C. Montgomery, and D. L. Raney, "Determination of stability and control derivatives using computational fluid dynamics and automatic Differentiation," in *Proceedings of the 17th AIAA Applied Aerodynamics Conference*, AIAA Paper, pp. 99–3136, Norfolk, Va, USA, June 1999.

[9] M. A. Park and L. L. Green, "Steady-state computation of constant rotational rate dynamic stability derivatives," in *Proceedings of the 18th Applied Aerodynamics Conference*, AIAA Paper 20004321, Denver, Colo, USA, 2000.

[10] V. Selmin and L. Formaggia, "Unified construction of finite element and finite volume discretizations for compressible flows," *International Journal for Numerical Methods in Engineering*, vol. 39, no. 1, pp. 1–32, 1996.

[11] L. Djayapertapa, N. Ceresola et al., "Time accurate methods," Tech. Rep. AG38/TP154, GARTEUR AD, 2006.

[12] J. M. Weiss and W. A. Smith, "Preconditioning applied to variable and constant density flows," *AIAA Journal*, vol. 33, no. 11, pp. 2050–2057, 1995.

[13] A. Hellsten, "New advanced k-ω turbulence mode for high-lift aerodynamic," in *Proceedings of the 42nd AIAA Aerospace Sciences Meeting and Exhibit*, Reno, Nev, USA, January 2004.

[14] S. Wallin and A. V. Johansson, "An explicit algebraic Reynolds stress model for incompressible and compressible turbulent flows," *Journal of Fluid Mechanics*, vol. 403, pp. 89–132, 2000.

[15] D. Biermann and E. P. Haetman, "Wind tunnel tests of four- and six-blade single- and dual-rotating tractor propellers," Tech. Rep. 747, NACA, 1942, http://nix.nasa.gov/search.jsp?R=19930091825&qs=N%3D4294129243%2B17%2B4294448544.

16

Numerical Simulation of the Static and Dynamic Aerodynamics of a UAV under Wake Flows

Bai-gang Mi[1] **and Hao Zhan**[2]

[1]School of Aerospace Engineering, Tsinghua University, Beijing 100084, China
[2]School of Aeronautics, Northwestern Polytechnical University, Xi'an, Shaanxi 710072, China

Correspondence should be addressed to Bai-gang Mi; mibaigang@163.com

Academic Editor: Andrea Monteriù

Frequent flight conflicts will be observed as the number of aircrafts increases, and such conflicts will cause unprecedented challenges in flight safety; thus, the flight characteristics of small aircrafts under the wake flow of a large airliner should be thoroughly analyzed. Combined with the sliding mesh technique, a computational fluid dynamics (CFD) method is proposed in this paper to simulate three wake flow patterns, i.e., wingtip vortex, jet flow, and propeller slipstream, and then, the static and dynamic derivatives that represent the stability of the fly wing under the wake flow are identified by using the least squares method. The results demonstrate that both the steady and unsteady aerodynamics of the fly wing are affected by wake flows: wingtip vortices increase the lift-to-drag ratio and considerably change the dynamic damping; jet flow reduces both the static and dynamic damping; and propeller slipstream leads to slow variations in the dynamic damping and decreases in the lift-to-drag ratio.

1. Introduction

Improvements in aeronautical techniques and increased demand have led to an increasing number of flight vehicles in various fields. According to forecasts by Boeing, over 50000 large jets will be required in the coming decades. Thus, air traffic systems will be tasked with addressing this increased congestion, and frequent flight conflicts will place unprecedented pressure on control systems and threaten flight safety. One of the negative phenomena related to flight congestion is the impact of the wake of the front aircraft on the rear aircraft, and this phenomenon has caused several air crashes in the history of aircraft. Moreover, the wake effect [1] should also be considered for military aircraft, such as during aerial refueling, because accurate control of the refueling aircraft in the wake flow of a large aerial tanker is a challenging task. Thus, to ensure flight safety and improve the task completion efficiency, the effect of a wake on the aerodynamics of small aircraft should be analyzed in detail.

The wake flows of large aircraft can be divided into the following aspects: wingtip vortex [2], jet flow [3], propeller slipstream [4], and wing-body turbulence. Wingtip vortices are circular patterns of rotating air that trail behind a wing as it generates lift, and they eventually roll up into large vortices near the wingtip. The jet flow generated by a jet engine could present high temperatures and speeds and thus could have a considerable impact on the rear airplane. The propeller slipstream is highly complex because of the self-induced velocities produced by the propeller vortex system; thus, the slipstream tends to deform and roll up to produce a so-called slipstream tube, which presents strong gradients in various directions. The turbulence of the wing-body is mainly caused by the spanwise flow on the surface of the wing and the body of the aircraft. Among all the wake flow patterns, the aerodynamic impacts of the first three types are the most important and will be analyzed in detail in this paper.

In recent decades, many studies have been conducted to simulate wingtip vortices. Dacles-Mariani [5] used the one-equation Baldwin-Barth turbulence model to calculate the behavior of wingtip vortex flow. The results indicated that the model could describe the basic aerodynamic characteristics but could not accurately capture the detailed structures of the vortex. Li [6] adopted a Large Eddy Simulation with a filtered-structure-function subgrid model to simulate the wingtip vortex of a rectangular half-wing with a NACA 0012 airfoil section and a rounded wingtip. The calculations demonstrated that this model was much more accurate than traditional Reynolds Average Navier-Stokes (RANS) methods and could better reveal the detailed characteristics of vortices. Matthew [7] investigated the accuracy of current state-of-the-art turbulence models and a compressible RANS solver model at simulating the formation of a wingtip vortex in the near field, and he found that these models predicted the mean flow accurately; however, none of them could accurately capture the magnitudes of the turbulence quantities or the lag of the Reynolds stress components underlying the corresponding strain-rate components. Moreover, many scholars have also studied the influence of wingtip vortices in formation flight. Inasawa [8] investigated the static aerodynamic interactions of wingtip vortices in close-formation of two wings that consist of the NACA 23012 airfoil section. The results showed that the tip vortices of the leading and trail wings interacted closely with each other to form a pair of counterrotating vortices for smaller overlap conditions of less than 5% with zero vertical offset. Kasper [9] had used both experimental and computational approaches to analyze in detail the upwash generated from the wingtip vortices of the formation flight for UAVs. The results revealed that formation flying can benefit flight performance in many ways and that altering the formation can lead to different benefits depending on the configuration.

A series of studies have also been conducted on the jet flow. Rudnik [10] evaluated the influence of increasing the bypass ratio and varying the position on the ALVAST narrow body wing/fuselage model using a numerical method. The investigations showed that the lift loss is proportional to the engine size and alleviated by the upstream movement of the engine position. Ritter [11] used the RANS-CFD method to investigate the strong impacts of Ultra-High-Bypass-Ratio (UHBR)-Engines on the aerodynamics of high-lift wings. The results of the study mainly focused on the effects of engine position on the achievable maximum lift coefficient and lift-to-drag ratio to identify an optimal engine position related to high-lift performance. Melber-Wilkending [12] considered the numerical solutions of the aerodynamics of two different configurations with an open and a closed thrust-gate by using RANS equations. The results indicated that, with an increased angle of attack, a flattening of the jet hull occurred at the upper side, and an increase in sidewise spreading was detected. The closed thrust-gate caused a more distinctive effect, with the inboard flap being partially passed by the engine jet.

Studies on propeller slipstream have been conducted for a long time. Schroijen [13] developed a tool to estimate the propeller slipstream effects on directional control in the critical case of a failed engine. Based on vortex theory, this simple model can quickly evaluate the sign and magnitude of the slipstream effects on the engine-out yawing moment. Moens [14] presented two numerical studies of the propeller/wing interactions for transport aircraft. The first one used the Euler conditions to compute a three-element high-lift section in take-off configuration and analyzed the details of the complex slipstream flow interactions on the flap system, and the second case evaluated the global performance of a four-propeller generic transport aircraft with a N-S solver and correctly predicted the blowing effect. Similar work was performed by Lenfers [15] and Xu [16], who used a computational method to study the propeller and high-lift wing interaction with different RANS and grid forms. Khan [17] presented a slipstream model that considered both the acceleration and diffusion phenomena via simple analytical and semiempirical equations to accurately predict the induced velocity at up to 8-10 propeller diameters downstream of the propeller plane.

Although recent research on wingtip vortices, jet flow and propeller slipstream mainly focused on the impacts of these phenomena on the airplane itself, few works have investigated the wake flow effects on the static and dynamic aerodynamics in aircraft formation.

The present study proposes a method that combines high-fidelity computational fluid dynamics (CFD) and a sliding mesh technique to assess the static and dynamic aerodynamics of the fly wing configuration [18, 19] under the effects of wingtip vortices, jet flow, and propeller slipstream, and the corresponding derivatives are also identified to evaluate the stability of the aircraft. The methods and conclusions obtained in this paper can provide detailed explanations of the aerodynamic characteristics of aircraft in wake flows and additional technical support for flight control system design.

2. Numerical Method

2.1. Governing Equation. The 3D Reynolds average N-S equation used to describe the flow field is expressed as

$$\frac{\partial}{\partial t}\iiint_V \mathbf{W} dV + \iint_{\partial V} \mathbf{F} \cdot \mathbf{n} ds = \frac{1}{\mathrm{Re}} \iint_{\partial V} \mathbf{F_v} \cdot \mathbf{n} ds \quad (1)$$

where V indicates the control volume, $\mathbf{W}$ denotes the vector of conserved variables, $\mathbf{F}$ is the inviscid component of the flux vectors, $\mathbf{F_v}$ represents the viscous flux vector, and Re is the Reynolds number.

The two-equation turbulence model called the shear-stress transport (SST) $k-\omega$ model is used in this paper to calculate the cases. Developed by Menter [20], this model takes the transport of the turbulent shear stress into the Wilcox $k-\omega$ model to give highly accurate predictions of the onset and the amount of flow separation under adverse pressure gradients. A sliding mesh technique is adopted to simulate the dynamic motion in all the unsteady cases. The sliding mesh is a special case of a general dynamic mesh motion in which all the cells move rigidly in a specific dynamic zone. This technique employs two or more cell zones, and each cell zone is bounded by at least one interface zone, where it meets the opposing cell zone. The interface zones of adjacent cell zones are associated with one another to form a mesh interface. During the unsteady calculation, the cell zones slide relative to one another along the interface in discrete steps.

2.2. Static and Dynamic Derivative Identification. Static derivatives are generally obtained by traditional methods via interpolations of steady aerodynamic forces and moments, whereas dynamic derivatives are calculated with the single-point or integration method by imposing a forced sinusoidal motion around the aircraft center of gravity. However, the artificial separation of static and dynamic derivative identifications may lead to substantial errors, especially in cases with nonlinear characteristics.

In this paper, we use a new method to simultaneously compute static and dynamic derivatives [21, 22] that is based on the aerodynamics model and incorporates unsteady aerodynamic forces and moments of forced sinusoidal motion.

The unsteady aerodynamic model of forced motion at small angles of attack (AoAs) can be expressed as

$$C_i = C_{i0} + C_{i\alpha}\Delta\alpha + C_{i\dot{\alpha}}\hat{\dot{\alpha}} + o(\alpha) \quad (2)$$

where $i = L, D, M$ (indicating the lift, drag, and moment, respectively); C_i represents the unsteady force or moment coefficient; $C_{i\alpha}$ denotes the static derivatives and is used to evaluate the trend that the aircraft moves back to original balanced state with restoring moments; $C_{i\dot{\alpha}}$ is the dynamic derivative focused on the time response process with disturbance; and $\hat{\dot{\alpha}} = \dot{\alpha}c/2V$ denotes the nondimensional rate of the AoA.

When discarding the high-order terms $o(\alpha)$, the frequently used aerodynamic derivatives model is given as

$$C_i = C_{i0} + C_{i\alpha}\Delta\alpha + C_{i\dot{\alpha}}\hat{\dot{\alpha}} \quad (3)$$

Because the rate of the AoA $\dot{\alpha}$ and angular velocity q have the same form when the freestream remains unchanged, they are difficult to separate and are often denoted as

$$C_{i\dot{\alpha}} = C_{i\dot{\alpha}} + C_{iq} \quad (4)$$

This term is referred to as the combined dynamic derivatives.

The model is fit to most cases at a small AoA with linear or weak nonlinear aerodynamic characteristics. Moreover, we use the least squares method to identify the static and dynamic derivatives.

2.3. Method Validation with SACCON Configuration. A generic UCAV called SACCON (Stability and Control CONfiguration), which is used by an Applied Vehicle Technology Task Group (AVT-161) established by the NATO Research and Technology Organization (RTO), is adopted to verify the method, and the relevant static and dynamic aerodynamics are measured in the Low Speed Wind Tunnel Braunschweig (DNW-NWB) [23].

The model contains a lambda wing with a leading-edge sweep angle of 53°, as shown in Figure 1. The root chord is approximately 1 m, the wing span is 1.53 m, the mean aerodynamic chord is 0.479 m, and the reference wing area is 0.77 m^2. The moment reference point (MRP) is located 0.6 m from the head of the wing, and the point of rotation (POR) for dynamic testing is 0.255414 m behind the MRP.

We firstly evaluate the static cases of the single fly wing to understand its aerodynamic performance. Comparisons of lift and pitching moment coefficient are, respectively, shown

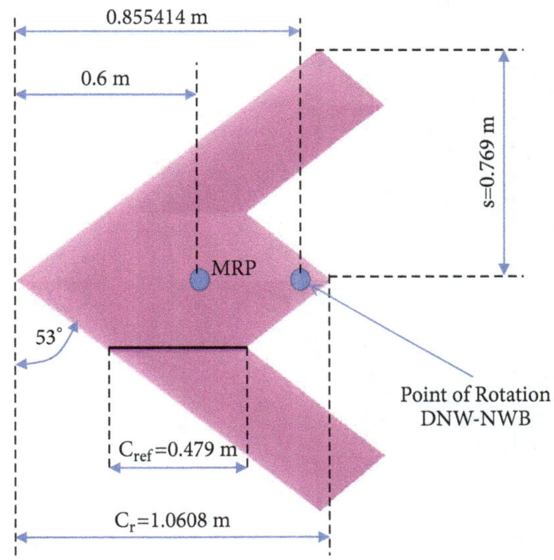

FIGURE 1: Geometric parameters of the SACCON UCAV configuration.

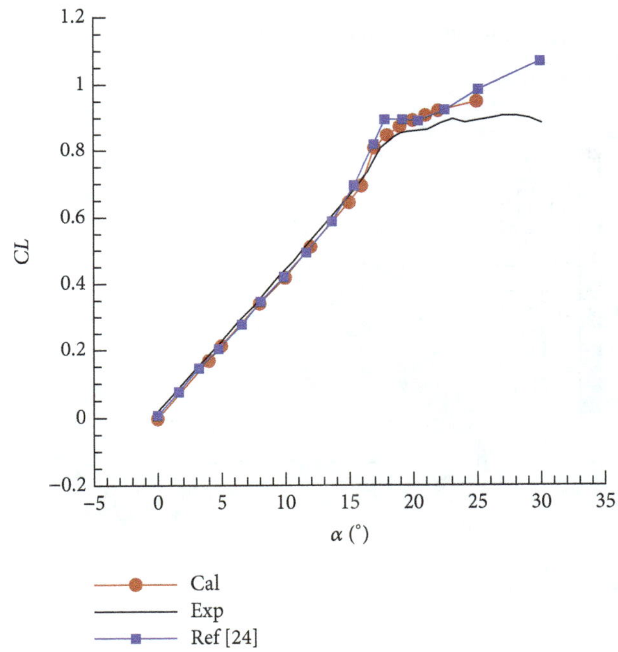

FIGURE 2: Lift coefficients.

in Figures 2 and 3. The lift coefficients have good agreement with the wind tunnel data or the calculating data in [24] below 17 degrees, while both the lift coefficients calculation and reference data diverge when ranging up to larger angle of attack. Therefore, none of the simulations well match the test data. The changing process of pitching moment coefficients are more complicated with the increment of the angle of attack. First of all, both the calculation and reference data have bad agreement with the experiment results when the angle of attack is below nearly 15 degrees, including the magnitude and slope of profiles. The possible reason may be that the sting and wind tunnel wall effects are not considered

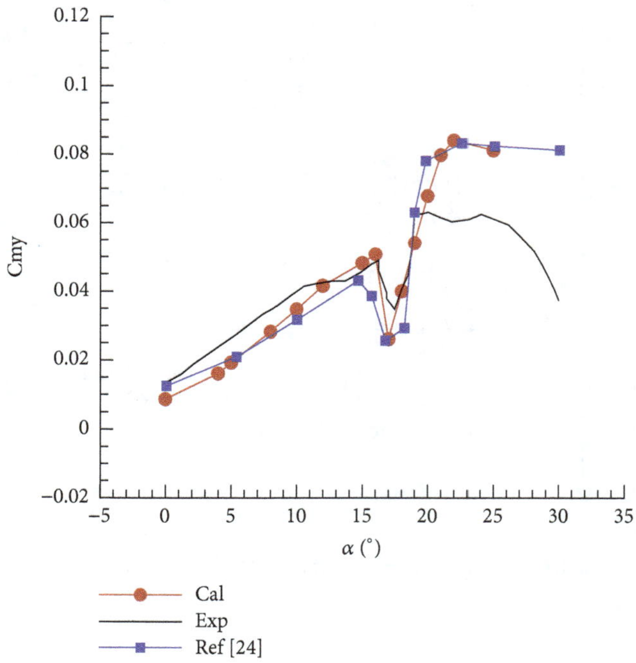

FIGURE 3: Pitching moment coefficients.

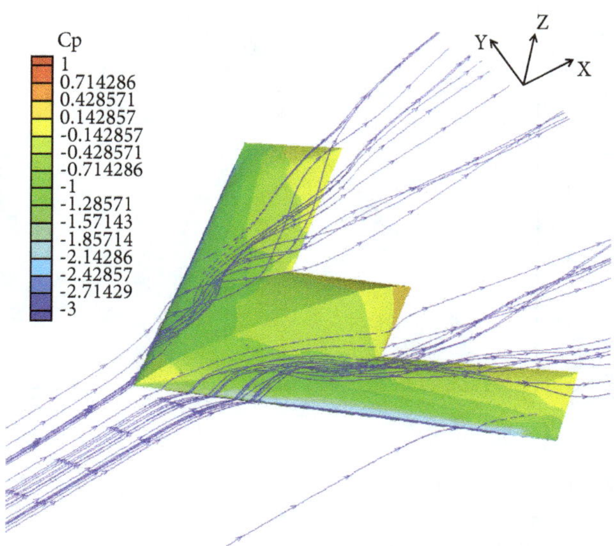

FIGURE 4: Leading-edge vortex at AoA=15°.

in the simulation. The predictions become better when the angle of attack increases further. All the methods capture the sharp reduction in nose-up pitching moment, with a sharp increase afterwards in positive moment between 17 and 20 degrees. This indicates that all the methods can well describe the basic vortex variations during the static process, shown in Figure 4.

Then we calculate the dynamic motion with sliding mesh technique to identify the stability derivatives. The validation case of a single fly wing without a sting is considered using the same conditions as those of the wind tunnel test (Table 1). Based on the method in this paper, the static and dynamic

TABLE 1: Simulation conditions.

Parameters	Values
Mach number *Ma*	0.146
Reynolds number based on reference chord c	1.13×10^6
Initial AoA α_0	0°
Reduced frequency *k*	0.090289397
Additional angle of attack $\Delta\alpha$	$1^\circ \sin(6\pi t)$

TABLE 2: Calculation results of static and dynamic derivatives.

Stability derivatives	Calculation	Wind tunnel test	Error
Static derivative $C_{z\alpha}$	1.49487	1.3752	8.7%
Dynamic derivative $C_{m\dot{\alpha}} + C_{mq}$	-0.76599	-0.82	6.59%

derivatives are calculated and shown in Table 2, and they are consistent with the wind tunnel test data [24].

3. Computational Models and Grids

Three wake flow types are considered here: wingtip vortices, jet flow, and propeller slipstream. The following computational models (Figure 5) are used: a combined airliner and fly wing model is used to simulate the effect of wingtip vortices; the power-on wing-body and fly wing models are used to calculate the jet flow; and a wing-body model with a propeller is applied to evaluate the interference on the fly wing. Except for the spanwise and chordwise locations, the fly wings have the same flight attitude with the large aircrafts ahead in three models. Moreover, the sizes of the fly wings have been adjusted to match the large aircrafts. Compared to the standard SACCON configuration, the spans for wingtip vortex and jet flow simulations are extended to 7*m*, while the value is changed to 4.5*m* due to the relatively small size of the propeller aircraft. It is noted that the three models in this paper are just selected to analyze the static and dynamic aerodynamics of a small UAV under the wake flows of large aircrafts, they may be something different from actual cases, but it is similar in methods and results.

The commercial software ANSYS ICEM CFD (Pittsburgh, Pennsylvania, USA) is used to generate the computational mesh. Based on the sliding mesh technique, the flow field can be divided into two zones: the inner dynamic zone with the fly wing and the outer static zone with the civil aircraft. An interface is set up to connect the two zones. The unstructured computational grids are shown in Figure 6. The mesh cells are clustered around the leading edges. As the reason that the wake flows are difficult to capture due to numerical dissipation of CFD methods, we set a mesh refined zone to improve the accuracy of solutions, seen in Figure 5, where adequate volume meshes have been enhanced to describe the true flow fields.

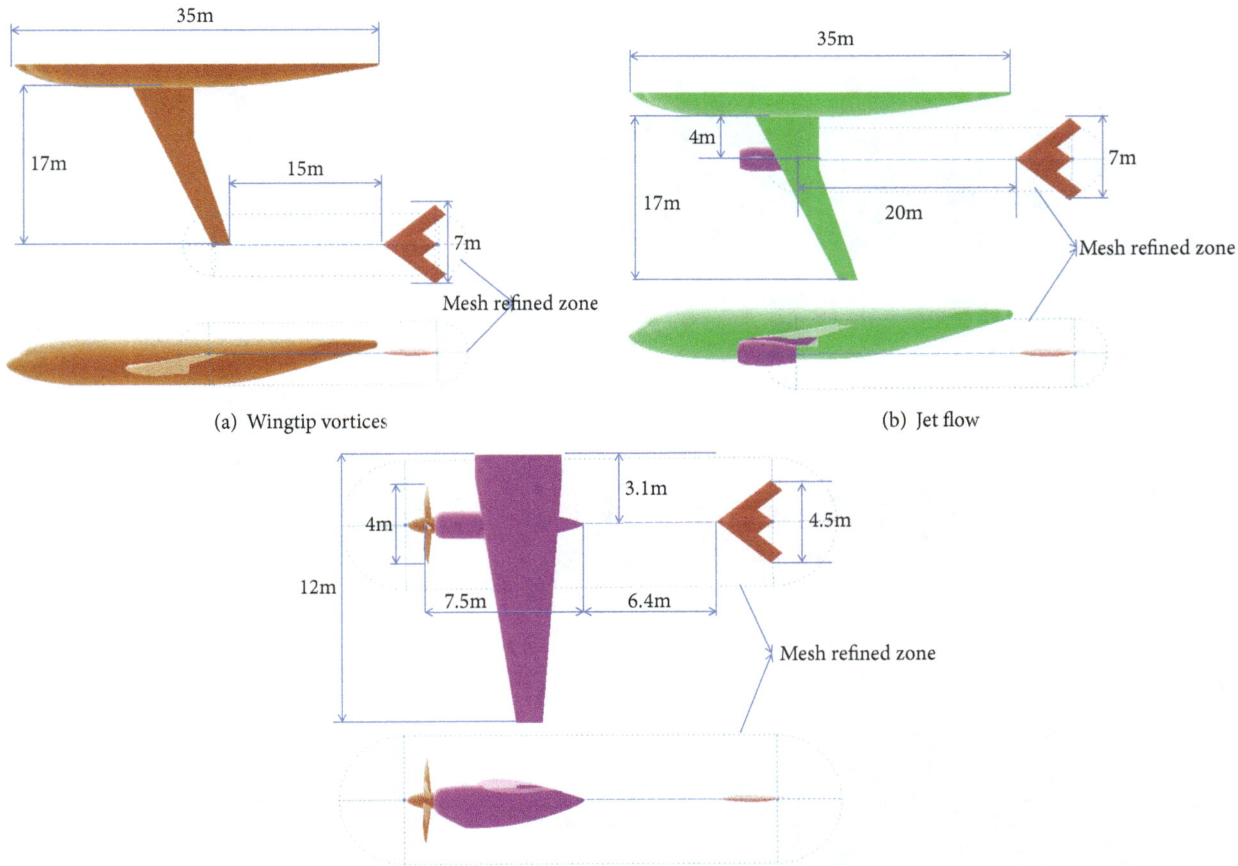

FIGURE 5: Computational models.

4. Calculation Results

Using the method discussed earlier in this paper, we initially perform steady case simulations, and these results can be used to further calculate unsteady motions. The static and dynamic derivatives can be then identified via the unsteady aerodynamic forces and moments.

4.1. Effect of Wingtip Vortices. For a large airliner, the wingtip vortices are apparently strong and cause a wide range of effects. To improve the accuracy and reduce the numerical dissipation, the computational grids around and behind the wingtip were refined. We choose the representative cruise mode with a Mach number of 0.7 and altitude of 11 *km* as the simulation conditions, and the attitude of the airliner remains unchanged while the fly wing oscillates around its center of gravity with three different AoAs: 0, 5, and 10 degrees.

Figure 7 shows the vorticity values of the airliner's wingtip vortex, which indicates that the wingtip can extend a long distance to affect the aerodynamic characteristics of the flight vehicles under this flow. Figures 8 and 9 display a comparison of the steady aerodynamics between the wingtip vortices and the single fly wing cases. The results show that both the lift and drag coefficients have been enhanced under increases of the AoA, with slightly greater increases in the lift, which indicates that the lift-to-drag ratio can be improved via increases in the AoA. These results explain why the formation of fly wings is always used during the actual flight. Moreover, the pitching and yawing moment coefficients do not show significant changes compared with the single fly wing cases. However, because of the shearing velocity of the wingtip vortices, unbalanced forces develop on two sides of the fly wing and cause positive rolling moments to appear.

The static and dynamic derivatives obtained by using the least squares method are shown in Figures 10 and 11. Because the steady pitching moments remain unchanged at small AoAs, the static derivatives are nearly the same as those of single fly wing cases, indicating that the static stabilities are weakly affected by the wingtip vortices of the airliner. However, the dynamic derivatives are heavily affected by this wake flow, and although the values are negative and the fly wing still exhibits dynamic damping, the vortices cause greater damping as well as local dynamic stabilities at 0° and 5° AoAs compared with normal cases. When the AoA increases to 10°, the effect is weakened via the small flow separation on surface of the fly wing.

4.2. Effect of Jet Flow. To simulate the inner and outer flows in a real flight, the effect of engine power must be considered. A jet flow with high temperatures and speeds affects not

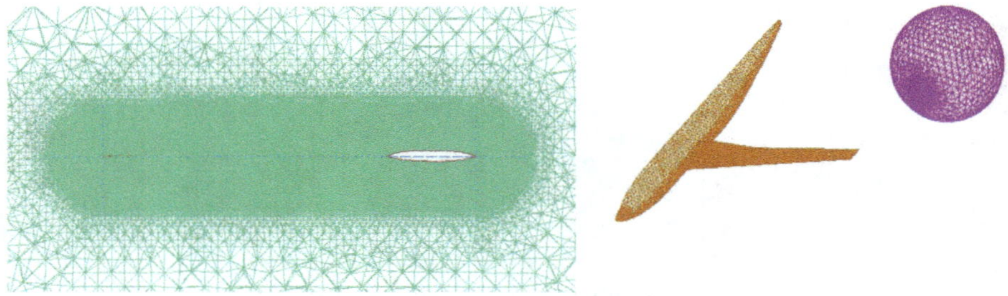

(a) Wingtip vortices

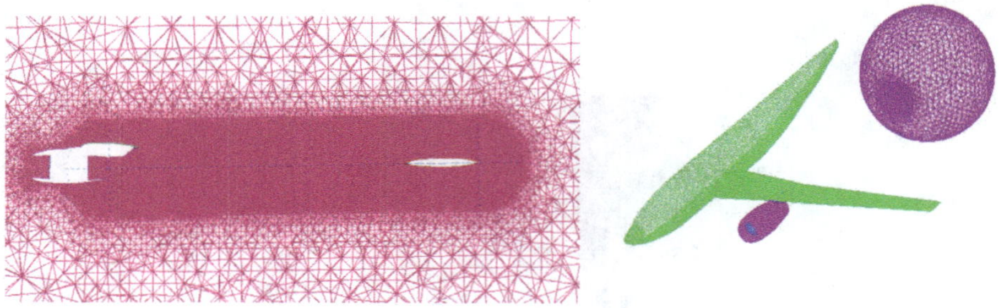

(b) Jet flow

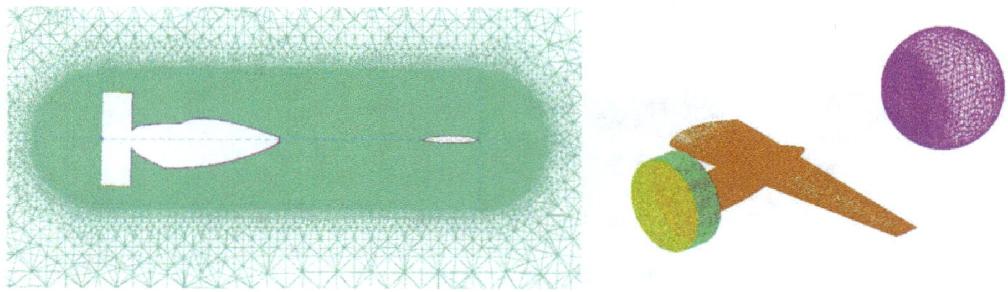

(c) Propeller slipstream

FIGURE 6: Computational grids.

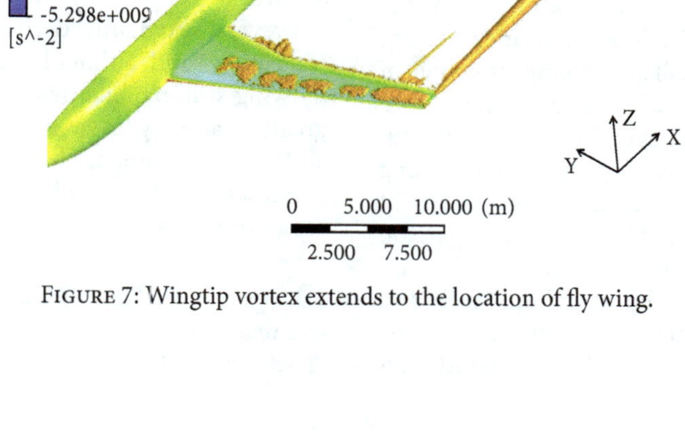

FIGURE 7: Wingtip vortex extends to the location of fly wing.

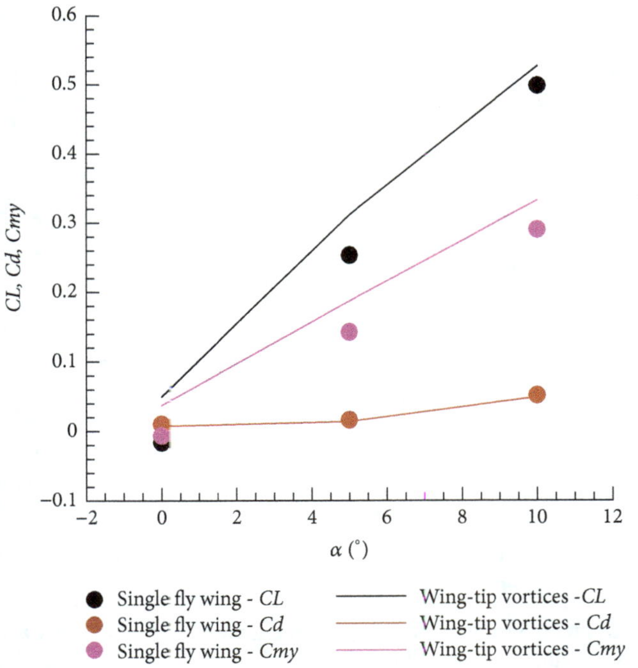

FIGURE 8: Steady aerodynamics-lift coefficients *Cl*, drag coefficients *Cd*, and pitching moment coefficients *Cmy*.

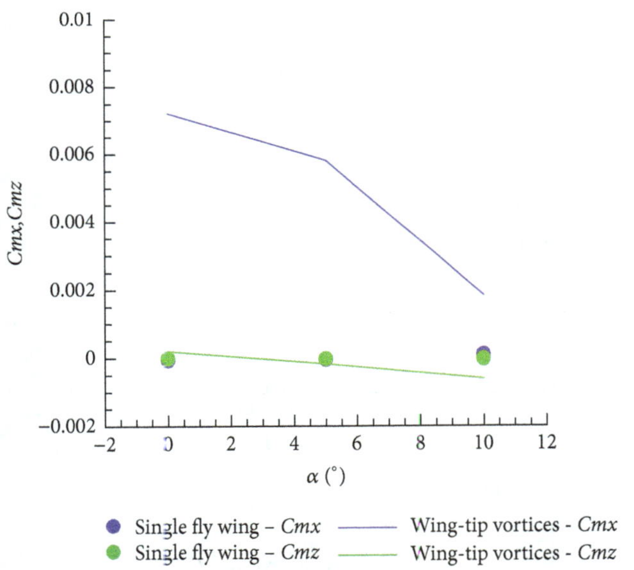

FIGURE 9: Steady aerodynamics, rolling moment coefficients *Cmx* and yawing moment coefficients *Cmz*.

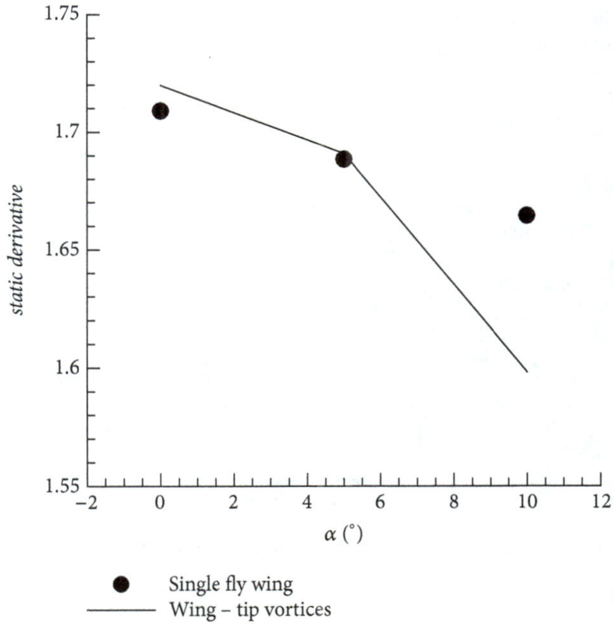

FIGURE 10: Static derivatives $C_{z\alpha}$.

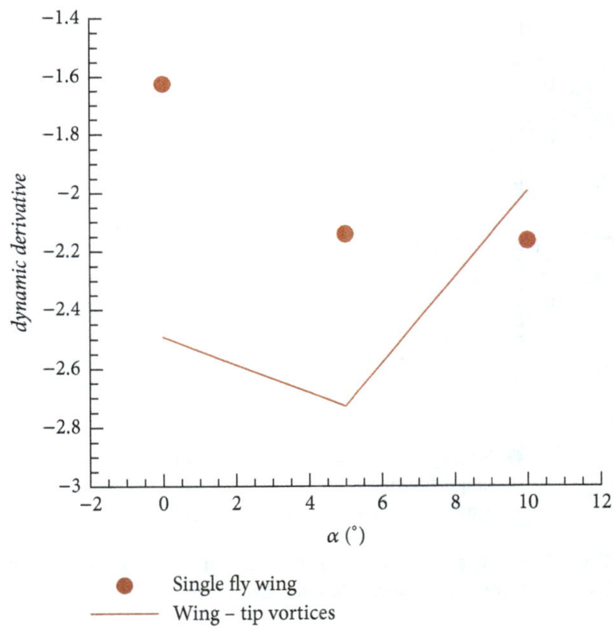

FIGURE 11: Dynamic derivatives $C_{m\dot{\alpha}} + C_{mq}$.

only the aerodynamic and structure characteristics of the flap (mainly refer to the wing flap close to the jet nozzle) but also the flight vehicle behind the flow. We adopt the numerical method and include the inlet and outlet boundary conditions instead of the real engine model to simulate the influence of engine power because this method has been shown to accurately solve this problem. The simulation conditions are the same as those of the wingtip vortices. We still choose the representative cruise mode with a Mach number of 0.7 and altitude of 11 *km*. The attitude of the airliner remains unchanged while the fly wing oscillates around its center of gravity at AoAs of 0°, 5°, and 10°. Unsteady motion can be expressed as $\Delta\alpha = 1^{\circ}\sin(8.6269311t)$ with a corresponding reduced frequency of $k = 0.05$.

Figure 12 displays the flow flied of the airliner with power-on engine, the high temperature and speed jet flow can still downward to the fly wing even it is 20*m* away behind the nozzle, therefore, the local flow flied around the fly wing is changed. The steady aerodynamic coefficients of forces and moments are presented in Figures 13 and 14. Compared with the shear effect of wingtip vortices, the jet flow changes the

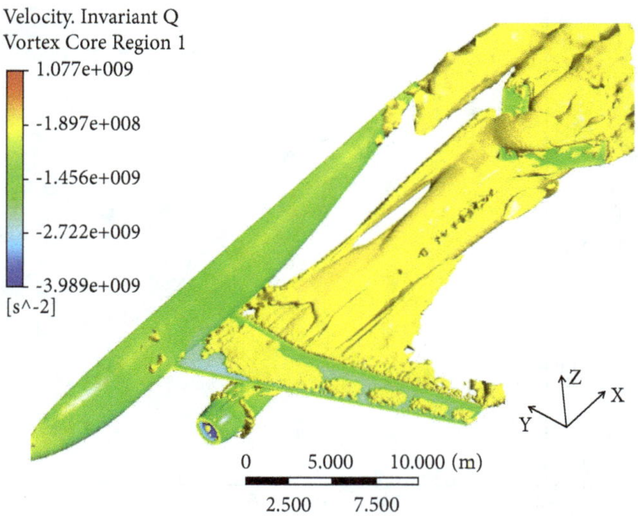

FIGURE 12: Influence of the jet flow on the fly wing.

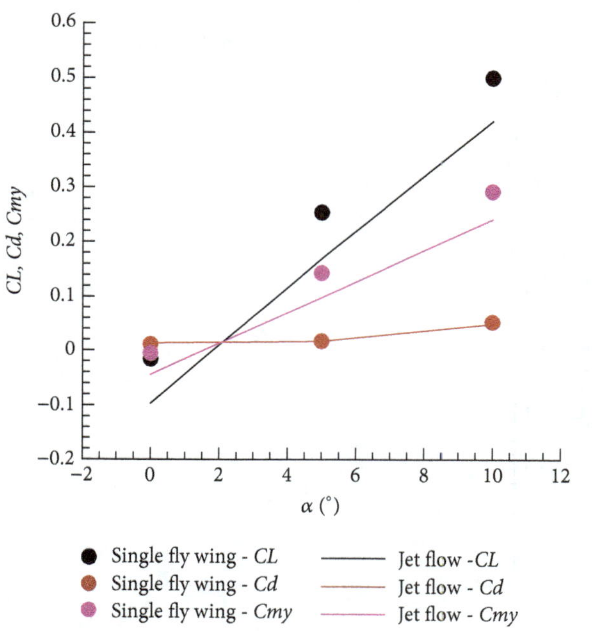

FIGURE 13: Steady aerodynamics-lift coefficients *Cl*, drag coefficients *Cd*, and pitching moment coefficients *Cmy*.

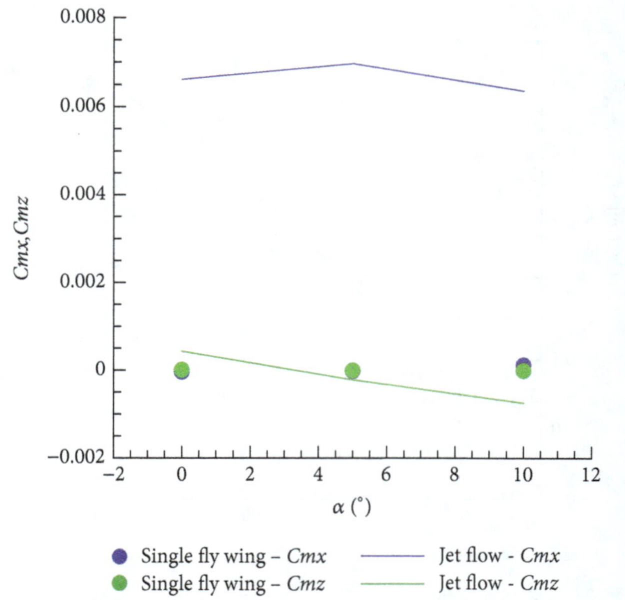

FIGURE 14: Steady aerodynamics, rolling moment coefficients *Cmx* and yawing moment coefficients *Cmz*.

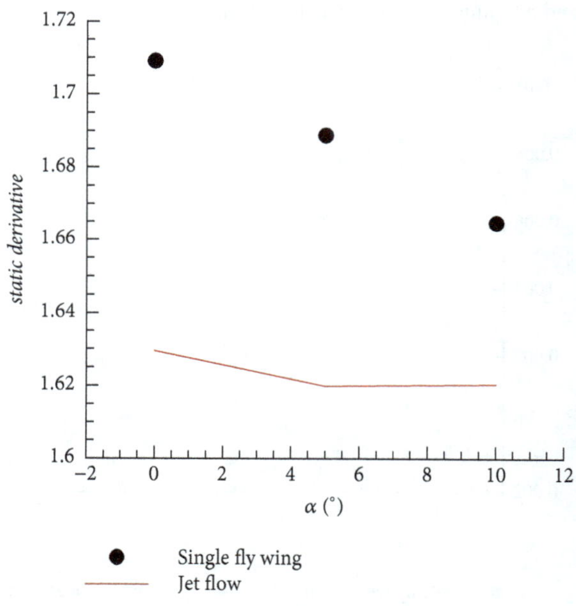

FIGURE 15: Static derivatives $C_{z\alpha}$.

local velocity and direction of the flow around the fly wing, which leads to a decrease in drag and a greater decrease in lift. Thus, the lift-to-drag ratios are reduced at all AoAs. Similar to the effect of wingtip vortices, the pitching and yawing characteristics do not show considerable differences compared with normal cases, whereas the asymmetric aerodynamics on two sides of the fly wing cause a huge change in the rolling moment. However, note that the change in the rolling moment is closely related to the position of the fly wing behind the engine.

Figures 15 and 16 show the static and dynamic derivatives of the fly wing, respectively. Compared with normal flight, the jet flow has an obvious effect on the static and dynamic characteristics. The static stability remains weakly unstable because of the smaller positive values of derivatives, and the absolute values of the dynamic derivatives with increments of the AoA are less than those of the single fly wing cases, indicating that the dynamic damping is diminished with the jet flow. All the changes of the static and dynamic characteristics should be determined in detail, and the attitude of the fly wing should be adjusted in real time.

4.3. Effect of Propeller Slipstream. The propeller of a plane has its own characteristics, and the simulation conditions are different from those of the previous calculations. The

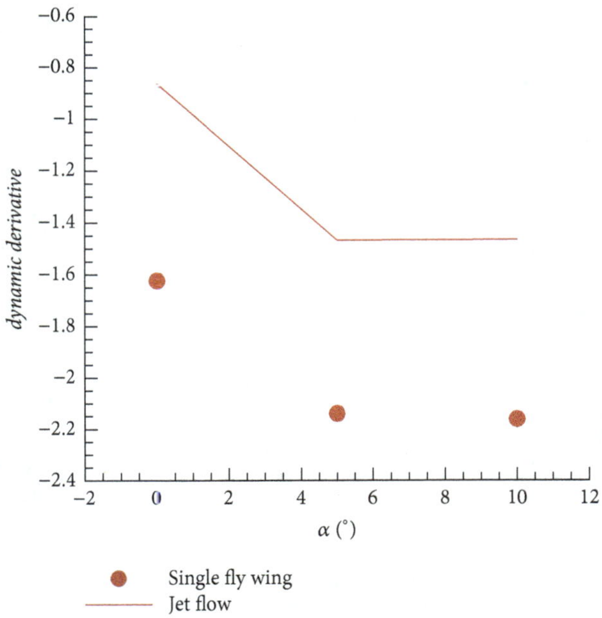

FIGURE 16: Dynamic derivatives $C_{m\dot{\alpha}} + C_{mq}$.

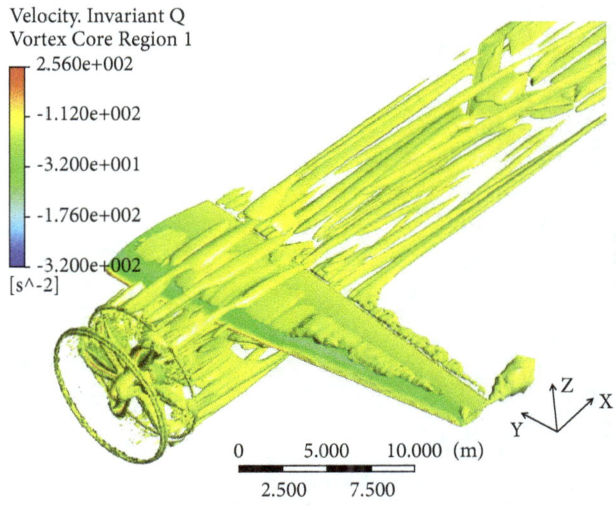

FIGURE 18: Propeller slipstream effect on the fly wing.

FIGURE 17: The counterclockwise rotation of the propeller.

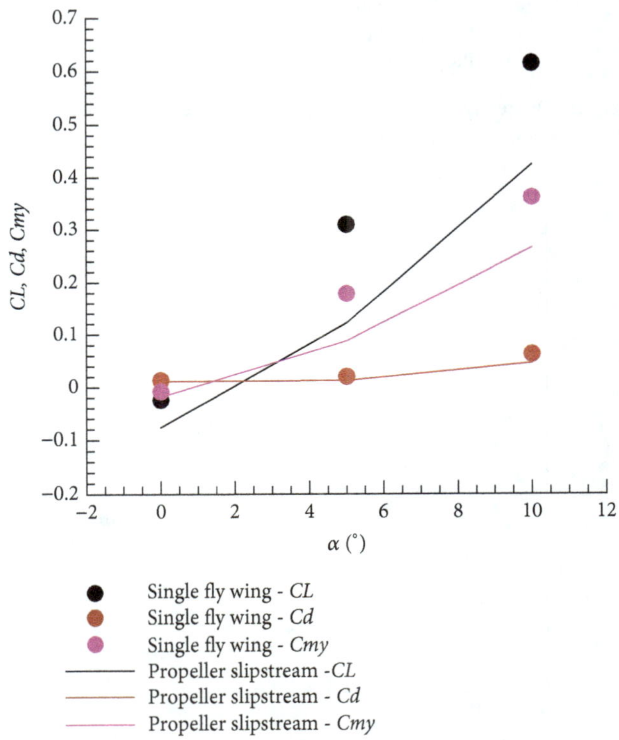

FIGURE 19: Steady aerodynamics-lift coefficients *Cl*, drag coefficients *Cd*, and pitching moment coefficients *Cmy*.

Mach number of the freestream is assumed to be Ma=0.55, the cruising altitude is 8 km, and the unsteady motion is described as $\Delta\alpha = 1^\circ \sin(7.075595t)$ at the three AoAs considered. The counterclockwise rotation speed of the propeller with 6 blades is n_s=1080 rev/min (Figure 17), and the sliding mesh is used to simulate the unsteady motion of the propeller.

The propeller slipstream is much more serious than the wingtip vortices and the jet flow because the long range of the flow region with high speed air rotation behind the propeller substantially affects the aerodynamics of wingman flight. Figure 18 describes the vorticity of the propeller slipstream. As the interruption of the wing and the engine nacelle, the slipstream is weakened but still moves to downstream and affects the aerodynamics of the fly wing. Figures 19 and 20 present the steady forces and moments of the fly wing. The corresponding results indicate that the lift and drag as well as the lift-to-drag ratio decrease at these AoAs, whereas the pitching and yawing moments have no significant difference relative to those of normal flight. Furthermore, the rolling moment suffers heavy interference from the synthesis speed of the rotating flow.

The static and dynamic derivatives are shown in Figures 21 and 22. The static derivatives at these three AoAs remain positive and the values become larger than those of the case

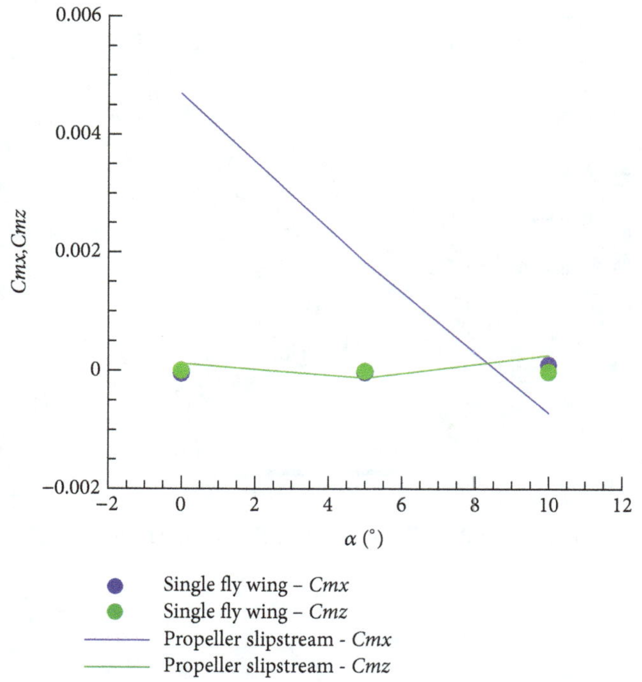

FIGURE 20: Steady aerodynamics, rolling moment coefficients *Cmx* and yawing moment coefficients *Cmz*.

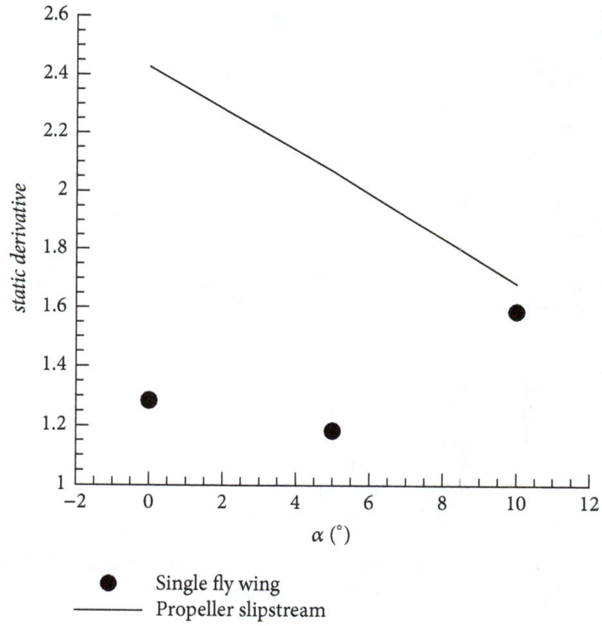

FIGURE 21: Static derivatives $C_{z\alpha}$.

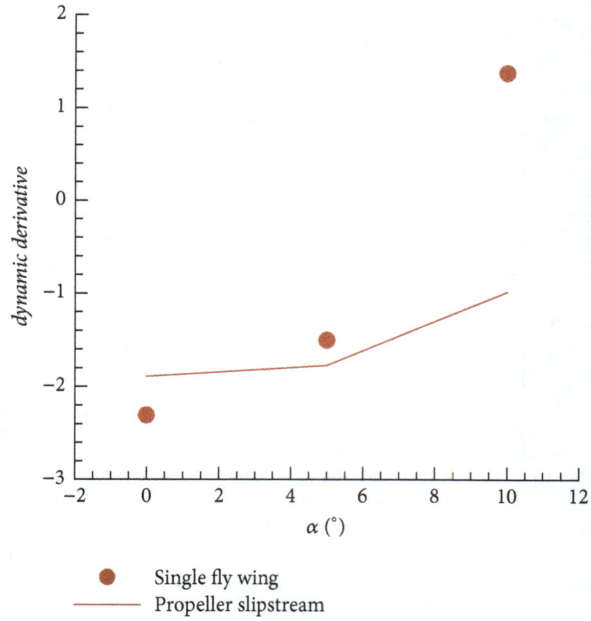

FIGURE 22: Dynamic derivatives $C_{m\dot{\alpha}} + C_{mq}$.

of no wake flow. However, the trend of the static stability appears to be fairly different. The dynamic characteristics are much more complex, although the trend is the same as that observed for the normal cases. Because of the effect of propeller slipstream, the values of the dynamic derivatives are negative, indicating that the unsteady motions are always accompanied by dynamic damping. The absolute values decrease as the AoA increases because the flow begins to separate from the leading edge of the fly wing at high AoAs and the dynamic aerodynamics before and after the aerodynamic center are substantially changed. Compared with a single fly wing, the propeller inhibits the change in dynamic stability.

5. Conclusions

In this paper, a novel CFD method based on a sliding mesh technique was proposed to calculate the static and dynamic aerodynamics of three wake flow patterns, and the static and dynamic stability derivatives were identified by using the least squares method. The conclusions can be summarized as follows:

(1) The least squares method can accurately obtain the static and dynamic derivatives.

(2) Wingtip vortices have a strong effect on the fly wing behind the airliner, which enhances the lift-to-drag ratio and rolling moment. The static stabilities in the presence of wingtip vortices have no apparent difference compared with those of the normal flight, whereas the dynamic damping is heavily affected by the wingtip vortices.

(3) The jet flow changes the local velocity and direction of the flow around the fly wing; thus, the lift-to-drag ratio is reduced at all the evaluated AoAs, and the rolling moments are caused by the asymmetric aerodynamics on two sides of the fly wing. The static damping and dynamic damping are both weakened under the effect of jet flow.

(4) The propeller slipstream has a long-range effect behind the aircraft and seriously interferes with the

aerodynamics of the fly wing, which causes the lift-to-drag ratios as well as the lift and drag to decrease at the given AoAs. Moreover, the trend in static stability is different from that of normal flight, although the positive values of the static derivatives become larger, and the propeller slipstream inhibits the change in dynamic stability.

In conclusion, although the static and dynamic aerodynamics of the fly wing under the wake flow patterns were calculated and the derivatives identified, real cases are much more complex. Therefore, high-fidelity CFD methods must be further developed to capture the detailed characteristics of the wake flow, and additional research must be conducted on the influencing factors between aircrafts, such as the distance, position, and altitude.

Data Availability

The data used to support the findings of this study are available from the corresponding author upon request.

Conflicts of Interest

The authors declare that we have no financial and personal relationships with other people or organizations that can inappropriately influence our work; there is no professional or other personal interest of any nature or kind in any products, service, and/or company that could be construed as influencing the position presented in, or the review of, the manuscript entitled.

Acknowledgments

The authors would like to acknowledge the support of National Natural Science Foundation of China (Grant no. 11672236) and Project funded by China Postdoctoral Science Foundation (Grant no. 2018M641381). Part of the previous work was finished by Mrs. Lu Sisi; we also gratefully acknowledge her contribution.

References

[1] C. Hwang and W. Pit, "Aircraft wake flow effect and horizontal tail buffet," *Journal of Aircraft*, vol. 16, no. 4, pp. 282–284, 1979.

[2] I. G. Sheldon, *Wing Tip Vortices*, Springer Publishing, Germany, 1995.

[3] D. P. Rizzetta, M. R. Visbal, and M. J. Stanek, "Numerical investigation of synthetic-jet flowfields," *AIAA Journal*, vol. 37, no. 8, pp. 919-927, 1999.

[4] E. W. M. Roosenboom, A. Stürmer, and A. Schröder, "Advanced experimental and numerical validation and analysis of propeller slipstream flows," *Journal of Aircraft*, vol. 47, no. 1, pp. 284–291, 2010.

[5] J. Dacles-Mariani, S. Rogers, and D. Kwak, "A computational study of wingtip vortex flowfield," in *Proceedings of the 24th Fluid Dynamics Conference, AIAA93-3010*, Orlando, Fla, USA, 1993.

[6] L. Jiang, J. Cai, and C. Liu, "Large-eddy simulation of wing tip vortex in the near field," *International Journal of Computational Fluid Dynamics*, vol. 22, no. 5, pp. 289–330, 2008.

[7] M. J. Churchfield and G. A. Blaisdell, "Numerical simulations of a wingtip vortex in the near field," *Journal of Aircraft*, vol. 46, no. 1, pp. 231–243, 2009.

[8] A. Inasawa, F. Mori, and M. Asai, "Detailed observations of interactions of wingtip vortices in close-formation flight," *Journal of Aircraft*, vol. 49, no. 1, pp. 205–213, 2012.

[9] S. Kasper, T. Burke, J. Rapski, and N. Saneholtz, "Aerodynamics of formation flight for UAVs," in *Proceedings of the AIAA Atmospheric Flight Mechanics Conference, AIAA 2016-1285*, San Diego, Calif, USA, January 2016.

[10] R. Rudnik, C. Rossow, and H. F. V. Geyr, "Numerical simulation of engine/airframe integration for high-bypass engines," *Aerospace Science and Technology*, vol. 6, no. 1, pp. 31–42, 2002.

[11] S. Ritter, "Impact of different UHBR-engine positions on the aerodynamics of a high-liftwing," *Notes on Numerical Fluid Mechanics and Multidisciplinary Design*, vol. 131, pp. 367–380, 2016.

[12] S. Melber-Wilkending, R. Wilhelm, and H. Geyr, "RANS soluations for a complex high-lift configuration of a transport aircraft with engine including improved resolution of the nearfield," in *Proceedings of the 22nd Applied Aerodynamics Conference and Exhibit, AIAA 2004-5081*, Providence, RI, USA, 2004.

[13] M. J. T. Schroijen and R. Slingerland, "Propeller slipstream effects on directional aircraft control with one engine inoperative," in *Proceedings of the 45th AIAA Aerospace Sciences Meeting and Exhibit, AIAA 2007-1046*, pp. 12592–12604, Reno, Nev, USA, January 2007.

[14] F. Moens and P. Gardarein, "Numerical simulation of the propeller/wing interactions for transport aircraft," in *Proceedings of the 19th AIAA Applied Aerodynamics Conference, AIAA 2001-2404*, Anaheim, Calif, USA, June 2001.

[15] C. Lenfers, N. Beck, and M. Bauer, "Propeller and active high lift wing interaction in experiment and simulation," in *New Results in Numerical and Experiment Fluid Mechanics X*, Springer International Publishing, 2014.

[16] H. Xu, Z. Ye, and A. Shi, "Numerical study of propeller slipstream based on unstructured dynamic overset grids," *Journal of Aircraft*, vol. 49, no. 2, pp. 384–389, 2012.

[17] W. Khan, R. Caverly, and M. Nahon, "Propeller slipstream model for small unmanned aerial vehicles," in *Proceedings of the AIAA Modeling and Simulation Technologies (MST) Conference, AIAA 2013-4907*, Boston, Mass, USA, August 2013.

[18] R. Paul and W. L. Garrard, "Dynamic and control of tailless aircraft," in *Proceedings of the 22nd Atmospheric Flight Mechanics Conference, AIAA1997-3776*, August 1997.

[19] S. Esteban, "Static and dynamic analysis of an unconventional plane: flying wing," in *Proceedings of the AIAA Atmospheric Flight Mechanics Conference and Exhibit, AIAA Paper 2001-4010*, Montreal, Canada, August 2001.

[20] F. R. Menter, "Two-equation eddy-viscosity turbulence models for engineering applications," *AIAA Journal*, vol. 32, no. 8, pp. 1598–1605, 1994.

[21] S. H. Park, Y. Kim, and J. H. Kwon, "Prediction of dynamic damping coefficients using unsteady dual-time stepping method," in *Proceedings of the 40th AIAA Aerospace Sciences Meeting and Exhibit, AIAA 2002-0715*, Reno, Nev, USA, January 2002.

[22] L. L. Green, Spence A. M., and P. C. Murphy, "Computation methods for dynamic stability and control derivatives," in *Proceedings of the AIAA Atmospheric Flight Mechanics Conference, AIAA 2004-0015*, Reno, Nevada, USA, 2004.

[23] D. V. Dan, D. L. Thomas, and S. Andreas, "SACCON forced oscillation tests at DNW-NWB and NASA Langley 14×22-foot tunnel," in *Proceedings of the 28th AIAA Applied Aerodynamics Conference, AIAA 2010-4394*, Chicago, Ill, USA, July 2010.

[24] J.-F. Le Roy, S. Morgand, and D. Farcy, "Static and dynamic derivatives on generic UCAV without and with leading edge control," in *Proceedings of the 32nd AIAA Applied Aerodynamics Conference, AIAA 2014-2391*, Atlanta, Ga, USA, June 2014.

17

On Unsteady Effects in WIG Craft Aerodynamics

Nikolai Kornev

Mechanical Engineering Dept., University of Rostock, A. Einstein Str. 2, Rostock 18059, Germany

Correspondence should be addressed to Nikolai Kornev; nikolai.kornev@uni-rostock.de

Academic Editor: Jose Carlos Páscoa

The paper presents the analysis of unsteady forces and their influence on the aerodynamics and motion of a wing-in-ground (WIG) effect craft. Two-dimensional and three-dimensional aerodynamic models based on the potential flow are coupled with time domain simulations in the longitudinal plane. A special attention is paid to the explanation of the dynamic ground effect on both the sink and pitching motions. The influence of unsteady and quasi-steady forces on the dynamic ground effects and the craft motion is analyzed for different heights of flight.

1. Introduction

WIG (wing-in-ground) crafts are still considered as a very promising means of transport since they possess amphibian properties and can attain speeds which are unattainable for waterborne vessels. A very strong advantage of a WIG craft is the high lift to drag (L/D) ratio which theoretically can attain values over 30 and even more. For comparison, the L/D ratio of modern airplanes is less than twenty and the possibilities to improve it have been sufficiently exhausted. Unfortunately, to the knowledge of the authors, the L/D ratios over 20 have never been attained for real WIG configurations due to the following reasons. First, a very high L/D ratio is documented at extremely low clearance between the craft and the ground when the ratio of the clearance under the trailing edge h referred to as the mean aerodynamic chord $\bar{h}$ is less than 0.05. In reality, the minimum height of flight near the free surface is limited by waves and safety considerations. Therefore, only a large craft can fly at small relative heights of flight and completely utilize the ground effect attaining a high L/D ratio. All crafts known to the authors both designed and tested have the cruise height of flight $\bar{h}$ over 0.15. Second, since the fuselage of the WIG craft is designed to reduce the dynamic loadings during the takeoff and landing, its form is far from the aerodynamically optimal one. For instance, one utilizes very often the stepped bottom to reduce the drag in the takeoff mode. Third, to solve the problem of the longitudinal stability, most of the WIG crafts have big tail units located relatively high and far from the central main wing (see wing arrangements 1, 2, and 3 in Figure 1). The tail unit area of the best craft is varied between 22 and 30 percent of the main wing area. Such an arrangement requires the big fuselage which reduces the L/D ratio. Fourth, the majority of the WIG crafts built so far are small-sized ones and have relatively low speed U. To achieve the required lift, the cruise angle of attack is relatively large (from four to six degrees) and does not correspond to the optimal angle of attack at which the L/D ratio is maximal. Fifth, the main wing has usually small aspect ratio AR. It is well known from aerodynamics that the increase of the aspect ratio is the most efficient way to raise the L/D ratio. There is an opinion that the small aspect ratio is sufficient for WIG because the L/D ratio is high enough just due to the ground effect. Additional reasons speaking for small AR are the possibility to perform sharp turning maneuvers at big roll angles without water surface contact and increase of the efficiency of the power augmentation regime (PAR) (the most recent study, see, for instance, [1]). However, the moderate L/D ratio attained for the existing craft motivated designers to come back to the well-tried recipe to improve the L/D ratio by increase of AR. The last WIG Russian designs like SKB and marine passenger ekranoplan MPE designed by D. Synitsin (see the configuration 2 in Figure 1) in the Bureau of Alexeyev or SKP designed by Volkov and Ganin at KSRI have side wings of high aspect ratio located above the main wing plane by using V-angle

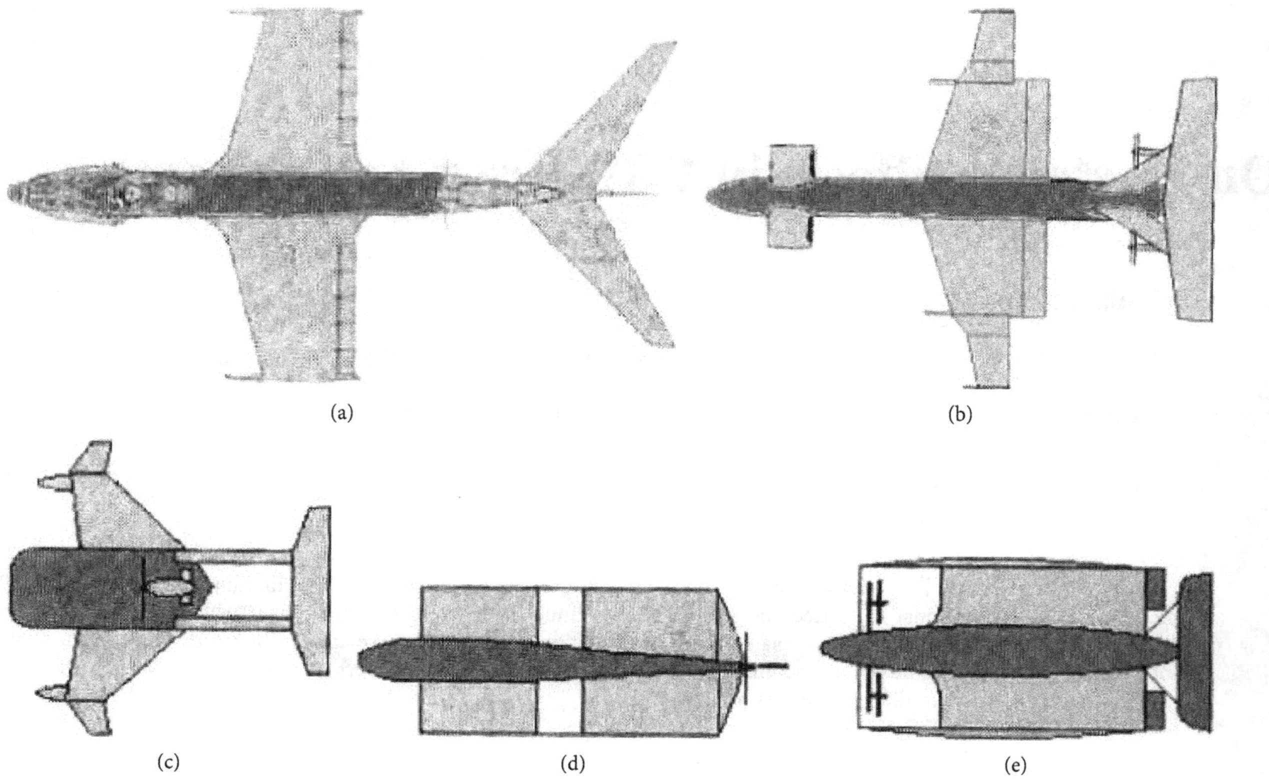

FIGURE 1: Typical wing arrangements of the wing in the ground effect craft [3]: (a) airplane type, (b) composite wing, (c) Lippish type, (d) tandem, (e) Volga type.

either for the whole side wing (SKP) or for its part (SKB and MPE). Due to this arrangement change, the overall AR was increased from the usual values of two or three to five and higher. A strong limitation for AR increase is the raise of the derivative dh/dU which reduces the "binding" between the craft and ground [2]. This can cause jumps of the craft in a vertical direction due to wind gusts or thrust increase. With the growth of craft size many of these problems can be overcome and *L*/*D* can be sufficiently increased for ultra large WIG craft which are getting probable with the appearance of modern light materials.

At present, there is a large body of literature on the ground effect aerodynamics although some of the key works were published in various reports and are still being not available to readers. For instance, according to the author's opinion, the most comprehensive and informative experimental study of the PAR regime was performed by the group of Prof. Epstein from the Central Aerohydrodynamic Institute (TsAGI). This group studied the effect of the wing planform, aspect ratio, distance between the engine nozzles and the leading edge of the wing, the blowing angles in horizontal and vertical directions, jet gas temperature, and the free surface (Froude number (Fn) effects) on the PAR efficiency. It was shown particularly that the optimal PAR efficiency determined as the ratio of the lift to thrust is attained at a small aspect ratio of the main wing around 0.7 (see Figure 48.14 in [3]). This explains the concept of the composite wing developed by many designers in Russia and Germany (see, for instance, configurations 2 and 3 in Figure 1 and in [4]). The wing is subdivided into two sections along the span, one of which is used for PAR whereas the second one is to generate lift in cruise regime. The free surface effect has a dramatical influence at Fn~0.7 similarly to common air cushion ships. At cruise Fn numbers, no influence of the free surface deformation on WIG aerodynamics was documented in the PAR regime.

Most of physical phenomena in ground effect aerodynamics have been already described in various papers and monographs [5–7]. In this introduction, we would like to mention just three facts which are often not sufficiently interpreted in the literature. First, it concerns the behavior of the drag when the height h decreases. Very often, it is said that the drag also decreases due to reduction of the induced resistance. In reality, this effect depends on the wing arrangement. The drag can increase whereas the lift to drag ratio increases almost in all cases when the lift gets higher. Second, as correctly mentioned in many papers, the lift decreases when the Venturi effect comes into play. A less known fact is the lift reduction at high lift conditions. This effect which is documented, for instance, in the experimental study by Gadezky [8], can be explained from the analysis of the pressure distribution. The overpressure is strongly increased on the pressure side at small *h* whereas the under pressure slightly decreases on the suction side. At small angles of attack without flaps, the first effect is proved to be much stronger resulting in the substantial lift increase. For strong flap deflections or at high pitch angles and small clearances, the pressure under the wing cannot become higher than the stagnation one whereas the under pressure on the upper side keeps the tendency for reduction. As a result, the lift decreases when

h goes to zero for high lift conditions in extreme ground effect. Third, there is a popular opinion that reversed delta wing arrangement 3 (Figure 1) proposed by Lippisch is the only one that has proven to be inherently stable in ground effect [9]. Stability of the WIG craft is determined by the distance between the aerodynamic centers in height X_h and in pitch X_ϑ, position of the center of gravity, damping properties, and binding criterion.

Numerical analysis shows that there exist a few other wing arrangements with common trapeze-like planform and tail unit which have similar stability properties as the reversed delta wing. The reader can easily prove it using the software Autowing [10] (http://www.lemos.uni-rostock.de/cfd-software/).

The problems mentioned above are not the main barrier for a sustainable development of WIG technology. The amphibian property of WIG is the sufficient factor for their use in the transport system despite of the moderate L/D ratio. The main problem of WIG is the stability and the flight safety. Most detailed description of the dynamics, stability, and control of the WIG craft is given by Zhukov [11] and Diomidov [12] who made serious contribution to the design of all famous Soviet ekranoplans. An overview of some results can be found in the book of Rozhdestvensky [6] and in the SNAME handbook [3]. In fact, WIG crafts designed by Russian and German specialists were thoroughly proved using the linear stability theory and nonlinear motion simulations. Although all famous crafts satisfy the criteria of both static and dynamic stabilities, the trial tests and operation of WIG crafts have been accompanied by stability problems. A possible reason is the fact that the classical stability theory is valid for the case of small perturbations and it is based on the linear representation of forces which are strongly nonlinear near the ground. Avvakumov [13] introduced the term of the stability "in small" meaning the stability predicted by the linear stability analysis to differentiate it from the stability "on the whole." The realistic perturbations cannot be considered as small ones, and the conclusions drawn from the linear stability theory are not sufficient to guarantee safe flight.

As shown in our recent paper [14], the motion of the WIG craft which is optimal from the point of the classical stability theory including Irodov's criterion [15], position of the center of gravity, and binding criterion can become unstable under relatively weak perturbations like wind gust. Therefore, the nonlinear simulation of the motion under perturbations is the inevitable part of the flight safety analysis to study the stability "on the whole."

Full 3DOF time domain simulation of the WIG motion coupled with the force determination is the task which can be solved using the modern CFD (computational fluid dynamics) codes. However, it requires big computational resources and for study of WIG dynamics with variation of craft and perturbations parameters in reasonable time is invisible. A more practical way is the decoupled consideration of dynamics and aerodynamics. Aerodynamic calculations are performed as preprocessing providing the coefficients which are then used for integration of the motion equations. Such a procedure is sufficiently based on the way of force representation in the form of Taylor or Fourier series on kinematic parameters. The set of kinematic parameters representing the WIG motion is still being a discussion subject. This problem is addressed in Section 2.1 of the present paper. Representation (6) was widely used in the Russian WIG community. It is based on the assumption of linear representation of forces caused by unsteady effects. The coefficients of unsteady forces are usually determined under the assumption that the craft performs harmonic oscillations with small amplitude and small frequency. This raises the following questions:

(i) Is this representation accurate enough to simulate nonlinear unsteady WIG dynamics in the ground effect

(ii) How is calculation of the coefficients in this representation done

(iii) Is the quasi-steady approach suitable for motion simulations

These three questions are in the focus of the present paper. Before proceeding further, we need to give the definition of the unsteady forces. Under unsteady force, we imply the fraction of the total aerodynamic force arising due to change of kinematic parameters and caused by the following vortex wake effects:

(i) Creation of unsteady vortex wake in the form of vortices with dominating transversal orientation

(ii) Change of the tip vortex circulation along its length

(iii) Oscillations of the tip vortex in horizontal and vertical directions

The second and the third effects which are typical only for three dimensional flows have influence on the upwind velocities and forces arising on the tail unit.

Contrary to unsteady forces, the quasi-steady time-dependent forces are calculated using the assumption that the aerodynamics is fully determined by instantaneous values of kinematic parameters neglecting unsteady vortex wake effects.

The motivation for this paper is the so-called dynamic ground effect mentioned in some recent papers (see, for instance, [16–18]). Surprisingly, this effect has never been explicitly discussed in the community involved in the development of WIG crafts. To understand this fact, this phenomenon is analyzed in Section 3.2 of this paper. The analysis is performed on the basis of the potential flow models from the following reasons. First, separate analysis of the quasi-steady and unsteady force components, necessary to interpret the dynamic ground effect, is only possible within the potential theory. Second, at moderate pitch angles and very high Reynolds numbers, the flow around the WIG is free of flow separations and influence of the viscosity is negligible. Third, as our viscous simulation experience shows [19], deformation of grids and their topology during the motion close to the ground makes the interpretations of the result difficult. It is necessary to separate the grid change effects from the real aerodynamic ones.

2. Mathematical Models

2.1. Force Representation in WIG Aerodynamics. The aerodynamic model is based on the representation of forces in a form of truncated Taylor series. This idea taken from aviation has sufficiently been modified by Prof. Treshkov in the late sixties. The original force coefficient C representation used in aviation is formulated as

$$C(\alpha, t) = C^0(\alpha_0) + \left.\frac{\partial C}{\partial \alpha}\right|_{\alpha_0} \cdot (\alpha(t) - \alpha_0) + \left.\frac{\partial C}{\partial \bar{\dot{\alpha}}}\right|_{\alpha_0} \cdot \frac{\dot{\alpha} b}{U} + \cdots. \quad (1)$$

Here, α is calculated in radians and $\bar{\dot{\alpha}} = \dot{\alpha} b/U$. In the proximity to the ground, the force coefficient depends additionally on the height of flight h. As Treshkov noted, representation (1) is not convenient for WIG crafts because of ambiguous definition of the angle of attack. In aviation, the angle of attack can be produced either by the craft pitching ϑ or by the vertical motion h. Thus, it can be stated that

$$\alpha = \vartheta \text{ or } \alpha = -\frac{\dot{h}}{U}. \quad (2)$$

Therefore, there is no difference in the derivatives between two ways in the change of the angle of attack:

$$\left.\frac{\partial C}{\partial \vartheta}\right|_{\vartheta_0} = -\left.\frac{\partial C}{\partial \bar{\dot{h}}}\right|_{\alpha_0}, \quad (3)$$

where $\bar{\dot{h}} = \dot{h}(t)/U$ is the nondimensional vertical velocity.

The motion is steady if

$$\dot{\vartheta} = 0, \quad \ddot{h} = 0. \quad (4)$$

In proximity to the ground, the vertical motion is principally unsteady and it is not similar to the pitching rotation. Thus, relation (3) is not valid and representation (1) has to be extended for motion in the proximity to the ground. For aerodynamic coefficients, a linear representation can be used:

$$C(\vartheta, h, t) = C^0(\vartheta_0, h_0) + \left.\frac{\partial C}{\partial \vartheta}\right|_{\vartheta_0, h_0} \cdot (\vartheta(t) - \vartheta_0) + \left.\frac{\partial C}{\partial \bar{h}}\right|_{\vartheta_0, h_0} \cdot \frac{h(t) - h_0}{b} + \left.\frac{\partial C}{\partial \bar{\dot{\vartheta}}}\right|_{\vartheta_0, h_0} \cdot \frac{\dot{\vartheta}(t) b}{U} + \left.\frac{\partial C}{\partial \bar{\dot{h}}}\right|_{\vartheta_0, h_0} \cdot \frac{\dot{h}(t)}{U} + \cdots, \quad (5)$$

where the angle of attack was excluded and replaced by the pitch angle to escape the ambiguous definition. $\bar{\dot{\vartheta}}$ is the nondimensional angular velocity $\dot{\vartheta}(t)b/U$. Since the steady force fraction described by the second and fourth terms in (5) is strongly nonlinear, it is used further in the nonlinear form:

$$C(\vartheta, h, t) = C^0(\vartheta, h) + \left.\frac{\partial C}{\partial \bar{\dot{\vartheta}}}\right|_{\vartheta, h} \cdot \frac{\dot{\vartheta}(t) b}{U} + \left.\frac{\partial C}{\partial \bar{\dot{h}}}\right|_{\vartheta, h} \cdot \frac{\dot{h}(t)}{U} + \cdots. \quad (6)$$

The last two terms on the r.h.s. in (6) contain both quasi-steady and unsteady fractions of the force.

For the study of unsteady force effects, we use a hierarchy of two models. In the next subsection, we present the coupled two-dimensional aerodynamic and motion models with two degrees of freedom based on the potential formulation. Since the resistance and induced drag are not captured, the change of the motion speed is neglected. To take the three-dimensional effects into account, we use the nonlinear vortex lattice method to calculate the coefficients in representation (6) which are then used for 3DOF motion simulation.

In what follows, for the sake of brevity, we use designations $C^{\vartheta}_{l,m}$, $C^{h}_{l,m}$, $C^{\dot{h}}_{l,m}$, $C^{\dot{\vartheta}}_{l,m}$, and $C^{\ddot{h}}_{l,m}$ meaning the derivatives on the pitch angle in radians and nondimensional parameters $\bar{h}$, $\bar{\dot{h}}$, $\bar{\dot{\vartheta}}$, and $\bar{\ddot{h}}$. Determination of these derivatives discussed in this paper is of a big importance because they are used as input parameters for the design of WIG control systems (see details in [11, 12]).

2.2. Two-Dimensional Coupled Aerodynamic and Motion Models with Two Degrees of Freedom (2D + 2DOF Model). Simulation of WIG motion coupled with the simultaneous determination of aerodynamic forces is a laborious problem requiring big computational resources. If the primary aim of the study is the qualitative analysis, it is worth to work first with simplified fast models retaining main typical aerodynamic features and containing minimum free parameters typical for CFD (e.g., grid and turbulence model). In this work, we studied influence of unsteady aerodynamic effects on the WIG dynamics using the two-dimensional potential flow model incorporated into two degrees of freedom motion simulation. The lifting system consists of the main airfoil with the chord length of $b = 1.0$ and tail unit with the chord of 0.25 (see Figure 2). Both lifting elements have the same setup angle of attack. The distance between the trailing edge of the airfoil and the leading edge of the tail unit in the horizontal direction is 0.65 and 0.5 in the vertical direction. The flow is assumed to be inviscid and has potential. The no-penetration condition is enforced on airfoils. The ground effect is modeled using the mirror image. The condition of zero circulation around contours C_1 and C_2 (see Figure 2) is satisfied according to the Thomson theorem. The airfoils and vortex wake are modeled using the method of discrete vortices [20]. The vortex sheet is represented by point vortices both on airfoils and in the vortex wake. The no penetration condition is satisfied at collocation points located between

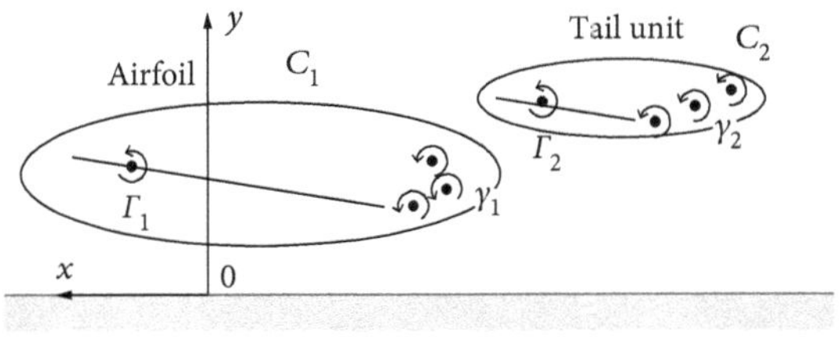

FIGURE 2: Sketch of the lifting system, the coordinate system, and contours C_1 and C_2.

point vortices on the profile. The first free point vortex in the wake lies at the line tangential to the airfoil. The trailing edge is exactly in the middle between the first free wake vortex and the last point vortex closest to the trailing edge on the profile. Such an arrangement of point vortices and collocation points secures fulfillment of the Kutta condition. Motion of free vortices in the wake is determined from the equations of fluid particle trajectories.

The pressure is calculated from the Bernoulli equation written for unsteady potential flow in the reference system moving with the speed of airfoil $U^*(t)$:

$$\frac{\partial \Phi}{\partial t} + \frac{U_a^2}{2} - \vec{U}_a \vec{U}^* + \frac{p}{\rho} = \frac{p_a}{\rho}, \qquad (7)$$

where $\vec{U}_a = \vec{U}_r + \vec{U}^*$ is the absolute velocity, $\vec{U}_r$ is the relative velocity, and Φ is the potential. Introducing the nondimensional quantities $\varphi = \Phi/Ub$, $u_a = U_a/U$, $u^* = U^*/U$, $\tau = tU/b$, and $C_p = p - p_a/(\rho U^2/2)$, one obtains after some transformations the difference of the pressure coefficient between the pressure and the suction sides

$$\Delta C_p = C_{p-} - C_{p+} = -2\frac{\partial}{\partial \tau}(\varphi_- - \varphi_+) - u_{r-}^2 - u_{r+}^2. \qquad (8)$$

The relative velocities on the suction and pressure sides are

$$u_{r-} = u_{00} + \frac{\gamma}{2}, \qquad u_{r+} = u_{00} - \frac{\gamma}{2}, \qquad (9)$$

where γ is the strength of the vorticity sheet and u_{00} is the direct value of the velocity on the airfoil (see Figure 2). Introducing circulation $\Gamma = \varphi_+ - \varphi_-$, finally, we have

$$\Delta C_p = C_{p-} - C_{p+} = -2\left(\gamma u_{00} - \frac{\partial \Gamma}{\partial \tau}\right). \qquad (10)$$

Integrating ΔC_p and $x/b\Delta C_p$ along the profile, one obtains the force and moment coefficients. The two-dimensional code was successfully verified by comparison with analytical data for a single airfoil at $h = \infty$.

The motion is calculated in the semiconnected coordinate system xoy (see Figure 2). The horizontal speed is assumed to be constant. Two degrees of freedom motion equations read

$$\begin{cases} m\ddot{y} = F_N \cos\vartheta - mg, \\ J_{zz}\ddot{\vartheta} = M_z, \end{cases} \qquad (11)$$

where F_N is the normal force, M_z is the pitching moment, and y is the distance between the center of gravity and the ground. In the nondimensional form, the system reads

$$\begin{cases} \ddot{\bar{y}} = \dfrac{d^2\bar{y}}{d\tau^2} = \mu C_N \cos\vartheta - Fn^{-2}, \\ \ddot{\bar{\vartheta}} = \dfrac{d^2\vartheta}{d\tau^2} = i_z C_m, \end{cases} \qquad (12)$$

where $\mu = \rho b^2 L/2m$, $Fn^{-2} = gb/U^2$, $i_z = \rho b^4 L/2J_{zz}$, $C_N = F_N/(\rho U^2/2)bL$, and $C_m = M_z/(\rho U^2/2)b^2 L$. Here, $L = 1$ is the wing span size.

This model is nonlinear with respect to the vortex wake influence and is applicable for arbitrary nonlinear dynamics with the only restriction of the constant horizontal speed.

2.3. Simulation of Motion on the Basis of Aerodynamic Coefficients Obtained from the Vortex Lattice Method (3D + 3DOF Model). Within this model, the aerodynamic coefficients are precomputed using the vortex lattice method [10] and then saved in the form of lookup tables. The data from the tables approximated with B-splines are used then for motion simulations. The vortex lattice method assumes the potential inviscid flow. The ground effect is modeled by the mirror image method. The lifting surfaces are modeled by vortex sheets represented by a set of M discrete horseshoe vortices with unknown circulations Γ_k determined from the no-penetration condition transformed to a system of linear equations with respect to Γ_k:

$$\sum_k^M \Gamma_k(\mathbf{w}_{jk}\mathbf{n}_j) = \mathbf{F}_j \boldsymbol{n}_j, \quad j = 1, M, \qquad (13)$$

where $\mathbf{w}_{jk}$ is the velocity induced by kth horseshoe vortex with unit circulation at the jth collocation point and $\mathbf{n}_j$ and $\mathbf{F}_j$ are the normal vector and the vector of incident flow speed, respectively, at the jth point. All time-dependent quantities are supposed to be harmonically changing with small frequencies and small amplitudes. $\Gamma_k(t)$, $\mathbf{w}_{jk}(t)$, and $\mathbf{F}_j(t)$ can be represented in the form of truncated Fourier series:

$$\Gamma_k(t) = \Gamma_k^0 + \sum_i^N \left(\Gamma_k^{q_i} q_i + \Gamma_k^{\dot{q}_i}\dot{q}_i\right) + O(\omega_i^2, q_i^{*2}, \omega_i q_i^{*2}), \qquad (14)$$

where $q_i = q_i^* \cos\omega_i t$ is the ith kinematic parameter changing with the small amplitude q_i^* and frequency ω_i, $\dot{q}_i = dq_i/dt$. The resting small terms of orders higher than one are neglected. In the limit of small frequencies and amplitudes, system (13) is reduced to three systems of linear equations solved sequentially:

$$\sum_k^M \Gamma_k^0\left(\mathbf{w}_{jk}^0\mathbf{n}_j\right) = \mathbf{F}_j^0\mathbf{n}_j, \tag{15}$$

$$\sum_k^M \Gamma_k^q\left(\mathbf{w}_{jk}^0\mathbf{n}_j\right) = \mathbf{F}_j^q\mathbf{n}_j - \sum_k^M \Gamma_k^0\left(\mathbf{w}_{jk}^q\mathbf{n}_j\right), \tag{16}$$

$$\sum_k^M \Gamma_k^{\dot{q}}\left(\mathbf{w}_{jk}^0\mathbf{n}_j\right) = \mathbf{F}_j^{\dot{q}}\mathbf{n}_j - \sum_k^M \Gamma_k^0\left(\mathbf{w}_{jk}^{\dot{q}}\mathbf{n}_j\right) - \sum_k^M \Gamma_k^q\left(\mathbf{w}_{jk}^{\text{unsteady}}\mathbf{n}_j\right). \tag{17}$$

The first system (15) describes the steady problem and the second one (16) provides the derivatives on kinematic parameters q_i whereas unsteady derivatives on $\dot{q}_i$ are calculated from the last system (17). System (16) is independent of (17) only in the limit of small frequencies. Function $\mathbf{w}$ is a pure geometrical function depending on coordinates of horseshoe vortices. For surge, heave, and pitch motions, the latter does not depend on $\dot{q}_i$, i.e., $\mathbf{w}_{jk}^{\dot{q}} = 0$. Each discrete vortex element produces unsteady vortices to satisfy the theorem of Thomson which is fulfilled for each element separately. As a result, the vortex sheet shed from each horseshoe element moves along the lifting surface and propagates further downstream in the wake. The intensity of such a vortex sheet is proportional to Γ_k^q. The last term in (17) describes the contribution of the unsteady vortex sheet to the normal component of velocity. Within this work, the position of all free vortices is prescribed. They oscillate along with the lifting surfaces as rigid constructions connected to lifting surfaces. The longitudinal vortices are straight lines being tangential to local chords. The transversal vortices of the vortex sheet are parallel to elements which they shed from. In summary, the aerodynamic model is linearized with respect to the angle of attack, wake form influence, and airfoil thickness effect. However, the following very important factor of the nonlinearity in the ground effect is considered, namely, the no penetration condition is satisfied at the actual position of the lifting surface at each time moment not being projected to any plane parallel to the ground as it is the case in fully linear wing theories. Consideration of instantaneous orientation of lifting surfaces with respect to the ground is the main source of nonlinearity of aerodynamic coefficients on the height of flight and pitch angle. The importance of this factor is discussed in [10]. Once Γ and its derivatives are known, the forces and moments are calculated from the Joukowsky theorem. Further details can be found in [10].

Although representation (6) is usually used in the frequency domain simulation formulation, it is utilized here within the nonlinear three DOF time domain simulation based on the system of three ordinary differential equations:

$$\begin{aligned} m\dot{U}_{xcg} &= T - \frac{\rho U_a^2 S}{2} C_d, \\ m\dot{U}_{ycg} &= \frac{\rho U_a^2 S}{2} C_l - mg, \\ J_{zz}\dot{\omega}_z &= \frac{\rho U_a^2 S b}{2} C_m, \end{aligned} \tag{18}$$

where $\dot{h}_{cg} = U_{ycg}$, $\dot{\vartheta} = \omega_z$, $U_y = U_{ycg} - w_y(t)$, and $U_a = U_{xcg} - w_x(t)$. Index cg denotes the center of gravity. The possible lift component and pitching moment due to the propulsion are neglected. Aerodynamic coefficients C_l, C_d, and C_m are represented in the form of (6).

The Vortex lattice method was implemented in the software Autowing [10] (http://www.lemos.uni-rostock.de/cfd-software/). Description of the method and its thorough validation are given by Kornev and Matveev [2], Kornev and Treshkov [10], and Benedict et al. [21].

The 3D + 3DOF model described above was used to simulate the dynamics of a WIG configuration of the Lippisch type (Figure 3). The fuselage and the side rudders were neglected. The configuration consists of the central wing which models the aerodynamic influence of the fuselage, main wing, tail unit, endplates, and winglets (Table 1). All sizes are referred to the root chord of the central wing $b = 3.85$ m. All lifting elements have the profile Clark Y. The endplate is modeled as the thin plate. The Venturi effect was not documented at pitch angles larger than minus three degrees at the minimal height of flight of 0.1 m. The mass of the craft is 1.7 tons when the craft flies with a zero pitch angle at height $h = 0.5$ m. The inertia moment is $4694\,\text{kg}\cdot\text{m}^2$ for these conditions. The craft is faultless from the point of the classical stability analysis. It possesses both static stability and dynamic stability up to the height of flight of 2.5 meters at a zero pitch angle. At the height of 0.5 m, Irodov's criterion referred to b is about 0.13. The speed of the craft is $U = 41.7$ m/s if not explicitly mentioned further. The binding criterion (see [2, 11])

$$\frac{dh}{dU} = \frac{2}{U}\frac{C_l}{\left(\partial C_l/\partial \bar{h}\right)}\frac{X_\vartheta}{X_h - X_\vartheta}, \tag{19}$$

calculated at this speed and a height of flight of 0.5 m, is 0.05 and can be considered as quite satisfactory. In contrast to CFD viscous methods, the VLM approach possesses a fast convergence reaching the final result at a few thousands of panels at heights of flight larger than $h/b \geq 0.1$. The total number of vortex elements utilized for the calculations of the hypothetical WIG craft was around two thousand using square panels.

3. Results

3.1. Validation Tests. To the author's knowledge, there are no well-tried benchmark tests proven in different independent experiments for validation of numerical simulations of wing arrangements in the close proximity to the ground. Very often, the available data raise the doubts at small heights of

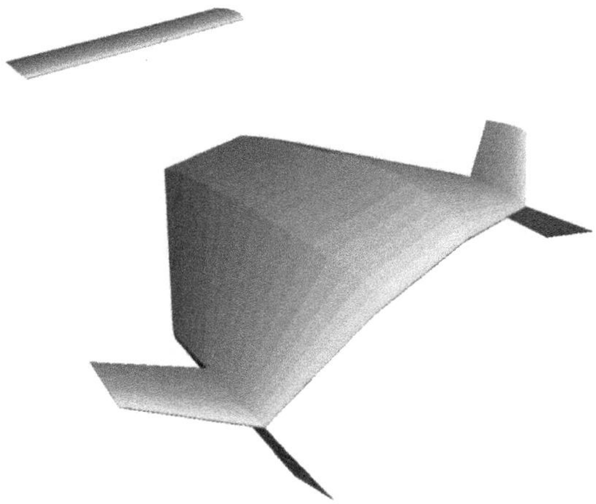

FIGURE 3: Hypothetical WIG craft used for the study.

TABLE 1: Main geometrical characteristics of the WIG craft used in the calculations.

Part	Chord	Aspect ratio	Taper ratio	Sweep angle	Pitch angle	V-angle
Central wing	1.0	0.3	1.0	0°	6°	0°
Main wing	1.0	0.5	0.3	−20°	6°	−12°
Tail unit	0.2	4.0	1.0	0°	2°	0°
Winglet	0.2	1.4	0.7	25°	5°	35°

flight because of inaccurate ground modeling especially if a simple plate without any suction of the plate boundary layer is utilized. This concerns both steady and unsteady aerodynamic parameters.

In this section, we evaluate the accuracy of the potential Vortex lattice method (VLM) by comparison of Autowing simulation data with available measurements and CFD calculations using the OpenFOAM code. Figure 4 presents the lift coefficient of a single wing with the aspect ratio of 1.5 and the profile NACA 6409 for two nondimensional heights of flight $h/b = 0.05$ and 0.3. The OpenFOAM simulations (see [19]) were performed using the limitedLinear scheme with the mix 90% CDS +10%UDS for the convection term. The Autowing computations show a very strong convergence when the number of vortex elements grows. The results were obtained with 60 panels along the span and 40 along the chord, i.e., with square panels. The ground is modeled by the mirror image method in VLM and the moving wall in OpenFOAM. As seen, the results of simulations are in very good agreement with measurements [22] for the moderate height $h/b = 0.3$. For very small $h/b = 0.05$, CFD shows a good accuracy whereas VLM results deviate sufficiently from experiments especially at moderate pitch angles. Higher lift in VLM can be explained by insufficient modeling of the Venturi effect. Indeed, the thickness influence in VLM is partly taken into account by the sources located on the mean wing surface. This approach models properly the potential displacement effect at moderate heights. At extremely low

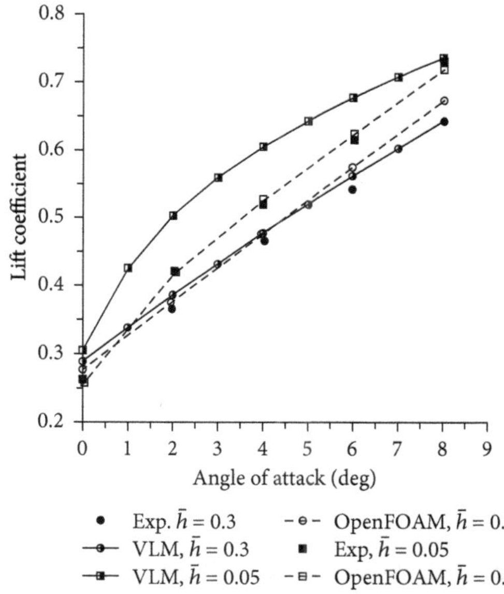

FIGURE 4: Lift coefficient of the wing with the aspect ratio of 1.5 and profile NACA 6409 versus the angle of attack (or the pitch angle). Comparison between VLM and OpenFOAM simulations and experiment [22].

heights, the profile form should be modeled more accurately. The second reason of the overestimation of the lift coefficient is the neglect of viscosity in the potential theory. As mentioned in Introduction, the heights of $\bar{h} = 0.05$ are unrealistic for WIG crafts because of wave limitations. Our experience shows that at more realistic heights larger than 0.15, such a thickness model is appropriate.

OpenFOAM calculations (dashed lines) are used to validate the VLM simulations (solid lines) for the configuration of the SeaFalcon WIG [23] with a schematic fuselage model (Figure 5). The results shown for different pitch angles ϑ are taken from [19]. It should be noted that the setup angle of the main wing is 5.5 deg, so that the resulting angle of the wing is 5.5 degrees larger. The VLM results are in good agreement with CFD for the lift coefficient (Figure 5) and for the Irodov criterion (Figure 6). The latter result is a convincing indication of the good VLM accuracy because the Irodov criterion depends on derivatives $X = C_m^h/C_l^h - C_m^\vartheta/C_l^\vartheta$ which proper computation is a more challenging problem than determination of coefficients. The pitching moment coefficient is small (Figure 6) because the reference point used for its calculation was located close to the pressure center. Therefore, a small difference between VLM and CFD for the moment coefficient is not important.

Validation results for the WIG craft Hoverwing 20 [24] presented in Figure 7 demonstrate an excellent agreement between VLM and measurements for the pitch angles of the craft of 0 and 2 degrees. The setup angle of the main wing is 6 degrees. The disagreement of approximately ten percent appears at $\vartheta = 4$ deg which corresponds to the pitch angle of 10 deg for the main wing. Additional measurements are necessary to clarify the reason for this growing discrepancy. Among others, it could be an improper modeling of the ground in experiments using plates without boundary layer

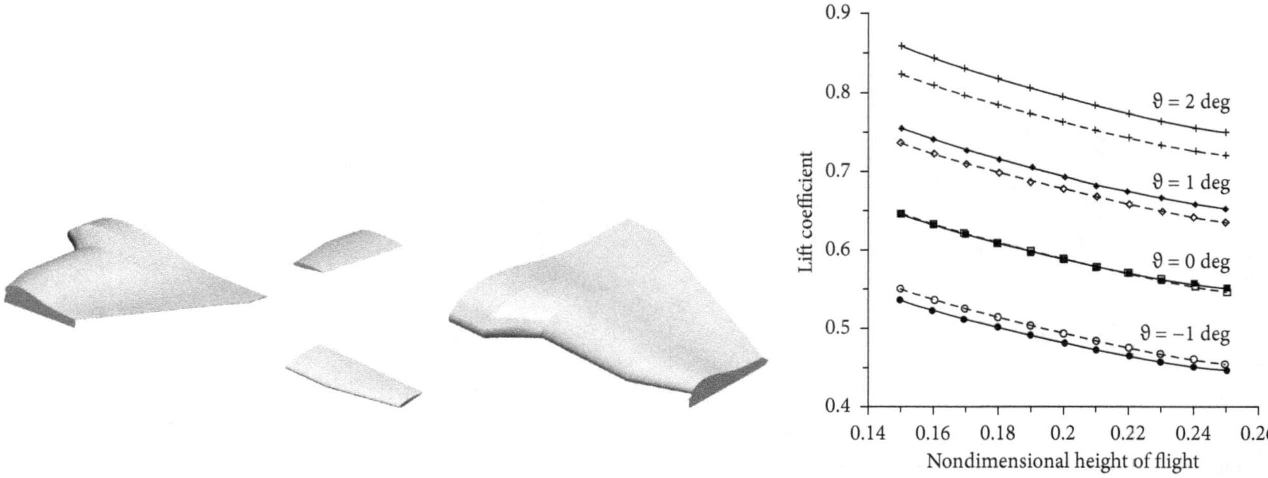

(a) (b) (c)

FIGURE 5: Half of the schematic configuration of the SeaFalcon WIG (a, b) and lift coefficient (c) versus the nondimensional height.

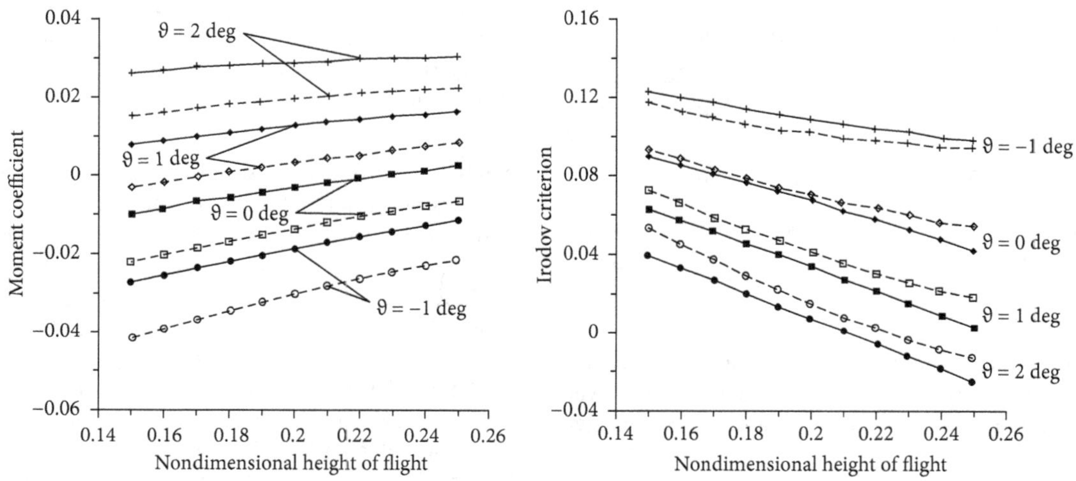

(a) (b)

FIGURE 6: Pitching moment coefficient (a) and the Irodov criterion (b) versus the nondimensional height.

suction. This results in a higher lift. Generally, in the past, any detailed validations of WIG craft simulations were seriously complicated by the lack of extensive experimental wind tunnel studies which were not available to small-design companies because of their high costs.

Validation for the unsteady forces is performed for a single wing with the aspect ratio of 3.0 and the profile NACA 0015 moving with prescribed harmonic vertical oscillations $\bar{h} = h^0 + 0.045 \sin \omega t$ with the dimensionless frequency $p = \omega c/V = 0.25$, where $V = 12$ m/s is the wing speed. VLM results are compared with unsteady OpenFOAM simulations obtained by Andreas Groß (unpublished material obtained by the research group lead by the author) using the morphing grid technology and $k - \omega SST$ turbulence model. Forces in VLM unsteady simulations are calculated in form (6). The ground is modeled by the mirror image method in VLM and the moving wall in OpenFOAM.

Figure 8 illustrates a good agreement between two approaches at a moderate height of flight $h^0 = 0.3$. The VLM provides the averaged lift coefficient of $C_l^0 = 0.768$ whereas OpenFOAM $C_l^0 = 0.756$. The time-dependent force part obtained from the VLM reads $\Delta C_l(t) = C_l^h h(t) + C_l^{\dot{h}} \dot{h}/V = -0.28h(t) - 4.39\dot{h}/V$. Closer to the ground at $h^0 = 0.15$, the discrepancy between two methods increases. While the averaged values are in good agreement $C_l^0 = 0.824$ for the VLM versus $C_l^0 = 0.838$ for the OpenFOAM, there is a certain deviation between unsteady parts: $\Delta C_l(t) = C_l^h h(t) + C_l^{\dot{h}} \dot{h}/V = -0.44h(t) - 4.91\dot{h}/V$ for the VLM and $\Delta C_l(t) = -0.88h(t) - 5.13\dot{h}(t)/V = Ah(t) + B\dot{h}(t)/V$ for the OpenFOAM. Since at small frequencies, $B \approx C_l^{\dot{h}}$ and both B are close to each other (−5.13 versus −4.91), one can conclude that the accuracy of the linear representation of unsteady parts in (6) for wing oscillations with relatively big oscillation amplitude and moderate frequencies is acceptable. The increase of derivative $C_l^{\dot{h}}$ with decreasing height (−4.39 at $h^0 = 0.3$ versus −4.91 at $h^0 = 0.15$) is due to the dynamic ground effect discussed in the next chapter. The difference between unsteady parts is

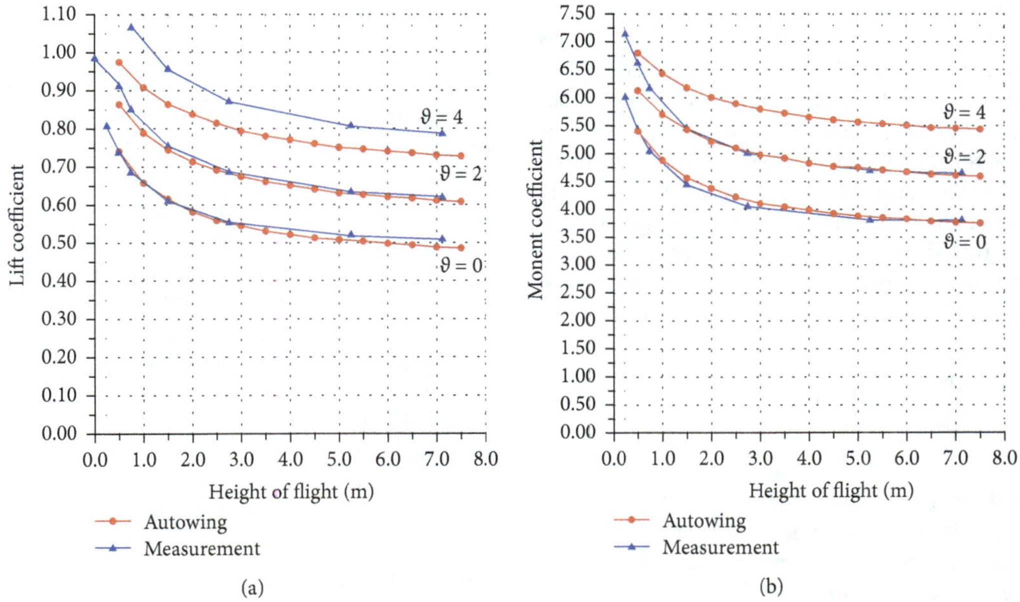

FIGURE 7: Lift (a) and pitching moment coefficient (b) for the WIG craft Hoverwing 20 [24] versus height of flight.

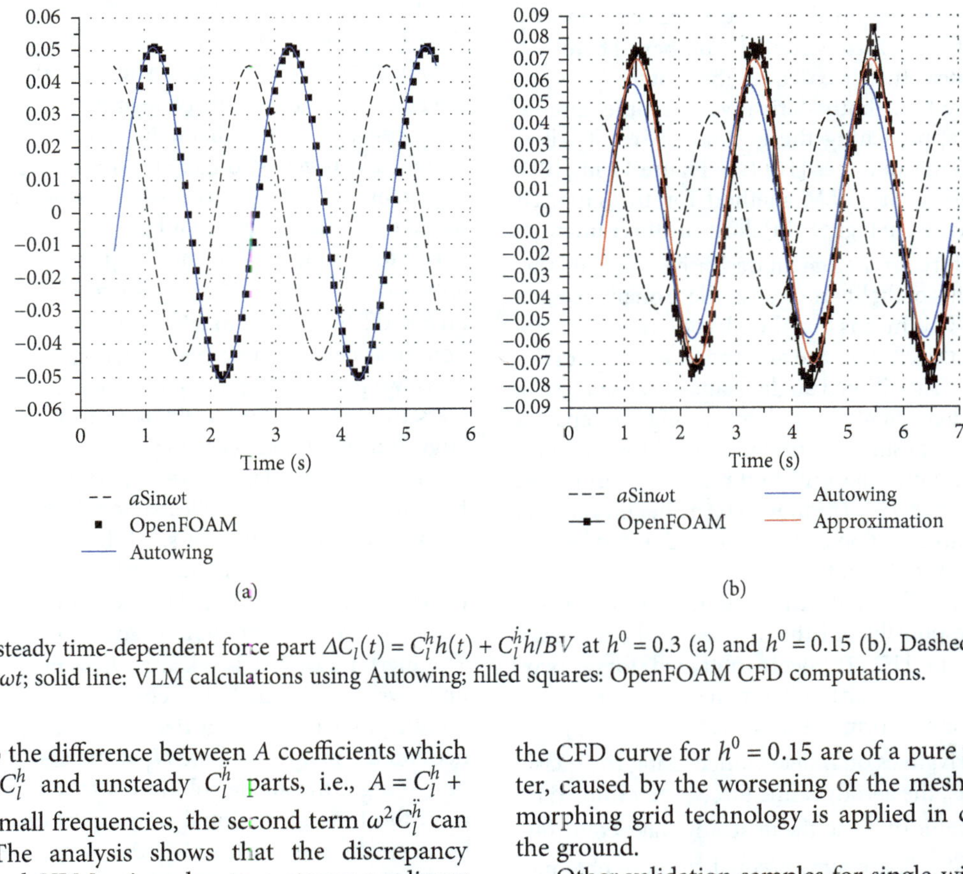

FIGURE 8: The unsteady time-dependent force part $\Delta C_l(t) = C_l^h h(t) + C_l^{\dot{h}} \dot{h}/BV$ at $h^0 = 0.3$ (a) and $h^0 = 0.15$ (b). Dashed line: change of the flight height $a \sin \omega t$; solid line: VLM calculations using Autowing; filled squares: OpenFOAM CFD computations.

obviously due to the difference between A coefficients which include steady C_l^h and unsteady $C_l^{\ddot{h}}$ parts, i.e., $A = C_l^h + (\omega c/V)^2 C_l^{\ddot{h}}$. At small frequencies, the second term $\omega^2 C_l^{\ddot{h}}$ can be neglected. The analysis shows that the discrepancy between CFD and VLM arises due to a strong nonlinear change of the steady coefficient C_l resulting in a large derivative C_l^h. At the small height of flight $h^0 = 0.15$ and amplitude $a/b = 0.045$, the instantaneous height of flight becomes around 0.1. In this case, the accuracy of the VLM computations is degraded due to insufficient modeling of the profile thickness effect and neglect of viscosity as mentioned above for the steady single-wing test case. The data fluctuations in the CFD curve for $h^0 = 0.15$ are of a pure numerical character, caused by the worsening of the mesh quality when the morphing grid technology is applied in close proximity to the ground.

Other validation samples for single-wing and WIG craft arrangements are presented by Kornev and Matveev [2], Kornev and Treshkov [10], and Benedict et al. [21].

Concluding, the validation results presented above can be considered as satisfactory and the VLM model can be utilized for the study in which the primary aim is the determination of the qualitative effects of the ground effect aerodynamics.

3.2. Phenomenon of the Dynamic Ground Effect. One of the ground effect phenomena attracting the attention of experts in the latest time is the so-called dynamic ground effect. For the review of works on the dynamic ground effect, the reader is referred to papers [17, 18]. The essence of this effect is a substantial increase of the total lift force when WIG sinks near the ground. Outside of the ground effect, the lift increase is purely due to an additional angle of attack caused by the sink rate of the craft. In the ground effect, the lift is proved to be much stronger. Within force representation (6), the additional dynamic force caused by the sink rate is described by the term

$$\left.\frac{\partial C_l}{\partial \dot{h}}\right|_{\vartheta,h} \cdot \frac{\dot{h}}{U} = C_l^{\dot{h}} \frac{\dot{h}}{U}. \quad (20)$$

The fact that derivative $C_l^{\dot{h}}$ gets much larger in the ground effect is not new. It is well known for a long time and can be found in some papers widely available to readers. For instance, in the monograph of Rozhdestvensky [6] (see Figures 3.14 and 3.15 in [6]), we find that this derivative on $\dot{h}$ multiplied by the height of flight $hC_l^{\dot{h}}$ is approximately 1.4 whereas the derivative on the vertical acceleration $hC_l^{\ddot{h}}$ is 0.5 in the limit of a zero Strouhal number. Therefore, according to the linear theory based on the matched asymptotic expansion method, both derivatives become infinite when the craft is approaching the ground. In reality, it is restricted as shown in our paper [10] (see Figure 11 in [10]).

Calculations using the VLM method for the Lippisch configuration presented in Figure 9(a) illustrate three interesting facts. First, the force and moment derivatives on $\dot{h}$ at large and moderate heights of flight are close to these on pitch angle ϑ (see Figure 9(a)). This means that the effect of the sink rate is almost the same as the effect of the pitch angle. Change of the force is almost due to additional angle of attack caused by the sink rate. At small heights of flight, there is a big discrepancy between two derivatives since the orientation of the craft with respect to the ground plays a significant role. This result is in accordance with studies [17, 18] in which the height range is subdivided into three regions. At high and intermediate heights, the effect of the sink rate is almost the same as that of the pitch angle. At small heights, the authors make compression work effects responsible for the increase of the unsteady force. The results in [17, 18] were obtained for a two-dimensional flow around an airfoil. The dynamic ground effect in the three-dimensional formulation was studied in [16].

Second, the derivatives on $\dot{h}$ determined using the quasi-steady approach, i.e., the aerodynamic force is determined by the instantaneous value of $\dot{h}$ and the unsteady wake contribution is zero $\mathbf{w}_{jk}^{\text{unsteady}} = 0$, are almost equal to these obtained using the unsteady vortex lattice method, i.e., with account for unsteady vortices. For this case, the equation for $\Gamma_k^{\dot{h}}$ can be reduced to that similar to steady problem system (15):

$$\sum_{k}^{M} \Gamma_k^{\dot{h}} \left(\mathbf{w}_{jk}^{0} \mathbf{n}_j \right) = \mathbf{F}_j^{\dot{h}} \mathbf{n}_j. \quad (21)$$

Therefore, $\Gamma_k^{\dot{h}}$ for the flat wing can be expressed through Γ_k^0 as $\Gamma_k^{\dot{h}} = \Gamma_k^0 (\mathbf{F}_k^{\dot{h}} \mathbf{n}_k)/(\mathbf{F}_k^0 \mathbf{n}_k) = \Gamma_k^0 \tan^{-1}\vartheta$. The obvious similarity between (15) and (21) points out that the reason for the so-called dynamic ground effect consisting in the increase of the lift at the sink rate is just the same as that in the increase of the lift in a steady flight, i.e., due to stagnation of the flow between the ground and the lifting surface. Unsteady aerodynamic effects connected to change of the lift and creation of free vortices play a minor role.

Third, the dynamic ground effect is strongly pronounced in the pitching motion much more than in the sink one. For this case, the unsteady aerodynamic effects have a strong impact on the derivatives on the pitching velocity $\dot{\vartheta}$ (see Figure 9(b)).

Concluding, the dynamic ground effect has been known in the ground effect aerodynamics for a long time although not discussed explicitly. For the simple sink rate motion, it is caused by the stagnation of the flow between the ground and the lifting surface.

3.3. Influence of Unsteady Aerodynamic Effects on Motion of a Lifting System Near the Ground. The ground effect has a substantial influence on mean forces and moments as well on their derivatives on h, ϑ, $\dot{h}$, and $\dot{\vartheta}$. Moreover, the unsteady aerodynamic effects increase when $h \rightarrow 0$ especially for derivatives on $\dot{\vartheta}$. The growth of unsteady derivatives in ground effect suggests an idea that their influence on motion and stability analysis gets stronger and the accuracy of their prediction cannot be satisfactory when using either the methods based on the assumption of harmonic change of parameters with small amplitudes and frequencies or a quasi-steady approach. Indeed, the unsteady part of forces in (6) depends on $\dot{h}$ and $\dot{\vartheta}$ linearly although the ground effect is an essentially nonlinear aerodynamic phenomenon. To clarify these questions, we perform a study of the sensibility of motion parameters to unsteady forces obtained from different models at large and small heights of flight.

The first calculations were performed using the 2D + 2 DOF model which is applicable to arbitrary motion and does not use the assumption of small harmonic oscillations. The mass m of the schematic craft with $b = 1$ m shown in Figure 2 and the position of the center of gravity were chosen from the force and moment balance equation for each height. The inertia moment is taken as $J_{zz} = m \cdot 10\,\text{kg} \cdot \text{m}^2$. Simulations were performed at a constant speed $U = 40$ m/s and middle pitch angle $\vartheta_0 = 5$ deg. The numbers of discrete vortices on the main airfoil are $N_1 = 100$ and $N_2 = 25$ on the tail unit. To get the stable solution with vortex sheet deformation, the time step should be proportional to the distance between discrete vortices on the airfoil, i.e., $\Delta\tau_a = b/N_1 = 0.25b/N_2 = 10^{-2}$ (see [20]). Unfortunately, this time step is too coarse to get the stable motion computation. Increase of N_1 was impossible due to numerical instability problems originating from the fact that the evolution of thin vortex sheets belongs to the ill-posed mathematical problems. To secure the stability of the overall simulation, two time steps are utilized. The determination of aerodynamic forces and motion parameters is carried out with the step $\Delta\tau = 10^{-3}$.

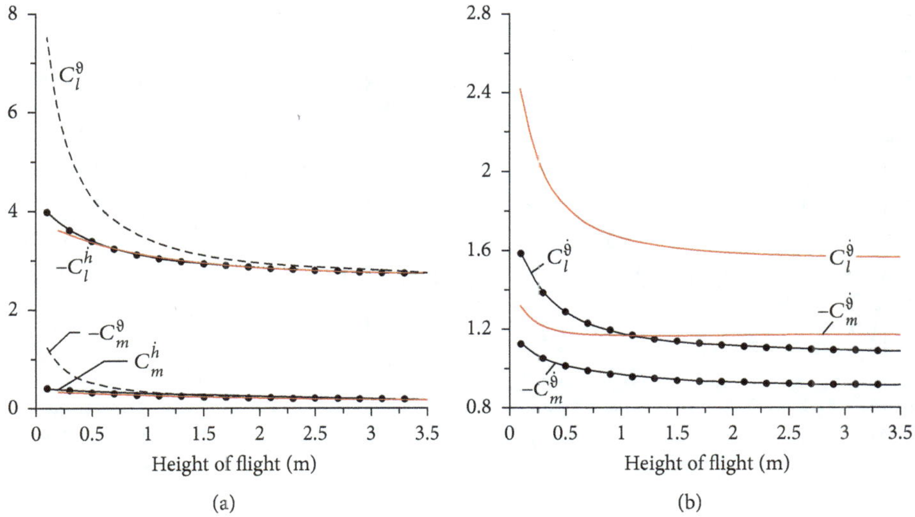

FIGURE 9: Influence of the ground effect on aerodynamic derivatives for the Lippisch configuration craft. Red lines: full unsteady calculations; black lines with circles: quasi-steady calculations.

Within the time step $\Delta\tau_a > \Delta\tau$, the circulation of the first wake free vortices closest to the trailing edges is changed whereas their positions remain constant with respect to the lifting system. The shedding of free vortices from the trailing edges into wake and motion of free vortices occurs with $\Delta \tau_a = 10^{-2}$. This two-level time step procedure allows one to hold the distance between free vortices proportional to N_1^{-1}, to avoid the Kelvin Helmholtz instability and to reduce the number of free vortices in the wake. The lifting system moves under step-like wind perturbations

$$\frac{w_x}{U} = \begin{cases} -0.1, & \text{if } 20 < \tau \leq 30, \\ 0, & \text{else.} \end{cases} \qquad (22)$$

Results are presented in Figure 10. The red lines correspond to full nonlinear unsteady simulation, whereas the black lines with circles present the simulation with quasi-steady determination of forces, i.e., without the last term in equation (10).

$$\Delta C_p = C_{p-} - C_{p+} = -2\gamma u_{00}. \qquad (23)$$

The most important conclusion from this analysis is that the difference between the unsteady aerodynamics formulation and quasi-steady one decreases drastically when the lifting system approaches the ground. This conclusion was reproduced in all simulations regardless of m, b, U, N_1, N_2, and $\Delta\tau$ and the method of motion equation integration.

The next motion simulation was carried out with the use of the 3D + 3DOF model. The perturbed motion was studied for two initial heights of flight 0.5 m and 3 m. The corresponding nondimensional values, referred to the mean aerodynamic chord are, respectively, 0.13 and 0.78. Motion equation (18) was integrated using the Runge-Kutta method of the 4th order with the time step of 10^{-4} s. The mass of the WIG craft and position of the center of gravity were chosen from the force and moment balance equation for each height.

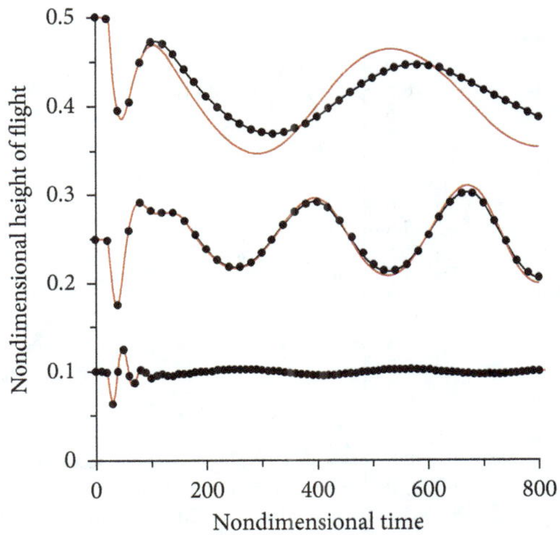

FIGURE 10: Trajectory of the schematic WIG craft under step-like wind perturbations. Black lines with circles correspond to the simulation with quasi-steady determination of forces.

The trajectory of the WIG craft (see Figure 3) under sinusoidal wind perturbations $w_x = w_{x0} \sin 2\pi t/T$, where $w_{x0} = -2$ m/s and $T = 10$ s is shown in Figure 11. Figure 12 shows the trajectory of the WIG craft under step-like wind perturbations

$$w_x = \begin{cases} -2 \text{ m/s}, & \text{if } 0 < t \leq 4 \text{ s}, \\ 0, & \text{if } t > 4 \text{ s}. \end{cases} \qquad (24)$$

We come to the same conclusion as this with 2D + 2DOF simulation. For calculation of motion, the importance of consideration of unsteady effects in the calculation of aerodynamic forces goes down drastically when the height of flight decreases. Moreover, the quasi-steady approach and the force representation are proved to be sufficient for application near the ground. This sufficiently simplifies the nonlinear motion simulation used to estimate the stability of the WIG craft "on

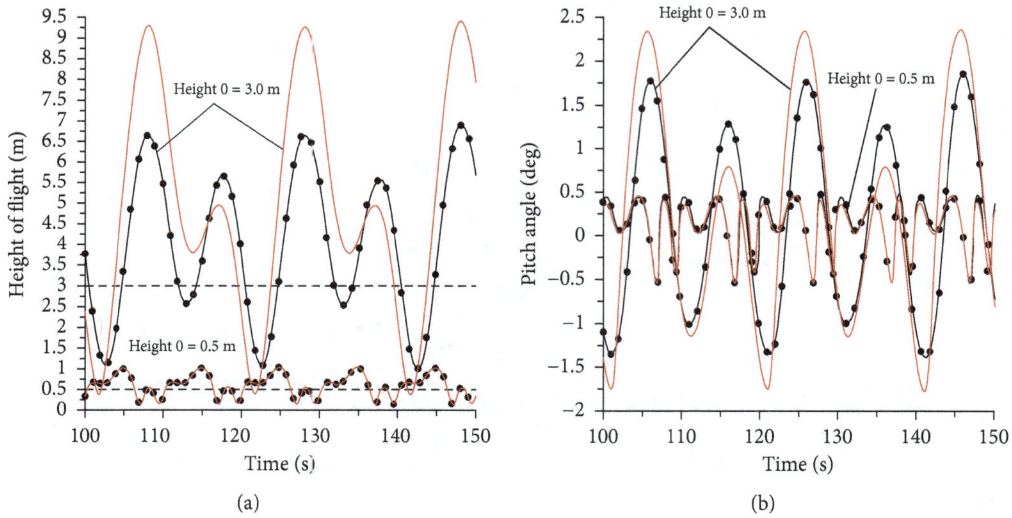

FIGURE 11: Trajectory of the WIG craft under sinusoidal wind perturbations $w_x = w_{x0} \sin 2\pi t/T$, where $w_{x0} = -2$ m/s and $T = 10$ s. Black lines with circles correspond to the simulation with quasi-steady determination of forces.

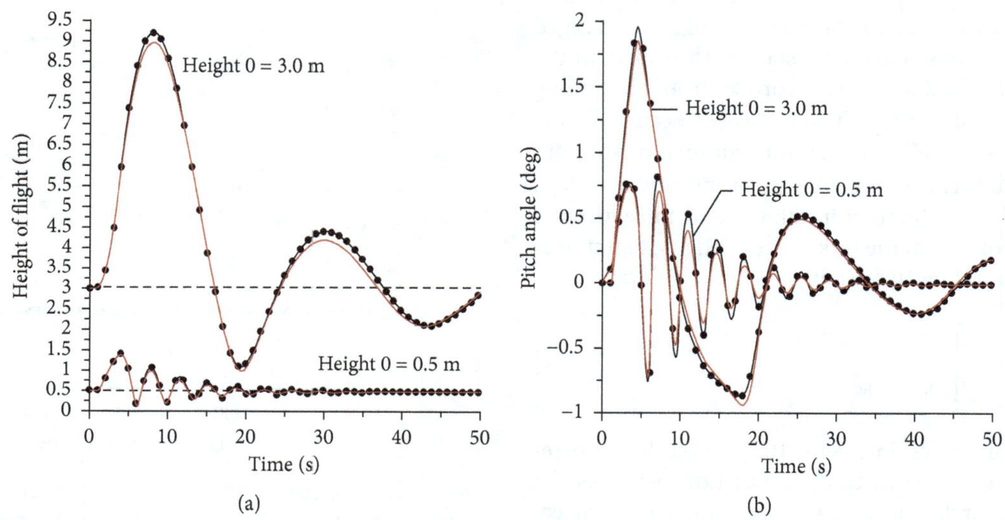

FIGURE 12: Trajectory of the WIG craft under step-like wind perturbations. Black lines with circles correspond to the simulation with quasi-steady determination of forces.

the whole" (see definition in Introduction) and to design the WIG control systems.

The explanation of this rather surprising result is the increase of restoring forces close to the ground. The increase of unsteady effects, i.e., effects caused by the formation of free vortices and their dynamics, when $h \to 0$, has no remarkable effect on WIG dynamics because of big restoring forces which also grow near the ground. The motion is mostly determined by instantaneous values of kinematic parameters. Far from the ground, in the area of weak stability, small differences in forces result in large deviation in trajectories. Note that the quasistatic approach ($\dot{h} = 0, \dot{\vartheta} = 0$) is fully nonapplicable regardless of height of flight.

The influence of unsteady effects on the position of the stability roots (see [4, 11]) is illustrated in Figure 13. Locus of the first three roots λ_i, $i = 1, 3$, was obtained for heights of 0.5 and 2.0 meters using quasi-steady and unsteady aerodynamic forces. The first root corresponds to the aperiodic motion $Im(\lambda_1) = 0$ whereas two others Re (λ_2) = Re (λ_3), $Im(\lambda_2) = -Im(\lambda_3)$ correspond to the long-period oscillatory mode. The roots corresponding to the short-period fast-decaying mode are not shown. The difference between unsteady and quasi-steady approaches for the height of 2.0 m is negligible, whereas two roots corresponding to the long-period oscillatory mode for the height of 0.5 m are slightly different. Since $-$Re $(\lambda_{2,3})$ is relatively large for $h = 0.5$, the oscillatory mode decays relatively quickly and the difference between unsteady and quasi-steady approaches is not remarkable in the direct time domain simulation of motion.

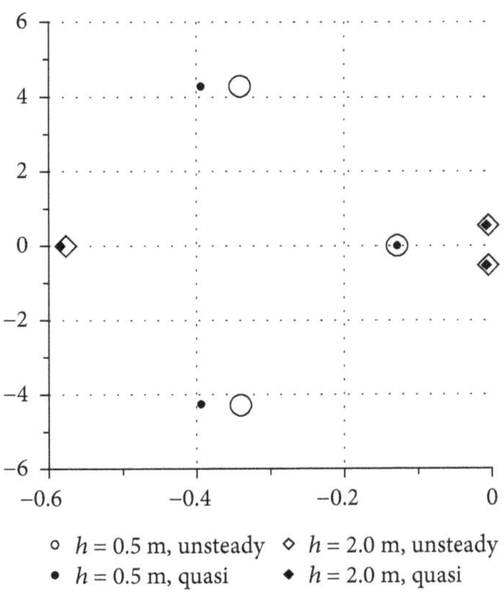

FIGURE 13: Locus of three first roots of the WIG craft at two heights of flight obtained under unsteady and quasi-steady aerodynamic forces.

4. Conclusion

Unsteady forces arising due to change of kinematic parameters and formation of the vortex wake have strong impact on the aerodynamical characteristics of lifting systems near the ground. They are responsible for the so-called dynamic ground effect consisting in a substantial increase of the total lift force when a wing sinks near the ground. Here, the lift at $h = \infty$ is compared with that at any small h and the same sink rate. Strictly speaking, this effect has been known in the ground effect aerodynamics for a long time although not discussed explicitly. At high and intermediate heights, the effect of the sink rate is almost the same as that of the pitch angle change. At small heights, the orientation of the craft with respect to the ground plays a significant role and the unsteady lift due to the sink rate is much larger than this of the pitch angle change. The present analysis based on the potential flow model shows clearly that the reason for the dynamic ground effect due to the sink is just the same as the increase of the lift in a steady flight, i.e., due to stagnation of the flow between the ground and the lifting surface. Unsteady aerodynamic effects connected to the change of the lift and creation of free vortices play a minor role. The dynamic ground effect is strongly pronounced in the pitching motion much more than that in the sink one.

Although the fraction of forces caused by unsteady effects increases near the ground, the influence of unsteady aerodynamic effects on the motion of a lifting system gets smaller when the height of flight decreases. The motion simulation can be predicted quite properly using the quasi-steady approach. The explanation of this rather surprising result is the increase of restoring forces close to the ground. The increase of unsteady effects, i.e., effects caused by the formation of free vortices and their dynamics, when $h \to 0$, has no remarkable effect on WIG dynamics because of big restoring forces which also grow near the ground. The motion is mostly determined by instantaneous values of kinematic parameters. Far from the ground, in the area of weak stability, small differences in forces result in large deviation in trajectories.

Nomenclature

b:	Chord (m)
C_l:	Lift coefficient (-)
C_m:	Pitching moment coefficient (-)
C_p:	Pressure coefficient (-)
h:	Height of flight measured from the trailing edge (m)
J_{zz}:	Moment of inertia (kg·m^2)
L/D:	Lift to drag ratio (-)
m:	Mass (kg)
p:	Pressure (N/m^2)
q:	Kinematic parameter
q^*:	Amplitude of the kinematic parameter
S:	Characteristic area (m^2)
T:	Thrust (N)
t:	Time (s)
U:	Craft speed (m/s)
U_a:	Craft speed wrt air (m/s)
X_ϑ, X_h:	Aerodynamic centers in pitch and height (m)
x, y, z:	Streamwise or axial, transversal, and spanwise directions (m)
α:	Angle of attack (deg)
Δt:	Time step (s)
Γ:	Circulation (m^2/s)
λ:	Stability root
ρ:	Density (kg·m^{-3})
ϑ:	Pitch angle (deg)
$\dot{\vartheta} = d\vartheta/dt = \omega_z$:	Angular velocity (1/s)
ω:	Oscillation frequency (1/s)
w_x, w_y:	Wind fluctuations (m/s).

Data Availability

The data used to support the findings of this study are available from the corresponding author upon request.

Conflicts of Interest

The author declares that they have no conflicts of interest.

Acknowledgments

The support of the German Research Foundation (Deutsche Forschungsgemeinschaft) under grant KO3394/3-1 is gratefully acknowledged.

References

[1] Y. Zhgang and Y. Wei, "Complex flow for wing-in-ground effect craft with power augmented RAM engine in cruise," *Chinese Journal of Aeronautics*, vol. 23, no. 1, pp. 1–8, 2010.

[2] N. Kornev and K. Matveev, "Complex numerical modeling of dynamics and crashes of wing-in-ground vehicles," in *41st*

Aerospace Sciences Meeting and Exhibit, pp. 1–9, Reno, NV, USA, January 2003.

[3] K. Fach, H. Fischer, N. Kornev, and U. Petersen, "Wing in ground effect craft," in *Ship Design and Construction*, vol. 48, SNAME, 2004.

[4] K. Rozhdestvensky, "Wing-in-ground effect vehicles," *Progress in Aerospace Sciences*, vol. 42, no. 3, pp. 211–283, 2006.

[5] A. Maskalik, B. Kolyzaev, and V. Zhukov, *Ekranoplans: Pecularities of Theory and Design*, Sudostroenie, 2000.

[6] K. Rozhdestvensky, *Aerodynamics of a Lifting System in Extreme Ground Effect*, Springer, Berlin, Heidelberg, 2000.

[7] L. Yun, A. Bliault, and J. Doo, *WIG Craft and Ekranoplan. Ground Effect Craft Technology*, Springer, 2009.

[8] V. Gadezky, "Study of aerodynamic characteristics of profile with flaps near the ground," in *Proc. Central Aerohydrodynamic Institute(TsAGI)*, vol. 1256, Moscow, 1970.

[9] H. Wang, C. J. Teo, B. C. Khoo, and C. J. Goh, "Computational aerodynamics and flight stability of wing-in-ground (WIG) craft," *Procedia Engineering*, vol. 67, pp. 15–24, 2013.

[10] N. Kornev and V. K. Treshkov, "Numerical investigation of nonlinear unsteady aerodynamics of the WIG vehicle," in *Proc. Conf. High Performance Marine Vehicle Conference HPMV-92*, pp. 38–48, Arlington VA, USA, 1992.

[11] V. I. Zhukov, *Peculiarities of the WIG's Aerodynamics, Dynamics and Stability*, TsAGI Press, 1997.

[12] V. B. Diomidov, *Automatic Control of Ekranoplan Motion*, Elektropribor Publisher, 1996.

[13] M. N. Avvakumov, "Requirements for aerodynamic characteristics of ekra- noplan from viewpoint of stability, linear estimates of dynamic properties of the craft and modeling of its motion," in *Proceedings of Conference GEM*, pp. 25–34, St. Petersburg, Russia, 2000.

[14] N. Kornev and A. Gross, "Investigations of the safety of flight of WIG craft," *Ship Technology Research*, vol. 61, no. 2, pp. 80–92, 2014.

[15] R. D. Irodov, "Criteria of longitudinal stability of ekranoplan," *Technical Notes of the Central Aerohydrodynamic Institute*, vol. 1, no. 4, 1970.

[16] S. Schmid, T. Lutz, and E. Kraemer, "Impact of modelling approaches on the prediction of ground effect aerodynamics," *Engineering Applications of Computational Fluid Mechanics*, vol. 3, no. 3, pp. 419–429, 2009.

[17] Q. Qu, J. Xi, W. Wang, and P. Liu, "Numerical simulation of the flowfield of an airfoil in dynamic ground effect," *Journal of Aircraft*, vol. 51, no. 5, pp. 1659–1662, 2014.

[18] Q. Qu, Z. Lu, P. Liu, and R. K. Agarwal, "Numerical study of aerodynamics of a wing-in-ground-effect craft," *Journal of Aircraft*, vol. 51, no. 3, pp. 913–924, 2014.

[19] N. Kornev and A. Gross, "Influences on safety of flight of WIG craft," in *52nd Aerospace Sciences Meeting*, pp. 1–17, National Harbor, MD, USA, 2014.

[20] S. M. Belotserkovsky and I. K. Lifanov, *Method of Discrete Vortices*, CRC Press, 1993.

[21] K. Benedict, N. Kornev, M. Meyer, and J. Ebert, "Complex mathematical model of the WIG motion including the take-off mode," *Ocean Engineering*, vol. 29, no. 3, pp. 315–357, 2002.

[22] K. H. Jung, H. H. Chun, and H. J. Kim, "Experimental investigation of wing-in-ground effect with a NACA6409 section," *Journal of Marine Science and Technology*, vol. 13, no. 4, pp. 317–327, 2008.

[23] http://www.seafalcon.net/technology.html.

[24] K. Matjasic and H. Fischer, "WIG (wing in ground effect) craft in military and paramilitary applications: the Howerving concept. A new generation of high speed marine craft," in *Nato Conference RTO-AVT Symposium. Novel and Emerging Vehicle and Vehicle Technology Concept*, p. 11, Brussels, Belgium, 2003.

Permissions

All chapters in this book were first published in IJAE, by Hindawi Publishing Corporation; hereby published with permission under the Creative Commons Attribution License or equivalent. Every chapter published in this book has been scrutinized by our experts. Their significance has been extensively debated. The topics covered herein carry significant findings which will fuel the growth of the discipline. They may even be implemented as practical applications or may be referred to as a beginning point for another development.

The contributors of this book come from diverse backgrounds, making this book a truly international effort. This book will bring forth new frontiers with its revolutionizing research information and detailed analysis of the nascent developments around the world.

We would like to thank all the contributing authors for lending their expertise to make the book truly unique. They have played a crucial role in the development of this book. Without their invaluable contributions this book wouldn't have been possible. They have made vital efforts to compile up to date information on the varied aspects of this subject to make this book a valuable addition to the collection of many professionals and students.

This book was conceptualized with the vision of imparting up-to-date information and advanced data in this field. To ensure the same, a matchless editorial board was set up. Every individual on the board went through rigorous rounds of assessment to prove their worth. After which they invested a large part of their time researching and compiling the most relevant data for our readers.

The editorial board has been involved in producing this book since its inception. They have spent rigorous hours researching and exploring the diverse topics which have resulted in the successful publishing of this book. They have passed on their knowledge of decades through this book. To expedite this challenging task, the publisher supported the team at every step. A small team of assistant editors was also appointed to further simplify the editing procedure and attain best results for the readers.

Apart from the editorial board, the designing team has also invested a significant amount of their time in understanding the subject and creating the most relevant covers. They scrutinized every image to scout for the most suitable representation of the subject and create an appropriate cover for the book.

The publishing team has been an ardent support to the editorial, designing and production team. Their endless efforts to recruit the best for this project, has resulted in the accomplishment of this book. They are a veteran in the field of academics and their pool of knowledge is as vast as their experience in printing. Their expertise and guidance has proved useful at every step. Their uncompromising quality standards have made this book an exceptional effort. Their encouragement from time to time has been an inspiration for everyone.

The publisher and the editorial board hope that this book will prove to be a valuable piece of knowledge for researchers, students, practitioners and scholars across the globe.

List of Contributors

Zhu Jianyang, Jiang Lin and Hou Yu
School of Machinery and Automation, Wuhan University of Science and Technology, Wuhan 430081, China

Yue Liu, Xiaorong Guan and Cheng Xu
School of Mechanical Engineering, Nanjing University of Science and Technology, Nanjing 210094, China

Tae-Yong Park and Hyun-Ung Oh
Space Technology Synthesis Laboratory, Department of Aerospace Engineering, Chosun University, 309 Pilmun-daero, Dong-gu, Gwangju City 61452, Republic of Korea

Su-Hyeon Jeon and Su-Jeong Kim
LIG Nex1 Co. Ltd., 207 Mabuk-ro, Giheung-gu, Yongin City, Gyeonggi-do 16911, Republic of Korea

Sung-Hoon Jung
Agency for Defense Development (ADD), Yuseong-gu, Daejeon City 34186, Republic of Korea

Hao Cheng and Hua Wang
School of Astronautics, Beihang University, Beijing 100191, China

Zhaoke Xu and Jian Xia
College of Aerospace Engineering, Nanjing University of Aeronautics and Astronautics, Nanjing 210016, China

Yang Yu and Zhengjie Wang
School of Mechatronical Engineering, Beijing Institute of Technology, Beijing 100081, China

Shijun Guo
Aerospace Engineering, Cranfield University, Cranfield MK43 0BL, UK

Du Siliang, Tang Zhengfei, Xu Pei and Ji Mengjiang
National Key Laboratory of Science and Technology on Rotorcraft Aeromechanics, Nanjing University of Aeronautics and Astronautics, Nanjing 210016, China

A. R. Ansari
Department of Mechanical and Aerospace Engineering, Science and Research Branch, Islamic Azad University, Tehran, Iran

A. R. B. Novinzadeh
Faculty of Aerospace Engineering, K. N. Toosi University of Technology, Tehran, Iran

Zhenqing Wang, Mingfang Shi, Lidan Xu and Xiaokun Sun
College of Aerospace and Civil Engineering, Harbin Engineering University, Harbin 150001, China

Xiaojun Tang
China Academy of Space Technology, Beijing 100094, China

L. T. Liu and Y. X. Hao
College of Mechanical Engineering, Beijing Information Science and Technology University, Beijing 100192, China

W. Zhang
College of Mechanical Engineering, Beijing University of Technology, Beijing 100124, China

J. Chen
Architecture and Civil Engineering Research Center, Shenzhen Research Institute, City University of Hong Kong, Shenzhen, China

Shan Pang and Xiaofeng Zhang
College of Information and Electrical Engineering, Ludong University, Yantai 264025, China

Xinyi Yang
Department of Aircraft Engineering, Naval Aeronautical and Astronautical University, Yantai 264001, China

Selin Aradag, Kubra Asena Gelisli and Elcin Ceren Yaldir
Department ofMechanical Engineering, TOBB University of Economics and Technology, Sogutozu Cad. No. 43, 06560 Ankara, Turkey

Tao Zhang, Bo Dong, Xun Zhou, Linan Guan, Weizhong Li and Shengqi Zhou
Key Laboratory of Ocean Energy Utilization and Energy Conservation of Ministry of Education, Dalian University of Technology, Dalian 116024, China

Qi'an Peng, Sanmin Wang, Bo Li, Changjian Zhi and Jianfeng Li
School of Mechanical Engineering, Northwestern Polytechnical University, Xi'an, China

Hong-hua Cai, Wan-sheng Nie, Xin-lei Yang, Rui Wu and Ling-yu Su
Department of Space Equipment, Equipment Academy, Beijing 101416, China

Hua Su, Chunlin Gong and Liangxian Gu
Shaanxi Aerospace Flight Vehicle Design Key Laboratory, Northwestern Polytechnical University, Xi'an, Shaanxi 710072, China

Index

www.ingramcontent.com/pod-product-compliance
Lightning Source LLC
LaVergne TN
LVHW080917210926
848485LV00004B/664